Wireless Communication Signals

Wireless Communication Signals

A Laboratory-based Approach

Hüseyin Arslan
University of South Florida
Tampa, FL, USA
Istanbul Medipol University
Istanbul, Turkey

This edition first published 2021

Registered Office
John Wiley & Sons, Inc., 111 River Street, Hoboken, NJ 07030, USA

Editorial Office
111 River Street, Hoboken, NJ 07030, USA

For details of our global editorial offices, customer services, and more information about Wiley products visit us at www.wiley.com.

Wiley also publishes its books in a variety of electronic formats and by print-on-demand. Some content that appears in standard print versions of this book may not be available in other formats.

Library of Congress Cataloging-in-Publication Data

Names: Arslan, Hüseyin, 1968- author.
Title: Wireless communication signals : a laboratory-based approach / Hüseyin Arslan, University of South Florida, Tampa, FL, USA, Istanbul Medipol University, Istanbul, Turkey.
Description: First edition. | Hoboken, NJ, USA : Wiley, 2021. | Includes bibliographical references and index.
Identifiers: LCCN 2020050695 (print) | LCCN 2020050696 (ebook) | ISBN 9781119764410 (cloth) | ISBN 9781119764427 (adobe pdf) | ISBN 9781119764434 (epub)
Subjects: LCSH: Wireless communication systems–Textbooks.
Classification: LCC TK5105 .A76 2021 (print) | LCC TK5105 (ebook) | DDC 621.384–dc23
LC record available at https://lccn.loc.gov/2020050695
LC ebook record available at https://lccn.loc.gov/2020050696

Cover Design: Wiley
Cover Image: © Andrey Suslov/Shutterstock

Set in 9.5/12.5pt STIXTwoText by SPi Global, Chennai, India

Contents

Preface

Wireless technologies have been evolving enormously over the last couple of decades. This ever-changing communication industry is in severe need of a highly skilled workforce to shape and deploy future generation communication systems. Therefore, well-designed training is needed to satisfy the telecommunication industry's thirst for resilient professionals. Distinguished telecommunication engineers must have a solid understanding of the fundamental theory, an ability to translate this theoretical knowledge to numerical implementation, and the capability of implementing them in hardware. Also, they must be equipped with independent thinking and creative problem-solving skills to pioneer future telecommunication systems. Furthermore, these prodigious engineers have to express their ideas clearly and function effectively on a team.

In this book, wireless communication systems and concepts are introduced from a practical and laboratory perspective with a goal to provide readers valuable experience to analyze wireless systems (along with wireless circuits) using modern instrumentation and computer-aided design software. The book is aimed to provide readers the knowledge to understand basic theory, software simulation, hardware test and modeling, system and component testing, and software and hardware interactions and co-simulations.

Broad Topical Coverage

This book aims to cover the following main categories:

- Wireless systems.
- Basic digital communications theory.
- Signal and waveform analysis.
- Digital transceiver algorithm design.
- Modeling, designing, testing, and measurement.
- Components (hardware) used for communication.
- Test-equipment and software defined radios.
- Component testing/modeling/evaluation.
- Advanced computer-aided simulation techniques.
- Interaction of simulation with hardware and test equipment.
- Understanding the real wireless channel environment.
- Wireless testbeds and hands-on experiments.

The objectives and overarching goals of the book can be listed as:

- Providing a platform to master the design, implementation, and experimentation of a communication system.
- Understand the impact of hardware on the performance of communication algorithms.
- Enable trainees to be able to send and receive actual communication waveforms; design, analyze, test, and measure these signals.
- Connect different disciplines of communication and teach cross-layer aspects of communication systems so that trainees can understand the relations and interactions between various layers.
- Provide basics of different tools (hardware emulation and software simulation) and utilize low cost and/or high-end SDR platforms.
- Enable learning the concepts with hands-on experiments.
- Allow instructors to develop their version of laboratory-based teaching of wireless and digital communication.
- Provide online laboratory teaching experience for those who do not have access to the equipment and wireless components.
- Provide reference Matlab codes (with very simple and basic isolated simulations) so that trainees can test the theory in a simulation environment.
- Developing link-level and system-level simulation tools that can allow students to test, evaluate, visualize, confirm, and interpret communication systems in multilayer and cross-layer domain.
- Providing trainees the ability to think/design a complete (and basic) communication system, and to design/conduct experiments in wireless communication systems.
- Shed light for future research and enable insights for interdisciplinary research.

Audience

This book is intended to provide both introductory theoretical aspects for beginners and advanced technical overview of practical aspects intended for university graduate and undergraduate students, technical professionals in the communications industry, technical managers, and researchers in both academia and industry. The book addresses concepts that are useful in several disciplines including digital baseband signal processing, radio frequency (RF) and microwave engineering, wireless communication systems, digital communications, antennas and propagation, radio frequency integrated circuits, machine learning-aided wireless communications, etc. In this respect, this is one of the rare books that bring many disciplines together.

Basic background of wireless communications and digital communications is preferable for a full understanding of the topics covered by the book.

Course Use

The book is structured in such a way that it can be used in support of various wireless courses at all levels and can serve as a reference for research projects of both undergraduate and graduate students. This book complements traditional theoretical textbooks by introducing some practical aspects and hands-on experiments.

The book provides an organic and harmonized coverage of wireless communication systems from radio hardware and digital baseband signal processing, all the way to testing and measurement of the systems and subsystems. Within this framework, the book chapters are quite independent from one another. Even though each chapter is self-contained, there is very good harmony between chapters and all the chapters complement each other. Therefore, different options are possible according to different course structures and lengths, as well as targeted audience backgrounds. The topics are covered in both descriptive and technical manners and can therefore cater to different readers needs. For each chapter we expect that a reader may skip the advanced technical description and still greatly benefit from the book.

The author of the book has been teaching a wireless laboratory course since 2007. This book is the result of the experience gained through this laboratory course. Hence, the book is structured in a way that it can be used as a text material for this course and for similar courses. The book can also be used to teach practical aspects of digital communication courses. Several laboratory-based teaching (in-lab or online) implementations can be applied utilizing the book. The following are some of the options among many:

- A lab course that contains theory and simulation.
- Theory, simulation, and RF components can be used with high-end test and measurement equipment like vector signal generators and vector signal analyzers in classical laboratory test benches.
- Similar to above, low-cost software-defined radio kit (with required RF components) can be utilized for home and remote experimentation in conjunction with the online teaching of the theory and simulation.
- Theory and simulation can be complemented with the real data captured and provided. We will provide large sets of data that are captured with the equipment. However, the data set can be enriched with others as well.
- Students can access the laboratory equipment remotely and carry out the experiments online either with or without partners (through remote equipment access).

Chapters

The book is structured with the following chapters:

- Chapter 1 Hands-On Wireless Communication Experience
- Chapter 2 Performance Metrics and Measurements
- Chapter 3 Multidimensional Signal Analysis
- Chapter 4 Simulating a Communication System
- Chapter 5 RF Impairments
- Chapter 6 Digital Modulation and Pulse Shaping
- Chapter 7 OFDM Signal Analysis and Performance Evaluation
- Chapter 8 Analysis of Single Carrier Communication Systems
- Chapter 9 Multiple Accessing
- Chapter 10 Wireless Channel and Interference
- Chapter 11 Carrier and Time Synchronization
- Chapter 12 Blind Signal Analysis
- Chapter 13 Radio Environment Monitoring

Acknowledgements

This book was made possible by the extensive support of numerous individuals and organizations throughout the career of Dr. Arslan. The adventure started with Ericsson research where Dr. Arslan has learned applied and practical research upon completing his PhD. Ericsson research laboratory has taught him the real meaning of communication theory, advanced digital baseband algorithm design, and wireless communication. Even though the time at Ericsson was extremely rewarding, something was missing. He was not able to connect the digital world with the RF front-end. He has learned hardware and RF first through interactions with his colleagues at the Electrical Engineering Department of USF, special thanks to wireless and microwave group and its members. But the significant leap came with his interaction with Anritsu Company. He had the pleasure of working with Anritsu for several years, and during this period he was able to connect the digital world with the RF world. He has also learned the importance of the test equipment in the design of wireless communication systems. Still something was missing, he was always a part of a subsystem. He did not design a complete end-to-end system. He has finally learned how to design an end-to-end communication system during his interactions with the great research institutions of Tubitak. He had the opportunity of designing several standard based and propriety systems. Dr. Arslan is deeply indebted to all these companies and institutions for giving him the opportunity to learn. One of the best parts of being a university professor is the opportunity for lifelong learning. He was very lucky to have great graduate students and he has learned a lot from them. This book and many things that he has achieved would not have happened without them. Some of them are actively involved in the writing of some of the chapters in this book. Others who are not actively involved also contributed one way or another, at the minimum by taking the wireless laboratory course and being involved in the course. The teaching assistants over the span of 13 years deserve a big credit for making this book a reality. Especially, he would like to thank Alphan Şahin (who is now a professor in USC) and Ali Fatih Demir for their great feedback. Also, all the students who have taken the class and provided extensive feedback deserve his appreciation. Finally, he would like to thank his family for their love, support, and constant encouragement.

Hüseyin Arslan
University of South Florida
Tampa, FL, USA
Istanbul Medipol University
Istanbul, Turkey

List of Contributors

Musab Alayasra
Department of Electrical and Electronics Engineering
Istanbul Medipol University
Istanbul
Turkey

Hüseyin Arslan
Department of Electrical Engineering
University of South Florida
Tampa, FL, USA

and

Department of Electrical and Electronics Engineering
Istanbul Medipol University
Istanbul
Turkey

Mehmet Ali Aygül
Department of Electrical and Electronics Engineering
Istanbul Medipol University
Istanbul
Turkey

Abuu B. Kihero
Department of Electrical and Electronics Engineering
Istanbul Medipol University
Istanbul
Turkey

Ahmed Naeem
Department of Electrical and Electronics Engineering
Istanbul Medipol University
Istanbul
Turkey

Saira Rafique
Department of Electrical and Electronics Engineering
Istanbul Medipol University
Istanbul
Turkey

Mehmet Mert Şahin
Department of Electrical Engineering
University of South Florida
Tampa, FL, USA

Muhammad Sohaib J. Solaija
Department of Electrical and Electronics Engineering
Istanbul Medipol University
Istanbul
Turkey

Halise Türkmen
Department of Electrical and Electronics Engineering
Istanbul Medipol University
Istanbul
Turkey

Armed Tusha
Department of Electrical and Electronics Engineering
Istanbul Medipol University
Istanbul
Turkey

Acronyms List

3GPP	3rd Generation Partnership Project
4-QAM	four-state quadrature amplitude modulation
4G	fourth generation
5G	fifth generation
6G	sixth generation
8-PSK	eight-state phase shift keying
16-QAM	16-state quadrature amplitude modulation
32-QAM	32-state quadrature amplitude modulation
64-QAM	64-state quadrature amplitude modulation
128-QAM	128-state quadrature amplitude modulation
ACF	auto-correlation function
ACI	adjacent channel interference
ACLR	adjacent channel leakage ratio
ACPR	adjacent channel power ratio
ADC	analog-to-digital converter
ADSL	asymmetric digital subscriber line
AFC	adaptive frequency correction
AFH	adaptive frequency hopping
AIC	Akaike information criteria
AI	Artificial Intelligence
ALRT	average likelihood ratio test
AM	amplitude modulation
AM-AM	amplitude-to-amplitude
AM-PM	amplitude-to-phase
AoA	angle-of-arrival
AoD	angle-of-departure
AP	access point
ARQ	automatic repeat request
AR	auto-regressive
AWGN	additive white Gaussian noise
BER	bit-error-rate
BE	best effort service
BLER	block-error-rate
BPF	band pass filter
BPSK	binary phase shift keying

BS	base station
BSA	blind signal analysis
BT	bandwidth-time
BW	bandwidth
BWA	broadband wireless access
CAD	computer aided design
CAF	cyclic autocorrelation function
CCDF	complementary cumulative distribution function
CCI	co-channel interference
CDF	cumulative distribution function
CDL	clustered delay line
CDMA	code division multiple access
CDP	code domain power
CFC	channel frequency correlation
CFR	channel frequency response
CINR	carrier-to-inteference-plus-noise ratio
CIR	channel impulse response
CLT	central limit theorem
CMA	constant modulus algorithm
CNN	convolutional neural networks
CP	cyclic prefix
CPI	coherent processing interval
CRC	cyclic redundancy check
CRLB	Cramér-Rao lower bound
CR	cognitive radio
CSD	cyclic spectral density
CSIT	channel state information at the transmitter
CSI	channel state information
CSMA	carrier sense multiple accessing
CTS	clear to send
CW	continuous wave
DAB	digital audio broadcasting
DAC	digital-to-analog converter
DC	direct current
DDC	digital down conversion
DFE	decision feedback equalizer
DFRC	dual function radar communication
DFS	dynamic frequency selection
DFT-S-OFDM	DFT-Spread-OFDM
DFT	discrete Fourier transform
DL	deep learning
DNN	deep neural network
DoF	degrees of freedom
DS-CDMA	direct spread code division multiple access
DSP	digital signal processor
DSSS	direct-sequence spread-spectrum
DUC	digital up conversion

DUT	device under test
DVB-T	terrestrial digital video broadcasting
DTFT	discrete-time Fourier transform
DWT	discrete wavelet transform
ED	energy detection
EDGE	enhanced data rates for GSM evolution
EGC	equal gain-combining
EM	electromagnetic
eMBB	enhance mobile broadband
EVM	error vector magnitude
FB	feature-based
FCH	frame control header
FDD	frequency division duplexing
FDE	frequency domain equalization
FDMA	frequency division multiple accessing
FEC	forward error correction
FER	frame-error-rate
FFT	fast Fourier transform
FHMA	frequency hopped multiple access
FHSS	frequency-hopping spread spectrum
FH	frequency hopping
FMCW	frequency modulated continuous wave
FPGA	field-programmable gate array
FSK	frequency shift keying
G-REM	generalized radio environment monitoring
GB	grant-based
GFSK	Gaussian frequency shift keying
GF	grant-free
GLRT	generalized likelihood ratio test
GMSK	Gaussian minimum shift keying
GNSS	global navigation satellite system
GoF	goodness-of-fit
GOS	grade of service
GPP	general purpose processor
GPS	global positioning system
GSM	global system for mobile communications
HMM	hidden Markov model
IM	index modulation
ICI	inter-carrier interference
IDFT	inverse discrete Fourier transform
IDMA	interleave division multiple access
IDW	inverse distance weighting
IEEE	Institute of Electrical and Electronics Engineers
IFFT	inverse fast Fourier transform
IF	intermediate frequencies
IMD	inter-modulation distortion
INI	inter-numerology interference

IoT	Internet of Things
IP	internet protocol
I/Q	in-phase and quadrature-phase
ISI	inter-symbol interference
ITU	International Telecommunication Union
IUI	inter-user interference
IWI	inter-waveform interference
JRC	joint radar and communication
K–L	Kullback–Leibler
LAN	local area network
LCR	level crossing rate
LDPC	low density parity coding
LDS-MA	low density spreading multiple access
LFM	linear frequency modulation
LFSR	linear feedback shift register
LHS	left-hand side
LLR	log-likelihood ratio
LMMSE	linear minimum mean-square error
LMS	least-mean-square
LNA	low noise amplifier
LOS	line-of-sight
LO	local oscillator
LS	least squares
LSTM	long short-term memory
LTE	long term evolution
LTE-A	LTE-advanced
LU	licensed user
MACA	multiple access collision avoidance
MAC	medium access control
MAI	multi access interference
MANET	Mobile Ad-Hoc Network
MC-CDMA	multi-carrier code division multiple access
MCS	modulation and coding scheme
MCD	measurement capable device
MDL	minimum descriptive length
MI	mutual information
MIMO	multiple-input multiple-output
MLa	machine learning
MLE	maximum likelihood estimation
MLSE	maximum likelihood sequence estimation
MMSE	minimum mean-square error
mMIMO	massive MIMO
mMTC	massive machine type communication
mmWave	millimeter wave
MPA	message-passing algorithm
MPC	multipath component
MSE	mean-squared-error

MSK	minimum shift keying
MS	mobile station
MUI	multi user interference
NBI	narrow-band interference
NBF	narrow-band filter
NLOS	non LOS
NN	neural networks
NOMA	non-orthogonal multiple access
NR	new radio
OFDM-CSMA	OFDM carrier sense multiple accessing
OFDM-FDMA	OFDM frequency division multiple accessing
OFDM-TDMA	OFDM time division multiple accessing
OFDMA	orthogonal frequency division multiple access
OFDM	orthogonal frequency division multiplexing
OK	Ordinary Kriging
OMA	orthogonal multiple access
OOBE	out-of-band emission
OQPSK	offset quadrature phase shift keying
OSI	open systems interconnection reference
OSR	oversampling rate
OTA	over-the-air
PA	power amplifier
PAPR	peak-to-average power ratio
PDF	probability density function
PDMA	pattern division multiple access
PDP	power delay profile
PER	packet-error-rate
PHY	physical layer
PICR	peak interference-to-carrier ratio
PLL	phase-locked loop
PLS	physical layer security
PMF	partial match-filtering
PMP	point to multi-point
PN	pseudo-noise
PPP	Poisson point process
PRBS	pseudo-random binary sequence
PRI	pulse repetition interval
PRF	pulse repetition frequency
PSD	power spectral density
PSK	phase shift keying
PTS	partial transmit sequences
PU	primary user
PUSC	partially used sub-channeling
QAM	quadrature amplitude modulation
QoS	quality of service
QPSK	quadrature phase shift keying
RACH	random access channel

RAT	radio access technology
RA	reconfigurable antenna
radar	radio detection and ranging
RCS	radar-sensing and communication
RC	raised cosine
REM	radio environment map
RF	radio frequency
RF-REM	radio frequency REM
RHS	right-hand side
RIS	reconfigurable intelligent surface
RL	reinforcement learning
RLS	recursive least-squares
RMS	root mean squared
ROC	receiver operating characteristic
RRC	root raised cosine
RS	Reed–Solomon
RSM	range sidelobe modulation
RSSI	received signal strength indicator
RTS	request to send
RT	ray-tracing
RU	rental user
RVC	reverberation chamber
Rx	receiver
SAW	surface acoustic wave
SC-FDE	single carrier frequency domain equalization
SC-FDMA	single carrier frequency division multiple access
SCMA	sparse code multiple access
SCS	sub-carrier spacing
SC	selection combining
SCF	spectral correlation function
SDMA	space-division multiple access
SDR	software defined radio
SDU	service data unit
SER	symbol error rate
SIC	successive interference cancellation
SINR	signal-to-interference-plus-noise ratio
SIR	signal-to-interference ratio
SISO	single-input single-output
SM	selected mapping
SNR	signal-to-noise ratio
SPNIR	signal-to-phase-noise-interference-ratio
SS	subscriber station
STC	space time coding
STDCC	swept time-delay cross correlation
STFT	short time Fourier transform
SVD	singular value decomposition
SU	secondary user

TDD	time division duplexing
TDL	tapped delay line
TDM	time division multiplexing
TDMA	time division multiple access
TFA	time-frequency analysis
TH	time hopping
TP	transmission point
TVWS	television white space
Tx	transmitter
UAV	unmanned aerial vehicles
UE	user equipment
UMTS	universal mobile telecommunications service
URLLC	ultra-reliable, low-latency communication
USRP	universal software radio peripheral
UWB	ultra wide band
V2V	vehicle-to-vehicle
VANET	vehicular ad-hoc network
VCO	voltage-controlled oscillator
VNA	vector network analyzer
VR	virtual reality
VSA	vector signal analyser
VSG	vector signal generator
WCDMA	wide band code division multiple access
Wi-Fi	wireless fidelity
WiMAX	worldwide interoperability for microwave access
WLAN	wireless local area network
WMAN	wireless metropolitan area network
WPAN	wireless personal area network
WRAN	wireless regional area network
WSSUS	wide-sense stationary uncorrelated scattering
WSS	wide-sense stationary
WT	wavelet transform
XPR	cross polarization power ratio
ZP	zero-padding
ZPSK	zero-padded phase shift keying

1

Hands-on Wireless Communication Experience

Hüseyin Arslan

Department of Electrical Engineering, University of South Florida, Tampa, FL, USA
Department of Electrical and Electronics Engineering, Istanbul Medipol University, Istanbul, Turkey

Due to the remarkable surge in wireless technologies as well as the introduction of new concepts like cognitive and software defined radios, a strong need for developing a flexible laboratory platform to teach a wide variety of wireless techniques has emerged. Laboratory benches equipped with flexible radio transmitters and receivers can address this goal. In this chapter, a laboratory-based approach for better understanding of wireless communication is described. Model for a practical test bench is provided and its application toward developing various laboratory-based experiments is discussed. The experiments are not only useful for better understanding of wireless communication concepts but also for research and development of the new communication technologies.

1.1 Importance of Laboratory-Based Learning of Wireless Communications

The high demand for communications anywhere, anytime and now with anything has been the driving force for the development of wireless services and technologies. These technologies have evolved significantly over the last couple of decades, from simple paging to real-time voice communication further to very high rate data communications, and recently to a wide variety of applications. This dramatic change has affected society in many aspects, enabling people and machines to communicate in ways unimaginable in the past and contributing to the quality of life that is enjoyed today.

In parallel to the development of new wireless standards, recently software defined radio (SDR) [1] and cognitive radio (CR) [2] concepts gained significant interest among wireless communication community. One of the main characteristics of CR is the adaptability where the radio parameters (including frequency, power, modulation, and bandwidth) can be changed depending on the radio environment, user's situation, network condition, geolocation, and so on. SDR can provide very flexible radio functionality by avoiding the use of application-specific fixed analog circuits and components. Therefore, SDR can be considered as the core enabling technology for cognitive radio. With this ever growing wireless communication technologies and standards along with the introduction of new concepts like CR and SDR, a strong desire to develop a flexible laboratory platform to teach wide variety of wireless techniques has emerged. Laboratory benches that are equipped with SDR capable transmitters and receivers can address this goal.

Comprehending certain subjects such as telecommunications is very challenging due to their highly abstract nature. The classical theoretical knowledge, unless accompanied by practical and interactive experience, is destined to perish since it prevents observation of the various cause and effect relationships between different aspects of a communication system. Therefore, numerical

Wireless Communication Signals: A Laboratory-based Approach, First Edition. Hüseyin Arslan.

models are widely used to assist in the evaluation and visualization of such complex systems. Trainees can manipulate the system and subsystem parameters independently and grasp their individual effects using various numerical tools. These skills are especially critical in their professional career since troubleshooting problems with such tools is easier than fixing them in the hardware prototyping stage. Nonetheless, the validity of the aforementioned models is limited by the assumptions. Therefore, theoretically and numerically verified designs must also be implemented in hardware and tested under various channel conditions and realistic scenarios. The hardware implementation can be achieved practically using flexible radio platforms. It should also be pointed out that the numerical and hardware implementations provide a convenient way for trainees to design telecommunication systems by themselves and assist reinforced learning.

In this chapter, we introduce a wireless communication systems lab that will provide readers with the experience to design, test, and simulate wireless systems (along with wireless circuits) using modern instrumentation and computer-aided design (CAD) software. The described approach enhances understanding of the wireless and microwave concepts significantly and builds a bridge between concepts on radio frequency (RF) circuits/devices and wireless systems and networking. The described laboratory-based approach is developed in such a way that it can promote the understanding of other related wireless concepts at all levels: wireless systems, wireless networks, cognitive radio and SDR, digital baseband signal-processing algorithm development and hardware implementation, wireless circuits, wireless devices, and components. In addition, the lab approach can serve as a resource for several research projects for both undergraduate and graduate students. As a result, this new lab approach represents an opportunity to capture the momentum of the current wireless activity worldwide and dramatically enhance the critical area of communications systems. The relation of the lab approach with various components of wireless and microwave curriculum is shown in a very high level graph in Figure 1.1.

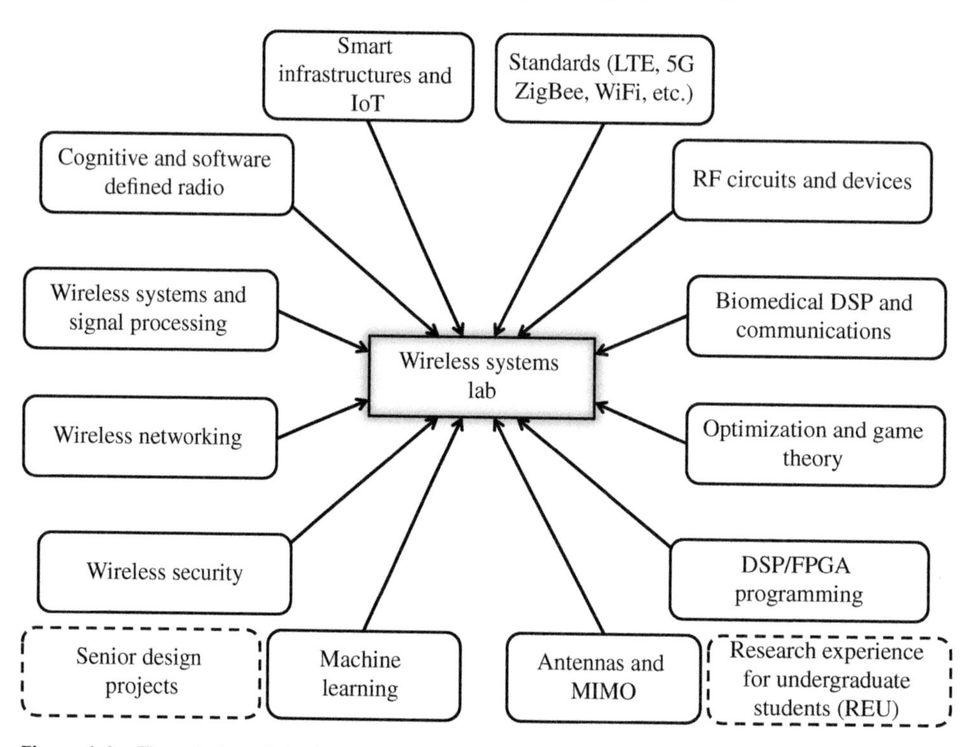

Figure 1.1 The relation of the lab approach with various components of wireless and microwave systems.

1.2 Model for a Practical Lab Bench

Figure 1.2 shows the laboratory bench setup along with the integration of various components and instruments. The four important elements of the model (theory, computer simulations and software, test and measurement instruments, and hardware) are shown in different rows. The top row shows the theory and concepts of various systems, subsystems, and components. Understanding the mathematics, statistical models, and the fundamental theories of wireless communication systems is extremely important. A solid design starts with a good understanding of the fundamentals. In our educational systems worldwide, most of the emphasis in teaching wireless communications is in the theory. However, the communication systems are getting more and more complex and exhibit nonlinear behavior, making it very difficult to analyze mathematically in closed form. Therefore, to help the understanding of these difficult concepts, simulations are also used. Simulations are especially important in the early design cycle as fixing mistakes is costly and time-consuming in the late stages of the wireless product development. The second row in Figure 1.2 presents the simulation world where the engineers (and students) can simulate and test various wireless communication systems (or subsystems). The simulation can help to get intuitive feeling of how theoretical knowledge is related to the real world; how to model and implement

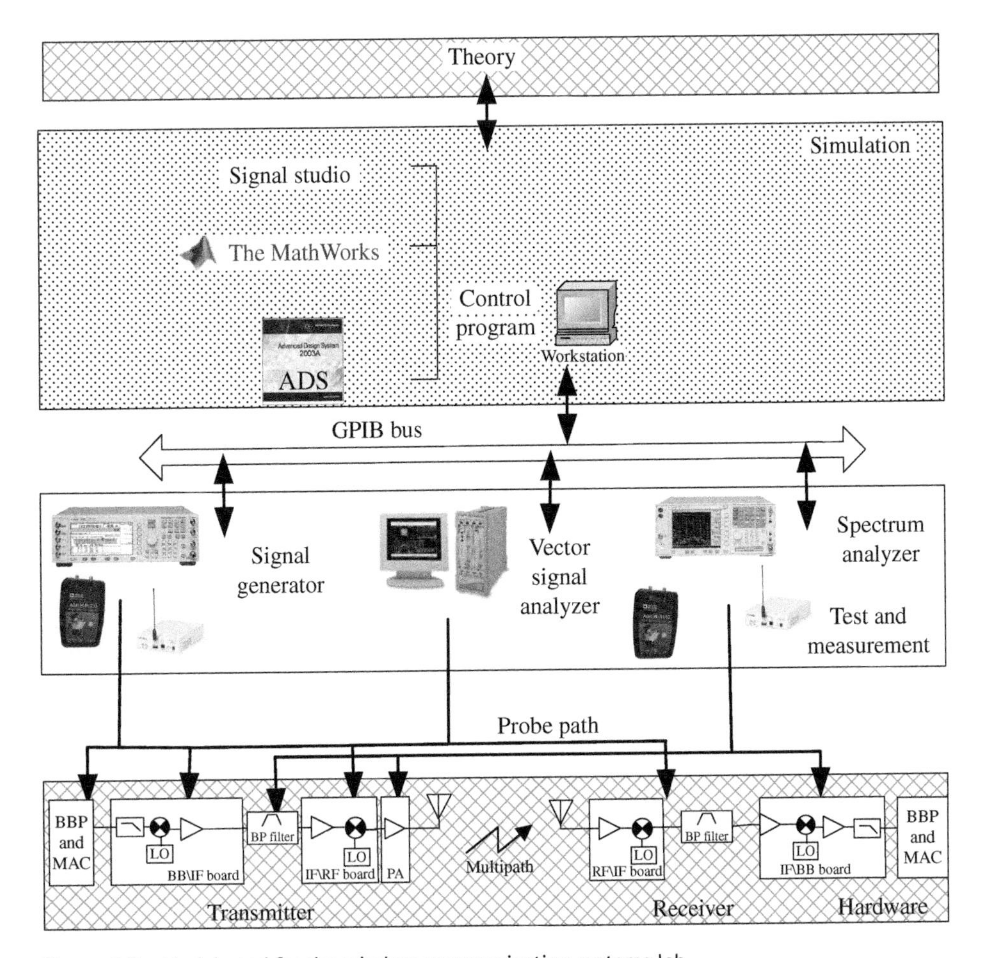

Figure 1.2 Model used for the wireless communication systems lab.

a realistic real-world wireless communication system using computer simulations; how various parameters affect the performance of the system, etc. Note that the traditional theoretical analysis and computer simulations are based on accurate modeling of the overall system, subsystems, and components.

The reliability and accuracy of the developed system and product depend on the accuracy of the assumptions and models used in the simulations and the theory. Hence, the system developers must be very careful in the assumptions and approximations they make in developing a system model. In the theoretical and simulation approaches, designs are at the algorithmic and behavioral level and hence often don't include any architectural or implementation details.

A step beyond the computer only simulation is the usage of the flexible radio hardware through a testbed. Testbed allows for further development and verification of the algorithms or ideas under real-world conditions. Additionally, testbeds provide an opportunity to employ mixed domain analysis. Especially, in developing SDR and cognitive radio systems, a full mixed domain design of the radio would be beneficial, where the baseband portions of the radio can be coupled with RF circuit level segments [3–6]. Test and measurement equipment are popularly used in developing testbeds and facilitating the mixed domain analysis. The third row in Figure 1.2 is related to the use of test and measurement equipment to connect the simulation world with the realistic hardware. These instruments are important elements of the wireless communication systems laboratory. The figure shows some of the important test and measurement equipment, i.e. vector signal generator (VSG), vector signal analyser (VSA), and spectrum analyzer (SA). The modern versions of spectrum analyzers with capabilities of providing in-phase and quadrature-phase (I/Q) samples through wide bandwidth digitizer can also be used as signal analyzer or the other way around. Therefore, a signal analyzer will be sufficient to study temporal, spectral, and waveform characteristics of the received signal.

The VSG can generate custom waveforms from various sources. Similarly, standard-based communication waveforms can also be generated using the VSG. These waveforms can either be integrated into the VSG, or they can be externally generated (using an external computer) and fed to the VSG. The simulation world in the second row can easily be connected to the VSG, as these devices are developed with such capabilities [7]. For example, a baseband signal can be developed using Matlab (or any specially designed Signal Studio software or advanced design system [ADS] simulator) and then downloaded to the VSG to create the physical signal. Physical signal generation is important in order to be able to feed the signal to the real hardware. Signal generators can also generate signals with noise and other impairments to test the receiver's ability to demodulate the signal in the presence of noise and impairments. Similarly, signals with fading and interference can be generated very easily. The baseband signal is then passed through the device under test (DUT) for study and characterization of the influence of different components, including intermediate frequency (IF) and RF up-converters, filters, power amplifiers, antennas, etc. The signal can also be passed through real radio channel to learn and understand the characteristics of realistic wireless channel. Alternatively, the signal can be transmitted through a multipath channel emulator. The channel emulator provides a controllable RF multipath channel model. Figure 1.3 shows a simplified block diagram of a VSG.

The wide frequency range of VSG can allow the study of RF signals up to tens of gigahertz range. Also, the wide bandwidth of the VSG can generate waveforms with high data rates. These flexibilities allow developers to generate a wide variety of custom and standard-based waveforms which can help them study and analyze current and possible future wireless communications systems and understand the related problems. Some of the important performance parameters of signal generators include: level accuracy (measure of how precisely the signal generator is able to output

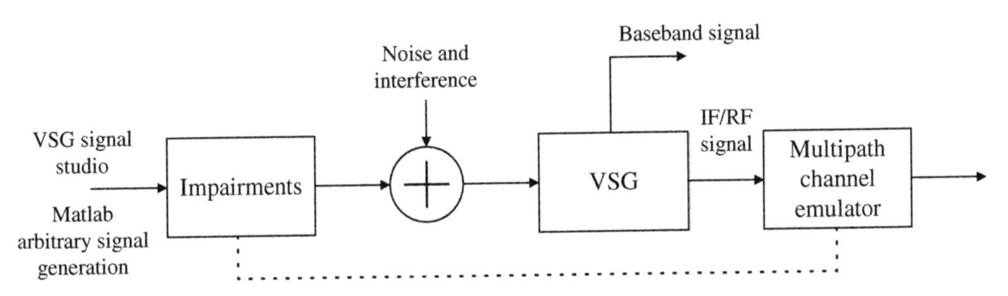

Figure 1.3 This simplified block diagram shows the main components of a vector signal generator.

the desired signal level); level repeatability (measure of how much the level output of the signal generator drifts); phase noise (random noise within the source that causes the power of a CW signal to be spread over a small range of frequencies); broadband noise (measure of noise over a wide frequency range); output power; etc.

The signal is received through the receiver antenna (or can be through direct cable) along with the interference sources that can be generated realistically in the testbed by another signal generator (or the interference model can be generated in the baseband signal that was transmitted along with the desired signal). Then, the received signal is passed through the receiver hardware and it is digitized with the VSA, the simplified block diagram of a VSA is shown in Figure 1.4. The VSA has the ability to demodulate the standard signals. However, it can also capture any arbitrary digital I/Q (in-phase and quadrature phase) samples and these samples can be processed with software components described in the first row. The simulation software (Matlab, ADS or any other possible software) can process the received data with the baseband receiver algorithms. The interaction between simulations and VSA is an excellent mechanism for teaching, studying and analyzing the current and possible future generation wireless systems. It allows a detailed study of the received waveform and development of advanced receiver algorithms considering realistic scenarios. Input to VSA can be a RF, IF, or baseband signal. Figure 1.3 shows a simplified block diagram of a VSA. Some of the important performance parameters of signal analyzers include: demodulation bandwidth (analysis bandwidth determines the maximum bandwidth that can be analyzed by the VSA); dynamic range (measure of how well a small signal can be analyzed in the presence of a big signal); I/Q memory (indication of how many samples can be captured and stored into memory, especially critical for wideband system measurements); residual error vector magnitude (EVM) (measure of how much the VSA contributes to the measured EVM); speed (measure of how fast the data can be processed); etc.

The final stage of the design cycle is the rapid prototyping of the complete system or subsystems. With rapid prototyping a sketch of the transmitter (or part of the transmitter) and receiver (or part of the receiver) hardware is developed. The prototyping is the stage which is closer to a final product.

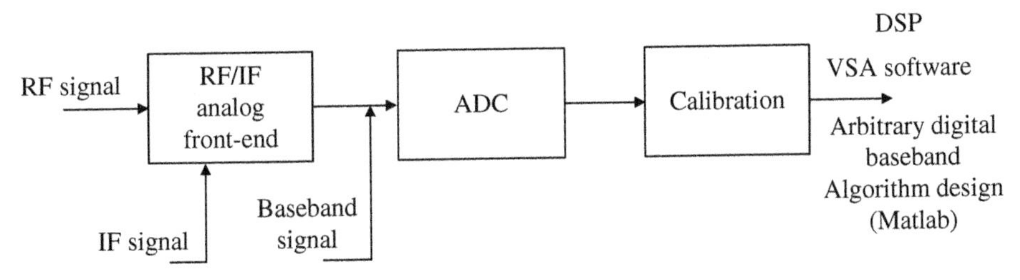

Figure 1.4 This simplified block diagram shows the main components of a VSA.

At this stage, and extensive testing and verification of the hardware that is prototyped is required. Test and measurement equipment along with the testbed described above can be used to evaluate the performance of the prototype as shown in the fourth row in Figure 1.2.

1.3 Examples of Co-simulation with Hardware

In the previous section, a model for a testbed is described. In this section, some examples for the co-simulation of system with hardware and test equipment, along with various DUT configurations, will be provided. In Figure 1.5 a limited set of test configurations is given. A more generic version is also given in Figure 1.6. For example, the radio channel can be characterized using the configuration in Figure 1.5a. A sounding sequence with good auto- and cross-correlation properties can be generated using the signal generator. This properly designed clean signal is then transmitted over the wireless channel. The perturbed received signal at the receiver side can be processed to obtain the channel characteristics. With such configuration, the radio channel can be measured and appropriate statistical models can be obtained to characterize the radio channel in the measured environment. The advantage of this approach is that the channel can be measured even if the transmitter and receiver are located far from each other. Note that in this configuration, the signal generator and signal analyzer can be controlled using CAD software (like Matlab) so that a user-defined sounding waveform can be used at the transmitter and the I/Q data can be processed at the receiver using a user-defined algorithm. The radio channel characteristics can be obtained at various carrier frequencies with different transmission bandwidth. The impact of carrier frequency, transmission bandwidth, the distance between transmitter and receiver, the operating physical environment can easily be observed with such a configuration.

In Figure 1.5b, a configuration to test power amplifier (PA) or low noise amplifier (LNA) is shown. In this case, the transmitter generates a clean modulated signal (or unmodulated carrier) with different input power levels and the signal is passed through the amplifier. The output signal is received at the receiver end and processed to characterize the impact of the amplifier at the transmitted signal. The measurement techniques to characterize the impact of the amplifier will be discussed in detail in the subsequent chapters. Note that the transmitted clean signal can be configured to generate a signal with different peak-to-average power ratio (PAPR) levels or with various other statistics. This capability will be especially critical to characterize the impact of the input signal and PA configuration together.

In Figures 1.5 (c) to (i), configurations to test various systems and subsystems are shown. For example, a complete transmitter can be tested by connecting the DUT (in this case it is the whole transmitter) to a signal analyzer. Note that in this case, the transmitter is not assumed to transmit a clean signal, rather we desire to measure how clean the transmitted signal is. If the signal analyzer is configured to receive the transmitted signal and appropriate algorithms to synchronize and demodulate the received data are implemented, testing the transmitter is pretty straight forward. In this case, the signal analyzer not only acts like an actual receiver tuned to receive the transmitted signal but also provides performance metrics to quantify the quality of the transmitted signal. These metrics will be discussed in detail in the subsequent chapters.

The signal analyzer and a well-designed analysis software can be used to evaluate and measure the reaction of DUT to various interference scenarios. The signal analyzer can also be used to observe the transient behavior of the DUT when the DUT changes the transmitted waveform as a reaction to interference or other stimulus. Hence, with a proper setup, we can control the interference environment with arbitrary signal generator, and we can observe and measure the

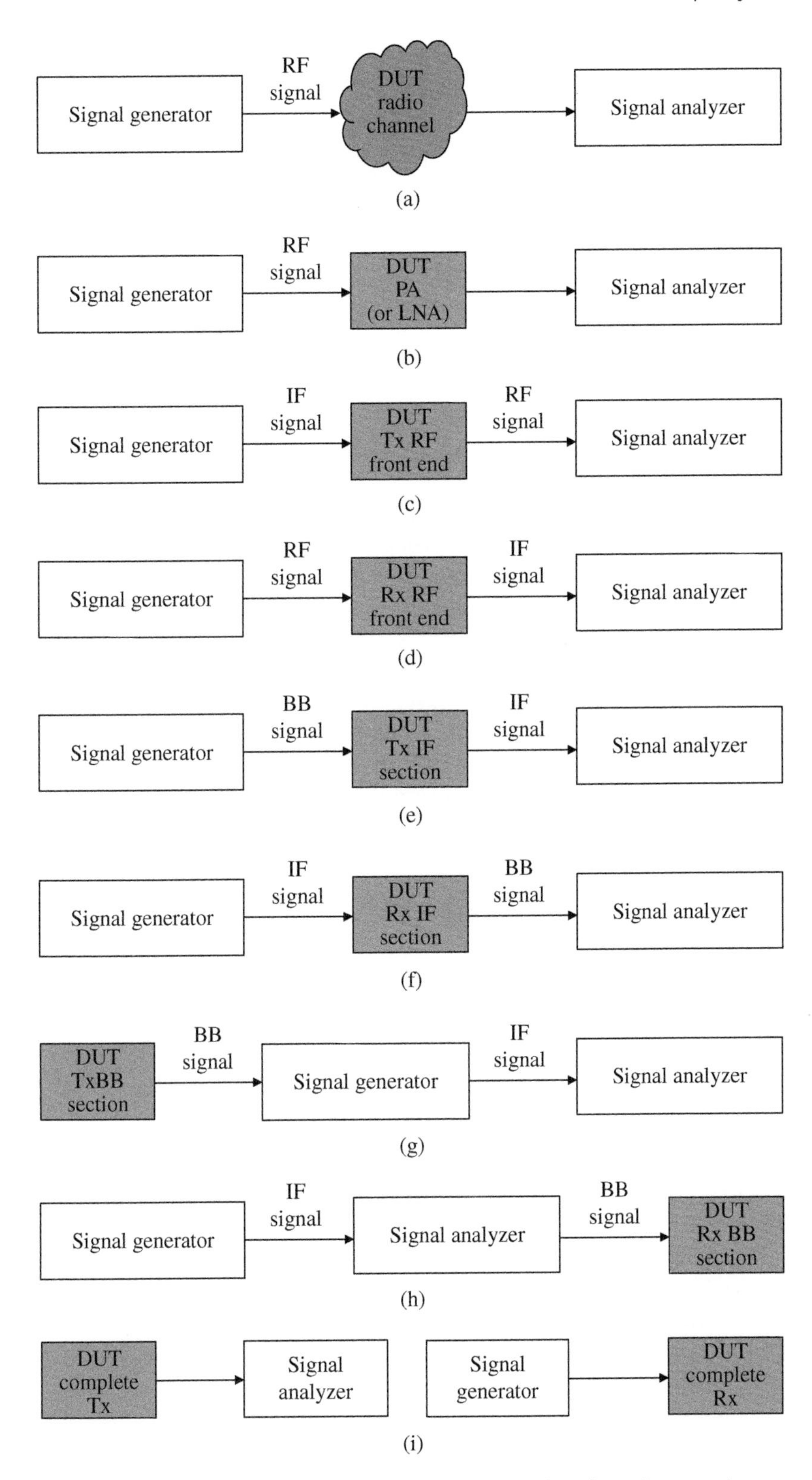

Figure 1.5 Testing systems and subsystems: various DUT configurations are shown.

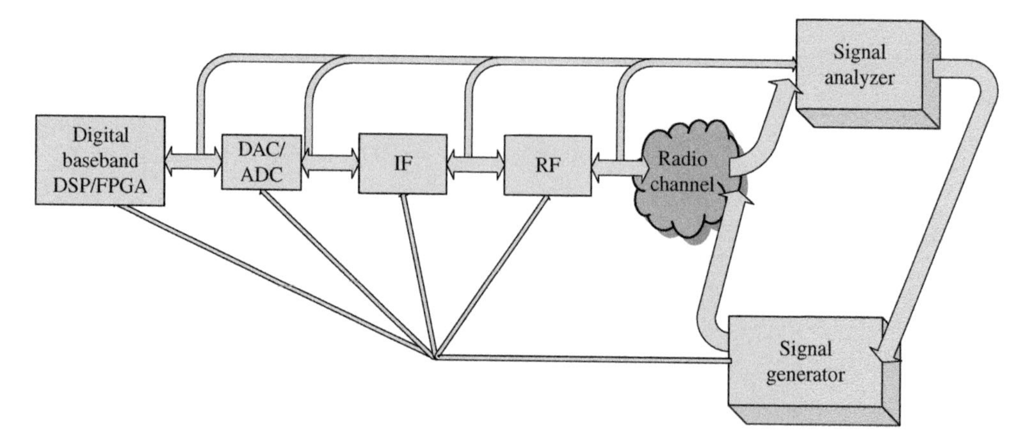

Figure 1.6 A more generic test and measurement set-up using signal generator and signal analyzer set.

DUT behavior and performance with signal analyzer along with enhanced test and measurement software. Such a setup can also be useful for testing link adaptation, in addition to other adaptive behaviors of the transmitter.

Although it is not shown in Figure 1.5, in measuring SDR based transmitter or receiver, it is important that the signal generator can apply test signals to the digital subsystem and signal analyzer can capture signals from the digital subsystem. For example, test signals directly from signal generator (instead of a real field-programmable gate array [FPGA]) can be used to drive digital-to-analog converter (DAC) (possibly after a digital signal interface module), or alternatively, a signal generated by field-programmable gate array (FPGA) can be fed to signal analyzer (after passing it through the digital signal interface module) to test the performance of the signal at the FPGA output. This is shown in a very basic form in Figure 1.6.

1.4 A Sample Model for a Laboratory Course

In this section, a model lab course will be provided as an example. Specifically, the Wireless Communications Systems Laboratory course that has been offered at the electrical engineering department of University of South Florida is given as the example. The Wireless Communications Systems Laboratory course is a three-credit course designed for graduate and senior-level undergraduate students. The theoretical aspects of the course are discussed in a weekly in-class session, which lasts approximately two hours. Principles of wireless communications, popularly used wireless communication technologies, standards and waveforms (single carrier and multicarrier), wireless communication receiver and transmitter hardware components, digital modulation, pulse-shaping filters, synchronization, and wireless propagation channel, multiple accessing are among the theoretical issues discussed. Experiments are designed in order to strengthen the trainee perspective on theoretical materials, and last approximately three and a half hours weekly.

The laboratory course is based on the successful completion of ten mandatory experiments and a design project. Team work and cooperation with the other trainees are strongly encouraged throughout the course. Postlaboratory reports, which are supposed to outline conclusions and outcomes derived from the experiments, are an essential part of the course. Trainees are also required to submit a report detailing the outcomes of their project and share their experiences with their classmates through a 30-minute demonstration at the end of the semester. Several pop

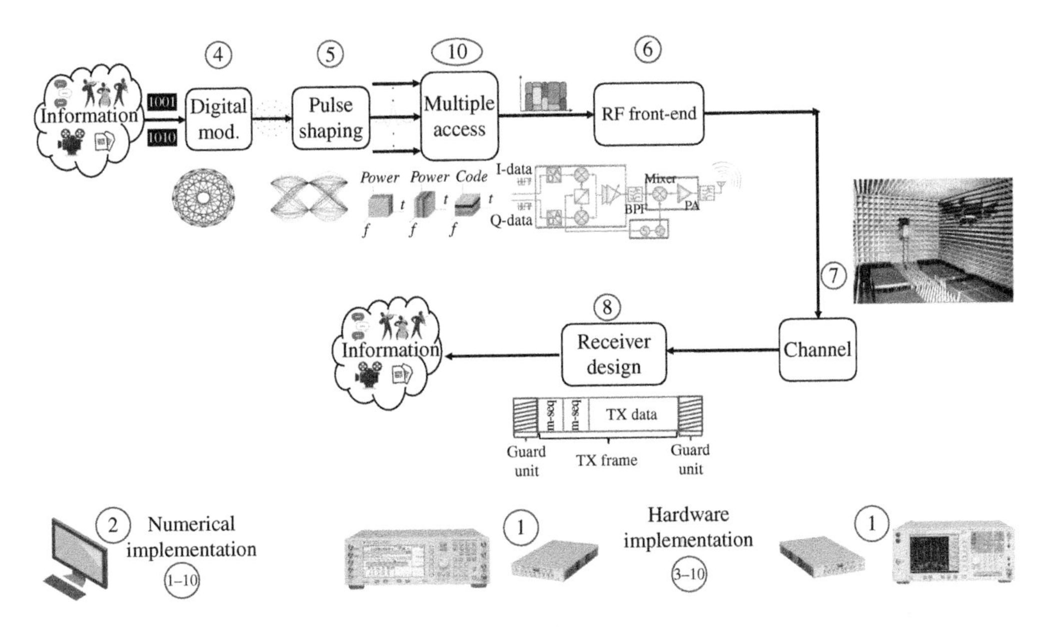

Figure 1.7 Functional model of the fundamental training.

quizzes and a final examination are also given to test trainees' understanding of the materials studied.

The course teaches telecommunications with an emphasis on physical, data link, and network layers. The training content presents the journey of bits through telecommunication systems as depicted in Figure 1.7. The theoretical content and numerical modeling can be taught either in a conventional classroom settings or computer laboratory or remotely. On the other hand, the hardware implementation requires a flexible radio platform that consists of SDR capable transmitter and receivers along with the modular RF front-end and configurable channel emulators. A basic SDR testbed includes a VSG, a VSA, and a computer that runs telecommunication system design and analysis tools as well as cables/antennas as shown in Figure 1.8. The signal generators and analyzer shown in the figure are high-end equipment with advanced capabilities. However, much lower cost versions are also available with relatively limited functionalities. In general, almost all the course contents can be delivered with high-end or low-cost equipment as all of them are SDR capable. A VSG can generate signals using various digital modulation techniques and baseband pulse shapes. Also, it can multiplex them in various domains such as time, frequency, and code to obtain standard and custom waveforms. These waveforms can either be generated internally using the VSG in standalone operation or externally using the computer and conveyed to VSAs. A VSA has the ability to demodulate the standard and custom signals in a standalone mode similar to a VSG. Also, it can convey the I/Q samples of the received signal to a computer for processing. The interaction between a computer and VSGs/VSAs is an excellent mechanism for teaching, studying, and analyzing the current and upcoming telecommunication systems.

The system design, scenarios, and channel conditions can be enriched further by extending the testbed with mobile transceivers, modular RF front-end, and channel emulation tools as presented in Figure 1.9. I/Q modems, DACs, analog-to-digital converters (ADCs), mixers, voltage-controlled oscillators (VCOs), power amplifiers (PAs), band-pass filters, cables, and antennas are some of the critical RF front-end components of telecommunication systems. Testing and measuring a telecommunication system using a modular RF front-end provides the ability to analyze the

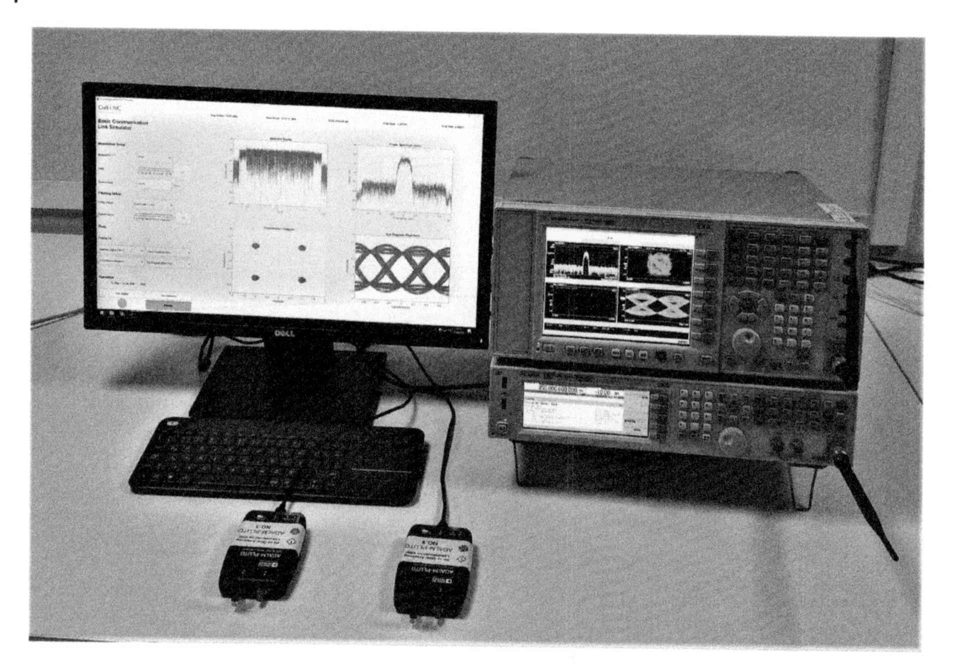

Figure 1.8 Basic SDR testbeds each consisting of a VSG, a VSA, a computer, and cables/antennas.

Figure 1.9 A flexible radio platform that features mobile SDR equipment, modular RF front-end, and configurable channel emulators.

function and effect of each component separately. In addition, mobile transceivers allow emulating various scenarios such as cellular hand-off and positioning. For high accuracy requirements, handheld VSGs and VSAs are preferable, whereas an abundance of low-cost SDR hardware is more convenient for experienced trainees to emulate applications involving networking and interference management. Furthermore, diverse channel conditions can be mimicked via reverberation and anechoic chambers, unmanned aerial vehicles (UAVs), and metal fans [8]. The frequency selectivity of the channel is controlled through the use of chambers, whereas time selectivity is managed using multispeed fans.

Basic telecommunication concepts can be taught in ten mandatory modules as numbered in Figure 1.7. Also, the final module (design project development) adapts trainees to contemporary telecommunication systems. A detailed description is provided on the training website,[1] and the key components of these modules are summarized as follows:

1.4.1 Introduction to the SDR and Testbed Platform

The primary objective of this module is to familiarize trainees with a basic SDR testbed. They learn how to use and control various SDR platforms (high-end vectors signal analyzers and generators, lower cost USRP, and even less expensive Adalm Pluto). Trainees learn how to generate and analyze signals utilizing the SDR platforms. They learn to generate basic single-carrier and multicarrier waveforms at the VSGs and assess their performances using the built-in functions of the VSAs along with the corresponding computer software. They also learn how to capture signals using SDRs and transfer these signals to the computers for further processing in Matlab. Similarly, they learn how to push Matlab generated signals to the signal generator hardware. Not only the signals that are generated by VSGs but also over the air signals such as FM radio, Wi-Fi, and Cellular signals are observed and analyzed using this module. Also, the received signal at VSA is downloaded to a computer, and spectro-temporal characteristics are analyzed using numerical tools. This part is especially important to illustrate the interconnection between the components of a basic SDR testbed.

1.4.2 Basic Simulation

In this module, trainees learn the simulation of a basic communication systems using Matlab. The importance of various models for simulation and a basic link setup is described and implemented. The performance of communication in additive white Gaussian noise is simulated and analyzed. Bit-error-rate (BER) versus signal-to-noise ratio (SNR) performance curves are obtained and interpreted. Basic modulation with rectangular pulse shaping (unlimited bandwidth), noise addition, optimal detector implementation are explained and implemented. Constellation diagram and polar diagrams are described and observed using the simulations. Observing the received signal and commenting on the quality of the signal under various noise levels are explained.

1.4.3 Measurements and Multidimensional Signal Analysis

Multi dimensional signal analysis and introduction to all performance measurement techniques are discussed in this module. The first part of this module mainly discusses signal characteristics in multiple domains such as time, spectrum, modulation, and code leading to the concept

1 http://wcsp.eng.usf.edu/courses/wcsl.html.

of multidimensional signal analysis. Multidimensional signal analysis is very important to understand challenges and trade-offs introduced by emerging wireless technologies. Learning multidimensional signal analysis helps trainees grasp difficult wireless concepts in laboratory environment. Then, trainees will learn all the link level measurements and visualization tools. The trainees will be familiar with the following items:

- Performance analysis tools and how to interpret them
- The spectral analysis and measurements (power spectrum, spectral masks, spectral flatness, frequency selectivity, bandwidth measurement
- Temporal analysis and time measurements (power versus time, transient behaviors, temporal selectivity of signals, histogram and time signal statistics)
- Spatial measurements (angular spread, space selectivity), etc.
- The relation between the multidimensional characteristics of the signals and exploring the dualities among various dimensions
- Correlation analysis
- Transient analysis of nonstationary signals (understanding and measuring transient behaviors of signals)
- Cyclostationarity analysis
- Code domain analysis
- Spectrogram
- Constellation and polar diagrams
- EVM, EVM versus time, EVM versus frequency
- PAPR, complementary cumulative distribution function (CCDF)
- Eye diagram
- Learn how to measure EVM, power spectrum, constellation, polar diagram, PAPR, SNR

1.4.4 Digital Modulation

Trainees commence the telecommunication system design by mapping bits to symbols after the overview in the first three modules. The theory part of the module teaches the purpose of digital modulation and its limiting factors such as SNR and PAPR. In the numerical and hardware implementation parts, various modulation schemes are demonstrated using built-in functions and characterized in terms of power efficiency, spectral efficiency, and ease of implementation. The analyses start with the throughput comparison of binary phase shift keying (BPSK) and higher-order quadrature-amplitude-modulation (QAM) and phase-shift-keying (PSK) modulation schemes as a function of SNR to demonstrate the trade-off between spectral and energy efficiency. Low order versus high order modulations and the trade-off between higher data rates and susceptibility to noise at higher orders of modulation are examined. This trade-off is made clear through the observation of EVM, I/Q polar diagram, eye diagram, and power CCDF curve. After initial characterization, alternative modulation schemes such as $\pi/2$-BPSK, $\pi/4$-DQPSK, Offset-QPSK, and minimum shift keying (MSK) that enhance PAPR to overcome hardware limitations at the expense of implementation complexity are demonstrated and examined. The importance of constant envelope modulation is explained and compared with nonconstant envelope modulation. After trainees comprehend the distinguishing characteristics of various digital modulation schemes, the instructor transmits a signal, and trainees attempt to figure out the digital modulation scheme blindly. Trainees understand the motivation behind the existence of various digital modulation schemes and link adaptation to channel conditions in this module.

1.4.5 Pulse Shaping

The objective of this module is to study the impact of baseband pulse shaping filters, which map digitally modulated symbols into waveforms, on telecommunication system performance. The discussions take off with orthogonal raised cosine (RC) filters. Trainees alter the roll-off parameter that controls the spectro-temporal characteristics, and observe changes in the spectrum, time envelope, power CCDF curve, I/Q polar diagram, and eye diagram. Furthermore, the performance of the RC pulse-shaping filter at the transmitter side is compared with using root raised cosine (RRC) pulse-shaping filter at both transmitter and receiver, and superiority of matched filtering is demonstrated. Also, an RRC pulse-shaped signal is transmitted without matched filtering at the receiver, and the presence of inter-symbol interference is pointed out. Once trainees are accustomed to the nonorthogonality in the filter design, the discussion continues with Gaussian pulse shaping filters that are preferred due to their spectral confinement. The similarity between the bandwidth-time (BT) product of Gaussian Pulse-shaping filters and the roll-off factor of RC pulse-shaping filters are pointed out. Global system for mobile communications (GSM) signals with different BT products are generated, and their performances are evaluated similar to RC pulse-shaped signals. After understanding the spectro-temporal characteristics of baseband pulse-shaping filters, the instructor transmits signals, and trainees predict the filter parameters. The fundamentals for spectral- and energy-efficient design are taught in this module, and robust filter design against RF front-end impairments and multiple access wireless channel are further elaborated in the following modules.

After completing this module, the trainee will have a full understanding of:

- Various filters and their effect on the communication system
- The effect of optimal match filtering under additive white Gaussian noise (AWGN)
- The effect of pulse shaping on eye diagram, bandwidth and time
- The effect of pulse shaping on CCDF
- Ambiguity function and time-frequency localization (time and spectrum analysis of the pulses)
- Localized versus nonlocalized pulses and their effects – inter-symbol interference (ISI)
- Details of raised cosine pulse shaping and impact of roll-off

1.4.6 RF Front-end and RF Impairments

Teaching the functionalities and characteristics of critical RF front-end components is the core objective of this experiment (see Figure 1.10). In the first stage of this module, the instructor informs trainees on I/Q modems and familiarize them with various impairments using the internal I/Q modulator on VSGs. I/Q gain imbalance, quadrature offset, and DC offset are intentionally introduced, and their effect on I/Q polar diagram, eye diagram, and EVM are exhibited. Following this stage, quantization effects are demonstrated by connecting an external DAC with various resolutions to VSGs and altering the dynamic range of VSAs. Trainees observe the results of quantization errors such as spectral regrowth, increase in EVM, interpolation issues in eye and I/Q polar diagrams. Afterward, the analog signal is conveyed to an upconverter unit that consists of a mixer and a VCO to shift the baseband signal to IF or RF. Trainees compare the stability of low-end standalone VCOs' output with that of VSGs. The reference signals generated by both VCOs are passed through mixers for upconversion. It is demonstrated that low-end standalone VCOs cannot output stable tones and produce phase noise.

Trainees analyze the effect of nonlinear behavior of mixers and PAs. First, they observe the spectrum while operating the PA in the linear region. Hence, the harmonics generated only by the mixer are presented. Afterward, the transmitted power is gradually increased, and power CCDF curve,

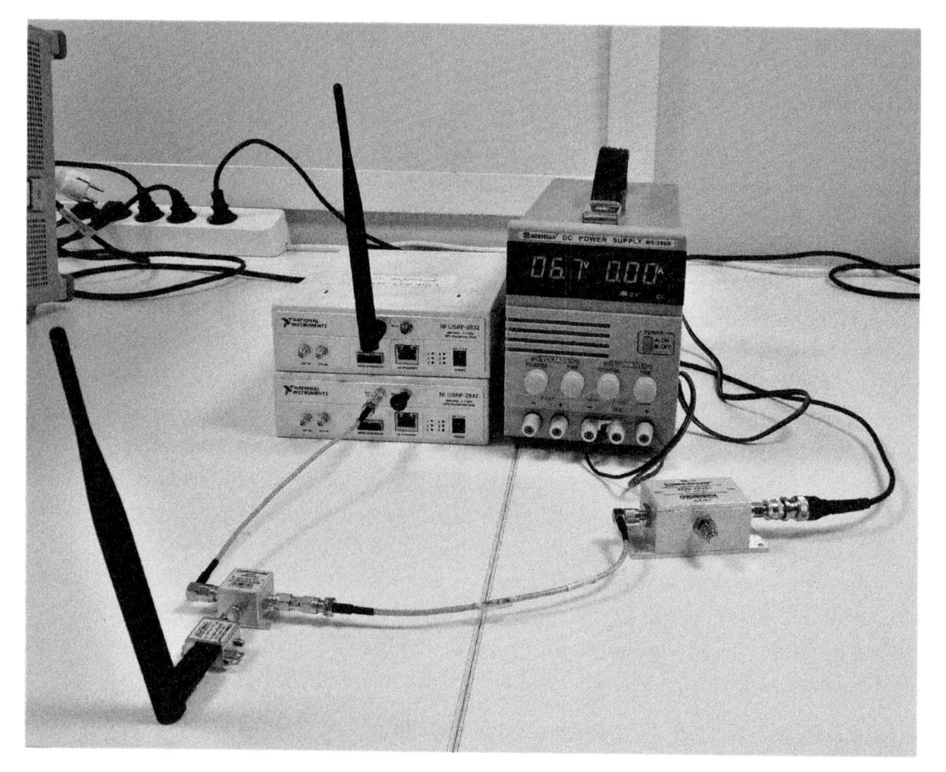

Figure 1.10 RF front-end design experiment.

EVM, I/Q polar diagram, and spectrum are observed. It is demonstrated that increasing the transmit power beyond the hardware limitations does not increase SNR. Band-pass filters are provided to suppress harmonics caused by both components. Furthermore, PAPR is reduced by using digital modulation techniques that omit zero-crossings and utilizing RRC pulse shapes with a higher roll-off factor. Thus, trainees understand both the inconveniences of these devices and techniques to cope with their disadvantages. Also, this module provides them an opportunity for building a complete RF front-end by connecting all the aforementioned components.

1.4.7 Wireless Channel and Interference

The purpose of this module is to present fundamental EM wave propagation concepts. The channel imposes various effects on EM waves such as distance and frequency-dependent path loss, large scale fading (i.e. shadowing), and small-scale fading (i.e. delay spread and Doppler spread). In the first stage, the distance- and frequency-dependent characteristics of the path loss are depicted by measuring the signal with different antenna separations at different frequencies. Trainees sweep the FM band and record the received power levels of various stations. They obtain the transmit power, antenna height, and distance information online. Afterward, a simple statistical path loss model is derived considering the measurements from the stations with the same parameters. Also, shadowing is demonstrated by pointing out the variation around the path loss.

The second stage of the experiment is concerned with small-scale fading, which originates from two phenomena, namely delay and Doppler spreads. This is illustrated in the lab using a reverberation chamber that provides control of the delay spread in an indoor environment. Trainees alter the transmission bandwidth and observe a frequency selective channel when the signal bandwidth exceeds the coherence bandwidth of the channel. It is challenging to demonstrate mobility in an indoor environment. Either the transceiver or the environment must be mobilized. A multispeed metal fan is utilized to create the variation in time. Trainees place a metal fan in between two antennas and operate it at different speeds while monitoring the spectrogram. Furthermore, a doubly-dispersive channel is emulated by moving the fan inside the reverberation chamber [8, 9]. Finally, trainees apply the concepts that are discussed in the previous modules to design telecommunication systems robust against channel effects. For example, the symbol rate is decreased to elevate immunity to delay spread, whereas it is increased to boost resistance to Doppler spread. Furthermore, the RRC roll-off factor is raised to improve the performance in both spreads with a penalty of degraded spectral localization.

1.4.8 Synchronization and Channel Estimation

The objective of this module is designing a complete digital baseband receiver. The instructor transmits a BPSK modulated time division multiple access (TDMA) frame. The frame consists of individualized messages following different repeated m-sequences for each testbed. Trainees capture the signal and record it for offline processing. They start with the coarse time synchronization step to pinpoint the position of their desired message. Since two m-sequences are appended back to back, their autocorrelation gives a rough estimate of the allocated slot beginning. Upon the coarse time synchronization step, the frequency offset amount is estimated using the constant delay between the consecutive m-sequences. The estimated frequency offset is compensated for the frequency synchronization. Then, fine time synchronization is performed by cross-correlating the local m-sequence copy with the frequency offset compensated signal. Trainees downsample the signal at the best sampling locations using the fine time synchronization information. Afterward, channel estimation must be carried out to compensate for the channel effect. The m-sequence of the received signal is compared with the local m-sequence copy in order to estimate the wireless channel. The phase and amplitude of the received signal are corrected upon estimation. In the final stage, digital demodulation is performed, and estimated bits are mapped to characters to reveal the transmitted information. Trainees are thrilled to be able to demodulate a physical signal similar to their daily devices. This module finalizes the modules that are pointed out in the functional diagram on Figure 1.7 and demonstrates a complete picture of a basic single-carrier communication system.

1.4.9 OFDM Signal Analysis and Performance Evaluation

The main features of orthogonal frequency division multiplexing (OFDM) waveform are studied in this module. The trainees learn to generate and receive a basic OFDM signal, and apply time and frequency domain analysis. Effects of impairments and synchronization errors on OFDM signal are investigated. Also, the effects of OFDM parameters (CP size, FFT size, etc.) on the system design considering hardware and channel impairments are experimented. The instructor supplies trainees with a cyclic prefix (CP)-OFDM transmission script along with the VSG and VSA connection libraries. In the first stage, trainees are expected to develop their OFDM baseband receiver algorithms similar to the previous module. Since there is no m-sequence in the frame, CP is used for time and frequency synchronization. Trainees repeat the hardware implementation steps of RF front-end and propagation channel modules for multicarrier communication after the receiver design. FFT size, number of active subcarriers, CP length, and modulation order are altered in each step, and their effect on the telecommunication system performance is analyzed through power

CCDF curve, constellation, and spectrum. Also, they compare the impairments for single-carrier and multicarrier communication schemes and point out the main differences. This module combines all the concepts that are covered in this training within the perspective of an advanced modulation scheme that is used in current cellular and Wi-Fi systems.

1.4.10 Multiple Accessing

The purpose of this module is to acquaint trainees with resource allocation among multiple users and to manage resulting interference. Previously, the testbeds share the channel in an frequency division multiple accessing (FDMA) manner. In this module, they get familiarized with other multiple accessing schemes such as TDMA, code division multiple access (CDMA), frequency hopping (FH), and orthogonal frequency division multiple access (OFDMA). Initially, GSM signals with varying relative burst powers for each slot are broadcasted by the instructor. Each slot is assigned to a certain testbed along with associated relative burst power. Trainees observe the TDMA frame structure and locate their slot. In the following stage, the instructor transmits various direct-sequence spread-spectrum (DSSS)-CDMA signals by assigning different codes to each testbed. As the number of active codes increases, trainees observe the power CCDF curve and conclude that an increased number of random variables degrades PAPR characteristics. Multiple CDMA signals are multiplexed into a time frame and trainees are asked their own data. Also, in this module, the trainees learn to obtain and interpret code domain power when multiple codes are active in the received signal. The trainees can identify active codes and their associated relative power levels by employing code domain power measurement. Similarly, trainees learn Rho measurement and understand the interpretation of this measurement for various code correlation and noise values. Also, a joint time-frequency analysis is performed on bluetooth signals from two smartphones using the spectrogram. The smartphones' distances to a VSA are utilized to identify the hopping structure of each frequency hopping spread spectrum code. If an arbitrary single user frequency hopping signal is generated, the trainees are asked to employ spectrogram measurement and identify the code sequence used at the transmitter. Finally, trainees learn OFDMA multiple accessing and understand the two-dimensional resource block concept and how to allocate these time-frequency resource blocks to different users.

Upon introducing conventional resource allocation techniques, multiple access interference is explained. Adjacent channel interference (ACI) is demonstrated by transmitting equipower RRC pulse-shaped signals in adjacent bands simultaneously without a guard band in between. Although the testbeds can demodulate their signals with slight performance degradation, they cannot demodulate each others' signals properly due to the near/far effect. In this step, trainees utilize their knowledge from the previous module and adjust the RRC roll-off factor to mitigate ACI. In the next stage, testbeds generate single-carrier signals at the same frequency and emulate a co-channel interference scenario. They observe that their communication performance significantly degrades since there is an insufficient spatial separation between the testbeds. Afterward, one of the testbeds switches to the DSSS scheme, and it is demonstrated that spread spectrum systems are more robust against narrowband interference. This module integrates previously learned concepts and improves multidimensional signal analysis skills.

1.4.11 Independent Project Development Phase

Trainees become ready to pursue independent projects upon the completion of the fundamental training phase. The independent project development phase requires trainees to acquire and apply new knowledge as needed without well-defined instructions and sharpens their analysis and synthesis skills. Also, it promotes teamwork and generates synergistic engineers. Trainees improve their presentation skills by demonstrating their projects effectively to a wide range of audience that includes undergraduate trainees, graduate trainees, faculty, and industry professionals. Discussions that take place during presentations and feedback received afterward are reportedly beneficial for professional interviews. The documentation of projects benefits technical writing skills. Furthermore, notable projects are encouraged for publication, and scientific interest among trainees is cultivated. Although novelty is not enforced, it is highly encouraged. Considering the last 50+ projects that are completed under the authors' supervision, the feasible and educational projects are categorically summarized to inspire prospective trainees as follows:

1.4.11.1 Software Defined Radio

The expertise acquired with the higher-end SDR platforms in the fundamental training phase equips trainees with the ability to build their own low-cost SDRs. These projects are especially suitable for embedded system developers seeking to implement the Internet of Things transceivers. The budget-limited SDR hardware comes with the price of excessive RF front-end impairments. Trainees must also develop advanced baseband algorithms to mitigate I/Q impairments, PA nonlinearities, and low-resolution DAC/ADC quantization issues.

1.4.11.2 Dynamic Spectrum Access and CR Experiment

In this project, trainees will learn how to scan a wide spectrum, estimate the spectrum usage (by other users or applications around), make a spectrum occupancy decision, and perform spectrum hand-off (i.e. move to another frequency band if the currently used spectrum is occupied). The project will be carried out in FM radio frequencies inside the laboratory environment. For carrying out the project 3 USRP boards will be used. Two of the USRP boards will be used for emulating a secondary transceiver (one board will be used for spectrum scanning and sensing and the other USRP board will be used for transmitting an FM radio broadcasting signal as a secondary basis). The third USRP board will be used by the instructor (or TA) to emulate the primary user (or interfering user). The secondary user will scan the spectrum, find an appropriate channel (the best channel or an interference-free channel) and start transmitting an FM broadcasting signal in an opportunistic manner. Other trainees in the room should be able to listen the broadcasting signal from their off-the-shelf FM radio by tuning to the band that the secondary user is transmitting. Hence, the trainees can confirm the transmission of secondary user signal and evaluate the quality of the FM transmission by the secondary user. While this operation is taking place, the instructor (or TA) will create another signal and transmit it over the band that the secondary user is utilizing (this will emulate the primary user or interfering signal). The secondary user should be able to detect the existence of primary user in this band through continual spectrum sensing operation, and make a spectrum hand-off to another band that is interference-free. Trainees will learn: How to use the USRP boards as a transceiver; how to control these boards from computer; spectrum sensing and measurement (real-time spectrum and interference awareness); FM signal generation and transmission via USRP board; dynamic spectrum access concept (real-time adaptation of the transmission or usage of channel for communication); Fourier analysis and spectral analysis; time-frequency analysis; FM modulation. Physical (PHY) layer, medium access (MAC) layer, and some network layer functionalities will be learned in this project. Trainees will be able to connect the knowledge of DSP, communication, and networking courses.

1.4.11.3 Wireless Channel

The fundamental training phase provides the ability to comprehend basic features of conventional cellular and Wi-Fi channels. Trainees with strong propagation engineering backgrounds may characterize and model different environments [10]. Furthermore, advanced channel features can be extracted to determine line-of-sight/nonline-of-sight conditions or to design an object identifier via machine learning. Location services driven by channel properties are also compelling projects. Triangular positioning, channel-based authentication, distance and angle of arrival estimation techniques are exemplary candidates. Moreover, the novel doubly-dispersive channel emulators [8, 9] used in the fundamental training phase were originally designed and implemented as trainee projects.

1.4.11.4 Wireless Channel Counteractions

After obtaining a well-understanding in channel characteristics, trainees can exploit them to improve telecommunication systems further. For example, multiple-input multiple-output (MIMO) systems are designed to take advantage of spatial diversity. Trainees with antenna and RF circuit design experience prefer MIMO front-end design and manufacturing, whereas trainees with solid communication theory background favor MIMO power allocation, channel estimation, and compensation projects. Furthermore, the millimeter wave channel, which will be deployed in 5G, can be studied as well. For instance, a trainee dealt with the millimeter wave blockage issue using an adaptive antenna beamwidth implementation as presented in [11].

1.4.11.5 Antenna Project

In this project, trainees will learn antenna patterns, directional and omnidirectional antennas and combine this knowledge with applications for sectorization, switched beam forming, angle of arrival-based locationing, spatial multiple accessing, adaptive antenna systems, and hand-off with sectored antennas (i.e. connect the theories they have learned in electromagnetics course with wireless communication systems, digital signal processing, and networking). Trainees will observe the radiation pattern of directional and omnidirectional antennas in the laboratory (with rich scattering environment) and in the corridors (with channel guide effects) and in open rooms and outside of the building. Portable laptop with USRP boards will be used for this purpose. Alternatively, handheld vector signal analyzer and generators can be used for the outdoor portion of this project. Once the trainees understand the beam patterns of various antennas, they will use the directional antennas for applications in communications. One of these applications is to employ hand-off process from one access point to another (this can be an emulation of one sector in cellular system, or an access point in Wi-Fi) as the transmitter antenna moves from one sector to another in the lab environment. For this project, three USRP boards are needed. One of the USRP boards will be used as mobile transmitter that is sending a signal to the access points. The other two USRP boards will be used as two different access points that are connected to directional antennas; each antenna is covering a sector. These two USRP boards will be controlled by a common computer (representing the mobile switching center or the common baseband controller). As the transmitter moves around, the baseband controller will receive signals from both access points, process the received signals using digital baseband signal processing, and decide which of the access points should be the point of contact.

1.4.11.6 Signal Intelligence

Trainees with digital signal processing (DSP) and communication theory backgrounds may utilize their extensive knowledge on multidimensional signal analysis, and communication channel to pursue signal intelligence applications. These applications include nondata aided blind receivers which estimate various parameters such as symbol rate, modulation type, and pulse shaping filter parameters in the presence of RF front-end and channel impairments. Furthermore, trainees can conceal the transmitted signal to prevent eavesdropping using various single- and multiantenna physical layer security techniques such as artificial noise transmission. Also, trainees may utilize channel-based authentication methods to identify the legitimate transmitter as well.

1.4.11.7 Channel, User, and Context Awareness Project

In this project, trainees will use one or multiple USRP boards to receive a signal transmitted from a transmitter and obtain some knowledge base from the received signals to identify information about the location of the transmitter, wireless channel between transmitter and receivers, and context of the transmitter (mobility, environment conditions, etc.). In other words, the receiver USRP boards will be used as radio sensors. Trainees will learn how to process received signal and how to extract useful information from multiple received signals. Trainees will be able to analyze received signals in multiple domains (time, spectrum, and angle) and also extract the channel correlation in multiple domains to identify the temporal, spectral, and angular variation of the channel. Then, this information will be used to identify some features of the transmitter like type of physical environment, location, mobility, etc. Trainees will also learn about the radio channel (path loss, shadowing, multipath) and relate the channel effect to the contextual information which promises further opportunities for the communication system. This project will be able to connect the knowledge learned from wireless communication, digital communication, digital signal processing, pattern recognition, electromagnetics, RF and microwave theory, and artificial intelligence courses.

1.4.11.8 Combination of DSP Lab with RF and Microwave Lab

In this project, the testbed can be integrated to the DSP lab and RF Microwave circuits lab for the design of an end-to-end communication system. Trainees will be able to connect their knowledge in DSP, RF and microwave circuits, and digital communications course. In many universities trainees take DSP lab (or microprocessor lab) course where they learn how to program a DSP processor (or General Purpose Processor). For example, they can develop a basic digital baseband transceiver in these lab courses without any problem. Similarly, in many universities, trainees take RF and microwave lab courses where they learn how to build RF front-end circuits and analyze these subsystems. However, both these courses are often disconnected and trainees cannot leverage the knowledge they have learned in these courses to develop a digital radio system. For example, a problem in an RF subsystem and its effect on the baseband signal processing is not considered properly. In this project, we will provide trainees this ability and also help them how to make use of the knowledge that they have learned from these courses to develop a complete system. The trainees will need a DSP (FPGA or general purpose processor [GPP]) board for implementing digital baseband algorithms. These boards need to have internal ADC and DAC. The RF front-end design needs to be integrated into the board. Therefore, trainees will need RF front-end components like oscillators, power supplies, filters, mixers, power amplifiers, low noise amplifiers, etc.

1.4.11.9 Multiple Access and Interference Management

The fundamental training phase mostly covers the centralized and orthogonal multiple accessing schemes. Alternative schemes such as cognitive radio (CR) and nonorthogonal multiple access (NOMA) attract trainees with data link layer interest and cultivate numerous projects. For example, the spectrum is utilized opportunistically in CR, and trainees avoid interfering the primary user. Furthermore, noncontiguous frequency resources can be aggregated in these projects. NOMA is another spectral-efficient resource utilization scheme and can be realized with multiuser detection, blind source separation, and interference cancellation algorithms. Moreover, trainees may manage self-interference and develop full-duplex communication schemes to improve the capacity of telecommunication systems.

1.4.11.10 Standards

Trainees can improve their expertise in telecommunication systems by partially implementing hardware layers of various standards such as 3G/4G/5G cellular, IEEE 802.11, Bluetooth, and stereo FM. In addition, those interested in network layer may realize vertical hand-off mechanisms to switch between different standards in a heterogeneous network scenario. Mobile SDR equipment is essential in these projects. Moreover, trainees can asses new technologies for future standards. For example, various waveforms might be implemented to evaluate their performances in realistic scenarios considering diverse channel conditions, multiple access interference, and RF front-end impairments.

1.5 Conclusions

The industry and academia require well-trained telecommunication engineers. The laboratory-based training is an excellent mechanism to educate outstanding engineers and to build a bridge between these two institutions by integrating theoretical proficiency, numerical modeling skills, and hands-on experience [12–22]. The provided array of skills as well as professional qualities help trainees to succeed in their professional journeys and to design future generations of telecommunication systems. Other institutions that are willing to setup a similar training can profit from the experiences shared in this chapter. Especially, the improvement of trainees confirm the effectiveness and make the proposed training a candidate for the flagship training of telecommunication education.

References

1 J. Mitola, "The software radio architecture," *IEEE Communications Magazine*, vol. 33, no. 5, pp. 26–38, May 1995.

2 ——, "III: Cognitive radio: an integrated agent architecture for software defined radio," Ph.D. dissertation, PhD thesis, KTH (Royal Institute of Technology), Stockholm, 2000.

3 M. D. L. Angrisani and M. D'Arco, "New digital signal-processing approach for transmitter measurements in third generation telecommunication systems," *IEEE Transactions on Instrumentation and Measurement*, vol. 53, no. 3, pp. 622–629, June 2004.

4 L. D. Vito and S. Rapuano, "A 3-D baseband signal analyzer prototype for 3G mobile telecommunication systems," *IEEE Transactions on Instrumentation and Measurement*, vol. 54, no. 4, pp. 1444–1451, Aug. 2005.

5 Agilent Technologies, Inc., "Agilent technologies application note: software defined radio measurement solutions," Printed in USA, 13 July 2007 [Online]. Available: http://cp.literature.agilent.com/litweb/pdf/5989-6931EN.pdf.

6 Tektronix, "Tektronix application note: software defined radio testing using real-time spectrum analysis," 14 Jan. 2007 [Online]. Available: http://www2.tek.com/.

7 Agilent Technologies, Inc., "Agilent E4438C ESG vector signal generator," Published in USA, 28 Feb. 2019 [Online]. Available: http://cp.literature.agilent.com/litweb/pdf/5988-4039EN.pdf.

8 S. Güzelgöz, S. Yarkan, and H. Arslan, "Investigation of time selectivity of wireless channels through the use of RVC," *Elseiver Journal of Measurement*, vol. 43, no. 10, pp. 1532–1541, 2010.

9 A. B. Kihero, M. Karabacak, and H. Arslan, "Emulation techniques for small-scale fading aspects by using reverberation chamber," *IEEE Transactions on Antennas and Propagation*, vol. 67, no. 2, pp. 1246–1258, Feb. 2019.

10 A. F. Demir, Q. H. Abbasi, Z. E. Ankarali, A. Alomainy, K. Qaraqe, E. Serpedin, and H. Arslan, "Anatomical region-specific in vivo wireless communication channel characterization," *IEEE Journal of Biomedical and Health Informatics*, vol. 21, no. 5, pp. 1254–1262, Sept. 2017.

11 S. Dogan, M. Karabacak, and H. Arslan, "Optimization of antenna beamwidth under blockage impact in millimeter wave bands," in *Proc. 2018 IEEE 29th Annu. Int. Symp. Personal, Indoor and Mobile Radio Commun.*, Bologna, IT, 9–12 Sept. 2018, pp. 1–5.

12 A. F. Demir, B. Pekoz, S. Kose, and H. Arslan, "Innovative telecommunications training through flexible radio platforms," *IEEE Communications Magazine*, vol. 57, no. 11, pp. 27–33, 2019.

13 M. Karabacak, A. H. Mohammed, M. K. Özdemir, and H. Arslan, "RF circuit implementation of a real-time frequency spread emulator," *IEEE Transactions on Instrumentation and Measurement*, vol. 67, no. 1, pp. 241–243, 2017.

14 A. Gorcin and H. Arslan, "An OFDM signal identification method for wireless communications systems," *IEEE Transactions on Vehicular Technology*, vol. 64, no. 12, pp. 5688–5700, 2015.

15 ——, "Signal identification for adaptive spectrum hyperspace access in wireless communications systems," *IEEE Communications Magazine*, vol. 52, no. 10, pp. 134–145, 2014.

16 S. Guzelgoz and H. Arslan, "Modeling, simulation, testing, and measurements of wireless communication systems: a laboratory-based approach," in *2009 IEEE 10th Annual Wireless and Microwave Technology Conference*, Clearwater, FL, 20–21 Apr. 2009, pp. 1–5.

17 S. Ahmed and H. Arslan, "Cognitive intelligence in UAC channel parameter identification, measurement, estimation, and environment mapping," in *OCEANS 2009-EUROPE*, Bremen, 11–14 May 2009, pp. 1–14.

18 M. E. Sahin and H. Arslan, "MIMO-OFDMA measurements; reception, testing, and evaluation of wimax MIMO signals with a single channel receiver," *IEEE Transactions on Instrumentation and Measurement*, vol. 58, no. 3, pp. 713–721, 2009.

19 A. M. Hisham and H. Arslan, "Multidimensional signal analysis and measurements for cognitive radio systems," in *2008 IEEE Radio and Wireless Symposium*, Orlando, FL, 22–24 Jan. 2008, pp. 639–642.

20 M. Sahin, H. Arslan, and D. Singh, "Reception and measurement of MIMO-OFDM signals with a single receiver," in *2007 IEEE 66th Vehicular Technology Conference*, Baltimore, MD, 30 Sept.–3 Oct. 2007, pp. 666–670.

21 H. Arslan, "IQ gain imbalance measurement for OFDM based wireless communication systems," in *2006 IEEE Military Communications Conference*, Washington, DC, 23–25 Oct. 2006, pp. 1–5.

22 H. Arslan and D. Singh, "The role of channel frequency response estimation in the measurement of RF impairments in OFDM systems," in *2006 67th ARFTG Conference*, San Francisco, CA, 16 June 2006, pp. 241–245.

2

Performance Metrics and Measurements

Hüseyin Arslan

Department of Electrical Engineering, University of South Florida, Tampa, FL, USA
Department of Electrical and Electronics Engineering, Istanbul Medipol University, Istanbul, Turkey

Measuring and evaluating the performance of wireless radio communication is a critical step toward developing a clear understanding the behavior of the whole system, subsystems, or components. Some of the measurements can take place offline during the development and design phases of the system and subsystems, while others need to be carried out during the operation of the actual communication in real time. For example, during the operation of communication, the quality of the link or network needs to be monitored continuously so that the necessary steps can be taken to improve the performance. These measurements can take place in different parts of the network (for example, in mobile stations or base stations in cellular networks). The network entities can also share these measurements between each other as the location of the measurement and the adaptation of the system parameters based on these measurements can take place in different parts of the network. In this chapter, various metrics and measurement techniques will be discussed to inspect the communication system architecture and its performance.

2.1 Signal Quality Measurements

Testing, trouble shooting, verification, design, and real-time adaptation of wireless communication systems need metrics to quantify the performance. Signal quality estimation is by far the most important measure that can be used in adaptive systems [1]. Different ways of measuring quality of the signal exist, and many of these measurements are done in the physical layer using digital baseband signal processing techniques [2–19]. In an actual receiver, the target quality measure is the frame-error-rate (FER) or bit-error-rate (BER), as these are closely related to higher level quality of service parameters like speech and video quality. However, these measurements do not provide enough information about the impairments and the quality of the signal. Therefore, other types of signal quality measurements, which are related to BER and FER, might be preferred. When the received signal is impaired only by white Gaussian noise, analytical expressions can be found relating the BER to other measurements. For other impairment cases, like colored interferers and RF front-end-related noise and impairment sources, numerical calculations and computer simulations that relate these measurements to BER can be performed. Therefore, depending on the system, the signal quality is related to the BER. Then, for a target BER (or FER), a required signal quality threshold can be calculated to be used with the adaptation algorithm.

The measurements can be performed at various points of a receiver, depending on the complexity, reliability, and delay requirements. Owing to the conflicting nature of different requirements, the latter two for instance, their simultaneous optimization might not be possible. System/network architects, therefore, need to make design choices keeping in mind the trade-offs for each user and

Wireless Communication Signals: A Laboratory-based Approach, First Edition. Hüseyin Arslan.

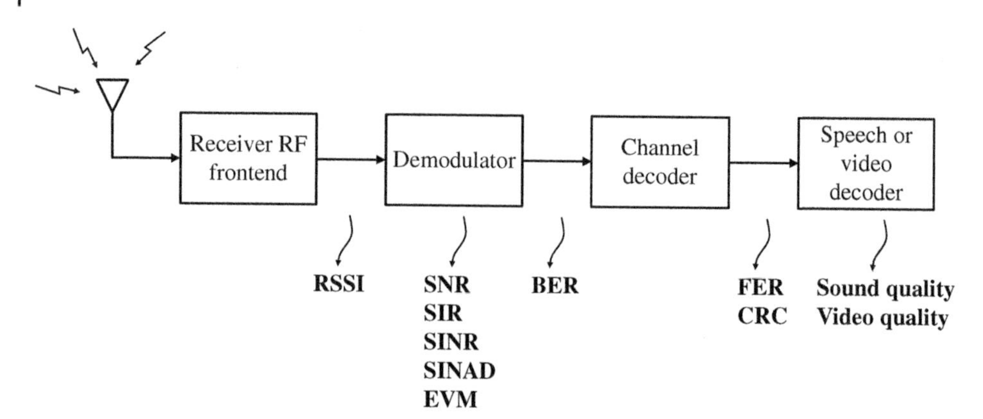

Figure 2.1 Various quality measurement techniques and possible locations to estimate these values in the receiver chain.

application. Figure 2.1 shows a simple example where some of these measurements can take place. In the following sections, these measurements will be discussed briefly.

2.1.1 Measurements Before Demodulation

Received signal strength (RSS) estimation provides a simple indication of how strong the signal is at the receiver front-end. If the received signal strength exceeds a threshold, then the link is considered to be good. The RSS measurement is simply done by reading samples from a channel and averaging these samples. However, averaging needs to be done carefully, especially for bursty and nonstationary signals. When the signal is discontinuous or nonstationary, long averaging intervals provide incorrect results. It is important that each segment is averaged separately, for example, in bursty transmission the averaging should be done when the signal is ON to get the RSS value of the signal during the transmission. Similarly, when the signal transmission is OFF, RSS measurement can be used to get an idea of the noise and interference power.

Compared to other measurements, RSS estimation is simple and computationally less complex, as it does not require processing and demodulation of the received samples. However, the received signal includes noise, interference, and other channel impairments. Therefore, receiving a high RSS measurement does not tell much about the channel and the signal quality. Instead, it gives an indication whether a strong signal is present or not in the channel of interest.

If the measurement is done in a wireless channel with a portable measurement device, the received signal power fluctuates rapidly due to fading. In order to obtain reliable estimates, the signal needs to be averaged over a time window to compensate for short term fluctuations. The averaging window size depends on the system, application, variation of the channel, etc. For example, if multiple receiver antennas are involved at the receiver, the window can be shorter compared to a single antenna receiver. For measurements with a cable connected between the device under test (DUT) and receiver, this is not an issue. Therefore, even short window of measurements can provide reliable received signal strength indicator (RSSI) values.

Most wireless receivers (like cell phones, Wi-Fi receivers) include a display element which reflects the RSS information. When there is an ongoing communication session, the RSS indicates measurements of the strength of a signal received over a communication channel. When a communication session is not in progress, the RSSI may reflect measurements of the strength of a control channel signal.

2.1.2 Measurements During and After Demodulation

Signal-to-interference-ratio (SIR),

$$SIR = \frac{P_{signal}}{P_{interference}}, \tag{2.1}$$

where P_{signal} is the signal power and $P_{interference}$ is the interference power, and signal-to-noise-ratio (SNR),

$$SNR = \frac{P_{signal}}{P_{noise}}, \tag{2.2}$$

where P_{noise} is the noise power, and also signal-to-interference-plus-noise-ratio (SINR) ,

$$SINR = \frac{P_{signal}}{P_{interference} + P_{noise}}, \tag{2.3}$$

are most common ways of measuring the signal quality during (or just after) the demodulation of the received signal. SIR (or signal-to-noise ratio [SNR], or SINR) provides information on how strong the desired signal is compared to the interferer (or noise, or interference plus noise). Most wireless communication systems are interference limited, therefore, SIR and SINR are more commonly used. Compared to RSS, these measurements provide more accurate and reliable estimates at the expense of computational complexity and additional delay. In measurement world, in addition to noise, distortion is also considered frequently. Therefore, signal-to-noise-and-distortion (SINAD) is also used as a performance measure,

$$SINAD = \frac{P_{signal}}{P_{noise} + P_{distortion}}, \tag{2.4}$$

where $P_{distortion}$ represents the distortion.

SIR estimation can be employed by estimating the signal power and the interference power separately and then by taking the ratio of these two. The training (or pilot) sequences can be used to obtain the estimate of SIR. Instead of using a training sequence, the data symbols can also be used for this purpose. For example, in [20], where signal-to-noise ratio (SNR) information is used as a signal quality indicator for rate adaptation, the cumulative Euclidean metric corresponding to the decoded trellis path is exploited for channel quality information. Another method for channel quality measurement is the use of the difference between the maximum likelihood decoder metrics for the best path and the second best path. In a sense, in this technique, some sort of soft information is used as channel quality indicator. However, this approach does not tell much about the strength of the interferer or the desired signal. There are several other SNR measurement techniques available in the literature, which can be found in [21].

Often, in obtaining the estimates, the impairment (noise or interference) is assumed to be white and Gaussian distributed in order to simplify the estimation process. However, in wireless communication systems, the impairment might be caused by a strong interferer that is colored. For example, in orthogonal frequency division multiplexing (OFDM) systems, where the channel bandwidth is wide and the interference is not constant over the whole band, it is very likely that some part of the spectrum is affected more by the interferer than the others. Hence, when the impairment is colored, estimates that take the color of the impairment into account might be preferable.

2.1.2.1 Noise Figure

Noise figure of a subsystem (or component) can be defined as the ratio of the SNR at the input to the SNR at the output of the system:

$$Noise\ Figure = \frac{SNR_{input}}{SNR_{output}}. \tag{2.5}$$

An ideal subsystem or component (let's say an amplifier) should have the same SNR at the input and output. Note that an amplifier boosts the power of both signal and noise at the input of the amplifier, therefore, the SNR will remain the same if the amplifier does not introduce additional noise. However, for practical amplifiers, the output SNR will be less since the amplifier will introduce additional noise and impairments into the system. The noise figure measurement helps the designers to identify the performance of the subsystem (or components). A low noise figure is desired as it means less noise is introduced to the system. The overall noise figure of a system can be calculated from the individual noise figures of the subsystems.

2.1.2.2 Channel Frequency Response Estimation

Channel frequency response (CFR) estimates provide information about the desired signals amplitude and phase variation across frequency. It is a much more reliable estimate than RSSI information, as it does not include the other impairments as part of the desired signal power. However, it is less reliable than SINR estimates, since it does not provide information about the noise and/or interference powers with respect to desired signals power. However, for white noise (like additive white Gaussian noise [AWGN]), channel frequency response estimate can also provide an idea about CINR expected at each frequency bin. For wireless measurements, CFR provides information about the dispersion (selectivity) of the medium. For measurements where the receiver is connected to the DUT with cable, CFR can provide an idea about the filter responses used at the transceivers. CFR is also useful for measuring spectral flatness which is a mandatory measurement required by many standards.

2.1.3 Measurements After Channel Decoding

Channel quality measurements can also be based on postprocessing of the data (after demodulation and decoding). BER, symbol-error-rate (SER), FER, and cyclic redundancy check (CRC) are some of the examples of the measurements in this category. BER (or FER) is the ratio of the erroneous bits (or frames) relative to the total number of bits (or frames) received during the transmission. The CRC indicates the quality of a frame, which can be calculated using parity check bits through a known cyclic generator polynomial. FER can be obtained by averaging the CRC information over a number of frames. In order to calculate the BER, the receiver needs to know the actual transmitted bits, which is not possible in practice. Instead, BER can be calculated by comparing the bits before and after the decoder. Assuming that the decoder corrects the bit errors that appear before decoding, this difference can be related to BER. Note that the comparison makes sense only if the frame is error-free (good frame), which is obtained from the CRC information. Note that in testing the DUT with the standard defined data (specified by the standard), the BER calculation is easy, as we know what is being transmitted, and we can compare it against what is received to get the BER performance.

Although these estimates provide excellent signal quality measures, reliable estimates of these parameters require observations over a large number of frames. Especially, for low BER and FER measurements, extremely long transmission intervals will be needed. Therefore, for some applications these measures might not be appropriate. Note also that these measurements provide information about the actual operating condition of the receiver. For example, for a given RSS or SINR measure, two different receivers which have different performances will have different BER or FER measurements. Therefore, BER and FER measurements also provide information on the receiver capability as well as the link quality.

2.1.3.1 Relation of SNR with BER

BER and SNR (or SINR or SINAD) are related under some situations. For example, when the noise is AWGN, then an analytical relation between SNR and BER can be derived. This relation depends on the modulation type and order. In many practical impairment conditions an analytical relation might not be possible, but, a loose relation between BER and SNR (SINR) still exists. We often use simulations to find out this relation. For example, in the following Matlab code, the BER performance of a binary phase shift keying (BPSK) modulated signaling is observed under AWGN channel for various SNR values. In the same code, a simple technique for calculating the SNR is also given. In obtaining the plots, these calculated SNR values are used instead of the desired valued. The corresponding performance curve is shown in Figure 2.2.

```
1   % A simple code to calculate BER vs SNR
2   N=1e6; % block size in terms of number of symbols
3   M=2;
4   SNR=[0:10]; % desired SNR values
5   symb=(randn(1,N) > 0)*2-1; % BPSK symbols
6   noise=(randn(1,N)+sqrt(-1)*randn(1,N))/sqrt(2);
7   sgnl_P=mean(abs(symb).^2); % calculated signal power
8   sgnl_N=mean(abs(noise).^2);% calculated noise power
9   for kk=1:length(SNR)
10      noise_var=10^(-SNR(kk)/10) ;
11      sgnl_rx=symb+noise*sqrt(noise_var);
12      symb_dtct=sign(real(sgnl_rx));
13      BER(kk)=(length(find(symb ≠ symb_dtct)))/N;
14  end
15  SNR_calculated=10*log10(sgnl_P/sgnl_N)+SNR;
16  semilogy(SNR_calculated, BER)
17  grid
18  xlabel('SNR (dB)')
19  ylabel('BER')
```

2.1.4 Error Vector Magnitude

While BER and FER are useful as conceptual figure of merits, they suffer from a number of practical drawbacks that compromise their value as a standard test in manufacturing or maintenance [22]. Calculation of BER has limited diagnostic value. If the BER value measured exceeds accepted limits, it offers no clue regarding the probable cause or source of signal degradation. Similarly, over a measurement interval with finite number of bits (even if the number of bits tested is very large), having a zero BER does not mean the system is perfect. For measurements and testing devices, error vector magnitude (EVM) is a viable alternative test method when looking for a figure of merit in nonregenerative transmission links. EVM can offer insightful information on the various

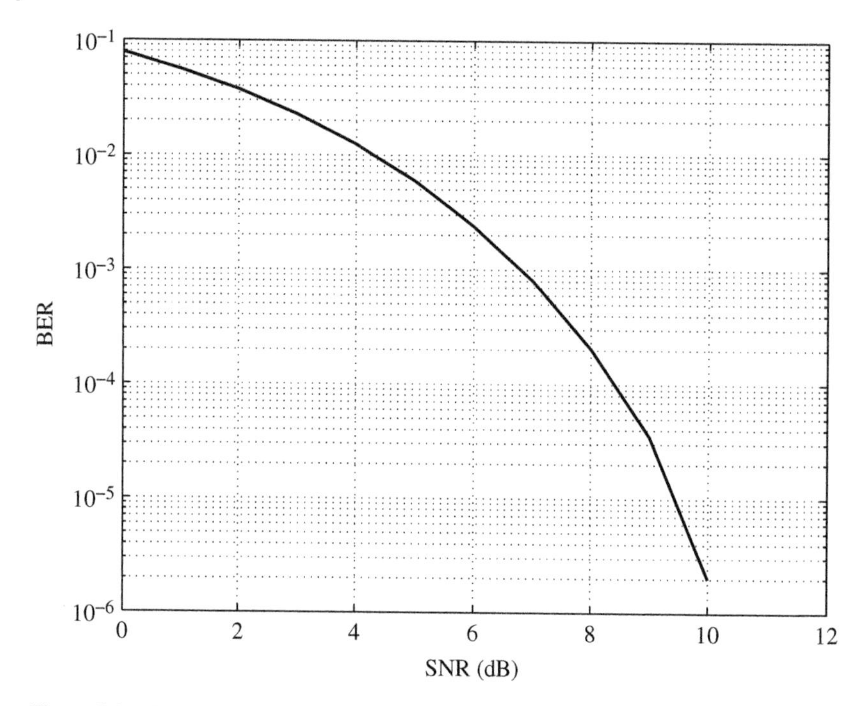

Figure 2.2 Relation with BER and SNR for BPSK modulation in AWGN noise.

transmitter imperfections, including carrier leakage, in-phase and quadrature-phase (I/Q) (in-phase and quadrature phase) mismatch, nonlinearity, local oscillator (LO) phase noise, and frequency error [23]. Requirements on EVM is already part of most wireless communications standards such as the IEEE 802.11 standard [24] and the IEEE 802.16 standard [25]. EVM measurements and simulations can be found in [22, 26, 27]. The impact of I/Q imbalance, as well as LO phase noise on EVM is investigated in [28] and EVM as a function of these impairments is derived. This work has been expanded in [23], where the effects of carrier leakage, nonlinearity and LO frequency error on EVM are considered. EVM can be defined as the root-mean-square (RMS) value of the difference between a collection of measured symbols and ideal symbols [29]. The value of the EVM is averaged over typically a large number of symbols and it is often expressed as a percentage (%) or in dB. The EVM can be represented as [30],

$$EVM_{RMS} = \sqrt{\frac{\frac{1}{N}\sum_{k=1}^{N}|s_r(k) - s_t(k)|^2}{P_0}}, \tag{2.6}$$

where N is the number of symbols over which the value of EVM is measured, $s_r(k)$ is the normalized received kth symbol which is corrupted by Gaussian noise, $s_t(n)$ is the ideal/transmitted value of the kth symbol, and P_0 is either the maximum normalized ideal symbol power or the average power of all symbols for the chosen modulation. For the remainder of this chapter, the latter definition of P_0 is used. In such case,

$$P_0 = \frac{1}{M}\sum_{m=1}^{M}|s_m|^2. \tag{2.7}$$

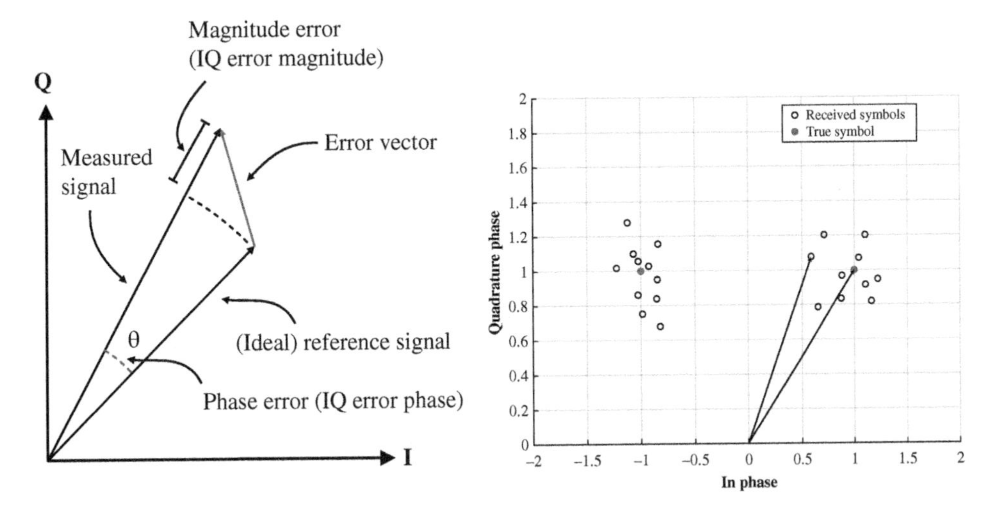

Figure 2.3 BPSK constellation plot that shows the ideal reference symbols and the actual received symbols. The vector difference between one of the received samples and the reference symbol is shown. The EVM values can be expressed in dB or as a percentage of the peak (or average) signal level. If the peak value is used, then, the corner state of the constellation diagram is used. For the average value, the average of all the constellation points is used. In BPSK (or quadrature phase shift keying [QPSK]) both average and peak values are the same.

The EVM value is normalized with the average symbol energy to remove the dependency of EVM on the modulation order. In Eq. (2.6), $s_r(n)$ is the detected symbol and $s_t(n)$ could be either known to the receiver (data-aided) if pilots or preambles are used to measure the EVM, or estimated from the received samples (nondata-aided) if data symbols are used instead. The following code calculates EVM for a BPSK modulated signals. Also, in Figure 2.3, the constellation diagram of the true BPSK symbols and detected samples can be seen. The error vector between one of these received samples and the true symbol is also shown in this figure.

```
1  % A simple code to calculate EVM and SNR
2  N=1e6; % block size in terms of number of symbols
3  SNR=15; % desired SNR in dB
4  symb=(randn(1,N) > 0)*2-1;
5  noise=(randn(1,N)+sqrt(-1)*randn(1,N))/sqrt(2);
6  noise_var = 10^(-SNR/10) ;
7  sgnl_rx=symb+noise*sqrt(noise_var);
8  scatterplot(sgnl_rx)
9  symb_dtct = sign(real(sgnl_rx));
10 EVM = sqrt(mean(abs(sgnl_rx-symb_dtct).^2));
11 SNR = 10*log10(1/EVM^2);
```

Special cases of EVM, phase error (sometimes also referred as I/Q error phase), and magnitude error (or sometimes referred as I/Q error magnitude) are also used as a performance metrics. In phase error, the error difference between ideal (reference) and actual (received) phases is calculated and averaged. On the other hand, in magnitude error, the magnitude difference between the actual and ideal signals are calculated. We refer these as special case of EVM, because in EVM, the vector error (both phase and magnitude) are involved.

2.1.4.1 Error-Vector-Time and Error-Vector-Frequency

As discussed in the previous section, to get a reliable EVM measurement, the individual error vector magnitude values are averaged over a large number of symbols. Sometimes, the EVM values change in time or over the frequency. Therefore, averaging EVM over both time and frequency might not provide the whole picture and misguide engineers. Therefore, there are two other popularly used EVM related measurements: error vector time (EVT) and error vector spectrum (EVS). Sometimes, we refer them as EVM versus time and EVM versus frequency.

EVT shows the EVM variations over time. This would be very useful if there are short transient noise (or interference) in the received signal. Similarly, the channel adaptation errors (channel tracking errors), sample clock error, and frequency offset can be detected easily by measuring EVT. EVT can also be useful to estimate some of these linearly varying impairments.

EVS provides the error vector magnitude variation over spectrum. Especially, this is a very important measurement for multicarrier modulation systems (like OFDM). Spectral flatness, narrowband interference, colored noise can be extracted from EVS measurement. In OFDM systems, sample clock error can also be observed from EVS measurement.

The above arguments on EVT and EVS can be expanded to other domains as well. For example, in multiple antenna systems, the EVM can be measured over multiple radio branches and EVM values can be obtained separately on each branch. Similarly, in adaptive beamforming and smart antenna system, EVM can be calculated with different beam formations, leading to a concept of EVM over angle (or error vector angle). Such measurements are not yet widely exploited in the literature, but, with the significant penetration of MIMO and smart antenna systems into the new generation of wireless standards, these measurements will also be gaining popularity.

2.1.4.2 Relation of EVM with Other Metrics

Relating EVM to other performance metrics such as SNR and BER is important [29, 31]. These relations are quite useful since it allows the reuse of already available EVM measurements to infer more information regarding the communication system. Moreover, using EVM measurements could reduce the system complexity by getting rid of the need to have separate modules to estimate or measure other metrics. When relating EVM to SNR, one assumption that has been made is that the EVM is measured using known data sequences (e.g. preambles or pilots) or that the SNR is high enough that symbol errors are negligible [29, 31]. Furthermore, assuming the presence of one or two dominant imperfections, the remaining ones can be modeled as Gaussian noise [23, 28]. However, the above analyses does not consider measuring EVM blindly for low SNR levels where symbol errors are possible. With new technologies for spectrum detection and utilization [32, 33], there is a need to detect signals at medium to very low SNR values. The signal quality metrics including EVM and SNR are possibly measured over unknown data sequences (nondata-aided) as well.

For data-aided EVM calculations, Eq. (2.6) is reduced to,

$$EVM_{RMS} = \sqrt{\frac{\frac{1}{N}\sum_{n=1}^{N}|\eta(n)|^2}{P_0}}, \tag{2.8}$$

where $\eta(n)$ is the instantaneous vector error for a given symbol n. If the EVM is measured over large values of N, then [29, 31, 34],

$$EVM_{RMS} \approx \sqrt{\frac{N_0}{P_0}} = \sqrt{\frac{1}{SNR}}, \tag{2.9}$$

where $N_0/2 = \sigma_n^2$ is the noise power spectral density.

The above EVM-SNR relation, however, only holds for data-aided receivers. For nondata-aided receivers, the transmitted symbols are estimated and those estimates are used to measure the EVM value. Thus, for low SNR values, errors are made when estimating the transmitted symbols. The measured EVM value in this case is expected to be less than its actual value as the symbol estimator tends to assign received symbols to their closest possible constellation point.

2.1.4.3 Rho

In CDMA systems, Rho is one of the key modulation quality metrics. Rho is the ratio of the correlated power in a multi coded channel to the total signal power. This measurement takes into account all possible impairments in the entire transmission chain. Rho provides the normalized correlation coefficient between the measured and ideal reference signals i.e.

$$\rho = \frac{P_{X_iX_m}}{P_{total}}, \tag{2.10}$$

where $P_{X_iX_m}$ is the correlated power of measured signal and ideal reference and P_{total} is the total signal power.

In other words, Rho indicates the fraction of the energy of the transmitter signal that correlates with the ideal reference signal. In the ideal situation (when there is only one active code), we obtain the maximum value of 1.0, which means the measured signal and reference signal are 100% identical. When there are noise and impairments in the system, then depending on the level of these noise and impairments sources the Rho value decreases. The distribution of power among the code channels can also be obtained by measuring Rho for different active codes, leading to code domain power (which will be discussed in Chapter 3). The following simple code relates the SNR with Rho measurement for a specific CDMA code obtained from Hadamard matrix. The related Rho is plotted in Figure 2.4 with respect to SNR.

```
1   %  Rho measurements
2   %  Observe the effect of noise in Rho
3   hdmrd=hadamard(1024);
4   sgnl_tx=hdmrd(518,:);
5   noise=randn(1,length(sgnl_tx));
6   SNR=[-3 0 3 6 9 12 15 20 25 30 35];
7   for ll=1:length(SNR);
8       noise_var = 10^(-SNR(ll)/10) ;
9       sgnl_rx = sgnl_tx + sqrt(noise_var) *noise;
10      sgnl_rx = sgnl_rx / sqrt(mean(abs(sgnl_rx).^2)) ;
11      Rho(ll) = sum(sgnl_rx .* conj(sgnl_tx)) / sum(abs(sgnl_tx).^2);
12  end
```

2.1.5 Measures After Speech or Video Decoding

The speech and video quality, the delays on data reception, and network congestion are some of the parameters that are related to user's perception. These measures are often related to the other measures mentioned in the previous sections. For example, speech quality for a given speech coder can be related to FER of a specific system under certain assumptions [35]. However, as discussed in [35], some frame errors cause more audible damage than others. Therefore, it is still desired to find ways to measure the speech quality more reliably (and timely), and to adapt the system parameters accordingly. Speech (or video) quality measures that takes the human perception of the speech (or video) into account would be highly desirable.

Perceptual speech quality measurements have been studied in the past. Both subjective and objective measurements are available. Subjective measurements are obtained from a group of people who rate the quality of the speech after listening to the original and received speech. Then, a mean opinion score (MOS) is obtained from these feedbacks. Although these measurements reflect

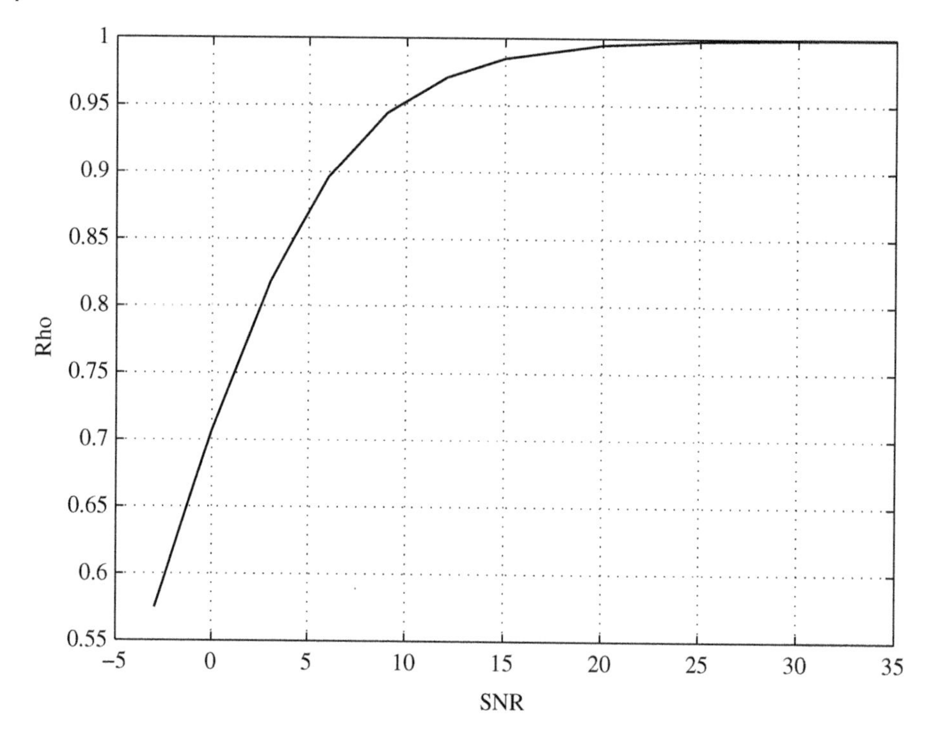

Figure 2.4 The impact of noise in Rho measurement is illustrated for a code of length 1024. The code is generated using Hadamard matrix.

the exact human perception that is desired for adaptation, they are not suitable for adaptation purposes as the measurements are not obtained in real-time. On the other hand, the objective measurements can be implemented at the receiver in real-time. However, these measurements require sample of the original speech at the receiver to compare the received voice with the undistorted original voice. Therefore, they are not applicable for many scenarios as well.

2.2 Visual Inspections and Useful Plots

In addition to various measurements to identify the quality of the signal, visual graphs are also very useful to get insight on impairments and other disturbances impacting the signal. For example, the points on a constellation diagram represent the modulated symbols. Constellation diagram (sometimes also referred as scatterplot) can be obtained by plotting the imaginary versus the real part of the complex signal. Received noisy (and possibly impaired) samples representing several symbols can be visualized as scatter diagram. With constellation diagram, it is very easy to see the I/Q impairments (like I/Q gain imbalance, I/Q quadrature error, I/Q offset), impact of noise and interference in the constellation points, effect of compression due to nonlinear devices, effect of phase noise, frequency offset, etc. When the system is working perfectly (ideal case), the constellation diagram shows the received samples at the exact position of the symbol alphabets. When the noise is introduced to the system, then the received samples form a cloud around the actual symbol alphabets. The size of cloud increases as the SNR decreases as shown in Figure 2.5.

Note that the constellation diagram is obtained at the symbol times (one sample per symbol). Polar diagram (or often referred as vector diagram) shows the signal trajectory (transition of signal

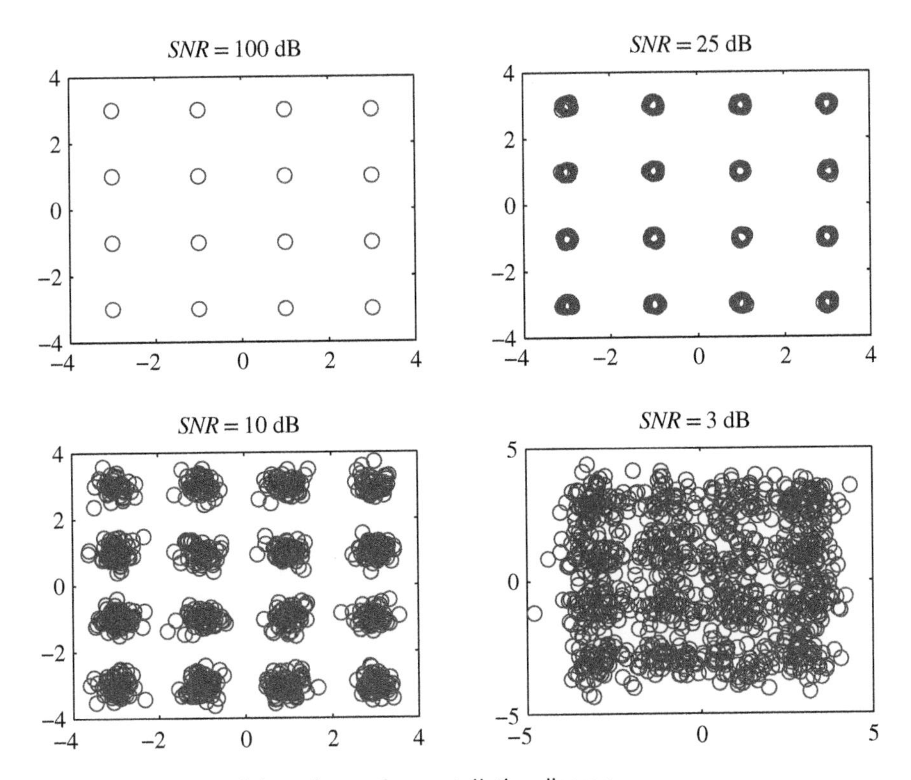

Figure 2.5 Impact of the noise on the constellation diagram.

from one state to the other). The values are shown not only at the symbol times but also during the transition from one symbol to another. You can see the effect of filters, modulation, channel, and effect of compression using the vector diagram.

Eye diagram, which displays a number of superimposed data patterns that resemble an eye, is another excellent visualization technique for digital communication systems. Intersymbol interference, sample clock error, and sample phase errors can be identified easily by using eye diagrams. A wide-open eye pattern corresponds to minimal signal disturbance. Optimum sampling time is the time at which the eye opening is the largest. The width of the eye provides information about tolerance to jitter (i.e. it provides information on the time interval over which the received signal can be sampled without error from noise and inter symbol interference). The height of the eye gives information about tolerance to additive noise. Effect of pulse shaping and channel characteristics on a digital modulation scheme can be studied by the use of eye patterns. Similarly, eye diagram allows visual inspection of the impact of impairments and how the samples are positioned with respect to the decision boundary, and hence providing the vulnerability of the samples to symbol errors.

Eye patterns can be obtained by wrapping the signal back onto itself in periodic time intervals and by retaining all traces. For a fixed time window equal to the symbol period, the I or Q component values are superimposed on top of one another to give the characteristic of an eye (see Figure 2.6). In binary modulation formats, there will be a single eye formation. When the order of the modulation increases, the number of eyes stacked vertically will also increase depending on the modulation and constellation shape. For example, in QPSK, we still have a single eye, but, one eye for real part of the signal and another eye for the imaginary part of the signal (i.e. we plot real and imaginary parts of the signal in separate eye diagrams). For rectangular shaped 16-state quadrature amplitude

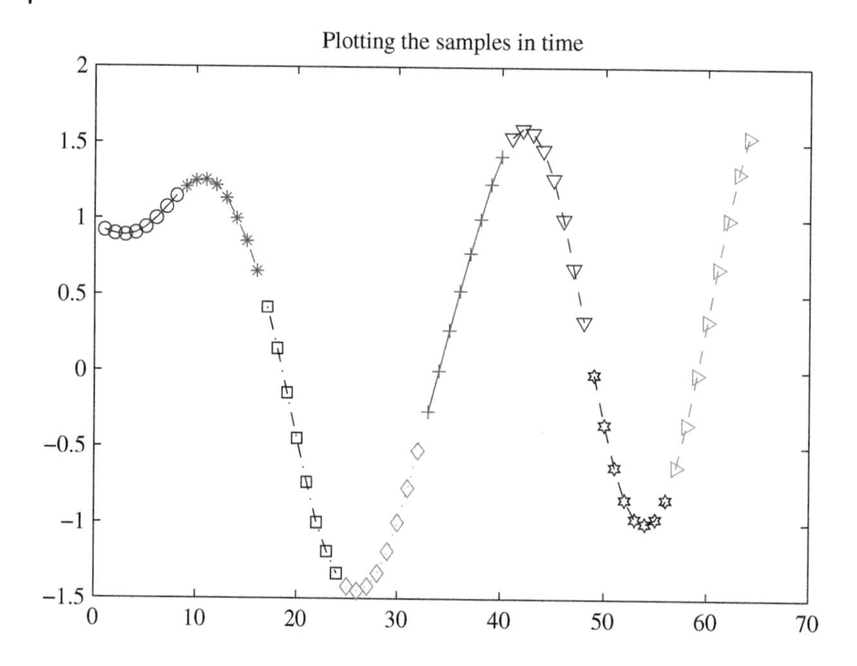

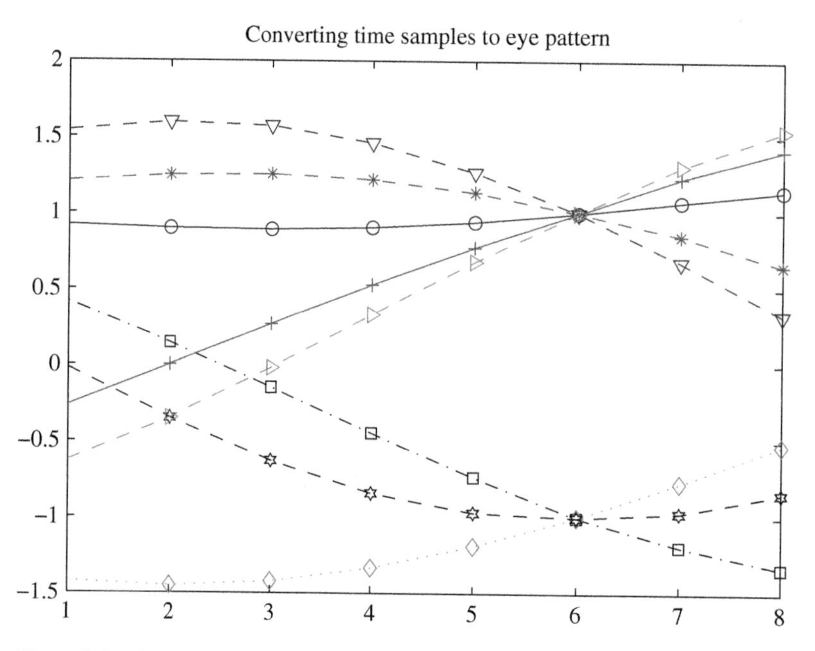

Figure 2.6 Construction of eye pattern.

modulation (16-QAM) modulation, there will be 3 eyes stacked vertically on the top of each other. In general, the number of eyes stacked vertically on the top of each other depends on the number of distinct I and Q values in the constellation diagram. If there are N points in the I-axis (or Q-axis), then, there will be $N - 1$ eyes stacked vertically on the eye diagram of the real (or imaginary) part of the signal (see Figure 2.7).

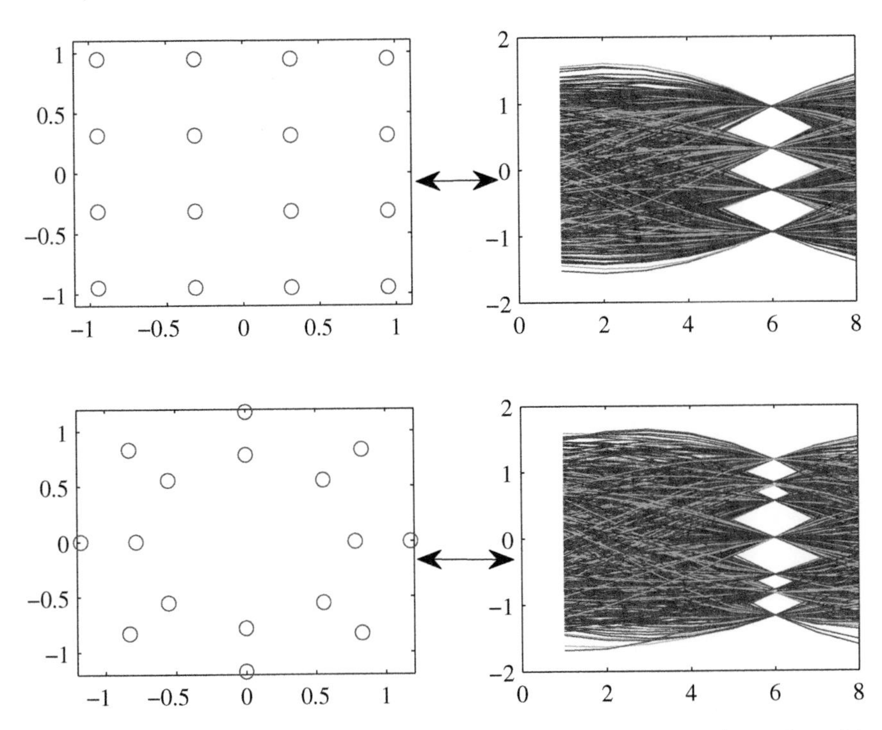

Figure 2.7 Eye diagram of 16-QAM with a popular rectangular constellation and an arbitrary (not optimized) circular constellation.

Following Matlab code provides examples for the above visual plots, and in Figure 2.8, the corresponding plots can be seen for a QPSK modulation.

```
1  % Digital modulation analysis measurements
2  % Observe the effect of:  SNR and filter roll-off on various plots
3  N=1e3; % block size in terms of number of symbols
4  sps=8; % oversampling rate (i.e. number of samples per symbol)
5  fltr_rolloff=0.3;
6  fltr=rcosine(1,sps,'normal',fltr_rolloff); % raised cosine filter
7  M=4; % modulation order
8  symb=qammod(floor(rand(1,N)*M),M); % QAM modulation
9  u_frame=upsample(symb, sps); % upsampling the modulated symbols
10 sgnl_tx = conv(fltr,u_frame); % filtering with raised cosine
11
12 SNR=25; % desired SNR in dB
13 noise=(randn(1,N)+sqrt(-1)*randn(1,N))/sqrt(2);
14 noise_var = 10^(-SNR/10) ; % desired variance calculated from SNR
15 sgnl_rx=symb+noise*sqrt(noise_var); % noise addition
16
17 subplot(2,2,1)
18 plot(real(symb),imag(symb),'*')
19 title('(a) Constellation Diagram')
20 subplot(2,2,2)
21 plot(real(sgnl_tx),imag(sgnl_tx))
22 title('(b) Vector Diagram')
23 subplot(2,2,3)
24 x=vec2mat(sgnl_tx(100:end-100),sps*2);
25 plot(real(x(1:end-1,:).'))
26 title('(c) Eye Diagram')
27 subplot(2,2,4)
28 plot(real(sgnl_rx),imag(sgnl_rx),'*')
29 title('(d) Scatter Plot')
```

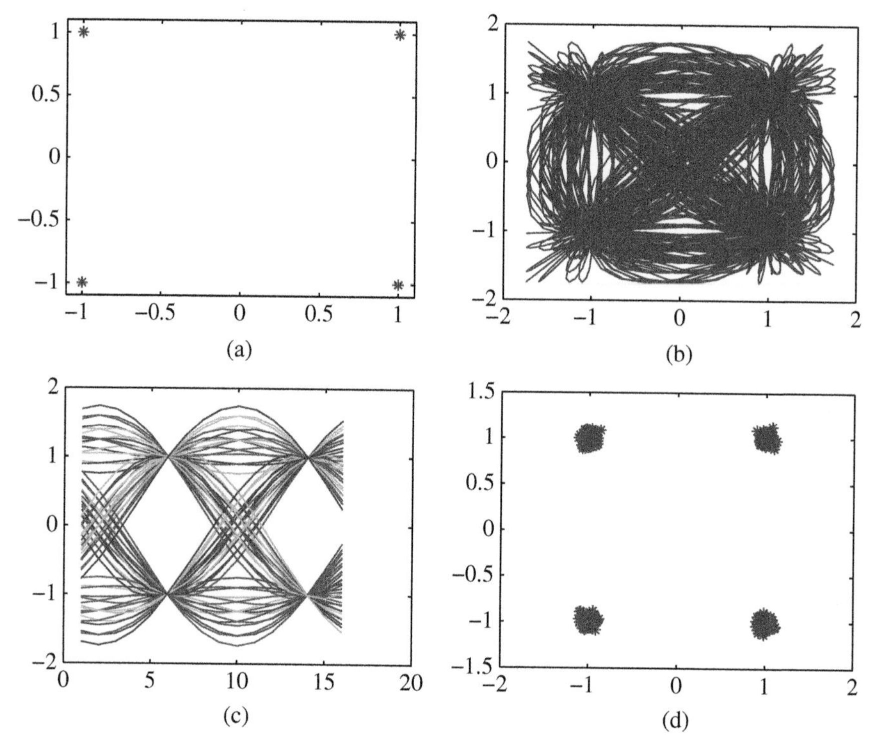

Figure 2.8 Constellation, vector and eye diagrams for QPSK modulation. (a) Constellation diagram. (b) Vector diagram. (c) Eye diagram. (d) Scatter plot.

In the following code, the visualization of several impairments using constellation diagrams is shown. The corresponding plots are given in Figure 2.9.

```
1  % Visualization of some RF impairments with constellation diagrams
2  % Observe the effect of:  SNR and other impairments
3  N=1e3; % Number of samples
4  M=4; % modulation order
5  symb= qammod(floor(rand(1,N)*M),M); % QAM modulation
6
7  SNR=30; % desired SNR in dB
8  noise=(randn(1,N)+sqrt(-1)*randn(1,N))/sqrt(2);
9  noise_var = 10^(-SNR/10) ;
10 symb=symb+noise*sqrt(noise_var); % noise addition
11 I_gain=1;Q_gain=0.5; % IQ gain imbalances
12 symb_g_imb = I_gain*real(symb) + j*imag(symb)*Q_gain ;
13 subplot(2,2,1)
14 plot(real(symb_g_imb),imag(symb_g_imb),'*')
15 title('(a) I/Q gain imbalance')
16 I_o=0.1;Q_o=-0.15; % DC offset
17 symb_DC_of = real(symb) + j*imag(symb)+I_o+j*Q_o;
18 subplot(2,2,2)
19 plot(real(symb_DC_of),imag(symb_DC_of),'*')
20 title('(b) I/Q offset')
21 po=pi/10; % phase offset
22 symb_ph_o = symb*exp(j*po);
23 subplot(2,2,3)
24 plot(real(symb_ph_o),imag(symb_ph_o),'*')
25 title('(b) Phase offset')
26 fo=1/4000; % frequency offset
27 symb_fo = symb.*exp(j*pi*[1:length(symb)]*fo);
28 subplot(2,2,4)
29 plot(real(symb_fo),imag(symb_fo),'*')
30 title('(b) Frequency offset')
```

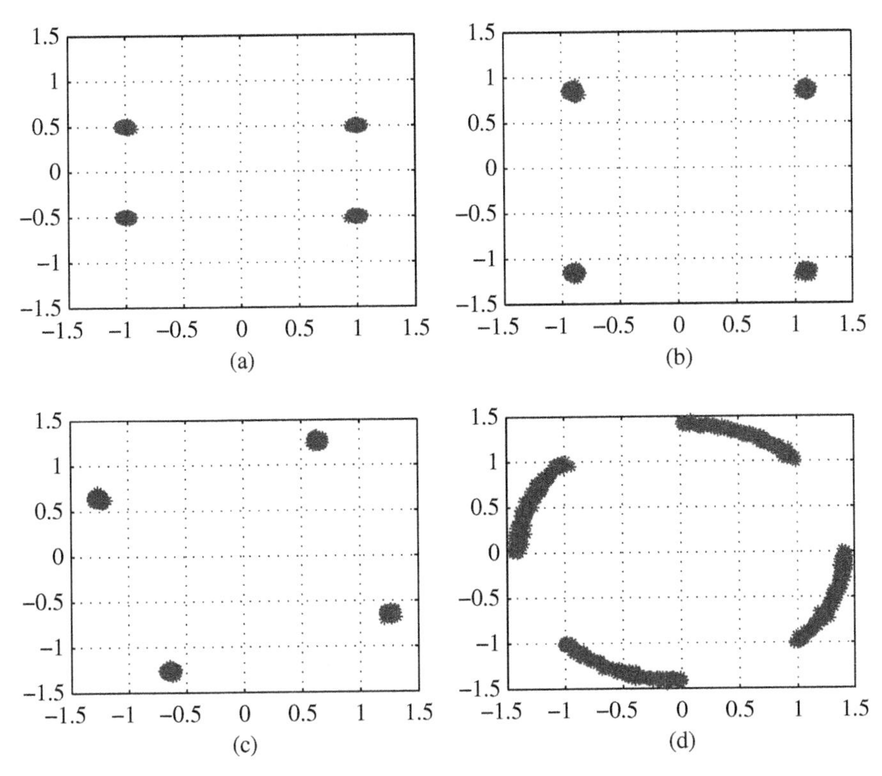

Figure 2.9 Effect of various impairments on the constellation diagram. (a) IQ gain imbalance. (b) IQ offset. (c) Phase offset. (d) Frequency offset.

In the following code, the visualization of intersymbol interference, sample clock error, and frequency offset is shown using eye diagrams. The corresponding plots are given in Figure 2.10.

```
1  % Eye diagram measurement. Visualization of intersymbol interference,
2  % sample clock error, and frequency offset using eye diagram
3  N=1e3; % Number of samples
4  sps=8;
5  fil1=rcosine(1,sps,'normal',0.3);
6  fil2=rcosine(1,sps,'sqrt',0.3);
7  M=4;
8  symb= qammod(floor(rand(1,N)*M),M);
9  u_frame=upsample(symb, sps);
10 sgnl_tx1 = conv(fil1,u_frame);
11 sgnl_tx2 = conv(fil2,u_frame);
12
13 subplot(2,2,1)
14 x=vec2mat(sgnl_tx1(100:end-100),sps*2);
15 plot(real(x(1:end-1,:).'))
16 title('(a) Using Nyquist Filter (raised Cosine)')
17 subplot(2,2,2)
18 x=vec2mat(sgnl_tx2(100:end-100),sps*2);
19 plot(real(x(1:end-1,:).'))
20 title('(b) Using non-Nyquist Filter (root-raised cosine)')
21 subplot(2,2,3)
22 sgnl_tx1_ce=resample(sgnl_tx1,10000,10003);
23 x=vec2mat(sgnl_tx1_ce(100:end-100),sps*2);
24 plot(real(x(1:end-1,:).'))
25 title('(c) With 300ppm clock error')
26 subplot(2,2,4)
27 sgnl_tx1_fo = sgnl_tx1.*exp(j*pi*[1:length(sgnl_tx1)]*1/8000);
28 x=vec2mat(sgnl_tx1_fo(100:end-100),sps*2);
29 plot(real(x(1:end-1,:).'))
30 title('(d) With Frequency offset')
```

Also, the effects of filter roll-off and noise on the eye diagrams are shown in Figures 2.11 and 2.12, respectively.

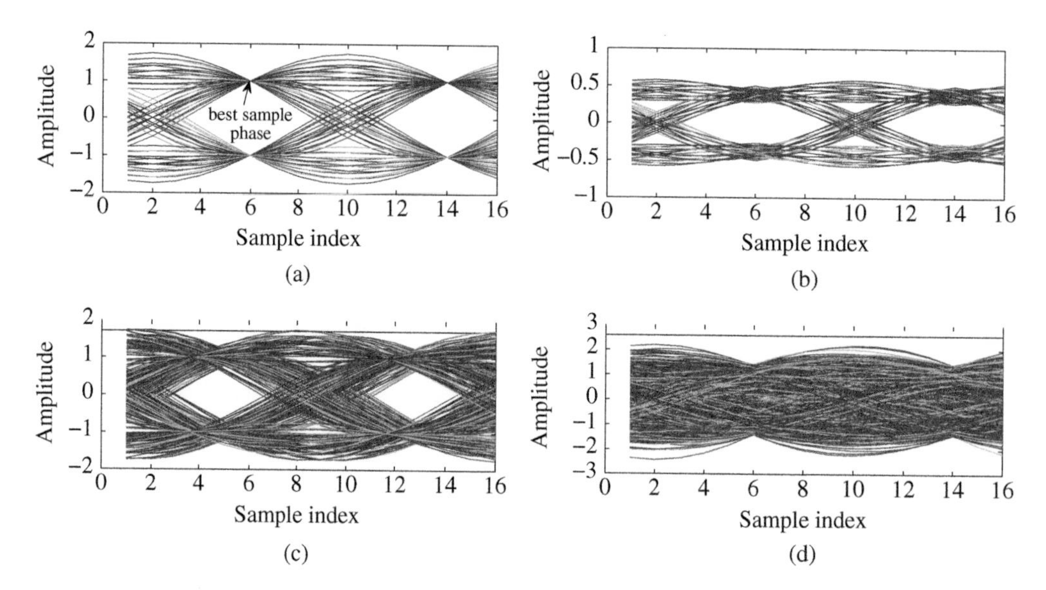

Figure 2.10 Effect of impairments on the eye diagrams. (a) Using Nyquist Filter (raised cosine). (b) Using non-Nyquist Filter (root-raised cosine). (c) With 300 ppm clock error. (d) With frequency offset.

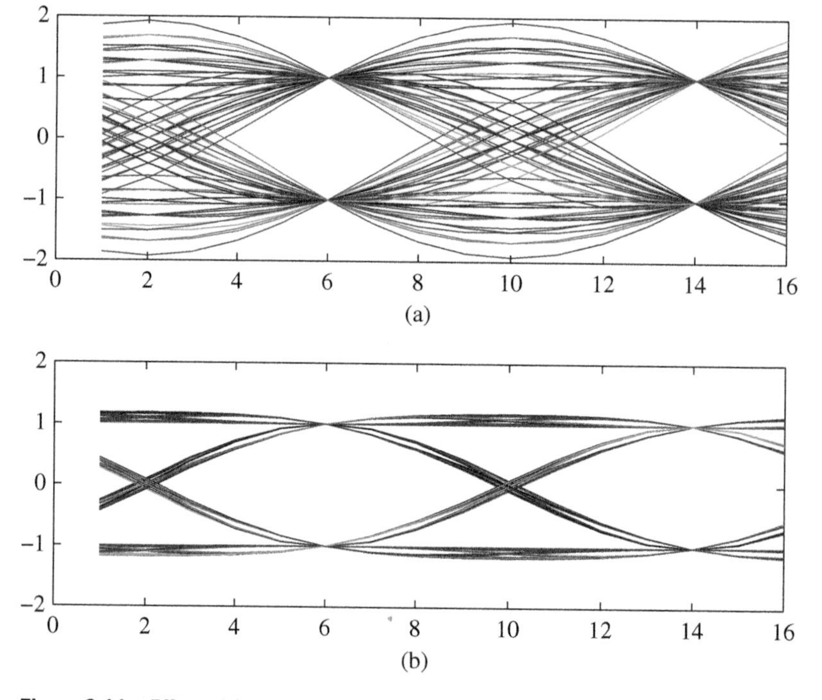

Figure 2.11 Effect of filter roll-off on eye diagram. (a) Roll-off = 0.1. (b) Roll-off = 0.8.

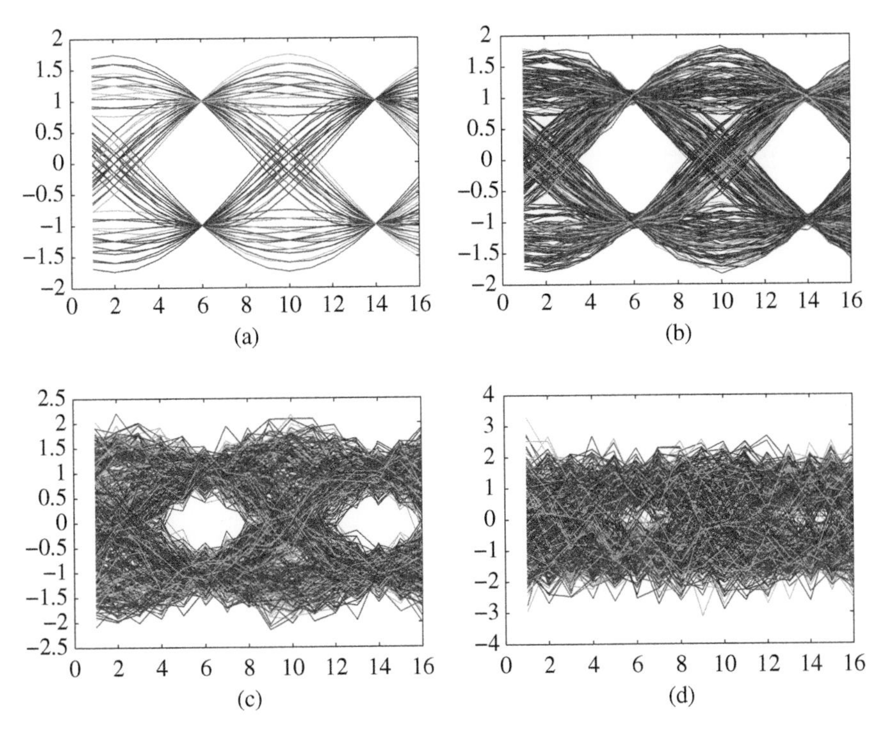

Figure 2.12 Effect of noise on eye diagram. (a) $SNR = 100$ dB. (b) $SNR = 25$ dB. (c) $SNR = 10$ dB. (d) $SNR = 3$ dB.

```
1  % 3-D visualization of constellations
2  % Observe the plots with various noise variance values
3  N=1e6;
4  M=4;
5  symb= qammod(floor(rand(1,N)*M),M);
6  sgnl_rx = symb + (randn(1,N) +j*randn(1,N))*0.5;
7  [n,x,y] = hist2d(real(sgnl_rx),imag(sgnl_rx),100);
8  subplot(2,2,1)
9  plot(real(sgnl_rx),imag(sgnl_rx),'*')
10 title('(a) Scatter plot')
11 subplot(2,2,2)
12 imagesc(x(1,:),-y(:,1),n);colorbar
13 title('(b) Scatter plot with color code')
14 subplot(2,2,3)
15 mesh(x(1,:),-y(:,1),n);
16 title('(c) 3-D Mesh')
17 subplot(2,2,4)
18 contour3(x(1,:),-y(:,1),n);
19 title('(d) 3-D Contour')
```

2.2.1 Advanced Scatter Plot

In the conventional constellation diagram, the signal is shown as a two-dimensional (2-D) scatter diagram in the complex plane. When the number of samples in this 2-D plane increases, it is hard to visualize the weight of samples in each point. Especially, for low SNR values, the plane will be completely filled with a lot of points making it very difficult to diagnose the impairment source and statistics. In order to see the weight in each point, a three-dimensional (3-D) plot would be useful. The following code provides how to obtain the 3-D scatter plot from the noisy received signals.

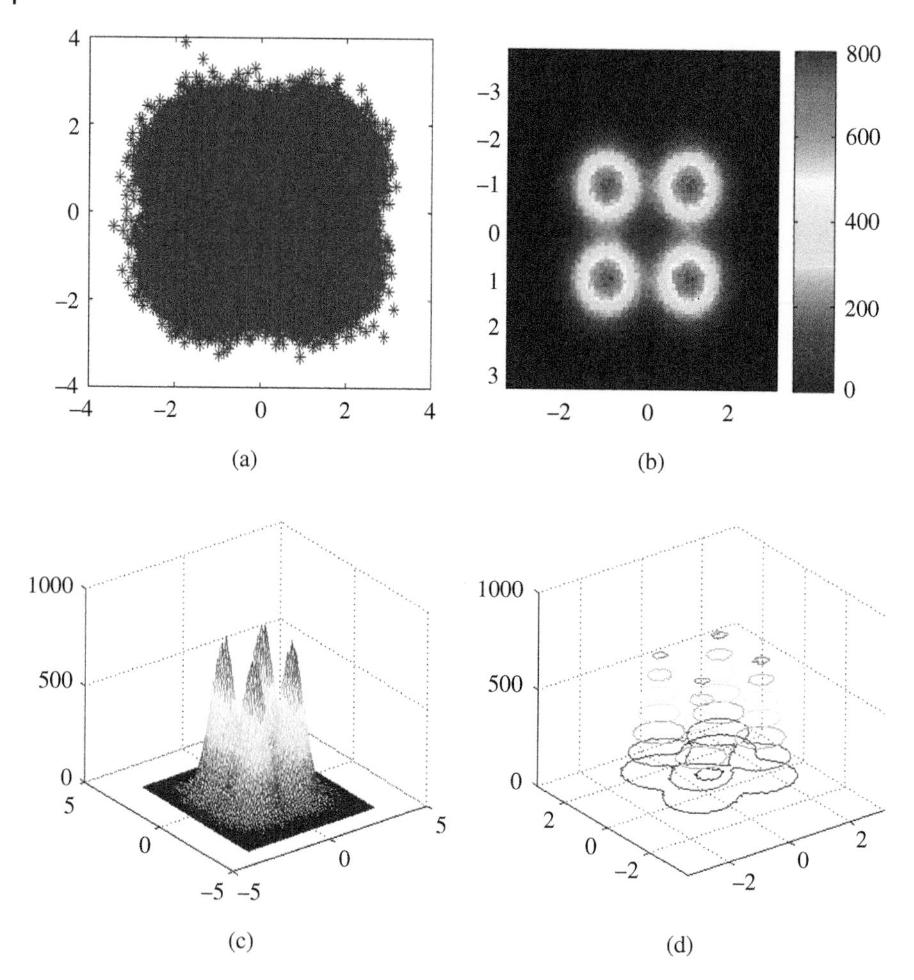

Figure 2.13 3-D visualizations of scatter plots. (a) Scatter plot. (b) Scatter plot with color code. (c) 3-D mesh. (d) 3-D contour.

Figure 2.13 shows the same samples in various plots. In Figure 2.13a, the regular 2-D scatter plot is shown. As can be seen, it is difficult to extract the statistics of the noise and the center of the constellations in 2-D plot. In Figure 2.13b–d, 3-D plots are shown. When they are printed in colored, these figures provide excellent visualization about the statistics of the noise.

2.3 Cognitive Radio and SDR Measurements

Software-defined radio (SDR) is envisioned to be a promising solution for interoperability, global seamless connectivity, multistandard, and multimode issues. SDR represents a very flexible and generic radio platform which is capable of operating with many different bandwidths over a wide range of frequencies and various different modulation and waveform formats. With SDR, the functionality of wireless devices is increased and they become more and more sophisticated.

Even though there is no consensus on the formal definition of cognitive radio (CR) as of now, the concept has evolved recently to include various meanings in several contexts. One main aspect of it is related to autonomously exploiting locally unused spectrum to provide new paths to spectrum access. Other aspects include interoperability across several networks; roaming across borders while being able to stay in compliance with local regulations; adapting the system, transmission, and reception parameters without user intervention; and having the ability to understand and follow actions and choices taken by their users, and learn to become more responsive over time. One of the most important elements of the CR concept is the ability to measure, sense, learn, and be aware of parameters related to the radio channel characteristics, availability of spectrum and power, interference and noise temperature, radio's operational environments, user requirements and applications, available networks (infrastructures) and nodes, local policies and other operating restrictions, etc. Since these parameters might change over time and over multitude of other dimensions, the radios need to be equipped with the proper mechanism to react these changes.

SDR and CR bring new challenges and opportunities to the test and measurement world. The testing, measuring, and verification of SDR capable devices is challenging not only in hardware (requiring wideband signal analysis and generation capabilities as well as high-speed ADC/DAC and fast processing capabilities), but also in software (requiring additional test vectors, performance metrics, user interface, etc.). The next-generation tools should have the ability to process wide variety of waveforms while being equipped with autonomous synchronization and detection capabilities. Especially, blind identification of the waveform parameters, modulation formats, and synchronization parameters is important. In addition, understanding the transient behaviors of the transmitted signals is crucial. The measurement devices should also have cognition capabilities which can enable the device to evaluate the ability of the radio being tested (device under test, or DUT) to react to the various factors.

Testing and measuring SDR/CR is a very difficult process, which requires many supporting functions, including advanced triggering mechanisms between the generator and analyzer. Signal intelligence and generic signal analysis (using blind estimation and detection techniques) could be extremely important for testing SDR transmitter, especially in doing government regulation, observing an operating network, qualification of a product, and to some extent R&D verification. Since SDR transmitters will be capable of transmitting various types of waveforms and will have the ability to change operating parameters (including modulation, bandwidth, frequency, power, time/frequency hopping pattern, etc.) during the transmission, the transient analysis capability is critical to be able to observe RF glitches, transients, and other short term abnormalities. Generic multiantenna and multidimensional (time/frequency/space/angle/code) signal analysis capability will be beneficial for providing detailed information about the transmitted signal. The SDR testing and measurement capability also requires new metrics to accurately evaluate the performance of the transceivers. In addition to the new metrics, advanced visualization of these metrics along with proper user interface and user control mechanisms is needed. Finally, it will be very useful if the SDR measurement analysis tool has the ability to demodulate multiple signals simultaneously (multiuser detection) and separate interference from desired signal. Multisignal separation and extraction of a signal of interest from multidimensional signal space helps to analyze each signal separately without the effect of other signals. This requires advanced windowing/filtering/correlation techniques.

In signal generation side, generic signal synthesis capability is critical. This should also include the ability to provide various transient behaviors and ability to change the waveform parameters during transmission. Additionally, it will be useful for the signal generator to be able to emulate various interference sources (co-channel, adjacent channel, narrowband, impulsive, etc.) and incorporate these into the signal of interest. Emulating various impairments (I/Q impairments, noise, frequency offset, sample time offset, etc.), and generation of multipath channel and incorporation of the channel into the signal are also critical to test and measure the behavior of the SDR system or component under various conditions.

In opportunistic spectrum usage concept of CR, spectrum sensing is a critical function. It is important that CR, sense the spectrum availability and spectrum occupancy accurately. Inability to perform the spectrum sensing properly leads to misdetection error and/or false alarm. Therefore, when testing the performance of a CR, these quantities need to be measured and methods to evaluate the cognitive radios performance need to be developed. Probability of false alarm and probability of misdetection rate as a function of the number of observed symbols, SNR, number of signals present (primary and secondary), level of cooperation (if a cooperative measurement is done with a number of CR devices) need to be measured accurately.

In addition to the items discussed above, we also need to develop techniques for quantifying the impact of the use of CR in the primary network. Further research is needed in this area to answer questions like the following: What kind of impact can a CR cause to the primary network and users? What types of measurements need to be developed to measure and quantify this impact in the primary users? Would the classical performance metrics be sufficient to identify the impact or do we need additional metrics?

2.4 Other Measurements

Wireless communication has evolved over the last four decades from voice-centric applications to wide variety of services. These different application classes have different requirements from a system design perspective, hence, wireless systems face several challenges, from increasing spectrum demand, to security threats, requiring novel and scalable solutions. One of the most critical challenge that wireless radio researchers are facing is the design of a flexible and adaptable communication system in every layer of the communication protocol stack. Even though at this time future 6G technologies look like an extension of their 5G counterparts, new user requirements, completely new applications/use-cases, and new networking trends of 2030 and beyond may bring more challenging communication engineering problems, which necessitate more flexible and cognitive communication paradigms in different layers, but particularly in the physical layer of future wireless systems. While wireless communication systems have enjoyed this enormous growth, another major phenomenon has gained significant interest, which is referred to as radio environment mapping (REM). Radio environment (RE) involves any radio and network-related parameter that might impact the operation of radio and network including radio spectrum, radio channel, power, angle of transmitter/receiver, interference, user location, mobility (individual and group), available networks and devices, user context (profile), RF front-end properties and impairments. Even though the estimation of some of these parameters, like speed, distance, and angle goes all the way back to the days of World Wars, and several techniques have been developed among radar signal processing community, the concept of REM is very recent and involves much more than what the radar literature presents. There remain essential challenges toward the development of this new REM framework complemented by new sets of critical sensing functions and mechanisms related to the complex, crowded, and multidimensional wireless environment. Developing novel and efficient algorithms to estimate important REM functions that are normally weak in a very convoluted and heterogeneous environment often requires techniques where the classical model-based approaches fails, leading to novel machine learning techniques to be used along with the basic domain knowledge.

Looking at the evolution of wireless communication technologies in time, one thing that can clearly be seen is that every new standard became more adaptive and more capable than the previous one. The adaptation is enabled by more awareness and measurements, and hence, the number of parameters that are measured and learned has increased significantly. Long term evolution (LTE) and 5G cellular standards already offer a wide variety of measurement capabilities utilizing a set of reference signals related to the signal strength, flexible frequency-dependent channel quality measurement, channel state information, interference measurements, etc. Also, there are measurements related to the operation of multiple antenna systems and beam-forming. Some of these measurements can be listed as:

- Channel state information (CSI)
- Channel quality indicator (CQI)
- Reference signal received power (RSRP)
- Reference signal received quality (RSRQ)
- RSSI
- CSI-IM (interference measurement)
- Rank indicator (RI)
- Precoding matrix indicator (PMI)
- Precoding type indicator (PTI)

These measurements are computed on the fly in LTE and 5G systems and used to optimize resource allocation among the various user equipments (UEs) that are requesting service. They can be used for link adaptation, adaptive scheduling the resources to the users, interference coordination, MIMO and beamforming, and user-cell association.

CSI is a kind of collective name of several different types of indicators that UE report. CQI is related to SINR. CQI value can be used to adapt the modulation and coding order. The frequency domain resolution in the CQI report can be varied. In addition to the wideband CQI, where the quality of the entire channel bandwidth is measured and averaged, there are different subband CQIs, each of which indicates the average transmission quality for a specific frequency subrange. As mentioned above, the CQI index reported to the BS by the UE is derived from the channel quality of the downlink signal (somewhat related to SINR). However, in contrast to the previous generations of mobile radio systems, the LTE CQI index is not directly associated with the measured SNR. Instead, it is also influenced by the signal processing in the UE. With the same channel, a UE featuring a powerful signal processing algorithm is able to forward a higher CQI index to the BS than a UE that has a weak algorithm. The precoding matrix determines how the individual data streams (called layers in LTE) are mapped to the antennas. Proper selection of the precoder maximizes the capacity. However, this requires knowledge of the channel quality for each antenna in the downlink, which the UE can determine through measurements. If the UE knows what the allowed precoding matrices are, it can send a PMI report to the BS and suggest a suitable matrix. The channel rank indicates the number of layers and the number of different signal streams transmitted in the downlink. The goal of an optimized RI is to maximize the channel capacity across the entire available downlink bandwidth by taking advantage of each full channel rank. CSI-IM measurement is used to obtain interference measurement from neighboring base stations.

RSSI is the linear average of the total received power observed only in symbols carrying reference symbols by UE from all sources, including co-channel nonserving and serving cells, adjacent channel interference and thermal noise, within the measurement bandwidth. RSRP and RSRQ are key measures of signal level and quality for modern LTE networks. In cellular networks, when a mobile device moves from cell to cell and performs cell selection/reselection and handover, it has to measure the signal strength/quality of the neighbor cells. In the procedure of handover, the LTE specification provides the flexibility of using RSRP, RSRQ, or both. RSRP is a cell-specific signal strength-related metric that is used as an input for cell reselection and handover decisions. For a particular cell, RSRP is defined as the average power (in Watts) of the resource elements (REs) that carry cell-specific reference signals (RSs) within the considered bandwidth. RSRP measurement, normally expressed in dBm, is utilized mainly to make ranking among different candidate cells in accordance with their signal strength. RSRQ measurement is a cell-specific signal quality metric. Similar to the RSRP measurement, this metric is used mainly to provide ranking among different candidate cells in accordance with their signal quality. This metric can be employed as an input in making cell reselection and handover decisions in scenarios (for example) in which the RSRP measurements are not sufficient to make reliable cell reselection/handover decisions.

It is expected that 6G will further bring new measurement and REM capabilities to the wireless world. It is expected that the REM concept will grow in the future, therefore, a separate chapter of this book is dedicated toward describing REM and radar measurement techniques in detail. In this chapter, without going into the details, some potential estimates that can be useful for adjusting the radio functionalities can be listed as:

- Dominant versus nondominant interference
- Location or direction of interference
- Time and frequency variation of channel
- Sparseness of the channel
- Sparse channel modeling and estimation
- Angular spread, delay spread, and Doppler spread of the channel
- Blockage estimation
- Mobility estimation (individual and group mobility)
- Line-of-sight (LOS) and non LOS (NLOS) estimation

Time scales of the measurements and system adaptation based on these measurements depend on several factors. *Very short term* time scales are those measurements that needs to be employed frequently, on the order of microsecond intervals. For example, multipath fading parameters, scheduling related measurements, CQI measurements, and other link adaptation-related measurements should be performed frequently, especially for fast varying channels. The measurement with *Short term* time scales do not need to be done as frequently. Examples for these measurements can be beamforming, tilt adaptation, user-cell association, blockage and shadowing compensation, path loss compensation (range extension)-related measurements. In *Long term* time scales, the measurements are done in longer terms like in the order of hours. A good examples for these would be base station ON and OFF mode, energy saving modes, low mobility user adjustment-related measurements.

2.5 Clarifying dB and dBm

In wireless communications, the signal power differences between what is transmitted and what is received could be significantly large. For example, a 1 W transmitter power can be received at the receiver with 0.000 000 001 W. Dealing with the linear values and interpreting them under various operations (especially multiplication and division) might be difficult. Therefore, use of logarithmic scale is useful when dealing with numbers that have a large range. The bel (B), which is named after the inventor of the telephone (Alexander Graham Bell), is the base 10 logarithm ($\log_{10}$) of the ratio of two power levels and it is a unitless measurement. When the **bel** value ($\log_{10}$) is multiplied with 10 (where the term **deci** comes from), then we obtain the decibel (dB), i.e. $10\log_{10}()$ dB. While dB is used to express the ratio of two values, the unit dBm (dB relative to a milliwatt) denotes an absolute power level measured in decibels and referenced to 1 milliwatt (mW). For example, when we want to express the SNR in logarithmic scale, we use $X = 10\log_{10}(S/N)$ dB, or say the SNR value is X dB. On the other hand, if we want to express the absolute transmit power of let's say 1 W, we say 30 dBm; ($10\log_{10}(1/0.001)$ dBm).

When expressing voltage ratios, we use $20\log_{10}()$ (20 instead of 10), because the power level is proportional to the square of the voltage level (and squaring is equivalent to doubling the logarithm). If we have two power values P_1 and P_2, the power ratios is $10\log_{10}(P_1/P_2)$ dB. If we re-write the power values in terms of voltages and resistance, $P_1 = V_1^2/R_1$ and $P_2 = V_2^2/R_2$, the power ratios can be re-written as $10\log_{10}((V_1^2R_2)/(V_2^2R_1))$ dB. When the resistance values are the same ($R_1 = R_2$), then, the result will be $10\log_{10}(V_1^2/V_2^2)$ dB, or alternatively $20\log_{10}(V_1/V_2)$ dB, which clearly explains the reason why 20 is used instead of 10 for voltage ratios.

2.6 Conclusions

Evaluation of the performance of communication offline and during real-time operation (online) is important. During offline testing, various measurements can provide insights about the performance of the system. These measurements are critical for the development phase of the systems. Test equipment are often used with such measurement capabilities during this phase. Online measurements are often used during the real-time operation for system adaptation. These online measurement are extremely important for robust and flexible system designs. CRs and networks depend heavily on these online measurements. Recently, the wireless standards are becoming more and more cognitive which provides a lot of flexibility in the operations of the systems in different layers of the protocol stacks. As a result, many new measurement techniques are introduced in recent standards. It is expected that the future standards will even have more things to measure about the quality of the channel, system, and networks.

References

1 S. Nanda, K. Balachandran, and S. Kumar, "Adaptation techniques in wireless packet data services," *IEEE Commun. Mag.*, vol. 38, no. 1, pp. 54–64, Jan. 2000.
2 A. F. Demir, B. Pekoz, S. Kose, and H. Arslan, "Innovative telecommunications training through flexible radio platforms," *IEEE Commun. Mag.*, vol. 57, no. 11, pp. 27–33, 2019.
3 M. Karabacak, A. H. Mohammed, M. K. Özdemir, and H. Arslan, "RF circuit implementation of a real-time frequency spread emulator," *IEEE Trans. Instrum. Meas.*, vol. 67, no. 1, pp. 241–243, 2017.
4 A. Gorcin and H. Arslan, "An OFDM signal identification method for wireless communications systems," *IEEE Trans. Vehicul. Technol.*, vol. 64, no. 12, pp. 5688–5700, 2015.
5 ——, "Signal identification for adaptive spectrum hyperspace access in wireless communications systems," *IEEE Commun. Mag.*, vol. 52, no. 10, pp. 134–145, 2014.

6 S. Guzelgoz and H. Arslan, "Modeling, simulation, testing, and measurements of wireless communication systems: A laboratory based approach," in *2009 IEEE 10th Annual Wireless and Microwave Technology Conference*, Clearwater, FL, 20–21 Apr. 2009, pp. 1–5.

7 S. Ahmed and H. Arslan, "Cognitive intelligence in UAC channel parameter identification, measurement, estimation, and environment mapping," in *OCEANS 2009-EUROPE*, Bremen, 11–14 May 2009, pp. 1–14.

8 M. E. Sahin and H. Arslan, "MIMO-OFDMA measurements; reception, testing, and evaluation of WiMAX MIMO signals with a single channel receiver," *IEEE Trans. Instrum. Meas.*, vol. 58, no. 3, pp. 713–721, 2009.

9 A. M. Hisham and H. Arslan, "Multidimensional signal analysis and measurements for cognitive radio systems," in *2008 IEEE Radio and Wireless Symposium*, Orlando, FL, 22–24 Jan. 2008, pp. 639–642.

10 M. Sahin, H. Arslan, and D. Singh, "Reception and measurement of MIMO-OFDM signals with a single receiver," in *2007 IEEE 66th Vehicular Technology Conference*, Baltimore, MD, 30 Sept.–3 Oct. 2007, pp. 666–670.

11 H. Arslan, "IQ gain imbalance measurement for OFDM based wireless communication systems," in *MILCOM 2006 – 2006 IEEE Military Communications Conference*, Washington, DC, 23–25 Oct. 2006, pp. 1–5.

12 H. Arslan and D. Singh, "The role of channel frequency response estimation in the measurement of RF impairments in OFDM systems," in *2006 67th ARFTG Conference*, San Francisco, CA, 16 June 2006, pp. 241–245.

13 T. Yucek and H. Arslan, "MMSE noise plus interference power estimation in adaptive OFDM systems," *IEEE Trans. Vehicul. Technol.*, vol. 56, no. 6, pp. 3857–3863, 2007.

14 B. Karakaya, H. Arslan, and H. A. Cirpan, "Channel estimation for LTE uplink in high Doppler spread," in *2008 IEEE Wireless Communications and Networking Conference*, Las Vegas, NV, 31 Mar.–3 Apr. 2008, pp. 1126–1130.

15 M. K. Ozdemir and H. Arslan, "Channel estimation for wireless OFDM systems," *IEEE Commun. Surv. Tutor.*, vol. 9, no. 2, pp. 18–48, 2007.

16 T. Yucek and H. Arslan, "OFDM signal identification and transmission parameter estimation for cognitive radio applications," in *IEEE GLOBECOM 2007 – IEEE Global Telecommunications Conference*, Washington, DC, 26–30 Nov. 2007, pp. 4056–4060.

17 H. A. Mahmoud and H. Arslan, "Improved channel estimation in OFDM systems with synchronization errors and back-off," in *IEEE Vehicular Technology Conference*, Montreal, QC, 25–28 Sept. 2006, pp. 1–4.

18 T. Yucek and H. Arslan, "MMSE noise power and SNR estimation for OFDM systems," in *2006 IEEE Sarnoff Symposium*, Princeton, NJ, 27–28 Mar. 2006, pp. 1–4.

19 H. Arslan, "Channel frequency response estimation under the effect of RF impairments in OFDM based wireless systems," in *IEEE Vehicular Technology Conference*, Montreal, QC, 25–28 Sept. 2006, pp. 1–5.

20 K. Balachandran, S. Kabada, and S. Nanda, "Rate adaptation over mobile radio channels using channel quality information," in *Proc. IEEE Globecom'98 Commun. Theory Mini Conf. Record*, 1998, pp. 46–52.

21 M. Türkboylari, and G. L. Stüber, "An efficient algorithm for estimating the signal-to-interference ratio in TDMA cellular systems," *IEEE Trans. Commun.*, vol. 46, no. 6, pp. 728–731, June 1998.

22 R. Hassun, M. Flaherty, R. Matreci, and M. Taylor, "Effective evaluation of link quality using error vector magnitude techniques," in *Proc. IEEE Wireless Commun. Conf.*, Aug. 1997, pp. 89–94.

23 R. Liu, Y. Li, H. Chen, and Z. Wang, "EVM estimation by analyzing transmitter imperfections mathematically and graphically," *Analog Integr. Circ. Signal Process.*, vol. 48, no. 3, pp. 257–262, 2006.

24 IEEE, *Supplement to IEEE standard for information technology telecommunications and information exchange between systems - local and metropolitan area networks-specific requirements. Part 11: Wireless LAN Medium Access Control (MAC) and Physical Layer (PHY) specifications: high-speed physical layer in the 5 GHz band*, The Institute of Electrical and Electronics Engineering, Inc. Std. IEEE 802.11a, Sept. 1999.

25 *IEEE Standard for Local and Metropolitan Area Networks Part 16: Air Interface for Fixed and Mobile Broadband Wireless Access Systems Amendment 2: Physical and Medium Access Control Layers for Combined Fixed and Mobile Operation in Licensed Bands and Corrigendum 1*, IEEE Std 802.16e-2005 and IEEE Std 802.16-2004/Cor 1-2005 (Amendment and Corrigendum to IEEE Std 802.16-2004) Std., 28 Feb. 2006, pp. 1–822. doi:10.1109/IEEESTD.2006.99107.

26 T. Nakagawa and K. Araki, "Effect of phase noise on RF communication signals," *Vehicular Technology Conference Fall 2000. IEEE VTS Fall VTC2000. 52nd Vehicular Technology Conference (Cat. No.00CH37152)*, vol. 2, Boston, MA, 24–28 Sept. 2000, pp. 588–591.

27 M. Helfenstein, E. Baykal, K. Muller, A. Lampe, P. Semicond, and S. Zurich, "Error vector magnitude (EVM) measurements for GSM/EDGE applications revised under production conditions," *2005 IEEE International Symposium on Circuits and Systems*, Kobe, 23–26 May 2005, pp. 5003–5006.

28 A. Georgiadis, "Gain, phase imbalance, and phase noise effects on error vector magnitude," *IEEE Trans. Veh. Technol.*, vol. 53, no. 2, pp. 443–449, 2004.

29 R. A. Shafik, M. S. Rahman, A. R. Islam, and N. S. Ashraf, "On the error vector magnitude as a performance metric and comparative analysis," in *International Conference on Emerging Technologies (ICET)*, Peshawar, 13–14 Nov. 2006, pp. 27–31.

30 S. Forestier, P. Bouysse, R. Quere, A. Mallet, J. M. Nebus, and L. Lapierre, "Joint optimization of the power-added efficiency and the error-vector measurement of 20-GHz pHEMT amplifier through a new dynamic bias-control method," *IEEE Trans. Microwave Theory Techol.*, vol. 52, no. 4, pp. 1132–1141, 2004.

31 K. M. Gharaibeh, K. G. Gard, and M. B. Steer, "Accurate estimation of digital communication system metrics-SNR, EVM and ρ in a nonlinear amplifier environment," *64th ARFTG Microwave Measurements Conference, Fall 2004*, IEEE, Orlando, FL, 2–3 Dec. 2004, pp. 41–44.

32 D. Cabric, S. M. Mishra, and R. W. Brodersen, "Implementation issues in spectrum sensing for cognitive radios," *Conference Record of the Thirty-Eighth Asilomar Conference on Signals, Systems and Computers*, vol. 1, Pacific Grove, CA, 7–10 Nov. 2004, pp. 772–776.

33 S. Haykin, "Cognitive radio: Brain-empowered wireless communications," *IEEE J. Select. Areas Commun.*, vol. 3, no. 2, pp. 201–220, Feb. 2005.

34 F. L. Lin and H. R. Chuang, "EVM and BER simulation of an NADC-TDMA radiophone influenced by the operator's body in urban mobile environments," *Wirel. Pers. Commun.*, vol. 17, no. 1, pp. 135–147, 2001.

35 K. Homayounfar, "Rate adaptive speech coding for universal multimedia access," *IEEE Signal Process. Mag.*, vol. 20, no. 2, pp. 30–39, Mar. 2003.

3

Multidimensional Signal Analysis

Hüseyin Arslan

Department of Electrical Engineering, University of South Florida, Tampa, FL, USA
Department of Electrical and Electronics Engineering, Istanbul Medipol University, Istanbul, Turkey

The human brain can perceive and analyze stimuli utilizing different organs. The multitude of sensory system, which is based on the sensory organs that include eyes, ears, skin, inner ear, nose, and mouth, has the capability of perceiving different features of the stimuli. As a result, in some stimuli, we use one organ to sense a feature that the others cannot, for some other stimuli our brain collects data through multiple organs. Similarly, wireless signals have features in different dimensions (like time, frequency, space, angle, code, modulation, etc.). Some of these features are very easy to sense (extract) in one dimension compared to others. This chapter discusses signal characteristics in multiple domains, leading to the concept of multidimensional signal analysis. Multidimensional signal analysis is very important to understand the challenges and trade-offs introduced by emerging wireless technologies. Learning multidimensional signal analysis helps trainees to grasp difficult wireless concepts in a laboratory environment. Understanding the multidimensional signal analysis is not only useful for better understanding the signals but also developing advanced receiver algorithms like synchronization, channel estimation, radio environment parameter estimation, blind signal analysis and reception. As a result, this chapter is relevant to almost all other chapters in this book.

3.1 Why Multiple Dimensions in a Radio Signal?

Signal analysis measurements are not directly related to the performance of a waveform. Instead, these measurements provide in-depth information about the statistics and characteristics of a signal. In blind signal detection, cognitive radio, adaptive system design, test and measurement, and for many other reasons signal analysis measurements are very useful. For example, in cognitive radio, signal analysis measurements can be used to enable the spectrum awareness, interference and primary user awareness, radio channel awareness, modulation and waveform awareness. By analyzing various dimensions of the received signal using advanced digital signal-processing techniques, information about the transmitted signal, operating radio channel medium, environmental conditions can be obtained, which are then used for improving the communication quality through adaptation of radio and network parameters. Similarly, in blind receiver design, signal analysis measurements allow the receiver to synchronize and demodulate the transmitted signal reliably, even if the transmitter does not send any known (training) bits to the receiver. In public safety and military applications, the signal analysis can be used to extract information regarding the location (or direction) of the signal that is received. Also, signal analysis along with software defined radio (SDR) can help to solve the interoperability issues in public safety radio systems. Last but not the least, signal analysis is a critical element for the test and measurement world.

Wireless Communication Signals: A Laboratory-based Approach, First Edition. Hüseyin Arslan.

The received signal can be analyzed and diagnosed to identify issues and impairments related to the signal.

Multidimensional signal analysis, in this chapter, is referred for analyzing the received radio signal from multiple dimensions jointly or separately as shown in Figures 3.1 and 3.2. The process involves multidimensional waveform awareness, signal extraction from this multidimensional signal space, and advanced signal analysis. For example, in Figure 3.2, a block diagram of joint time-frequency, two-dimensional (2-D), analysis process is shown. In this specific example, the first goal is to take the time samples (a mono-dimensional vector signal that is captured in time) and obtain a two-dimensional time-frequency representation of the signal. Time and frequency domain constitute two different ways of representing a signal. Some features of the signal are easier to represent in one domain while others are easier in the other. Traditional oscilloscopes and spectrum analyzers are capable of providing only single dimensional representation. However, wireless signals have multidimensional attributes with nonstationary characteristics over the transmission period and it is very difficult to provide detailed analysis using a single dimensional approach.

The time-frequency representation needs to be expanded to other dimensions. These other dimensions may include code domain, like code domain power used in code division multiple access (CDMA) measurements, angle domain (to provide direction of arrival, angular spread), waveform domain (to provide information about the type of signaling), polarization domain, etc. For instance, CDMA is a spread spectrum technology which employs a special coding scheme for multiple accessing. Therefore, analysis of the CDMA signals in code domain can provide very valuable information which cannot be extracted from other domains. Code domain power analysis provides the signal power projected on a code-space normalized to the total signal power. This is useful for composite signal analysis. The analyzed coded waveform is correlated with different codes to determine the correlation coefficients for each code. Once different channels are separated, the power in each code channel is determined. Similar techniques need to be

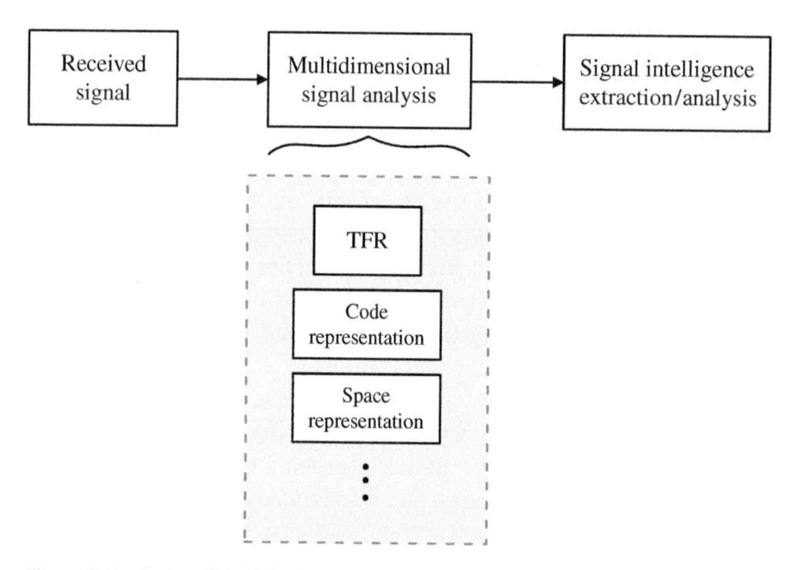

Figure 3.1 A simplified block diagram of multidimensional signal analysis. Joint time-frequency representation (TFR), code, space and dimensions can be employed in multidimensional signal analysis block.

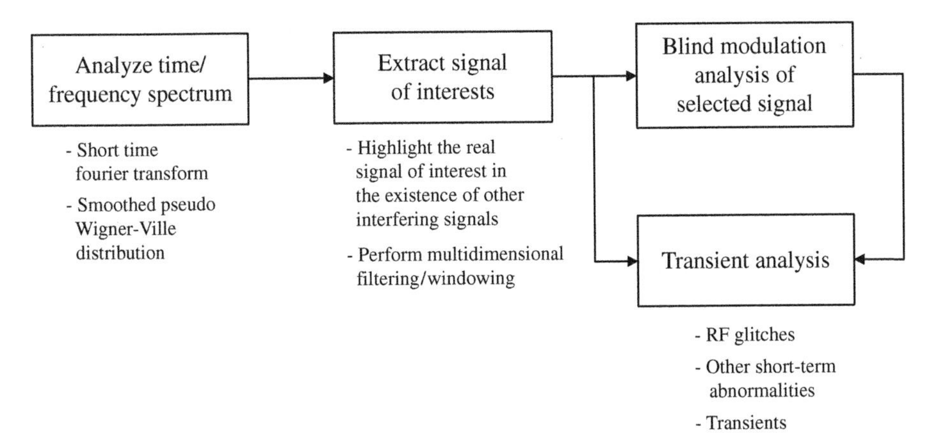

Figure 3.2 Use of joint time-frequency analysis to enable a better understanding of the signals and to allow further blind processing of the received signal.

investigated in other domains as well. The challenge is to represent all these dimensions jointly and to develop algorithms to optimally extract features in all these dimensions.

Once an accurate representation of the signal is obtained in multiple dimensions (2-D in the case of the example given in Figure 3.2), the next step is to be able to extract the desired information from this complex plane. It is very difficult to analyze the multidimensional signal. However, by extracting part of this signal (for example, the short transient part in time, or a part of desired frequency), a detailed analysis can be performed. Signal (or feature) extraction requires accurate windowing/filtering in multiple domains. In addition, advanced triggering and/or advanced user interfaces are also required to be able to extract the desired features of the signal.

The final stage is the analysis of the extracted signal. This includes processing the extracted data and providing accurate test and measurement results as well as statistics about the signal of interest. Blind modulation analysis is one example that can be employed within this stage.

In addition to the conventional measurements and analysis tools (like error vector magnitude, in-phase and quadrature-phase [I/Q] measurements, power versus time, power spectrum, frequency offset, sample timing offset, Rho) that provide information about the device under test (DUT), new generations of wireless systems will bring about new performance metrics and analysis options. Not only metrics or tools that can provide generic test and measurement capabilities, but also metrics that are unique for specific waveforms need to be developed. Given the vast number of possible waveforms in the future transceivers, providing all sorts of metrics and measures is quite challenging. In addition to the measurements that we discussed in Chapter 2, additional metrics and analysis tools need to be extracted from multidimensional signal space. Some of these measurements can be given as:

- Cross-coupling/cross-talk (among radio frequency [RF] components and high-speed digital components)
- Analog-to-digital converter (ADC) and digital-to-analog converter (DAC) impairments
- Predistortion/linearization impairments
- Time/frequency/space/angle statistics of interference and signal (like minimum bandwidth, maximum bandwidth, spectral occupancy rate, etc.)
- Cyclostationarity analysis
- Generic code domain analysis

- Waveform identification (whether the signal is multicarrier signal or single carrier signal; frequency hopping signal or not; spread spectrum signal or not)
- Automodulation detection
- Signal activity detection
- Direction of arrival (DoA)/angle of arrival (AoA) detection
- Protocol analysis
- Multiple antenna and multiple-input multiple-output (MIMO) measurements
- Interference awareness
- Advanced modulation domain analysis
- Transient behavior measures, like spectral re-growth during transient
- Power statistics, like peak power, average power, complementary cumulative distribution function (CCDF), peak-to-average power ratio (PAPR), average burst width

3.2 Time Domain Analysis

Time domain is the first step to the other domains. Many of the signal characteristics can be revealed from its time series. For instance, signal power (received signal strength), CCDF, eye diagram, pulse repetition intervals, on-time, off-time, duty cycle, peak power, average power, time interval (windowed) power, PAPR, signal-to-noise ratio (SNR), burst length, burst start, burst end, time selectivity (rate of change of time signal), and error vector magnitude (EVM) versus time are some of the signal attributes which can be extracted by time domain analysis. Energy detection, autoregressive modeling of a signal, and correlation techniques can also be performed in the time domain. Time domain analysis not only provides information about the temporal attributes of the signal, but also helps the receiver to synchronize (frame, symbol, and sample timing synchronization) to the received signal for detailed modulation analysis. Many of the blind and training-based receiver algorithms benefit from the time domain analysis directly or indirectly. In Figure 3.3, a time signal can be seen where a bursty signal along with background noise (with SNR value of 25 dB) and impulse noise (with 10 dB above the signal power level) is plotted. Burst start and burst end positions can be easily seen from the time domain analysis. By detecting the power ramp up and down

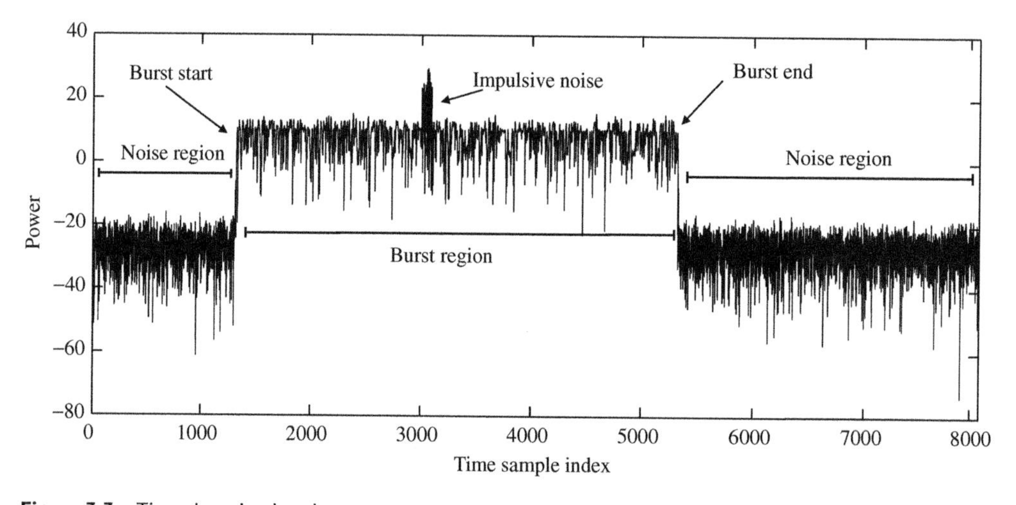

Figure 3.3 Time domain signal.

regions, the burst start and end positions can also be calculated in algorithmic level, providing a very easy coarse synchronization tool for the data burst of the received signal.

3.2.1 CCDF and PAPR

PAPR is the ratio of the peak power of a signal over a period of time samples to the average power level that occurs over the same time period:

$$PAPR = 10\log_{10}\left(\frac{P_{peak}}{P_{average}}\right). \tag{3.1}$$

On the other hand, CCDF is a statistical measure that characterizes the power statistics of a modulated signal. It represents the statistical probability of occurrence of signal peaks that are greater by a factor K in dB than the average value. It also represents the percentage amount of time that the signal power is above some value (for example, we can say the signal power level that is 6 dB above the average power value for 9% of the time, or alternatively, we can say the signal power exceeds the average by at least 6 dB for 9% of the time). Obtaining CCDF is quite simple. When we have the discrete data samples over a large period of time, we can calculate the discrete probability density function (PDF) which can be obtained using histogram function in Matlab. From the PDF discrete cumulative distribution function (CDF) can be obtained by employing cumulative sum (cumsum in Matlab). The CCDF is the complement of the CDF which can be obtained as $CCDF = 1 - CDF$. Following is a simple Matlab code for obtaining the CCDF curve from the received samples (in this figure random noise samples are used). Figure 3.4 shows the corresponding Matlab plots.

```
1  % A simple Matlab code for calculating CCDF
2  % of a random complex Gaussian noise.
3  N=1e7 ; % Number of samples
4  noise=(randn(1,N)+sqrt(-1)*randn(1,N))/sqrt(2);
5  noise_P=abs(noise).^2 ;
6  noise_P_av = mean(noise_P);
7  noise_P_norm = noise_P/noise_P_av;
8  subplot(2,2,1)
9  plot(10*log10(noise_P_norm));
10 xlabel('Sample index')
11 ylabel('Normalized Power in dB')
12 title('(a) Noise sample powers')
13 axis([0,1000,-30 10])
14 [x,y]=hist(10*log10(noise_P_norm),100);
15 subplot(2,2,2)
16 plot(y,x/sum(x));
17 xlabel('power in dB')
18 ylabel('probability')
19 title('(b) PDF')
20 axis([-40,15,0.000001 .08])
21 subplot(2,2,3)
22 plot(y,cumsum(x/sum(x)))
23 xlabel('power in dB')
24 ylabel('cumulative probability')
25 title('(c) CDF')
26 axis([-30,15,0 1.1])
27 subplot(2,2,4)
28 semilogy(y,1-cumsum(x/sum(x)))
29 xlabel('power in dB')
30 ylabel('cumulative probability')
31 title('(d) CCDF')
32 axis([-10,15,0.00000001 1.2])
```

Compared to PAPR, CCDF provides more in-depth analysis and more reliable measurements. Due to the random nature of the signals that we analyze in communication systems, the occurrence

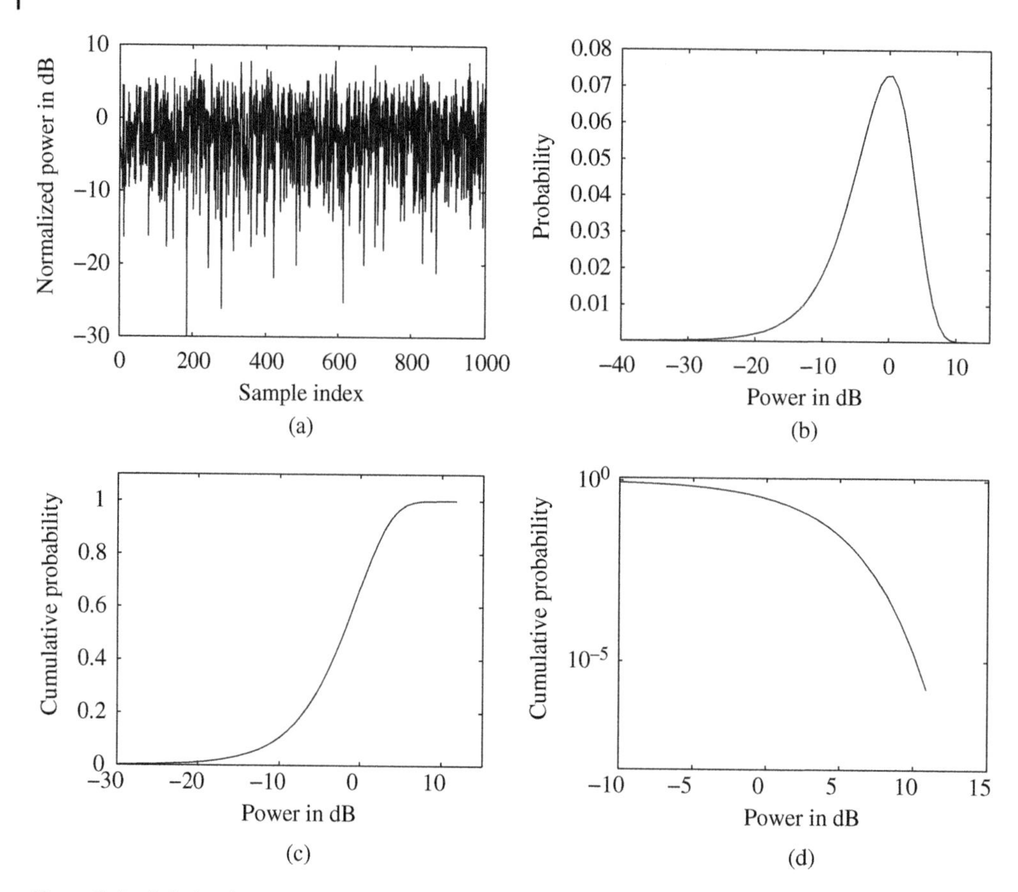

Figure 3.4 Relation between received signal power, PDF, CDF, and CCDF. (a) Noise sample powers. (b) PDF. (c) CDF. (d) CCDF.

of the highest peak value depends on the measurement time. Two separate measurements of the PAPR from two different sets of data windows might not give us the same results. Therefore, it might be better to look at the CCDF measurements.

By looking at the CCDF measurements, one can get some intuitive feelings of what type of modulation or filtering is used in the signal. For example, the CCDF curves for constant envelope modulation (let's say Gaussian minimum shift keying as used in global system for mobile communications [GSM]) and nonconstant envelope modulations (let's say 8-PSK mode as used in enhanced data rates for GSM evolution [EDGE]) will be different. Similarly, CCDF of various orders of quadrature amplitude modulation (QAM) will be different; (4-state quadrature amplitude modulation [4-QAM] will have different curves than 64-state quadrature amplitude modulation [64-QAM]). Therefore, by looking at the CCDF curves, it is possible to characterize various modulation and filtering types. In the following Matlab code, CCDF curves for two different QAM modulations are obtained and the corresponding plots are given in Figure 3.5.

It is also possible to characterize the multicarrier signals and multiuser CDMA signals from the CCDF curves. In fact, multiple digitally modulated signals combined together (and let's say transmitted from a single source like base stations) can also be characterized through CCDF plots. In general, superimposition of multiple signals (either coded with different Pseudo Random sequences like in CDMA or transmitted over multiple carriers like in orthogonal frequency

```
1  % A simple Matlab code for calculating CCDF of QAM modulations.
2  N=1e6; % block size in terms of number of symbols
3  sps=8; % oversampling rate
4  fltr=rcosine(1,sps); % raised cosine filter
5  M=4; % modulation order 4-QAM
6  alp=qammod(0:M-1,M);
7  norm=sqrt(mean(abs(alp).^2)) ;
8  symb= qammod(floor(rand(1,N)*M),M)/norm;
9  u_frame=upsample(symb, sps);
10 sgnl_tx_4 = conv(fltr,u_frame);
11 M=64; % modulation order 64-QAM
12 alp=qammod(0:M-1,M);
13 norm=sqrt(mean(abs(alp).^2));
14 symb= qammod(floor(rand(1,N)*M),M)/norm;
15 u_frame=upsample(symb, sps);
16 sgnl_tx_64 = conv(fltr,u_frame);
17 [x4,y4]=hist(abs(sgnl_tx_4).^2,100) ;
18 [x64,y64]=hist(abs(sgnl_tx_64).^2,100) ;
19 figure(1);clf
20 semilogy(10*log10(y4),1-cumsum(x4/sum(x4)),'-')
21 hold on
22 semilogy(10*log10(y64),1-cumsum(x64/sum(x64)),'—')
23 legend('4-QAM','64-QAM')
24 grid
25 xlabel('power in dB')
26 ylabel('cumulative probability')
27 title('CCDF')
28 axis([-10,8,0.000001 1.2])
```

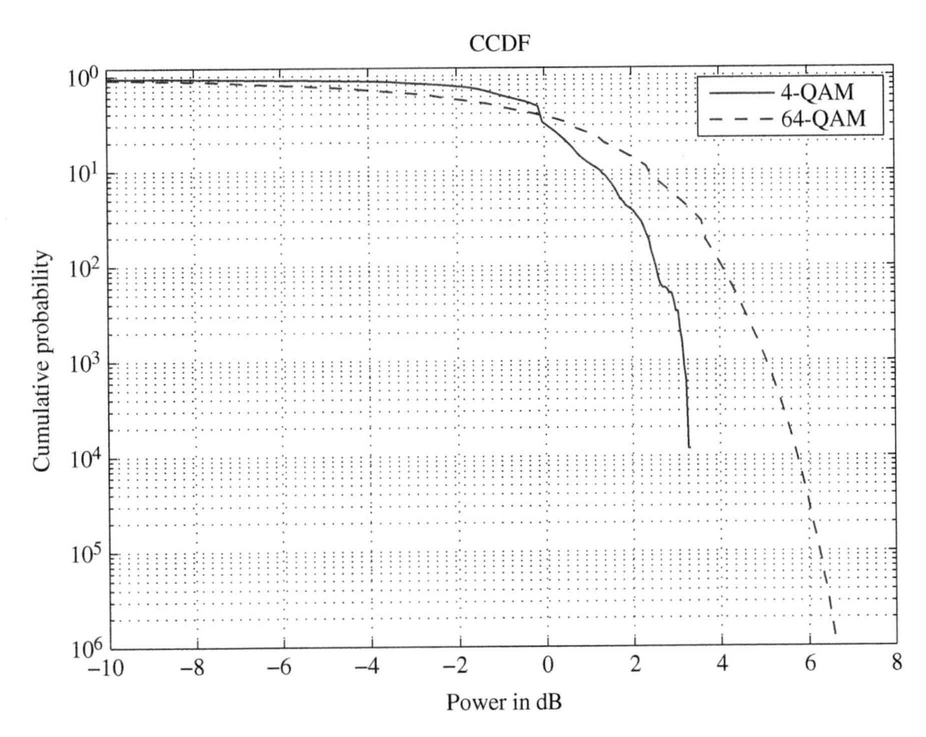

Figure 3.5 Comparison of CCDF curve for two different QAM modulations. As the modulation order increases, the PAPR also increases, shifting the CCDF curve to the right.

division multiplexing [OFDM] or transmitted in adjacent bands like in FDMA signals) increases the PAPR of the combined signal compared to the individual signals. For example, if a signal is made up of multiple (let's say N) superimposed tones that are of equal amplitude and that their phases are coherent in the sense that they add up constructively once over a certain period, then the PAPR of this N-tone signal can be estimated as $PAPR = 10\log_{10}(N)$. As can be seen here, as the number of tones (N) increases, the PAPR also increases. Such intuition, for example, can provide us rough information on the number of active codes used in a combined CDMA signal. Similarly, this can be used to get a rough idea on how many carriers are possibly used in an orthogonal frequency division multiplexing (OFDM) signal. On the other hand, the CCDF curves can also help us design the optimal number of carriers (or codes, or adjacent channel signals) so that the combined signal stays within the linear region of the nonlinear devices (like power amplifiers) to avoid signal distortion. Similarly, the CCDF curve helps us in designing efficient PAPR reduction techniques which are extremely important for OFDM based multicarrier modulation schemes.

CCDF curve can also provide us some idea about the compression of a signal by the nonlinear components like power amplifier, mixer, etc. For example, the PA compresses the signal when the power reaches into the saturation region. By looking at the CCDF characteristics of the signal before and after the PA, we can see whether the signal is compressed or not. If both the input and output signals exhibit similar CCDF characteristics, then we can say that the PA is not distorting the signal and amplification is linear (in other words the PAPR of the signal remains the same). On the other hand, if the input power levels are around the saturation region where the PA compresses the signal and the output signal is clipped, then the PAPR or the output signal will be less than the PAPR of the input signal. Similarly, the CCDF curves of the output signal and input signal will not be the same. In the case of the compressed signal, the CCDF curve will be shifted to the left. As can be understood from the above discussion, the CCDF plot can be a great tool to optimize the system design in various levels including the PA, input signaling type, input signal power, etc.

3.2.2 Time Selectivity Measure

CCDF is useful in obtaining statistics of the amplitude and power of the received signal. However, it does not provide information about the rate of change of the signal. Time selectivity measure is referred to as the variation of the channel that the signal is operating. The signal amplitude can also vary due to the modulation, multiple accessing, and frame format of the signal that is transmitted. These variations are not channel dependent, rather waveform dependent variations. In this subsection, time selectivity is referred for the variation of the operating channel (not the whole signal). In a later section of this chapter, we will discuss the variation of the whole signal (channel plus the transmitted waveform) using correlation and cyclostationary analysis.

Doppler shift is the frequency shift experienced by the radio signal when either the transmitter or receiver is in motion, and Doppler spread is a measure of the spectral broadening caused by the temporal rate of change of the mobile radio channel. Therefore, time-selective fading and Doppler spread are directly related. The coherence time of the channel can be used to characterize the time variation of the time-selective channel. It represents the statistical measure of the time window over which the two signal components have a strong correlation, and it is inversely proportional to the Doppler spread.

Time selectivity and Doppler spread estimation have been studied for several applications in wireless mobile radio systems. Correlation and variation of channel estimates, as well as correlation and variation of the signal envelope, have been used for Doppler spread estimation [1]. One simple method for Doppler spread estimation is to use *differentials* of the complex channel estimates. The differentials of the channel estimates are very noisy, hence require low-pass filtering. The bandwidth of the low-pass filter is also a function of the Doppler estimate. Another simple approach is based on the autocorrelation of complex channel estimates, which provides more reliable results compared to the differential. Channel autocorrelation can be calculated using the channel estimates over the known field of the transmitted data. Instead of using channel estimates, the received signal can also be used directly in estimating Doppler spread information. The Doppler information can be estimated as a function of the deviation of the averaged signal envelope. Similarly, level crossing rate of the average signal level can be used in estimating velocity and Doppler spread. Multiple antennas can also be exploited for Doppler spread estimation, where a linear relation between the switching rate of the antenna branches and Doppler frequency can be obtained.

3.3 Frequency Domain Analysis

Revealing signal spectral components is useful for many applications like center frequency and bandwidth estimation, in-band and out-of-band spurious emission detection, observing spectral re-growth, understanding intermodulation products, measuring the adjacent channel powers, monitoring spectral flatness, analyzing interference and noise sources, analyzing spectral contents of the desired and interfering signals, spectrum awareness (especially in the cognitive radio context), band power measurement (the total power between two selected frequencies), and calculating EVM versus spectrum (especially very important for OFDM systems). For example, adjacent channel power (ACP) measurement is an integral part of many wireless standards. The interference caused in the adjacent channel due to energy that spills from the desired communication is a critical parameter that needs to be controlled (often strictly regulated by wireless standards). Nonlinear devices, like PAs, introduce spectral re-growth (making the spectrum larger than it actually is), causing adjacent channel interference. Similarly, spurious and harmonics caused by nonlinear behaviors of the RF components can be measured using spectral measurements. These measurements are critical to monitor and control interference levels on communication systems operating in other bands.

Spectral measurements help the designer to monitor the spectrum and optimize the transmission parameters accordingly. Recently, in the context of cognitive radio, spectrum sensing has received a lot of interest to monitor primary signals (and interference temperature) for opportunistic spectrum usage.

In frequency domain analysis, the received time discrete samples are transformed to frequency by using discrete Fourier transform (DFT) which is implemented efficiently with fast Fourier transform (FFT). If the time domain samples are represented as $x(n),\ n = 1, \dots, N$, then the DFT of these samples will be:

$$X(k) = \sum_{n=0}^{N-1} x(n)e^{-\frac{j2\pi kn}{N}}, \tag{3.2}$$

where N is the FFT size and number of samples in the data block (after zero padding if necessary). The transformed frequency domain samples will have a resolution of $\frac{f_s}{N}$, where f_s is the sampling frequency. The frequency range of FFT starts from 0 Hz until $\frac{f_s}{2}$. The frequency resolution increases with the data record length N. In order to avoid aliasing, sample rate f_s should be above the Nyquist rate ($f_s \geq 2f_{max}$, where f_{max} is the highest frequency component of the signal). Hence, in spectrum analysis, the frequency span can be controlled by adjusting the sampling rate and for a given sampling rate the frequency resolution can be improved by increasing the data block size. If the signal is in higher frequencies, by employing band selectable analysis (BSA) the frequency resolution can be increased for a given data block size. The BSA allows us to reduce the frequency span by mixing the high-frequency signal to lower frequencies, digital filtering, and decimating the data. Since

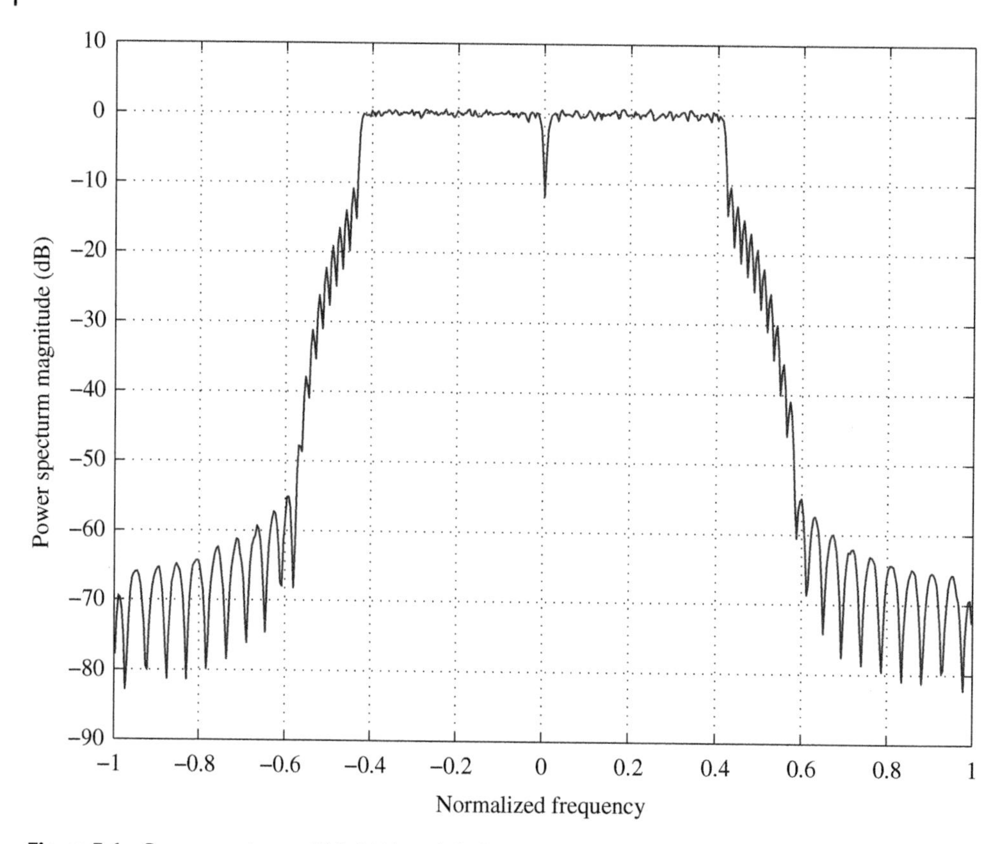

Figure 3.6 Power spectrum of 16-QAM modulation with raised cosine filtering ($\alpha = 0.3$).

the signal is shifted to lower frequencies, reducing the sampling rate is possible. In the following Matlab code, a 16-state quadrature amplitude modulation (16-QAM) modulated signal with raised cosine filtering (with roll-off factor of $\alpha = 0.3$) is generated. The 3 dB normalized bandwidth of the modulated and filtered signal will be $B_{3\text{ dB}} = 1 + \alpha$. The plot is shown in Figure 3.6.

```
1  % Power spectrum of raised cosine shaped QAM signal
2  % Observe the effect of:  roll-off and oversampling rate
3  N=1e4; % block size in terms of number of symbols
4  sps=4; % oversampling rate
5  fltr_rolloff=0.3;
6  fltr=rcosine(1,sps,[],fltr_rolloff);
7  M=16;
8  alp=qammod(0:M-1,M);
9  norm=sqrt(mean(abs(alp).^2));
10 symb= qammod(floor(rand(1,N)*M),M)/norm;
11 u_frame=upsample(symb, sps);
12 sgnl_tx_16 = conv(fltr,u_frame);
13 [Pxx,F]=pwelch(sgnl_tx_16,[],[],2048,sps,'twosided');
14 plot(F,10*log10(Pxx))
15 xlabel('Normalized (with symbol rate) Frequency (Hz)')
16 ylabel('Power (dB/Hz)')
```

In the following Matlab script, the impact of the sampling rate and data window in frequency resolution can be seen clearly. Two tones are generated with frequencies of 1000 and 1200 Hz (the spacing between the tones is 200 Hz). With a large data window size, the tones can be resolved from each other (see Figure 3.7a). When the data window length is reduced (300 samples in this specific scenario), then the frequency resolution reduces as well, making it difficult to resolve the

```
1  % Basics of spectrum analysis
2  % Observe: frequency resolution and span
3  f1=1000;
4  f2=1200;
5  frequency_max=max(f1,f2); % maximum frequency component
6  fs=6*frequency_max; % sampling rate
7  time=0:1/fs:10000/frequency_max;
8  x=cos(2*pi*f1*time);
9  y=cos(2*pi*f2*time);
10 z=x+y; % sum of two tones
11 subplot(2,2,1)
12 [Pxx,F]=pwelch(z(1:10000),[],[],2048,fs);
13 plot(F,10*log10(Pxx))
14 xlabel('Frequency (Hz)')
15 ylabel('Power (dB/Hz)')
16 title('(a) Large window size')
17 subplot(2,2,2)
18 [Pxx,F]=pwelch(z(1:300),[],[],2048,fs);
19 plot(F,10*log10(Pxx))
20 xlabel(' Frequency (Hz)')
21 ylabel('Power (dB/Hz)')
22 title('(b) Small window size')
23 subplot(2,2,3)
24 zz=z.*cos(2*pi*800*[1:length(z)]/fs);
25 [Pxx,F]=pwelch(zz(1:300),[],[],2048,fs);
26 plot(F,10*log10(Pxx))
27 xlabel('Frequency (Hz)')
28 ylabel('Power (dB/Hz)')
29 title('(c) After mixing the signal (without filtering)');
30 subplot(2,2,4)
31 frequency_max_new=400;
32 d_rate=frequency_max/frequency_max_new;
33 BB = fir1(150,0.12);
34 filtered=conv(BB,zz);
35 zd=downsample(filtered,d_rate);
36 [Pxx,F]=pwelch(zd(1:300),[],[],2048,fs/d_rate);
37 plot(F,10*log10(Pxx))
38 xlabel('Frequency (Hz)')
39 ylabel('Power (dB/Hz)')
40 title('(d) Band selective analysis');
```

tones (see Figure 3.7b). If the data window size is something we cannot control, by employing band selective analysis, we can increase the resolution. Since the tones were in higher frequencies, in the following code, we shifted them to further baseband. With the same data window size as in (b), with band selective analysis, the resolution is increased (see Figure 3.7d).

3.3.1 Adjacent Channel Power Ratio

Adjacent channel power ratio (ACPR) is a measure of how much interference a transmitter can cause on nearby channels. The measurement is a relative power with respect to the total transmitted power and expressed in dB, or the relative difference between the signal power in the main channel and the signal power in the adjacent or alternate channel. Depending on the particular communication standard, these measurements are often described as adjacent channel power ratio (ACPR) or adjacent channel leakage ratio (ACLR) tests. Many transmission standards, such as IS-95, CDMA, wide band code division multiple access (WCDMA), 802.11, and Bluetooth, contain a definition for ACPR measurements.

Wireless standards define spectral masks to set boundaries for the signal power spectrum. A transmit spectrum mask is the power contained in a specified frequency bandwidth, and wireless standards define how much power can be emitted from a transmitter at the center frequency and at the given frequency points (called offsets) on both sides of the center frequency. For example, in Figure 3.8, a spectral mask that is defined in the institute of electrical and electronics engineers (IEEE) specification 802.11a, 802.11b, and 802.11g is given.

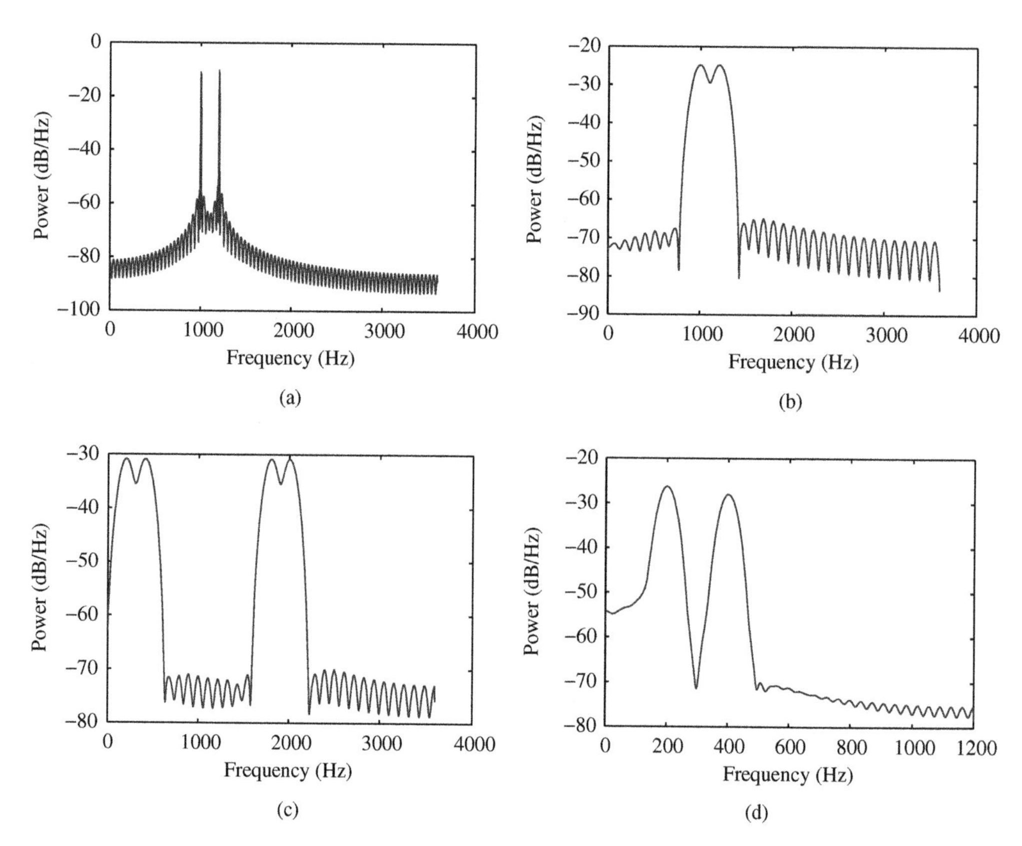

Figure 3.7 Spectrum analysis of two tone signal. (a) Large window size. (b) Small window size. (c) After mixing the signal (without filtering). (d) Band selective analysis.

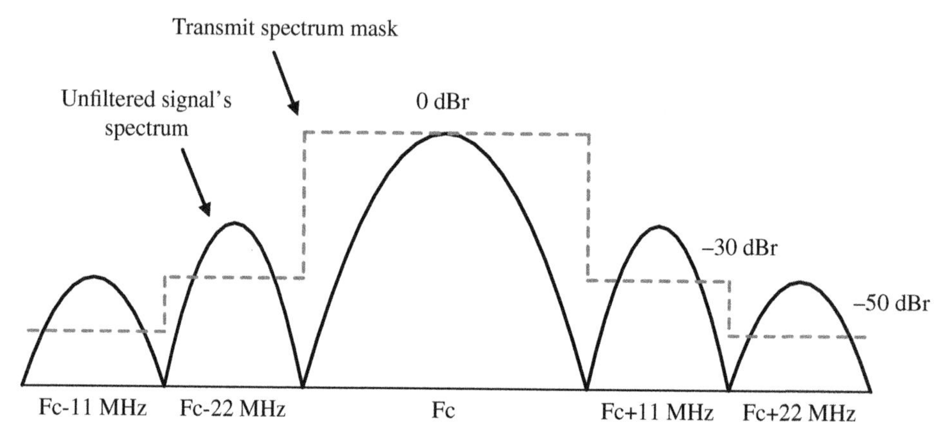

Figure 3.8 Spectrum mask defined in IEEE 802.11 specifications.

Similar to the ACPR measurement, out-of-band measurements can be done on signals outside of the desired signal's band. The signals (referred to as harmonics) that appear outside of the actual transmission band can interfere with other communication systems. The out-of-band measurement can be performed using spectrum analyzers with a wide frequency range and good dynamic range.

For ACPR measurements, the transmitter signal is modulated in normal operating conditions. For burst signals like in time division multiple access (TDMA), the measurements are to be made when the transmitter is active. All measurements are made at the transmitter output port. During the measurement, the transmitter power can be set to maximum output power. Then, the in-band power and out-of-band powers in adjacent and alternate frequencies can be measured by integrating the power within each band between the start and end frequency.

3.3.2 Frequency Selectivity Measure

Delay spread is one of the most commonly used parameters that describes the time dispersiveness of the channel, and it is related to the frequency selectivity of the channel. Frequency selectivity can be described in terms of coherence bandwidth, which is a measure of the range of frequencies over which the two frequency components have a strong correlation. The coherence bandwidth is inversely proportional to the delay spread [2].

Although dispersion estimation can be very useful for many wireless communication systems, it is particularly crucial for OFDM based wireless communication systems. OFDM, which is a multicarrier modulation technique, handles the ISI problem due to high bit rate communication by splitting the high rate symbol stream into several lower rate streams and transmitting them on different orthogonal carriers. The OFDM symbols with increased duration might still be affected by the previous OFDM symbols due to multipath dispersion. Cyclic prefix extension of the OFDM symbol avoids ISI from the previous OFDM symbols if the cyclic prefix length is greater than the maximum excess delay of the channel. Since the maximum excess delay depends on the radio environment, the cyclic prefix length needs to be designed for the worst case channel condition. This makes cyclic prefix as a significant portion of the transmitted data, resulting in reduced spectral efficiency. One way to increase spectral efficiency is to adapt the length of the cyclic prefix depending on the radio environment. The adaptation requires estimation of maximum excess delay of the radio channel, which is also related to the frequency selectivity of the channel. In HyperLAN2, which is a wireless local area network (WLAN) standard, a cyclic prefix duration of 800 ns, which is sufficient to allow good performance for channels with delay spreads up to 250 ns, is used. Optionally, a short cyclic prefix with 400 ns duration may be used for short-range indoor applications. Similarly, the wireless metropolitan area network (WMAN) standard, IEEE 802.16, defines several cyclic prefix options that can be used in different environments. Delay spread estimation allows adaptation of these various options to optimize the spectral efficiency. Other OFDM parameters that could be changed adaptively using the knowledge of the dispersion are the OFDM symbol duration and OFDM sub-carrier bandwidth.

Characterization of the frequency selectivity of the radio channel is studied extensively using level crossing rate (LCR) of the channel in frequency domain. Frequency domain LCR gives the average number of crossings per Hz at which the measured amplitude crosses a threshold level. An analytical expression between LCR and the time domain parameters corresponding to a specific multipath power delay profile (PDP) can be easily obtained. LCR is very sensitive to noise, which increases the number of level crossing and severely deteriorates the performance of the LCR measurement. Filtering the channel frequency response reduces the noise effect, but finding the appropriate filter parameters is an issue. If the filter is not designed properly, one might end up smoothing the actual variation of frequency domain channel response.

The instantaneous root mean squared (RMS) delay spread, which provides information about local (small-scale) channel dispersion, can be obtained by estimating the channel impulse response (CIR) in time domain. The absolute square of these CIR estimates can then be averaged in time to obtain the power delay profile and hence the average (long term) delay spread estimate. Channel frequency selectivity and delay spread information can also be calculated using the channel frequency correlation estimates, and analytical expressions between delay spread and coherence bandwidth can be obtained easily.

3.4 Joint Time-Frequency Analysis

In order to reveal the temporal and spectral components of a nonstationary signal, joint time-frequency analysis (TFA) could be performed. TFA combines time and frequency domain analysis which is especially useful to understand the characteristics of nonstationary, impulsive, and multicomponent signals. TFA gives a better representation of the dynamic usage of the spectrum over time. For example, in Figure 3.9, a time-frequency representation of a Bluetooth signal is given. For a signal like this, if we were to look at the power spectrum of the whole received signal, we can see as if the whole band is used all the time, misleading the communication engineer in the analysis of the signal. As can be seen from the figure, the whole bandwidth is not used all the time. The signal is frequency hopping and using only a fraction of the spectrum at each time interval.

Signal intelligence, blind signal identification, and CR/SDR (which have found applications in military and commercial communications including signal classification, interference identification and cancellation, spectrum management, surveillance, and electronic warfare) can benefit greatly from TFR-based signal-processing techniques. TFA can be used as an initial analysis tool to provide some characteristics about the signal (such as instantaneous frequency, instantaneous bandwidth, power statistics, and spurious emissions). Modulation domain, space domain, and code domain can then be evaluated as the next step after TFA and thus characterize different dimensions of the signal attributes for further processing such as signal classification and feature extraction. Alternatively, these domains can be evaluated jointly with TFA processing. For instance, mode

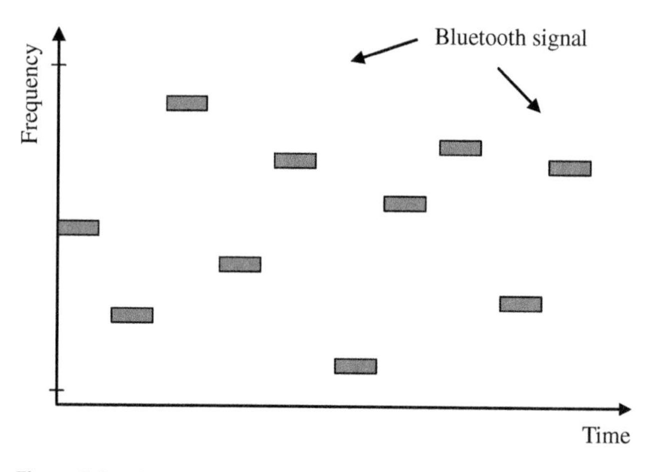

Figure 3.9 Illustration of bluetooth like signaling in time-frequency plane.

identification can use TFA for signal detection and feature extraction which is then used for modulation and mode identification.

In [3], TFA is combined with pattern recognition to identify superimposed transmission modes such as Bluetooth and 802.11b standards in an SDR terminal. In [4], TFA is used for better resolution in time-frequency planes for frequency-hopping spread spectrum (FHSS) parameters estimation in embedded noise. For instance, hop frequencies, hop duration, and time offset can be blindly estimated using TFA. TFA is used for modulation classification and symbol rate estimation in [5]. TFA is also exploited for interference suppression in wideband communication systems using direct-sequence spread-spectrum (DSSS) in applications such as CDMA, low probability of being intercepted (LPI) systems, communications over channels with multipath, resistance to intentional jamming [6]. Another important application for TFA is SDR testing and measurements as there is a need for software-based tests for the wide range of wireless communication standards. Recently, advanced signal analyzers for testing and measurement were released in the market. These equipment are capable of capturing and displaying RF signals in multidomain mode such as time, frequency, joint time-frequency, modulation constellation, error vector magnitude. Similar efforts for measurement and testing different technologies based on TFA are also provided in the literature to provide automated measurements such as estimation of instantaneous frequency, carrier frequency, instantaneous bandwidth, and RF power. TFA are exploited for generic measurement purposes in second and third generation telecommunication systems. The proposed techniques use a digital signal-processing approach for carrying the most needed measurements. For instance, TFA-based measurements for verifying global system for mobile communications (GSM) and universal mobile telecommunications service (UMTS) equipment were proposed in [7]. It can be concluded that TFA can provide better visualization of signal dynamics. This can be much beneficial compared to usual mono-dimensional analysis. Careful selection for the TFA method is needed to fully analyze the signal.

Many TFA methods exist in the literature [8]. One famous TFA techniques is short time Fourier transform (STFT) which is also referred to as spectrogram in many test equipments. In STFT, a prewindowing (in time) is applied before the transformation, providing spectral properties of the signal in short time periods (local spectrum). The performance of the spectrogram depends on the window type and size which often results in a trade-off in obtaining a good time resolution (at the expense of bad frequency resolution) or vice-versa. Decreasing the window size results in a better time resolution since the length of the signal which is processed at a time is shorter. Yet, this reduces the frequency resolution of the spectrogram. Recently, other techniques that provide a more accurate representation of time-frequency characteristics are developed. For example, it is reported that smoothed-pseudo-Wigner-Ville distribution (SPWD) can provide enhanced performance compared to STFT.

Most of the TFA techniques employ some kind of smoothing kernel, window, or filter to reduce cross-components [8]. Basically, TFA can be classified into two types: linear and quadratic TFAs [8]. Examples for the first type are STFT, Gabor expansion, and Wavelet transform. Whereas examples for quadratic TFA are Wigner-Ville distribution, Choi-Williams distributions, and time-frequency distribution series.

STFT of a signal can be given as

$$STFT(t, w) = \int_{-\infty}^{\infty} x(\tau)w(\tau - t)e^{-jw\tau}d\tau, \tag{3.3}$$

where $x(t)$ is the analyzed signal, and $w(t)$ is the window function. Due to the uncertainty principle, the window selection creates a trade-off between time and frequency resolutions and thus limits the use of STFT in applications where both resolutions are required. However, it was found that Gaussian windows achieve minimum time-frequency uncertainty.

Another well known TFA is wavelet transform (WT) which is a mathematical technique that correlates the signal with shifted and dilated versions of an analyzing function, called wavelets. WT thus provides multiresolution decomposition of the signal. The continuous WT (CWT) is given by

$$CWT(\tau, a) = \int_{-\infty}^{\infty} x(t)\psi\left(\frac{t-\tau}{a}\right) dt, \tag{3.4}$$

where a is the scale factor, τ is the translation factor (shift), and ψ is an arbitrary mother wavelet. The discrete wavelet transform (DWT) is then evaluated at discrete scales and translations which reduces computations. In practice, the implementation of a DWT suitable for finite-length discrete-time signals is based upon the multiresolution analysis which decomposes a signal into scales with different time and frequency resolutions. A major drawback for WT is that it has time resolution varying with the analyzed scale. This means that the time resolution for the small scale (high frequency) is better than frequency resolution, and the reverse is true for a low scale. Nevertheless, WT finds its common use in many applications as it provides multiband decomposition of the signal. This is beneficial for applications such as transient analysis, interference mitigation, signal classification, and many other applications.

A second class of the TFA is bilinear class. An example of this class is Smoothed-Pseudo-Wigner-Ville distribution (SPWD). SPWD is largely used because of its desired TFA properties Another TFA belonging to a bilinear category is Choi-Williams distribution (CWD). It provides high resolution in both time and frequency while minimizing the cross-terms but does not satisfy the finite support property in time and frequency. Comparison between some of the TFAs are given in [9].

In general, no one set of TFAs performs well in all applications. Smoothed representations provide high resolution, but they require a careful choice of filtering or windowing to match the signal characteristics and decrease interference terms. Adaptive algorithms provide better resolution, but they are either computationally expensive or require apriori data information. Suboptimal computational efficient algorithms exist for high time or frequency resolution. One possible solution is to use an adaptive algorithm like adaptive spectrogram as an initial TFA for unknown signals. After analyzing the signal, it may be possible to select a suitable TFA from the TF dictionary. For instance, in [4] adaptive STFT is used to analyze FH signals and then optimized TFA for a better analysis.

3.5 Code Domain Analysis

Code domain power analysis provides the signal power projected on a code-space normalized to the total signal power. It is the distribution of signal power across the set of code channels, normalized to the total signal power. This can be obtained from the correlation coefficient factor of each code. This is useful for composite signal analysis, allowing us to monitor the active channels with their individual channel powers. Ideally, inactive channels should have a zero level. The levels of the inactive channels can provide useful information about specific impairments. A related measurement, **Rho** provides the normalized correlation coefficient between the measured and ideal reference signals, i.e.

$$\rho = \frac{P_{X_i X_m}}{P_{total}}, \tag{3.5}$$

where $P_{X_iX_m}$ is the correlated power of measured signal and ideal reference and P_{total} is the total signal power.

In the ideal situation, we obtain the maximum value of 1.0, which means the measured signal and reference signal are 100% identical.

```
1  % Code domain power measurement
2  % observe the effect of: used codes, code powers, noise variance
3  codes=hadamard(16); % hadamard code generation
4  codes_used=[1 5 7 13]; % defines which codes are actively used
5  codes_Amp=[4 2 1 6]; % the code amplitudes
6  sgnl_tx = zeros(1,16);
7  noise_var=0.03;
8  for kk=1:length(codes_used)
9       sgnl_tx=sgnl_tx+codes(codes_used(kk),:)*codes_Amp(kk);
10 end
11 sgnl_rx=sgnl_tx+sqrt(noise_var)*(randn(1,16)+j*randn(1,16))/sqrt(2);
12 % code domain power calculation
13 for ll=1:16
14      code_corr(ll) = abs(sum(sgnl_rx .* codes(ll,:)))/16;
15 end
16 stem(code_corr); grid; xlabel('codes');ylabel('Code Power')
```

In the following code a simple code domain power measurement technique is given. For the sake of simplicity, 16 codes using Hadamard matrix is generated. A set of these codes (four of them) used to make up a signal. Additive white Gaussian noise (AWGN) noise added to the combined signal. The received noisy signal is then passed through code domain analysis. Figure 3.10 shows the result

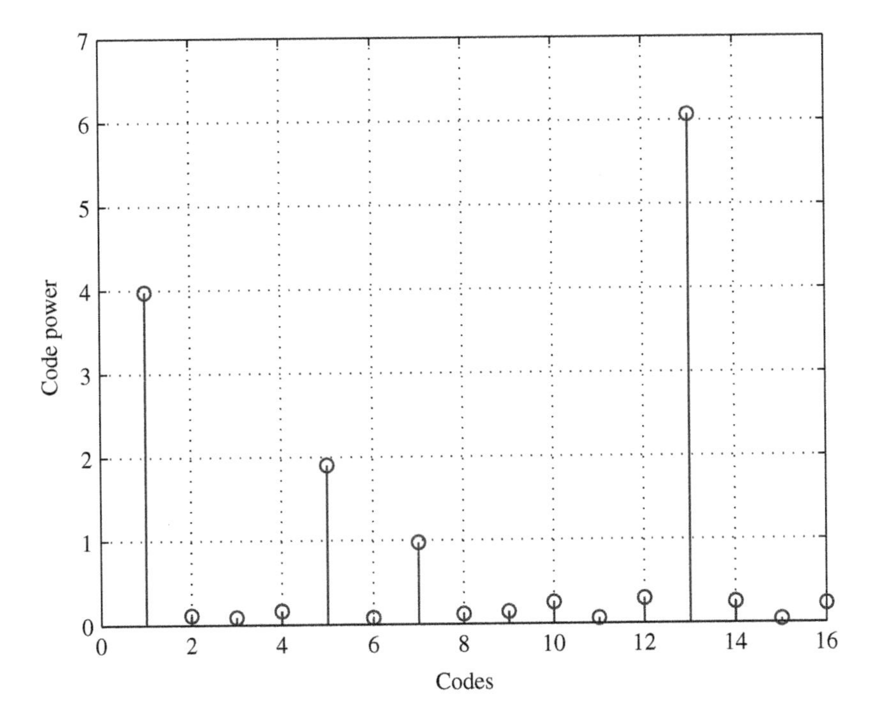

Figure 3.10 Code domain power plot. Four active codes from Hadamard matrix (of size 16) along with noise constitute a signal, and code domain power plot clearly identifies the active codes. The active codes have different power levels and as can be seen in the figure, code domain power measurement indicates the difference in power levels.

of the code domain analysis. As can be seen, the code domain power measurement displays the relative power in each channel. Note that due to the noise, some small power (depending on the SNR) appears on the codes that are not used, which is expected. When the SNR decreases, these undesired correlations in unused codes will also increase. In multiuser CDMA systems, each code (or often called channel) is assigned to a different user. Therefore, with code domain power measurements, we can view the presence of a signal in any code. Such feature can also be very useful for spectrum awareness measurement in cognitive radio if the primary user is employing CDMA type of transmission.

3.5.1 Code Selectivity

Code selectivity, like pseudo random (PN) codes in DSSS or time hopping (TH) codes in ultra wideband (UWB) or frequency hopping (FH) codes in FH systems, could be a strong measure for system design for future wireless radio systems. Many of the wireless systems are interference limited. Therefore, the capacity is determined by how much interference the system can tolerate. For example, the self interference (like intersymbol interference) which is caused by the nonzero autocorrelation sidelobes, and multi-access interference (MAI) due to the nonzero cross-correlations are major interference sources that are related to the code design. The effect of interference and near-far problem can be minimized by employing power control. Alternatively, decreasing sidelobes of the auto- and cross-correlation also reduces interference and increases spectral efficiency. Therefore, it is desirable to have sequences with ideal cross- and autocorrelation properties. However, it is proven that "perfect" sequences do not exist. Also, it is well known that there is a trade-off between obtaining good auto and cross-correlation properties, i.e. smaller ISI leads to larger MAI or vice-versa. In addition, the number of possible codes (and hence the capacity) can be increased by allowing some correlation (or interference) in code domain. In other words, the number of codes that have good correlation properties are limited. As a result, by allowing some correlation adaptively depending on the other system, channel, and transceiver parameters, the overall capacity of the system can be increased.

The correlation properties can also be changed adaptively to provide desired properties over a zone depending on other channel selectivity parameters. For example, there are several approaches in obtaining sequences with low and zero correlation zone properties. One common method of generating zero correlation zone (ZCZ) sequences is based on using complementary sequences [10]. In general, if there is a collection of K (K is even) mutually orthogonal complementary sets with each containing K sequences of length L_{ZCZ}, it is possible to construct a set of ZCZ sequences [10]. As described before, there is the same trade-off between the required L_{ZCZ} and the number of active users, given a fixed code length. Methods for generating low correlation zone (LCZ) sequences are also available in the literature. For example, LCZ sequences can be obtained from the Gold sequences that have the run property of cross-correlation with a consecutive value of -1 in a certain range. By the phase shift of Gold sequences, new Gold sequences that have run property around the origin can be obtained. There are also other techniques for generation LCZ sequences. However, here, we will not discuss them. The interested readers are referred to [11].

In summary, there are various ways to approach the interference problem in wireless communication systems. For example, one way to deal with it is based on allowing interference and employing receivers with interference cancelation capability. There has been considerable amount of research in this direction as it increases the spectral efficiency greatly. Another popular approach is to avoid interference by designing proper sequences as described above. The first approach usually requires complex receiver structures and might not remove interference completely.

Although the second approach avoids interference, it has a limited number of codes. A better approach is to employ a combination of those adaptively.

3.6 Correlation Analysis

Autocorrelation and cross-correlation are two popular correlation measurements. Correlation analysis provides the degree of similarity between two signals. The cross-correlation sequence between two discrete-time sequences, $x(n)$ and $y(n)$ is given by:

$$R_{xy}(l) = \sum_{-\infty}^{\infty} x(n)y(n+l). \tag{3.6}$$

Similarly, the autocorrelation sequence of a discrete-time sequence, $x(n)$, can be given by

$$R_{xx}(l) = \sum_{-\infty}^{\infty} x(n)x(n+l). \tag{3.7}$$

Correlation analysis is very widely used for the digital baseband algorithm design in wireless transmitters and receivers. Correlation properties of the noise and modulated signal are exploited in designing optimal algorithms. Similarly, the correlation of the wireless channel is used extensively for the optimization of the systems. The white noise process exhibits an autocorrelation of a delta function, $R_{xx}(l) = \delta(l)$, as the successive samples of a white noise sequence are uncorrelated. This property is often exploited in optimally detecting the modulated signal from the noise. Some of the applications of the correlation analysis can be given as:

- Speech encoding, especially vocoders
- Synchronization of signal
- Channel estimation algorithms
- Equalization algorithms
- Interference suppression algorithms
- Noise suppression algorithms
- Echo cancelation techniques
- Blind receiver design
- Spectrum sensing for cognitive radio

Cyclostationarity can be considered as an advanced version of correlation analysis [12–18]. Cyclostationary features are caused by the periodicity in the signal or its statistics like mean and autocorrelation. Instead of looking at the power spectrum, cyclic correlation function can be used for detecting signals present in a given spectrum. Cyclostationarity-based detection algorithms can differentiate noise from modulated signals. This is a result of the fact that noise is Wide Sense Stationary with no correlation while modulated signals are cyclostationary with spectral correlation due to the redundancy of signal periodicities [15].

The cyclic spectral density (CSD) function of the received signal $y(n) = x(n) + w(n)$, where $w(n)$ represents the noise and $x(n)$ represents the signal components, can be calculated as [19]

$$S(f,\alpha) = \sum_{\tau=-\infty}^{\infty} R_y^{\alpha}(\tau)e^{-j2\pi f\tau}, \tag{3.8}$$

where

$$R_y^{\alpha}(\tau) = E\{y(n+\tau)y^*(n-\tau)e^{j2\pi\alpha n}\}, \tag{3.9}$$

is the cyclic autocorrelation function, and α is the cyclic frequency. The CSD function outputs peak values when the cyclic frequency is equal to the fundamental frequencies of the transmitted signal. Cyclic frequencies can be assumed to be known [13, 18] or they can be extracted and used as features for identifying transmitted signals [16].

3.7 Modulation Domain Analysis

Digital modulation is one of the signal dimensions. Modulation domain analysis allows detection and recovering the digital data bits. Besides, modulation domain analysis provides powerful tools to identify and quantify impairments to the I/Q waveforms. For instance, modulation and signal quality measures through EVM, CCDF, I/Q vector and constellation, and eye diagrams are some of the critical measures in modulation domain analysis. We have discussed these measurements in Chapter 2.

Modulation analysis requires synchronizing to the signal and decoding the modulated symbols. Therefore, it requires the knowledge of the transmitted signaling features such as bandwidth, operating frequency, modulation type and order, pulse shaping, frame format, etc. Blind estimation of these features and applying modulation analysis subsequently is possible. However, blind estimation of all the waveform parameters would be tedious especially with the wide variety of signaling formats available today. For standard-based signaling, known patterns are usually utilized in wireless systems to assist synchronization or for other purposes. Such patterns include preambles, midambles, regularly transmitted pilot patterns, spreading sequences, etc. In the presence of a known pattern, synchronization can be simply performed by correlating the received signal with a known copy of itself [20, 21]. However, since there are a lot of wireless standards available currently, which standard-based parameters the receiver needs to be tuned to is another challenge. This problem can be avoided by employing a signaling type detection algorithm, i.e. by identifying the transmission technology used by the transmitter. For example, assume that the transmitter technology is identified as a Bluetooth signal. The receiver can use this information for extracting some useful information to lock into the signal and enable a detailed modulation analysis.

3.8 Angular Domain Analysis

Angular domain analysis can include finding the direction of the signal source and angular spread of the signal that is received.

3.8.1 Direction Finding

Radio direction finding has a long and rich history specifically in radar applications. As the name implies, direction finding refers to finding the direction of the transmitter of the signal received. Directional antennas (where the antenna receives or transmits signals only from certain directions) are often used to find the direction of the source. By physically rotating the directional antenna over all directions, the source of the direction can be found which is the direction where the received signal has most energy. Alternatively, the same goal can be achieved by employing multiple directional antennas and using a mechanical switch that can select antennas one-by-one. This is also called switched beam antenna system where signal strength is observed from one of several predetermined fixed beams (Figure 3.11). Both approaches given above correspond to

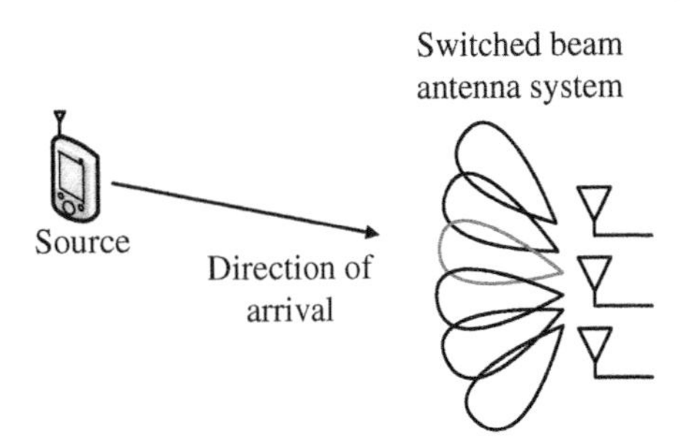

Figure 3.11 Use of directional antennas and switched beamforming to find the angle of arrival.

a single-channel directional finding as only one RF receiver chain is used to find the direction. Even though this is a quite simple technique, there is a potential of missing the source when the source is not transmitting continuously as all possible directions are not looked at simultaneously. Another potential problem is the impact of multipath fading. The signal strength might fade over the actual direction of the signal due to destructive combination of multipath components, leading to some potential errors in direction finding.

The direction of the source can also be found using Doppler-based approaches (Figure 3.12). Doppler-based direction finding can be achieved by physically spinning an antenna over a circular pattern. If the antenna is rotated very fast, it creates a Doppler frequency shift on the arriving signal depending on the speed of the rotation and the angle of the signal arriving to the antenna. The frequency increases when the antenna is moving toward the signal source and decreases when the antenna is moving away from the signal source. Maximum Doppler frequency shift as a function of antenna rotation can be written as

$$f_{max} = \frac{\omega r}{\lambda}, \tag{3.10}$$

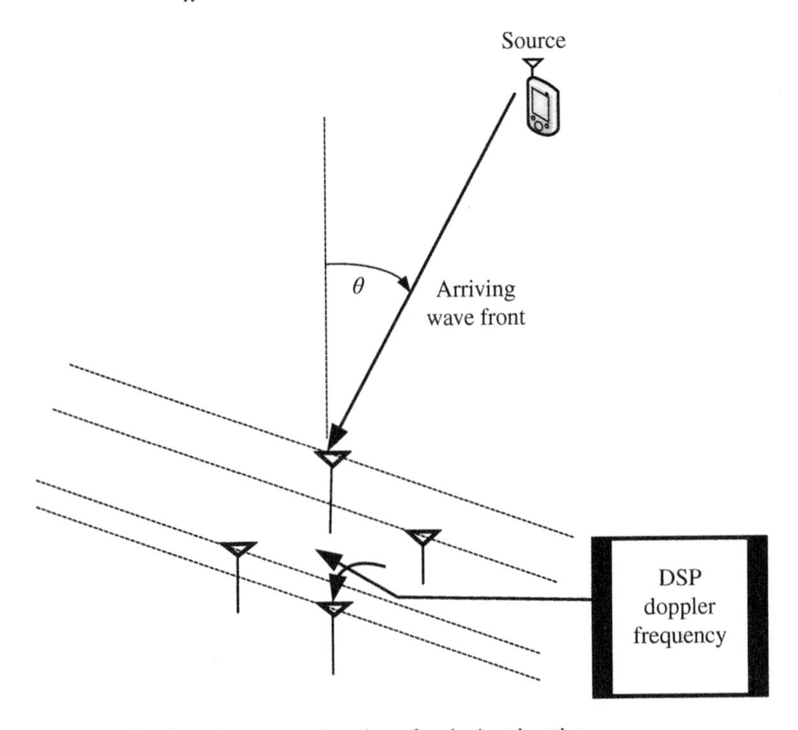

Figure 3.12 Doppler-based direction of arrival estimation.

where ω is the angular velocity of rotation in radians per second (i.e. $\omega = 2\pi v$ where v is the velocity of the rotation), r is the radius of the antenna rotation in meters, and λ is the wavelength ($\lambda = \frac{c}{f_c}$ where c is the speed of the light and f_c is the carrier frequency of the transmitted signal). The Doppler shift will be cycling between the positive and negative values depending on the angle of the antenna with respect to the angle of arrival of the signal. The magnitude of the Doppler shift is at a maximum as the antenna moves directly toward and away from the direction of the incoming wavefront. There is no apparent frequency shift when the antenna moves orthogonal to the wavefront.

Alternative to the rotation of the antenna, an approach that is called Pseudo-Doppler can be used with a multi-antenna circular array where each antenna is sampled in succession. In this case, rather than rotating an antenna, several antennas are sampled with a switch to emulate the Doppler effect. The Doppler-based direction finding methods work well even for short duration noncontinuous signals.

Smart antenna systems (which can be considered as multichannel directional finding) can also be used for estimating the direction of the source. Smart antenna systems are built upon an array of antenna elements and processing of the signals induced on the different antenna elements (Figure 3.13). The direction of arrival can be estimated by finding a spatial spectrum of the antenna array, where the peaks of this spectrum can provide the angles of arrivals from multiple sources. Algorithms that are used for spectral estimation such as multiple signal classification (MUSIC), estimation of signal parameters via rotational invariance techniques (ESPRIT) can also be used for finding the spatial spectrum.

3.8.2 Angular Spread

Angle spread is a measure of how multipath signals are arriving (or departing) with respect to the mean arrival (departure) angle. Therefore, angle spread refers to the spread of angles of arrival (or departure) of the multipaths at the receiving (transmitting) antenna array [22]. Angle spread is related to the spatial selectivity of the channel, which is measured by coherence distance. Like coherence time and frequency, coherence distance provides the measure of maximum spatial separation over which the signal amplitudes have a strong correlation, and it is inversely proportional to the angular spread, i.e. the larger the angle spread, the shorter the coherence distance. Local scattering in the vicinity of radio receivers (especially in indoor channels) results in larger angular spreads, as the received signals come from many different directions due to relatively richer local scattering environment. For a given receiver antenna spacing, this leads to weaker antenna correlations between the signals received by different antenna elements compared to that of antennas in open outdoor environments. Note that although the angular spread is described independent of the other channel selectivity values for the sake of simplicity, in reality, the angle of arrival can be related to the path delay. The multipath components that are arriving to the receiver with shorter delays are expected to have similar angle of arrivals (lower angle spread values).

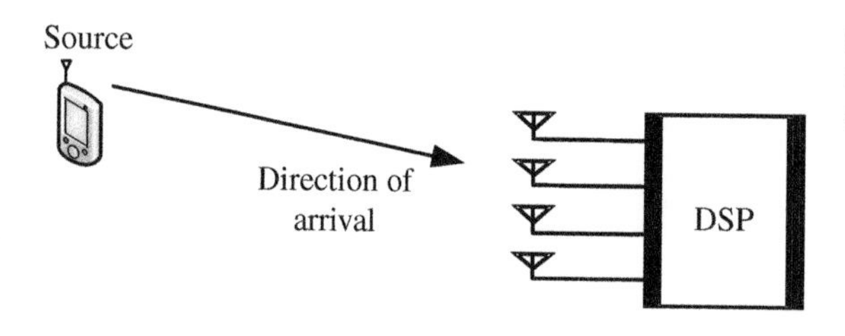

Figure 3.13 Use of smart antenna system with advanced DSP techniques to find the angle of arrival.

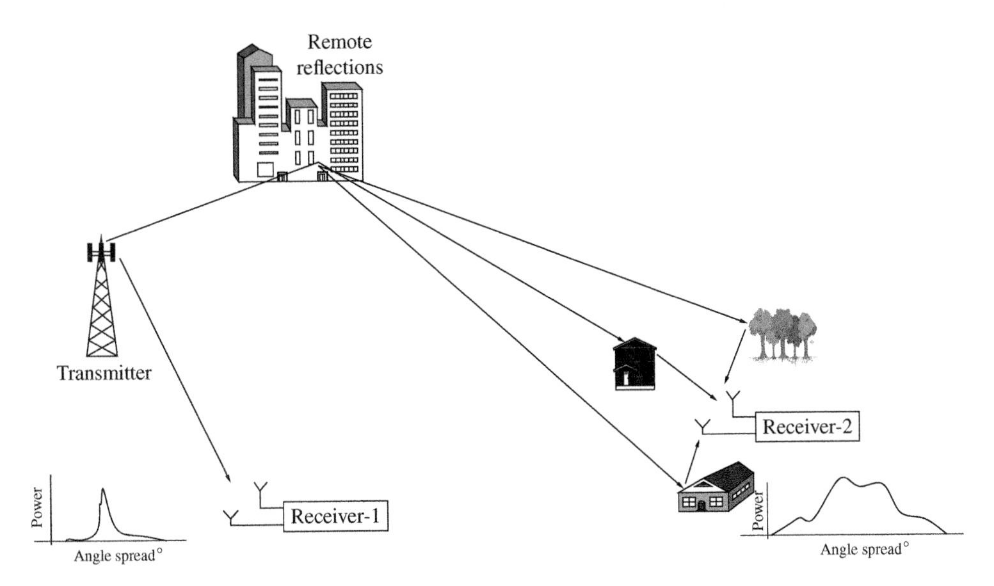

Figure 3.14 Illustration of spatial selectivity.

Compared to time and frequency selectivity, spatial selectivity has not been studied widely in the past. However, recently, there has been a significant amount of work in multi-antenna systems. With the widespread application of multi-antenna systems, it is expected that the need for understanding spatial selectivity and related parameter estimation techniques will gain momentum. Spatial selectivity will especially be useful when the requirement for placing antennas close to each other increases, as in the case of multiple antennas in mobile units (Figure 3.14).

Spatial correlation between multiple antenna elements is related to the spatial selectivity, antenna distance, mutual coupling between antenna elements, antenna patterns, etc. [23]. The spatial correlation has significant effects on multi-antenna systems. Full capacity and performance gains of multi-antenna systems can only be achieved with low antenna correlation values. However, when this is not possible, maximum capacity can be achieved by employing efficient adaptation techniques. Adaptive power allocation is one way to exploit the knowledge of the spatial correlation to improve the performance of multi-antenna systems. Similarly, adaptive modulation and coding, which employ different modulation and coding schemes across multi-antenna elements depending on the channel correlation, are possible. In MIMO systems, adaptive power allocation has been studied extensively by exploiting the knowledge of channel matrix estimate and by employing eigenvalue analysis.

3.9 MIMO Measurements

Multiple antenna systems are becoming an integral part of wireless transceivers. MIMO is the term that defines the use of multiple antennas at both the transmitter and receiver to improve communication performance, such as providing a significant increase in data throughput and expansion of the link range without additional bandwidth or transmit power.

Reliable MIMO implementation requires performing certain MIMO measurements on the system (Figure 3.15). Often, MIMO measurements are performed either by using multiple vector signal analysers (VSAs) or a VSA with multiple RF front-ends. Needless to say, this kind of a

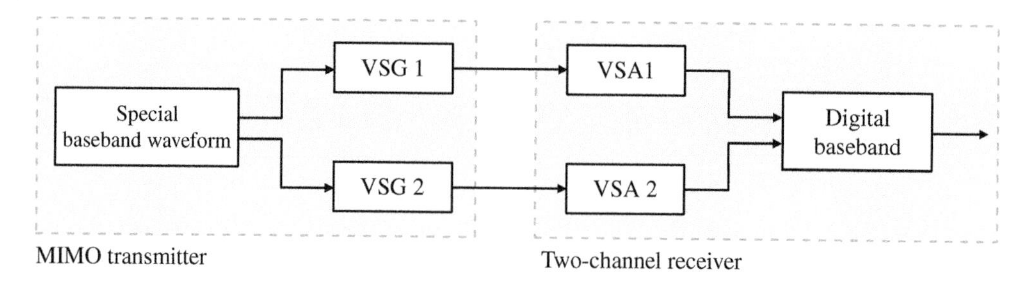

Figure 3.15 Two channel MIMO transmitter and receiver measurement set up. Two VSA and two VSA are used. Alternatively, a single VSA and vector signal generator (VSG) set with two channel transceivers can also be used.

measurement setup is extremely costly. Recently, measurement techniques using single VSA solutions are also proposed. The set of measurable parameters in MIMO systems comprise all parameters in single-input single-output (SISO) systems such as I/Q impairments, spectral flatness, frequency offset, and phase offset. However, there are additional measurements specifically critical for MIMO implementation. In the following, some of these measurements will be discussed briefly.

3.9.1 Antenna Correlation

The correlation between the antennas of a MIMO system is of vital importance for the system performance. A high correlation may substantially ruin the diversity and multiplexing gains targeted by using multiple antennas [24]. Antenna correlation can vary depending on the antenna separation and angular spread of the incoming wave [25]. Because of its significance on the achievable gains, antenna correlation has to be quantified while doing a system performance analysis. The correlation between the receiver (Rx) antennas can be measured with a simple setup. In the transmitter (Tx) side, only one branch is allowed to be active, whose signals are captured by all receiver antennas. By recording and correlating the signals received by each of the antennas, the receiver antenna correlations can be determined. If the complex correlation coefficient is higher than 0.7, a significant reduction in the targeted gains should be expected [26]. In such a case, the most reasonable solution would be to increase the distance between the antennas if it is possible.

3.9.2 RF Cross-Coupling

In general, the cross-coupling between the RF front-ends of separate branches of a MIMO system is not taken as seriously as the antenna correlation. However, signals at different front-ends may become correlated to each other because of the coupling between the front-ends. Hence, the negative effect of RF cross-coupling on the system performance can be significant, and it might be necessary to quantitatively measure it. The setup to measure the RF cross-coupling between two transmitter branches using a single VSA is shown in Figure 3.16. Two separate measurements are done. In each measurement, a known signal is transmitted from one of the branches while the inactive branch is directly connected to the VSA in the receiver part via a cable. Cross-coupling can be measured in this simple way also when the number of branches is more than two.

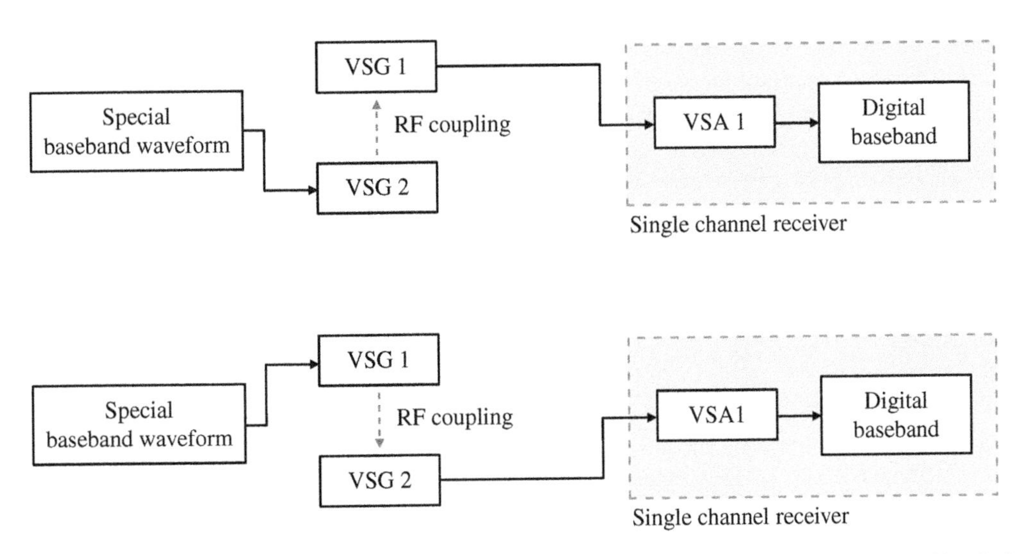

Figure 3.16 Measurement of MIMO cross-coupling between antenna branches using a single VSA solution.

3.9.3 EVM Versus Antenna Branches

EVM and its various versions (EVM versus time and EVM versus frequency) have been discussed in detail in the previous sections. In MIMO, EVM can also be measured across multiple antenna branches. Since in MIMO, each branch has its own radio, the EVM needs to be measured separately for troubleshooting each branch. Therefore, EVM versus the antenna branch could be a useful measurement to measure the quality of each antenna branch and compare it with other branches.

3.9.4 Channel Parameters

The measurement of the channel parameters constitutes an important initial step in the MIMO implementation [27]. A reliable separation of the signals received at each branch depends on correctly determining the channel fading coefficients. Other important channel parameters to be measured are the channel delay spread, channel coherence time, and the noise level in the channel. In order to get reliable statistics of channel parameters, the same measurements have to be repeated extensively, which constitutes the tough part of channel measurements. The method of performing channel estimation can vary from system to system. Once the channel parameters are determined, the real radio channel environment can be simulated in an RF test lab by using a channel emulator whose parameters are set according to the channel measurements. The use of a channel emulator can enable testing various channel conditions very conveniently, however this component considerably increases the hardware cost.

3.10 Conclusions

With the advances in wireless technologies, the dimensions of wireless signals are growing. Some dimensions that were not exploited before are now heavily utilized with the current standards [28–43]. A good example for this is the recent utilization of angular dimension through multiple antenna systems. It is expected that future standards and future digital signals will have more dimensions. Exploring, exploiting, understanding, and analyzing different dimensions of the signals are extremely important for wireless engineers and researchers. The basic concepts provided in this chapter are expected to shed some light to the researchers and engineers that are further exploring this important area. These concepts will especially be useful for blind signal analysis and reception. Also, there is a strong relation between what is discussed in this chapter and the communication performance parameters measurements that were discussed in Chapter 2 and radio environment measurement techniques that will be discussed in Chapter 13.

References

1 H. Arslan, L. Krasny, D. Koilpillai, and S. Channakeshu, "Doppler Spread Estimation for Wireless Mobile Radio Systems," vol. 3, Chicago, IL, 23–28 Sept. 2000, pp. 1075–1079.

2 H. Arslan and T. Yücek, "Delay spread estimation for wireless communication systems," in *Proc. The Eighth IEEE Symposium on Computers and Communications (ISCC'2003)*, Antalya, 3 July 2003, pp. 282–287.

3 M. Gandetto, M. Guainazzo, and C. S. Regazzoni, "Use of time-frequency analysis and neural networks for mode identification in a wireless software-defined radio approach," *EURASIP J. Appl. Signal Process.*, vol. 2004, no. 12, pp. 1778–1790, Jan. 2004.

4 W. Danzhi, L. Shujian, and S. Dingrong, "The analysis of frequency hopping signal acquisition based on Cohenreassignment joint time-frequency distribution," in *Proceedings of Asia-Pacific Conference on Environmental Electromagnetics*, Hangzhou, 4–7 Nov. 2003, pp. 21–24.

5 R. J. Mammone, R. J. Rothaker, and C. I. Podilchuk, "Estimation of carrier frequency, modulation type and bit rate of an unknown modulated signal," in *Proc. IEEE Int. Conf. Commun. (ICC)*, vol. 2, Seattle, WA, 1 Dec. 1987, p. 1006–1012.

6 D. S. Roberts and M. G. Amin, "Linear vs. bilinear time-frequency methods for interference mitigation in direct sequence spread spectrum communication systems," in *Proc. of ASILOMAR*, vol. 29, Pacific Grove, CA, Nov. 1995, p. 925.

7 M. D. L. Angrisani and M. D'Arco, "New digital signal-processing approach for transmitter measurements in third generation telecommunication systems," *IEEE Trans. Inst. Meas.*, vol. 53, no. 3, pp. 622–629, June 2004.

8 F. Hlawatsch and G. F. Boudreaux-Bartels, "Linear and quadratic time-frequency signal representations," *IEEE Signal Process. Mag.*, vol. 9, no. 2, pp. 21–67, Apr. 1992.

9 D. L. Jones and T. W. Parks, "A resolution comparison of several time-frequency representations," *IEEE Trans. Signal Process.*, vol. 40, no. 2, pp. 413–420, Feb. 1992.

10 X. Deng and P. Fan, "Spreading sequence sets with zero correlation zone,"*IEE Electron. Lett.*, vol. 36, no. 11, pp. 993–994, May 2000.

11 M. K. S. Kuno, T. Yamazato, and A. Ogawa, "A study on quasi-synchronous CDMA based on selected PN signature sequences," in *Proc. IEEE ISSSTA*, Oulu, pp.479–483, 4–6 July 1994.

12 D. Cabric, S. Mishra, and R. Brodersen, "Implementation Issues in Spectrum Sensing for Cognitive Radios," vol. 1, Pacific Grove, CA, 7–10 Nov. 2004, pp. 772–776.

13 M. Oner and F. Jondral, "Cyclostationarity-based air interface recognition for software radio systems," Atlanta, Georgia, Sep. 2004, pp. 263–266.

14 ——, "Cyclostationarity-Based Methods for the Extraction of the Channel Allocation Information in a Spectrum Pooling System,"Atlanta, GA, 22 Sept. 2004, pp. 279–282.

15 D. Cabric and R. W. Brodersen, "Physical Layer Design Issues Unique to Cognitive Radio Systems," Berlin, 11–14 Sept. 2005.

16 A. Fehske, J. Gaeddert, and J. Reed, "A New Approach to Signal Classification Using Spectral Correlation and Neural Networks," Baltimore, MD, 8–11 Nov. 2005, pp. 144–150.

17 S. Shankar, C. Cordeiro, and K. Challapali, "Spectrum Agile Radios: Utilization and Sensing Architectures,"Baltimore, MD, 8–11 Nov. 2005, pp. 160–169.

18 M. Ghozzi, F. Marx, M. Dohler, and J. Palicot, "Cyclostationarity-Based Test for Detection of Vacant Frequency Bands," Mykonos, 8–10 June 2006.

19 W. Gardner, "Exploitation of spectral redundancy in cyclostationary signals," *IEEE Signal Process. Mag.*, vol. 8, no. 2, pp. 14–36, 1991.

20 H. Tang, "Some physical layer issues of wideband cognitive radio systems," Baltimore, MA, 8–11 Nov. 2005, pp. 151–159.

21 A. Sahai, R. Tandra, S. M. Mishra, and N. Hoven, "Fundamental design trade-offs in cognitive radio systems," *Proceedings of the first international workshop on Technology and policy for accessing spectrum* Aug. 2006, pp. 2

22 A. Paulraj and B. Ng, "Space-time modems for wireless personal communications," *IEEE Pers. Commun.*, vol. 5, no. 1, pp. 36–48, Feb. 1998.

23 M. K. Özdemir, H. Arslan, and E. Arvas, "Mutual coupling effect in multi-antenna wireless communication systems," in *Proc. IEEE GlobeCom Conf.*, San Francisco, CA, 1–5 Dec. 2003.

24 A. Tulino, A. Lozano, and S. Verdu, "Impact of antenna correlation on the capacity of multi-antenna channels," *IEEE Trans. Inf. Theory*, vol. 51, no. 7, pp. 2491–2509, 2005.

25 D. Chizhik, F. Rashid-Farrokhi, J. Ling, and A. Lozano, "Effect of antenna separation on the capacity of BLAST in correlated channels,," *IEEE Commun. Lett.*, vol. 4, no. 11, pp. 337–339, 2000.

26 H. Sampath, S. Talwar, J. Tellado, V. Erceg, and A. Paulraj, "A fourth-generation MIMO-OFDM broadband wireless system: design, performance, and field trial results," *IEEE Commun. Mag.*, vol. 40, no. 9, pp. 143–149, 2002.

27 O. Fernandez, R. Jaramillo, and R. Torres, "Empirical analysis of a 2 x 2 MIMO channel in outdoor–indoor scenarios for BFWA applications," *IEEE Antennas Propag. Mag.*, vol. 48, no. 6, pp. 57–69, 2006.

28 A. F. Demir, B. Pekoz, S. Kose, and H. Arslan, "Innovative telecommunications training through flexible radio platforms," *IEEE Commun. Mag.*, vol. 57, no. 11, pp. 27–33, 2019.

29 M. Karabacak, A. H. Mohammed, M. K. Özdemir, and H. Arslan, "RF circuit implementation of a real-time frequency spread emulator," *IEEE Trans. Instrum. Meas.*, vol. 67, no. 1, pp. 241–243, 2017.

30 A. Gorcin and H. Arslan, "An OFDM signal identification method for wireless communications systems," *IEEE Trans. Vehicul. Technol.*, vol. 64, no. 12, pp. 5688–5700, 2015.

31 ——, "Signal identification for adaptive spectrum hyperspace access in wireless communications systems," *IEEE Commun. Mag.*, vol. 52, no. 10, pp. 134–145, 2014.

32 S. Guzelgoz and H. Arslan, "Modeling, simulation, testing, and measurements of wireless communication systems: A laboratory-based approach," in *2009 IEEE 10th Annual Wireless and Microwave Technology Conference*, Clearwater, FL, 20–21 Apr. 2009, pp. 1–5.

33 S. Ahmed and H. Arslan, "Cognitive intelligence in UAC channel parameter identification, measurement, estimation, and environment mapping," in *OCEANS 2009-EUROPE*, Bremen, 11–14 May 2009, pp. 1–14.

34 M. E. Sahin and H. Arslan, "MIMO-OFDMA measurements; reception, testing, and evaluation of wimax MIMO signals with a single-channel receiver," *IEEE Trans. Instrum. Meas.*, vol. 58, no. 3, pp. 713–721, 2009.

35 A. M. Hisham and H. Arslan, "Multidimensional signal analysis and measurements for cognitive radio systems," in *2008 IEEE Radio and Wireless Symposium*, Orlando, FL, 22–24 Jan. 2008, pp. 639–642.

36 M. Sahin, H. Arslan, and D. Singh, "Reception and measurement of MIMO-OFDM signals with a single receiver," in *2007 IEEE 66th Vehicular Technology Conference*, Baltimore, MD, 30 Sept.–3 Oct. 2007, pp. 666–670.

37 H. Arslan, "IQ gain imbalance measurement for OFDM based wireless communication systems," in *MILCOM 2006 – 2006 IEEE Military Communications Conference*, Washington, DC, 23–25 Oct. 2006, pp. 1–5.

38 H. Arslan and D. Singh, "The role of channel frequency response estimation in the measurement of RF impairments in OFDM systems," in *2006 67th ARFTG Conference*, San Francisco, CA, 16 July 2006, pp. 241–245.

39 J. Haque, M. C. Erturk, and H. Arslan, "Aeronautical ICI analysis and Doppler estimation," *IEEE Commun. Lett.*, vol. 15, no. 9, pp. 906–909, 2011.

40 H. Celebi, K. A. Qaraqe, and H. Arslan, "Performance comparison of time delay estimation for whole and dispersed spectrum utilization in cognitive radio systems," in *2009 4th International Conference on Cognitive Radio Oriented Wireless Networks and Communications*, Hannover, 22–24 June 2009, pp. 1–6.

41 F. Kocak, H. Celebi, S. Gezici, K. A. Qaraqe, H. Arslan, and H. V. Poor, "Time delay estimation in cognitive radio systems," in *2009 3rd IEEE International Workshop on Computational Advances in Multi-Sensor Adaptive Processing (CAMSAP)*, Aruba, 13–16 Dec. 2009, pp. 400–403.

42 S. Gezici, H. Celebi, H. V. Poor, and H. Arslan, "Fundamental limits on time delay estimation in dispersed spectrum cognitive radio systems," *IEEE Trans. Wirel. Commun.*, vol. 8, no. 1, pp. 78–83, 2009.

43 T. Yucek and H. Arslan, "Time dispersion and delay spread estimation for adaptive OFDM systems," *IEEE Trans. Vehicul. Technol.*, vol. 57, no. 3, pp. 1715–1722, 2008.

4

Simulating a Communication System

Muhammad Sohaib J. Solaija[1] *and Hüseyin Arslan*[1,2]

[1] *Department of Electrical and Electronics Engineering, Istanbul Medipol University, Istanbul, Turkey*
[2] *Department of Electrical Engineering, University of South Florida, Tampa, FL, USA*

Communication system design has undergone significant changes over the years. New technologies are required to be incorporated with exceeding regularity owing to the increasing user requirements. However, before any significant change is introduced into a practical system, it needs to be validated by modeling and simulation. This chapter looks at the basics of simulation of communication systems, covering topics such as strategy, modeling, general methodology, link and network-level categorization, error sources, performance evaluation, and practical issues.

4.1 Simulation: What, Why?

Communication systems have evolved drastically over recent years. This evolution is not merely limited to an increased number of users and higher data rates, rather it involves the introduction of various technologies, methods, and algorithms to support the higher demands associated with a wide variety of applications. Incorporation of new subsystems, technologies, and algorithms requires extensive validation and performance evaluation before they can be made part of the system.

There are three conventional approaches for a system's performance evaluation, namely, mathematical analysis, simulation, and hardware prototyping [1]. Analytical methods, while providing a direct relationship between the different parameters and system performance, become intractable for a complex system. Here, *intractability* refers to problems that can be solved but the solution takes too long for them to be useful in a practical setting [2]. Hardware prototyping, on the other hand, requires more time for development and provides limited flexibility to observe the effect of tweaking different system parameters. Simulation provides a trade-off in terms of accuracy and flexibility. While it does not provide exact closed-form expressions for different performance metrics, its invaluable advantage lies in offering a more economical and faster way to carry out the parametric studies in system design [3].

Figure 4.1 illustrates a communication system design cycle to highlight the importance of simulation in the design process. This process starts with the user requirements and corresponding system specifications. This is followed by setting up a simulation environment and verification of the parameters corresponding to the link/system. Once this verification is achieved, hardware development can begin. The hardware performance is then used to validate the simulation model.

The validated model can be used for various purposes, such as studying design trade-offs, evaluating system performance, or establishing test procedures. It gives the system designer the flexibility to see the effect of various design parameters on the system. Insights can be drawn regarding performance from a well-rounded simulation. Moreover, simulation also helps in identifying the

Wireless Communication Signals: A Laboratory-based Approach, First Edition. Hüseyin Arslan.

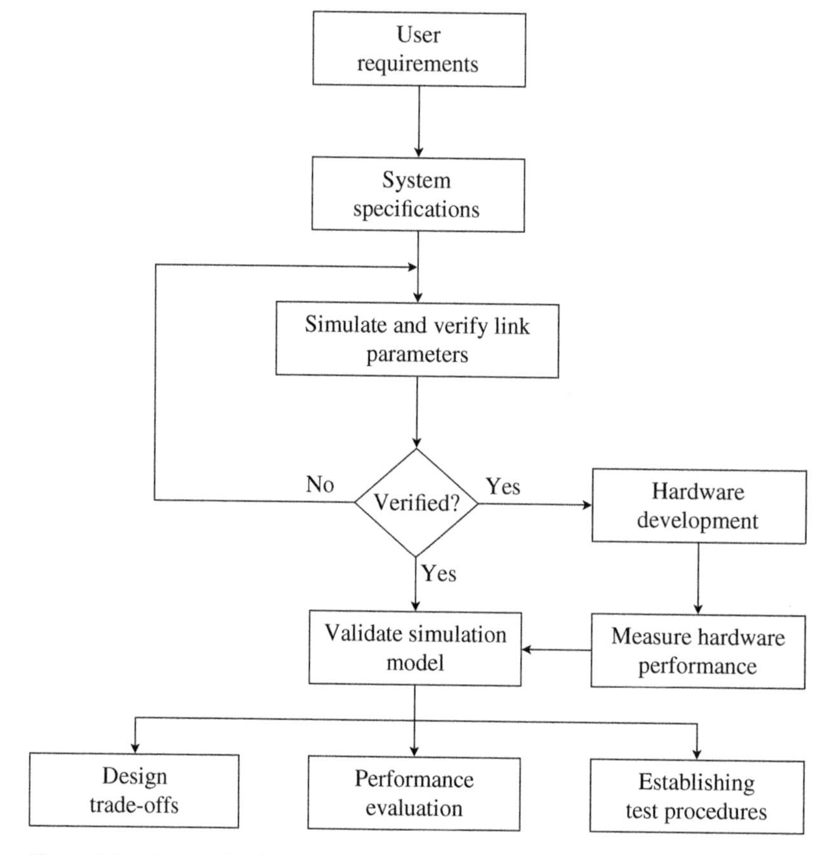

Figure 4.1 Communication system design process.

procedures that can be used to evaluate the real system in a more suitable manner. Other applications include studying the effect of aging on the system and end-of-life predictions [3].

Having established the need for simulation in any design process in general, we look at the strategy utilized in developing a simulation setup for any communication system or part thereof.

4.2 Approaching a Simulation

This section sheds light on some of the critical questions regarding the choices concerning a simulation setup. First, we discuss the strategies that can be opted for when developing such a system. Later, we provide the general steps to follow in solving any problem using a simulation approach.

4.2.1 Strategy

Before designing any simulation setup, it is necessary to decide on the general tactical approach. The most important question to answer is whether the setup is to be designed for a specific problem or needs to be generalized for various aspects of a system or a subsystem. The trade-off, in this case, is obvious; the former approach can provide results in a swifter manner but would require a new setup from scratch if a different problem were to be targeted. A generic setup, on the other hand, would require more time and effort to develop initially but can be adapted easily for analyzing various components of the system (or subsystem). This can be illustrated by considering a typical transmitter/receiver chain, as shown in Figure 4.2. It can be seen that the blocks on the receiver are counterparts of the transmitter blocks apart from the synchronization and equalization blocks. Assume a study related to the mapping/modulation operation which takes processed bits as its input. For this specific study, the implementation of source and channel coding blocks is

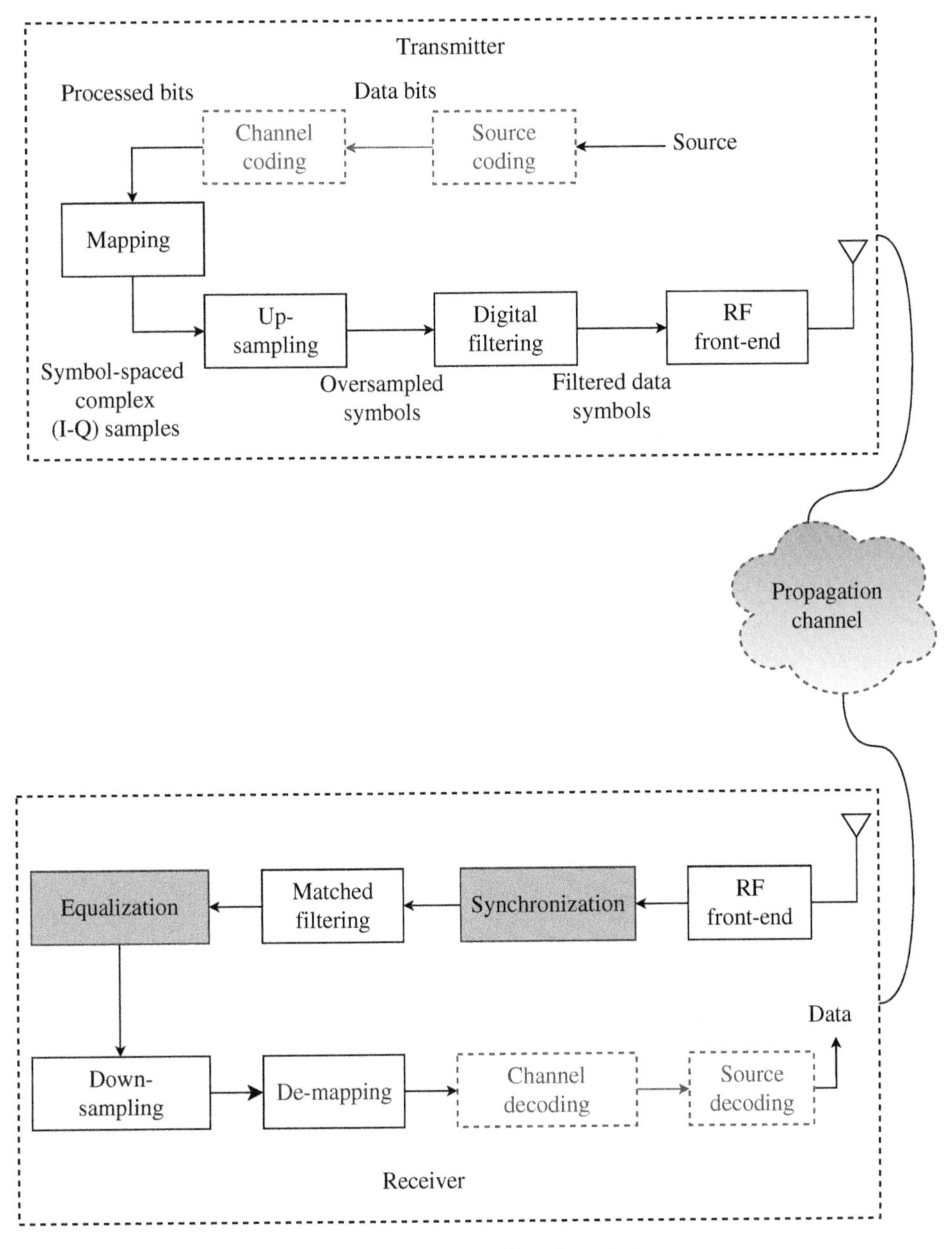

Figure 4.2 Block diagram of typical transmitter and receiver chains.

unnecessary and might be a computational burden. However, if different coding techniques were to be studied with the said mapping, the general approach comprising of all blocks would be more sensible. On the contrary, if a mapping scheme were to be studied without considering its impact on the radio frequency (RF) front-end of the communication system, it would be extremely misleading since the impairments depend significantly on the mapping used. This goes to show how critical it is to identify the various dependencies in a communication system and model them accordingly. It also needs to be ensured that all such blocks (which have these dependencies) are made part of the simulation setup.

In addition to the question about generic vs. specific approach, another issue that needs to be addressed is the error-time-complexity trade-off. The error can arise from selecting a simpler model, which provides faster results but introduces some error into the system. Balancing this trade-off is important since overly complex models cannot be used for larger systems, however, at the same time inaccuracies beyond a certain limit render the results unreliable. This trade-off is illustrated in Figure 4.3. It can be seen that practical operation tries to reduce both error and running time while keeping a moderate complexity.

The ultimate goal of any simulation is to provide performance results in an interpretable and palatable manner. This requires the selection of the appropriate performance metric and correct visualization tool. A simple example can be considered to see the effect of the latter's importance, assume signal quality measurement is of interest for a particular simulation and bit-error-rate (BER) is recorded for that. It is possible to look at the mean value of this metric for the given scenario and draw conclusions; however, the averaging process destroys the granularity that could be provided by measurement tools such as a probability density function (PDF) or cumulative distribution function (CDF). These functions can help identify different users (or categories thereof) looking at the probability distributions. That being said, in the case of research, performance results need to be validated and compared to state-of-the-art solutions in the literature, necessitating the use of corresponding metrics and tools.

4.2.2 General Methodology

While the methodology depends on the specific design process used, communication systems are generally designed utilizing a "top-down" approach, starting from a high-level, abstract depiction

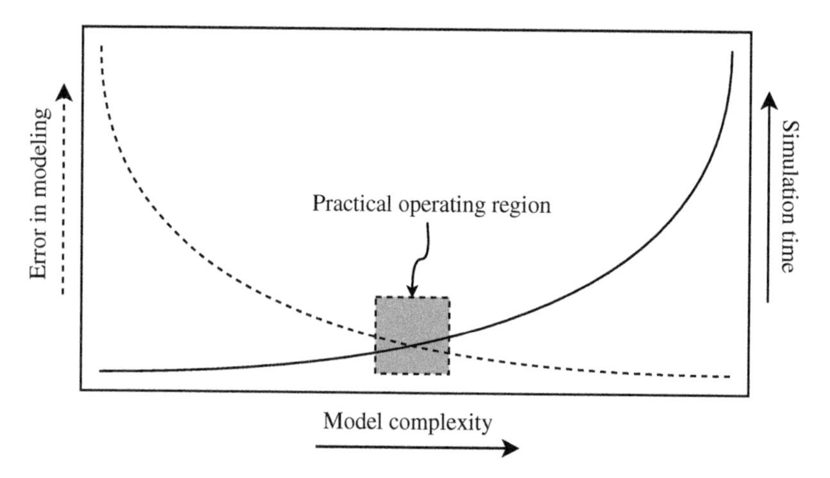

Figure 4.3 Illustration of error, time, and complexity trade-off for a simulation model.

of the system. This precedes the incorporation of more detailed models for different blocks of the system. The general simulation methodology is summarized below:

- Given a problem, the first step is to come up with an appropriate system model. This what certain authors/academicians refer to as "art" of simulation. While there are no concrete rules about what should constitute a system, dependencies in any subsystem under consideration can be used as a guideline for identifying other parts of the system model.
- Having identified the overall problem and the concerned system model, the next step is to decompose the problem into smaller subproblems. These subproblems are then mapped to the different blocks of the system model, where each block performs a specific signal processing operation. This modular approach helps in reducing the complexity of the system.
- This is followed by the selection of proper models for simulation of the various blocks. Since the real-world systems are too complex to be analytically tractable, it also renders their perfect representation in the virtual world of simulation impossible. This leads to designers and researchers making certain simplifying assumptions; however, these assumptions need to be realistic in order to provide practical solutions for the aforementioned subproblems.
- The results obtained for the identified subproblems can then be combined in a meaningful manner to solve the overall problem.

Here, it is important to mention that the models used in simulation generally follow analytical or mathematical derivations based on actual measurements. Moving from the hardware to simulation results in a higher abstraction with possibly more errors. This renders the modeling stage vital for any reasonable simulation study. It is, therefore, appropriate to get an overview of the basic modeling concepts at this stage.

4.3 Basic Modeling Concepts

Accurate modeling is necessary for both simulation and analytical approaches. Here it should be noted that accuracy refers not only to the models of the hardware under test but also the random processes being used in the system. Given below is a summary of the different levels of modeling that make up a simulation.

4.3.1 System Modeling

In the context of this book, a system generally refers to one or more communication link(s) under study. A typical link is represented by a block diagram consisting of multiple blocks, like the one shown in Figure 4.2. As mentioned earlier, certain assumptions can be leveraged to reduce the computational burden. This can include only using certain blocks of the system in simulations. However, the selection or removal of these blocks or subsystems needs to be backed by logical reasoning. A common example of this approach is the omission of the synchronization block which is assumed to be present in many analyses. Another such example is the analog-to-digital converter (ADC)/digital-to-analog converter (DAC) block which is rarely made part of the simulation. As a general rule, it is preferable to implement the sparsest block diagram which covers the subsystems of interest for any given problem.

4.3.2 Subsystem Modeling

A subsystem refers to a block in the system which carries out some specific (signal processing) task. While most of these subsystems or blocks have ideal models present, it is more realistic to incorporate meaningful deviations from the ideal behavior of the system. A good subsystem-level model should, therefore, provide tunable parameters. It is also possible that the available models of the subsystems are also obtained from real-world measurements (this is particularly relatable for channel models). In such a case, integrating uncertainty in the model is clearly understandable.

4.3.3 Stochastic Modeling

Stochastic processes form an imperative part of any communication system. These processes might be used to represent desired signals such as information and undesired ones like interference and/or noise. The realistic simulation of a system is therefore dependent on the proper use of stochastic models for these different aspects. The goal of these models is to correctly imitate the randomness of the system and signals.

Apart from noise, interference, and signals, the channel is also modeled as a stochastic process. This is particularly the case for multipath channels that are conventionally modeled with time-varying impulse response. The generation of random numbers is also discussed in Section 4.7.2, while further details related to channel models will be covered in Chapter 10.

Having gone over the general methodology and modeling concepts for communication system simulation, it is perhaps time to dive a little deeper into the world of simulations. We start off by looking at different possibilities of simulation at the link-level.

4.4 What is a Link/Link-level Simulation?

In the context of communication systems, a link consists of a transmitting part, a medium, and a receiving subsystem [1]. Again, we refer to Figure 4.2 where each block can be considered to be part of the link-level simulation. As mentioned in the last section, it is desirable to implement/simulate the sparsest possible system in any given situation, where the selected boxes are able to cover the dependencies. Alternatively, if a study compares two different designs which are equally affected by another component, the latter might not require implementation. Given below is a brief description of the different blocks that constitute a typical communication link from the simulation perspective.[1] Moreover, code snippets for different blocks of the communication system are provided considering our selected case study, i.e. communication in the presence of an additive white Gaussian noise (AWGN), which is discussed in the next section.

4.4.1 Source and Source Coding

Information sources in real life tend to have an analog nature. Simulations, on the other hand, are part of a digital world. A source, therefore, needs to be converted into an equivalent digital representation. However, this raises the question that whether it is necessary to simulate this part of the system at all. This includes simulation of the ADC and source coding blocks, the output of which is a compressed digital signal. The quick answer to this question is "no." Unless either of these blocks is the focus of the study, it is possible to replace them with a binary sequence [1].

1 Here we are only going to describe the blocks on the transmitter side since everything on the receiver side is pretty much an "undo" operation of these blocks. Synchronization and equalization blocks on the receiver side will, however, be described since they have no counterparts on the transmitter.

```
1  % An equiprobable binary sequence generation
2  N = 1e6; % length of the binary sequence
3  bin_seq = (randn(1,N)>0); % binary sequence extracted from a standard
4  %normal distribution
```

The code given above employs a standard normal distribution to generate a binary sequence using the condition of samples being greater than zero.

4.4.2 Channel Coding

Channel coding mechanisms are used to provide redundancy for the communication, reducing the average BER. However, this comes at the cost of increased bandwidth. There are two primary approaches for evaluating coding mechanisms in terms of simulation. The first focuses explicitly on the encoder/decoder pair (codec) and is suitable for studies which are directed toward analyzing the impact of different design parameters on the codec. The alternative is to separate the codec from the (discrete) channel. Here instead of implementing the oversampled channel and intermediate blocks between encoder and decoder, the discrete channel at symbol rate can be used, leading to increased computational saving.

Conventional schemes are categorized into block and convolutional coding approaches. In either case, the use of redundant bits leads to additional bandwidth being utilized. Modeling/simulations of codecs can be computationally prohibitive particularly for low BER scenarios (consider 10^{-6}), where the simulation length becomes extremely large (you need a sufficient number of errors where only one out 10^{-6} bits is erroneous on average). It is therefore more feasible to take advantage of some approximations or bounds in performance evaluation applications [1].

4.4.3 Symbol Mapping/Modulation

Modulation is a term that can refer to multiple operations in the context of a communication system. Classically, modulation is defined as transmitting a baseband signal over a passband centered at a carrier frequency, generally represented by f_c. While analog (angle, phase, and frequency) modulation schemes have a rich history, we will mostly focus on the digital schemes, also referred to as symbol mapping, within the scope of this book.

As the name suggests, symbol mapping involves mapping bits to different symbols or points in the in-phase and quadrature-phase (I/Q) domains. Depending on the order M, an M-ary scheme would map $\log_2 M$ bits to each of the M points in the I/Q domain. The points collectively form the constellation for any scheme.

```
1  % QPSK symbol mapping
2  M = 4; % Modulation order
3  % Setting up the modulator with pi/4 phase offset and Gray mapping
4  QPSKModulator = comm.QPSKModulator(M, 'BitInput', true,...
5  'PhaseOffset', pi/4, 'SymbolMapping', 'Gray') ;
6  constellation(QPSKModulator) % Show the constellation points
7  s_QPSK = QPSKModulator(bin_seq); % Mapping the binary sequence
```

The code snippet provided shows the mapping of the previously generated binary sequence to a quadrature phase shift keying (QPSK) constellation. There are different parameters such as phase offset and mapping scheme that can be controlled by the user. For this example, we use a $\pi/4$ phase offset (phase offset of 0 would lead to constellation points on the major axes) and Gray mapping. Gray mapping is used to reduce BER by ensuring no more than one bit is different between

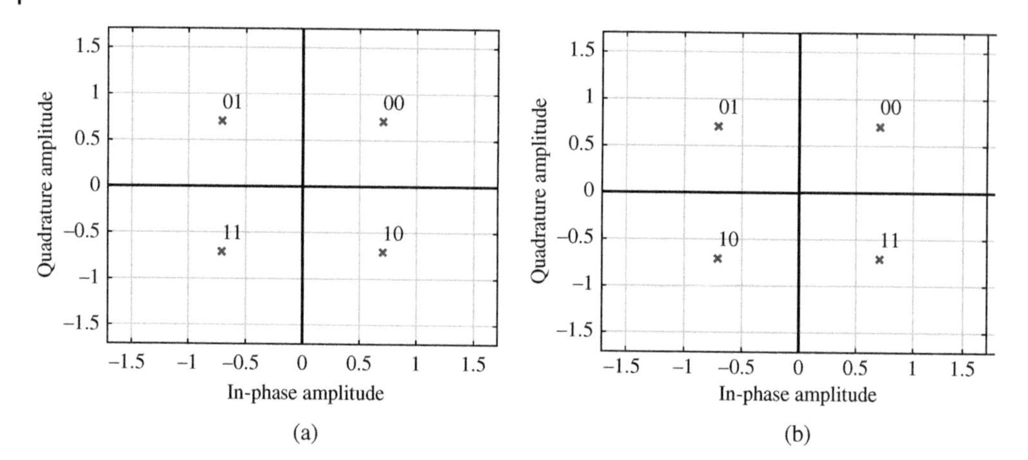

Figure 4.4 QPSK symbol mapping schemes. (a) Gray mapping. (b) Binary mapping.

adjacent constellation points (see Figure 4.4a). In order to use the conventional symbol mapping (see Figure 4.4b), the term "Gray" in the snippet provided above can be replaced by "Binary."

For a more detailed understanding of the modulation and mapping concepts, readers are referred to Chapter 6.

4.4.4 Upsampling

Upsampling allows conversion of data symbols to pulses, which can then be manipulated or shaped for different purposes. The upsampling also allows observing the transitions between different symbols using techniques such as eye and polar diagrams. It should be ensured that the upsampling factor corresponds to the rate of the subsequent pulse-shaping digital filter. Upsampling is carried out by inserting $N-1$ zeros for each symbol to achieve an upsampling by a factor of N.

```
1  % Upsampling the mapped sequence
2  sps = 8; % Samples per symbol
3  u_frame = upsample(s_QPSK, sps); % Upsampled signal
```

The snippet provided above oversamples the mapped symbols by a factor of 8, referred to as samples per symbol or "sps."

4.4.5 Digital Filtering

Digital filtering or pulse shaping translates the mapped symbols to physical signals for transmission over the communication link. The different parameters of these filters, such as roll-off for raised cosine (RC) or bandwidth-time (BT) product of Gaussian filters, can be manipulated to change the characteristics of the signal. The increase in bandwidth caused by increasing these parameters results in a smoother filter. While the bandwidth occupancy is increased, it leads to better localization of the signal in the time domain. This property enables the utilization of these pulse-shaping filters for reducing inter-symbol interference (ISI) by suppressing the sidelobes of time-domain signals.

An example of an ideal rectangular filter is given in the snippet below. The real part of (some) mapped symbols and their oversampled and filtered counterparts are also plotted as shown in Figure 4.5.

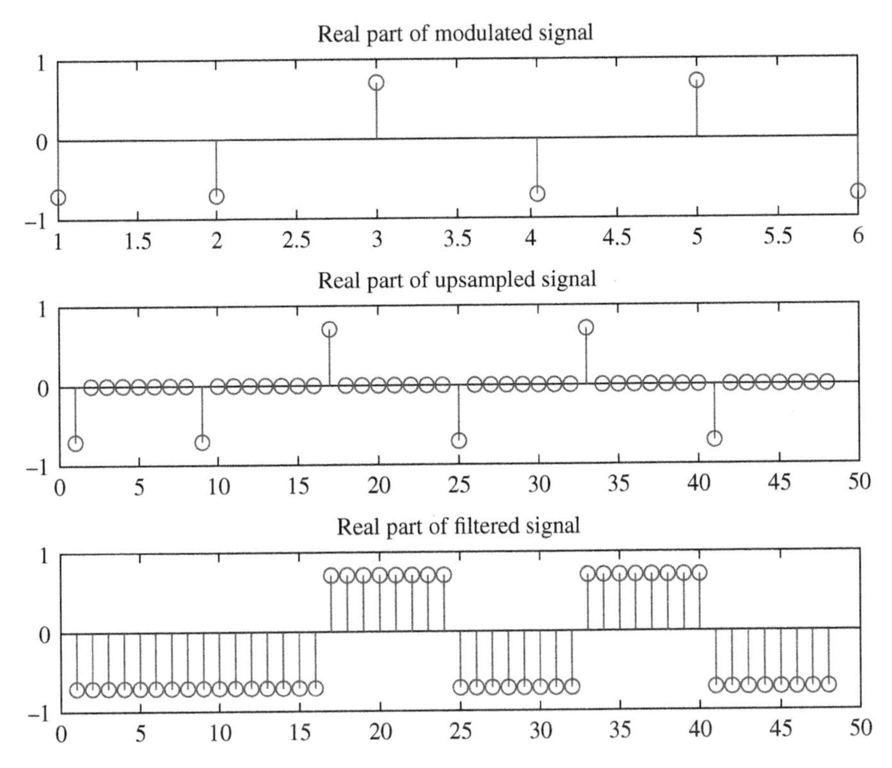

Figure 4.5 Comparison of the modulated/mapped symbol with its upsampled and (rectangular) filtered versions.

```
1  % Passing the oversampled signal through a rectangular filter
2  fltr = ones(1,sps); % Rectangular filter
3  sgnl_tx = conv(fltr,u_frame); % Signal after convolution with the filter
4
5  % Comparing the modulated, upsampled and filtered signals
6  syms_to_plot = 6;
7  subplot (3,1,1)
8  stem(real(s_QPSK(1:syms_to_plot)))
9  title('Real Part of Modulated Signal')
10
11 subplot (3,1,2)
12 stem(real(u_frame(1:syms_to_plot*sps)))
13 title('Real Part of Upsampled Signal')
14
15 subplot(3,1,3)
16 stem(real(sgnl_tx(1:syms_to_plot*sps)))
17 title('Real Part of Filtered Signal')
```

For a more detailed discussion regarding filtering and pros and cons of different filters, readers are referred to Chapter 6.

4.4.6 RF Front-end

The RF front-end of a communication system consists of ADCs/DACs, voltage-controlled oscillators (VCOs), power amplifiers (PAs), mixers, and filters. Imperfections in these pieces of equipment introduce issues such as carrier frequency offset, I/Q imbalance, and other nonlinear behavior. Despite its presence in all real communication systems, this block is often skipped from simulations unless the effect of different parameters such as modulation, filtering, or waveform is intended to be studied on the impairments introduced by its constituent components. This is partly due to the fact that the models for these blocks are considered a given from hardware design and there is no room for modification from a digital communication system designer's perspective.

For our particular scenario/case study, i.e. AWGN case, we will not consider the presence of any RF impairment in the system. However, readers are referred to Chapter 5 for more information related to this topic.

4.4.7 Channel

Channel in the context of wireless communication refers to the medium over which the exchange of information between different nodes takes place. The simplest channel used in these systems is the AWGN model, where the only impairment to the communication is the added noise. Other models consider the selectivity/dispersion in time and frequency domains, and their combination.

```
1  % Passing the signal through an AWGN channel
2  SNR = 10; % desired SNR in dB
3  noise = (randn(1,N) + 1j*randn(1,N))/sqrt(2);
4  noise_var = 10^(-SNR/10) ; % desired variance calculated from SNR
5  sgnl_rx = s_QPSK + noise*sqrt(noise_var); % noise addition
6
7  % plot the transmitted constellation
8  subplot(1,2,1)
9  plot(real(s_QPSK),imag(s_QPSK),'b*')
10
11 % plot the received constellation
12 subplot(1,2,2)
13 plot(real(sgnl_rx),imag(sgnl_rx),'b*')
```

The signal is passed through AWGN channel. To simplify the simulation, the signal power is kept unitary while the noise is added according to the desired signal-to-noise ratio (SNR) value.[2] Since the rectangular filter has no effect on the signal, the symbol-spaced signal, i.e. "s_QPSK" is used. The constellation diagrams for (part of the) transmitted and received signals are shown in Figure 4.6a and b. The clean constellation points present in the former are converted to a cloud in the latter due to the presence of noise.

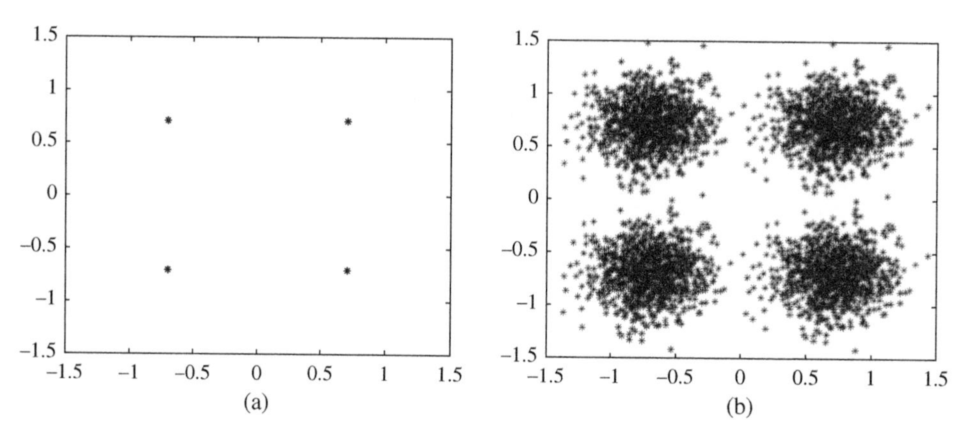

Figure 4.6 Transmitted and received constellations for AWGN channel at $SNR = 10$ dB. (a) Transmitted constellation. (b) Received constellation.

2 The power of the noise is given by its variance, hence the variance is calculated for the desired SNR.

Some more discussion regarding communication in the presence of AWGN can be found in Section 4.5, while Chapter 10 discusses in detail the different aspects of wireless channel and the corresponding models.

4.4.8 Synchronization and Equalization

As mentioned earlier, apart from synchronization and equalization, the other blocks on the receiver side perform the inverse operation of their corresponding blocks on the transmitter. Hence, these blocks will not be discussed here. Furthermore, in the context of AWGN case neither synchronization nor equalization is necessary. Therefore, only a conceptual introduction of both concepts is provided below.

The purpose of the synchronization block is to remove any mismatch in time and/or frequency domains between the receiver and transmitter sides. This mismatch is caused by impairments such as hardware imperfection (particularly of the VCO) and can lead to severe performance degradation. Furthermore, the oversampled data obtained at the receiver needs to be downsampled to the original symbol rate for further processing (keep in mind that the corresponding upsampling on the transmitter was done to allow pulse shaping). Synchronization is required to determine the particular sampling phase that ensures best SNR of the received signal.

The signal, when passing through the medium or channel, is affected in various ways. It may go through different paths experiencing varying degrees of reflection, refraction, diffraction, absorption, etc. The process of compensating for these channel effects is referred to as equalization. Once equalization is done, the samples can be used for detection of the received symbols/bits.

Both synchronization and equalization are discussed in Chapters 7 and 8 in the context of multi-carrier and single-carrier systems, respectively. Furthermore, Chapter 11 is dedicated to a more detailed study of synchronization in single-carrier systems.

4.4.9 Performance Evaluation and Signal Analysis

As mentioned earlier (in Section 4.2.1), selection of the appropriate performance metric and correspondingly proper visualization tool is an essential part of the simulation methodology. These metrics provide a way to measure the quality of the signal. Some examples include received signal strength indicator (RSSI), signal-to-interference-plus-noise ratio (SINR), BER, and packet-error-rate (PER) that can be measured at different stages of the receiver chain. RSSI, for instance, only provides information regarding the received power irrespective of its source, i.e. it is not capable of differentiating between a user's own signal and interference. SINR, on the other hand, allows this differentiation to provide an estimate of user's own received signal as compared to the interference and noise. This, however, comes at the cost of some additional processing. Therefore, depending on the application and motivation of simulation, the performance metric needs to be decided accordingly. Readers are referred to Chapter 2 for more detailed discussion regarding different metrics and visualization tools.

Apart from the performance evaluation afforded by simulation methodologies, they also enable designers to perform the analysis of the communication signals in different domains, including but not limited to, time, space and frequency. This facilitates the study of various trade-offs accompanying the algorithms and tools being analyzed. Chapter 3 of this book is dedicated to multidimensional signal analysis and readers are referred to it for more discussion on the topic.

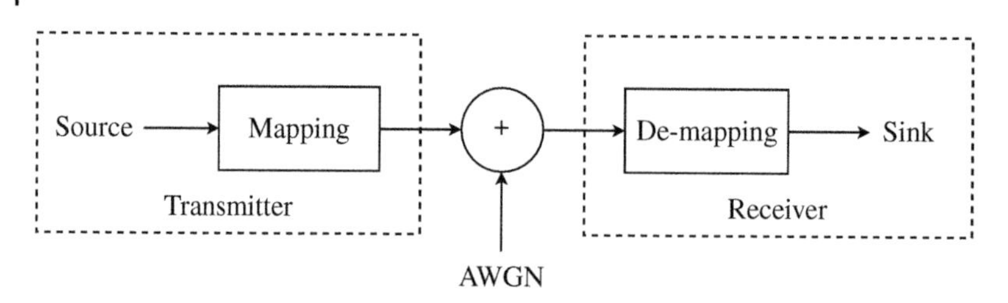

Figure 4.7 Block diagram for simulation model in AWGN.

4.5 Communication in AWGN – A Simple Case Study

From a wireless communication perspective, AWGN channel represents the most basic simulation setup. Figure 4.7 illustrates it in the form of a block diagram. This can be represented mathematically as:

$$r_k = b_k + z_k, \tag{4.1}$$

where r and b refer to the received and baseband signals, respectively, z is the additive noise and the subscript k represents the symbol index. It should be noted that the above-mentioned notation is the simplest possible representation of the AWGN phenomena which does not consider the effect of any pulse shaping or channel impairments (other than noise). The "sgnl_rx" obtained at the end of Section 4.4.7 is the equivalent of r_k in Eq. (4.1).

4.5.1 Receiver Design

This case study aims to analyze the effect of AWGN on communication under different noise levels. The performance can be characterized by the number of bit-errors or BER. For this purpose, the readers need to develop the appropriate simulation setup according to the guidelines given below:

1. Since only the impact of noise is studied, synchronization and filtering effects can be ignored. Consequently, upsampling/downsampling at the Tx/Rx can also be skipped.
2. De-mapping operation needs to be implemented. For this, users can calculate the distance of each received symbol, r_k, from the reference constellation points and map it to the nearest one. This is referred to as *maximum likelihood* detection.
3. A bit is in error when the detected bit is different from the one sent by the transmitter. This can be used to calculate BER as:

$$BER = \frac{number\ of\ bits\ in\ error}{total\ number\ of\ transmitted\ bits} \tag{4.2}$$

The simplicity of AWGN channel model raises the question regarding its applicability in today's communication systems. While it is true that these models are not suitable for the conventional (terrestrial) systems since they ignore fading, multipath effects, selectivity, dispersion, and interference, they are still applicable for scenarios involving satellite and space communication links. Moreover, the noise itself in the current systems is also modeled as AWGN.

4.6 Multi-link vs. Network-level Simulations

There are different terminologies used to categorize the scope of simulation. Some of the terms used include link, multi-link, system, and network-level simulations. Despite their common usage, no

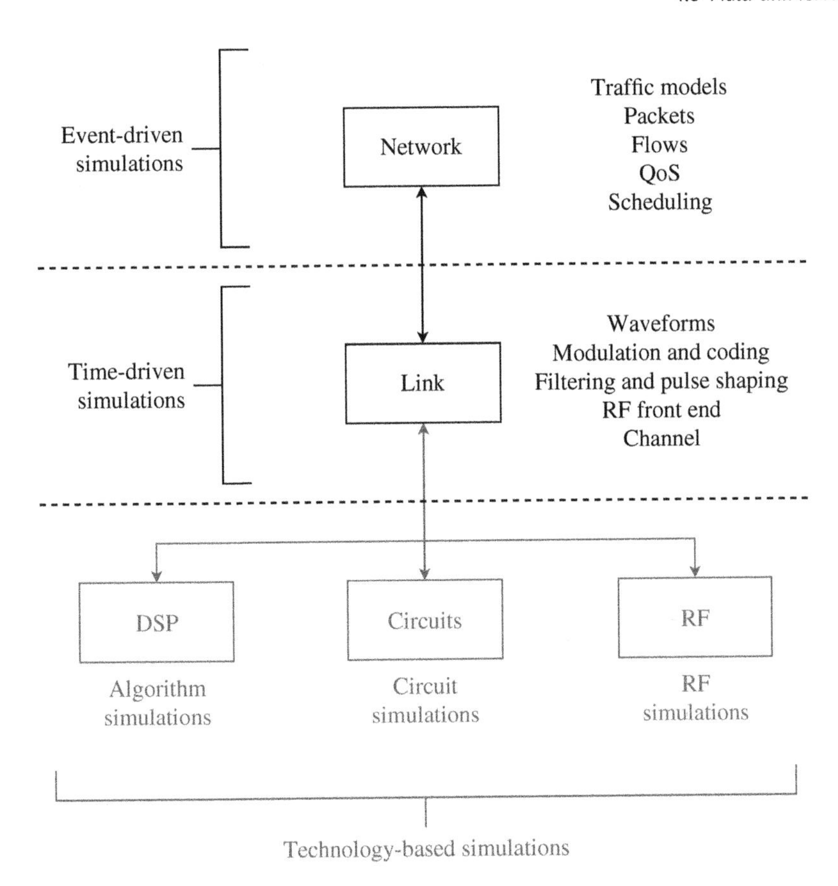

Figure 4.8 Simulation hierarchy.

clear definitions or distinctions are available in the literature. Therefore, we first try to clarify these terms before going on to establish the relationship(s) between them.

Using the bottom-up strategy, we have the link-level simulation first (technology-based simulation are not the focus in this book, therefore that portion is grayed out in Figure 4.8). To recap the discussion in the previous section, this level of simulation focuses on signal generation and manipulation (encoding, filtering, pulse shaping), its propagation over the physical medium, and its reception. The most common outputs are bits, symbol, or block error rates. In essence, these simulations are limited to the physical layer aspects of the communication system. This concept can be extended to multi-link simulations which effectively instantiates multiple link-level simulations simultaneously, allowing the effect of varying parameters and stochastic processes involved in the system to be observed.

Communication system designers/architects are not only interested in the individual link's performance but also the evaluation of the whole network. The simulations used for this purpose are referred to as system [4] or network-level [1] simulations. However, to avoid any confusion, we will use the term network-level simulation since system-level simulation is also used for the full transmit–receive chains of single links [3]. The network-level simulation incorporates issues such as network planning, user admission, mobility management, and resource management in general.

Owing to the large number of variables that constitute a network-level simulation, its computational complexity, time, and memory requirements can be a prohibitive factor. It is not possible

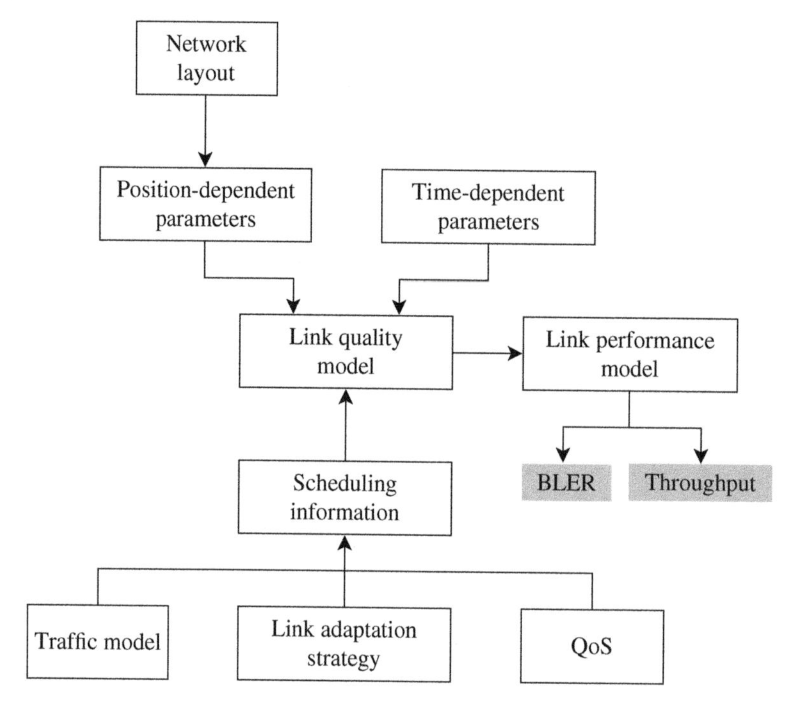

Figure 4.9 Link abstraction model.

to simulate all the characteristics of user equipments (UEs), base stations (BSs), and the links between them. This necessitates the use of some abstraction methods which simplify the simulation. Vienna's system-level simulator [5] is a popular example that performs link-to-network mapping by utilizing two blocks, namely, *link quality* and *link performance* models. A simplified illustration of this is provided in Figure 4.9. The link quality model takes into consideration the position-dependent parameters (such as path loss and shadowing) coming from the network layout, time-dependent small-scale fading, and scheduling information such as power and resource allocation. This model is responsible for providing the link performance after equalization and SINR metric is used in this particular case. The link performance model then translates the SINR to block error rate (BLER) and throughput values for the modulation and coding scheme (MCS) used. This is achieved using AWGN BLER curves which are obtained from link-level simulations [6].

While providing a complete network-level simulation is beyond the scope of this chapter, one aspect of it, i.e. network layout generation and some popular methods for it are described below.

4.6.1 Network Layout Generation

Network layout is one of the primary components of any given system model. It essentially dictates the communication channel parameters such as path loss, fading, shadowing, etc. Furthermore, in conjunction with user distribution, it characterizes the system performance from the perspective of interference between different BSs.

4.6.1.1 Hexagonal Grid

Hexagonal grid is one of the most popular cellular layouts used in communication systems. Even though it is not the most realistic since no cellular network has such a vivid hexagonal coverage, it enables an efficient coverage of a region without overlapping.[3] The following code snippet represents the construction of a homogeneous single-tier cellular network.

```
1  % Generating a 1-tier hexagonal grid
2  angles = 0:30:360; sin_angles = sind(angles); cos_angles = cosd(angles);
3  ISD = 250; % inter-site distance
4  size = 2*ISD/sqrt(3); %distance between the opposite vertices of a hexagon
5  t0_cellCntr = [0, 0];
6  scatter(t0_cellCntr(1), t0_cellCntr(2),  'rs', 'filled')
7  % plot the central BS
8  hold on
9  vrtcs_xCoord = t0_cellCntr(1) + (size/2).*cos_angles(1:2:end);
10 % x-coordinate of vertices
11 vrtcs_yCoord = t0_cellCntr(2) + (size/2).*sin_angles(1:2:end);
12 % y-coordinate of vertices
13
14 % Generating and plotting cell center coordinates for tier-1 cells
15 t1_cellCntr_xCoord = t0_cellCntr(1) + (ISD).*cos_angles(2:2:end);
16 t1_cellCntr_yCoord = t0_cellCntr(2) + (ISD).*sin_angles(2:2:end);
17 scatter(t1_cellCntr_xCoord, t1_cellCntr_yCoord, 'g^', 'filled')
18
19 % Generating users with uniform distribution in the central cell
20 usrLocs = unifrnd(-size/2, size/2, 2, 100); % random user generation
21 in = inpolygon(usrLocs(1,:),usrLocs(2,:),vrtcs_xCoord,vrtcs_yCoord);
22 % selecting the locations within the hexagon
23 plot(usrLocs(1,in), usrLocs(2,in),'b+', 'MarkerSize', 3)
24
25 % Generating and plotting the vertices for tier-1 cells
26 for index = 1:length(t1_cellCntr_xCoord)
27     t1_vrtcs_xCoord = t1_cellCntr_xCoord(index) + ...
28         (size/2).*cos_angles(1:2:end);
29     t1_vrtcs_yCoord = t1_cellCntr_yCoord(index) + ...
30         (size/2).*sin_angles(1:2:end);
31     plot(t1_vrtcs_xCoord, t1_vrtcs_yCoord, 'k-')
32 end
33 legend('Tier-0 Base Station', 'Tier-1 Base Station', 'User Location',...
34 'Cell Boundary')
```

The single-tier network is centered at the origin, with six hexagonal cells of the first tier surrounding the central cell, as shown in Figure 4.10. The variable "ISD" in the given code snippet represents the distance between centers of adjoining cells. Using the hexagonal geometry and angles, locations of vertices are calculated for each cell and plotted. Furthermore, the user locations represented by "+" signs in Figure 4.10 are generated from a uniform distribution. For this purpose, we have utilized the *rejection* method [7], where a number of points are generated in a square grid (easier to do with uniform distribution) and then checked against the boundaries of the central hexagon. Later, only the points lying within the hexagon's boundary are kept as user locations.

4.6.1.2 PPP-based Network Layout

A Poisson point process (PPP) is used to describe a random collection of points in a given space. The concept has been used in various domains, including telecommunications [8]. Referred to as *spatial point process*, this approach generalizes the PPP to 2-D and 3-D space and can be used to model the node locations in a wireless network [9]. Since the process is governed by a probability distribution, it allows the use of stochastic geometry methods to analyze the network behavior over its various spatial realizations.

Given below is the code snippet generating both BSs and UEs using a PPP. The only parameter that a PPP needs is λ, which represents the density of nodes in the unit area.

---

3 Square-shaped cells can also provide complete coverage without overlapping, but a larger number of them would be required.

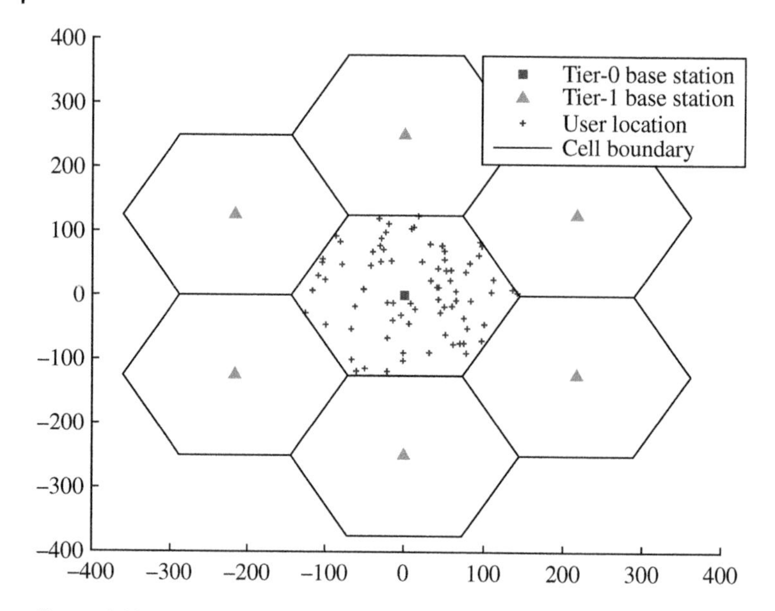

Figure 4.10 Hexagonal grid generation.

```
1  % Generating a Poisson BS and UE disribution
2  r_total = 0.4; %radius of disk (km^2)
3  areaTotal=pi*r_total^2; %area of disk
4
5  r_UEs = 0.1;                area_UEs = pi*r_UEs^2;
6
7  % Point process parameters
8  lambda_BS = 40; lambda_UE = 6000;
9
10 % Generating BS according to PPP with lambda_BS
11 numbPoints_bs = poissrnd(areaTotal*lambda_BS);% Poisson number of points
12 theta_bs = 2*pi*(rand(numbPoints_bs,1)); % angular coordinates
13 rho_bs = r_total*sqrt(rand(numbPoints_bs,1)); % radial coordinates
14 [Loc_BS(:,1),Loc_BS(:,2)] = pol2cart(theta_bs,rho_bs);
15 % Convert from polar to Cartesian coordinates
16 voronoi(Loc_BS(:,1),Loc_BS(:,2),'ks')
17 hold on
18
19 % Generating UEs according to PPP with lambda_UE
20 numbPoints_ue = poissrnd(area_UEs*lambda_UE);
21 theta_ue = 2*pi*(rand(numbPoints_ue,1));
22 rho_ue = r_UEs*sqrt(rand(numbPoints_ue,1));
23 [Loc_UE(:,1),Loc_UE(:,2)] = pol2cart(theta_ue,rho_ue);
24 plot(Loc_UE(:,1),Loc_UE(:,2), 'b    +', 'MarkerSize', 2);
25 legend('BS locations', 'Voronoi Boundaries', 'UE locations' )
26 axis([-0.4 0.4 -0.4 0.4])
```

Figure 4.11 shows the network layout generated as a result of the above-mentioned code. Here, the BSs are located within a disk of radius 0.4 km, centered at the origin, while the UEs are within a disk of 0.1 km radius. In certain network depictions [10], it is possible to have different UE distributions at the same time, e.g. a higher density of users can be used to represent a hotspot scenario with a lower density of users away from the center depicts the non-hotspot users. A *Voronoi* diagram is used to represent the cell boundaries such that the points inside each cell are closest to their own BS out of all BSs.

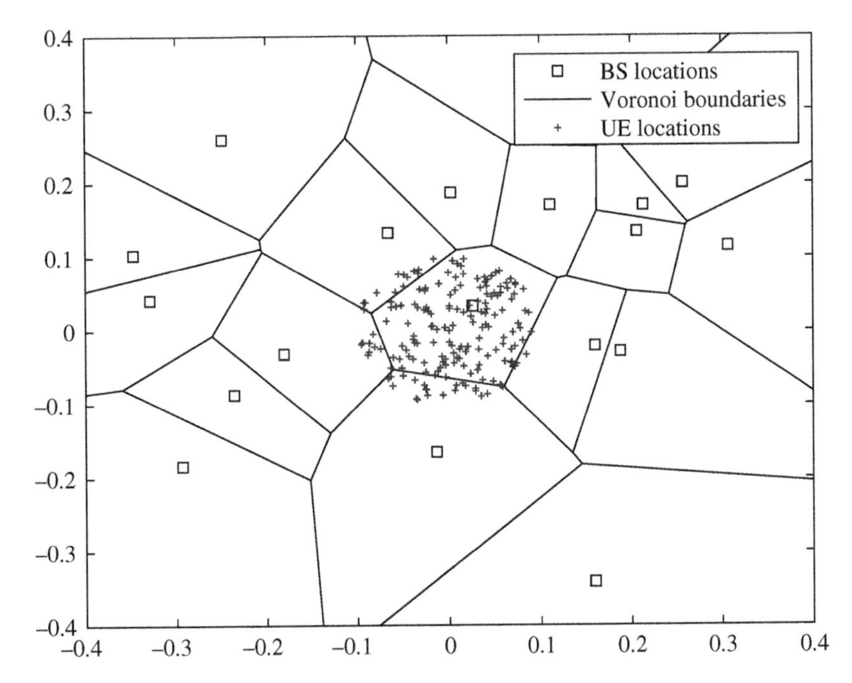

Figure 4.11 PPP-based network and user layout.

4.7 Practical Issues

This section gives an overview of some of the practical aspects of simulation. We start by looking at the concept behind the widely popular Monte Carlo methods. Later, we look at certain tips or rules-of-thumbs related to simulations, particularly about the sufficiency of statistics and corresponding stopping criteria.

4.7.1 Monte Carlo Simulations

Monte Carlo class of methods/techniques take their name from the notorious Monte Carlo casino and are used for applications that involve random processes. These methods form the basis of simulations where the input of a system is random. Simulation involving the calculation of BER, as discussed in Chapter 2, is a classical example of a Monte Carlo simulation. This can be generalized to a scenario where one or more random inputs are fed to a system (or any part of it) to get an output, as shown in Figure 4.12. $W(t)$ and $X(t)$ are random processes that form the input to a certain (sub)system.[4] As a result of the randomness of the inputs, even if the system itself is deterministic,

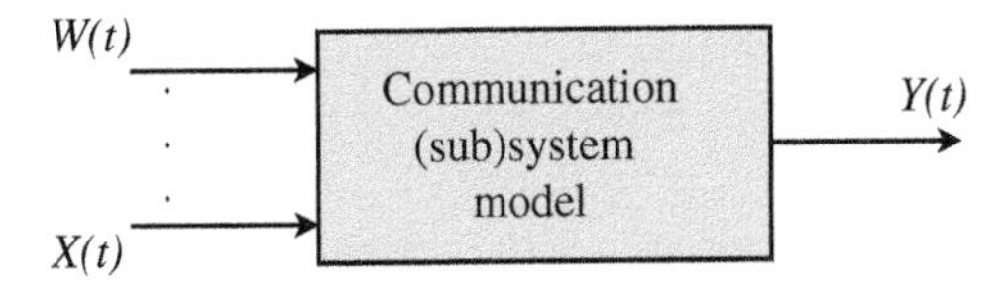

Figure 4.12 Illustration of the Monte Carlo simulation concept.

4 It is possible that only a subset of the inputs to the system are simulated as random while the others leverage analytical methods, in which case we have a "Quasianalytical Monte Carlo Simulation."

the output would have randomness in it. Hence, it is also modeled by a random process, denoted by $Y(t)$. Carrying out Monte Carlo simulations allows us to evaluate the statistical properties of the output or any function of it. Generally, the statistics of interest include mean, variance, correlation, and so on.

Since Monte Carlo methods are heavily dependent on the random numbers generated for various inputs, we take a quick look at this phenomenon next.

4.7.2 Random Number Generation

Random numbers or sequences are imperative in the simulation of communication systems. Their two most common applications include AWGN and random binary sequence, both of which are described below. The former serves as the simplest channel model with no impairments except the noise, while the latter is used to model the data bits after ADC conversion and coding. However, before we dive into their description it is important to note that in modeling and simulation a fundamental assumption often made for random processes is their ergodicity. This implies both wide and strict-sense stationarity, meaning the first- and second-order properties of the process are time-invariant.

4.7.2.1 White Noise Generation

As mentioned earlier, AWGN serves as the simplest channel model for communication, where (white) noise is the only impairment in the system. As the term *Gaussian* implies, this is modeled using the Gaussian or Normal PDF. The simplest implementation that can produce such numbers is the *sum-of-uniform* method which leverages the central limit theorem (CLT) [3].[5]

Assuming the availability of N_R independent uniform random variables, $U(i), i = 1, 2, \ldots, N_R$ in the interval $[0, 1]$, the Gaussian random variable can be calculated as:

$$Y = \sum_{i=1}^{N_R} (U(i) - 1/2) \tag{4.3}$$

A suitable value for N_R is 12, which provides a decent compromise between speed and accuracy [1]. However, a higher value is preferable if computational resources are not severely limited. Other methods include mapping a Rayleigh random variable to a Gaussian one using mathematical transform; however, we will not discuss it here.

4.7.2.2 Random Binary Sequence

As mentioned earlier, these random binary sequences are generally used to model the data bits that serve as the input for a communication system. Such a sequence ideally consists of an independent sequence of equiprobable 0's and 1's. However, achieving a truly random sequence via simulation is simply not possible. What we have are "pseudorandom" periodic binary sequences with an autocorrelation function similar to that of a truly random sequence.

These pseudorandom sequences are generally generated using feedback shift registers whose contents are logically combined to produce the sequence. Since this is a logical system, which is not truly random, it is in fact *periodic*. This period is dependent on the number of binary storage elements in feedback shift registers.

Note: An important thing to consider in random number generation is the correlation between different samples. While in cases such as AWGN, the samples are meant to be uncorrelated, there might be scenarios where an arbitrary amount of correlation is required such as when modeling fading in a network [11].

5 CLT states that the PDF of the sum of a large number of i.i.d. variables approaches a Gaussian distribution.

4.7.3 Values of Simulation Parameters

Proper selection of the simulation parameters is critical in ensuring the validity and usefulness of the obtained results. Considering the example given in Section 4.5, the BER performance can be studied as a function of the SNR. However, the assumed SNR values need to be realistic. The typical range of SNR values used in simulations is from 0 to 30 dB. Closely related with SNR are the consequent BLER or PER, which form part of the simulation parameters. For instance, the PER requirement for conversational voice (10^{-2}) varies significantly from what is needed for low-latency augmented reality applications (10^{-6}) [12]. Therefore, it is advisable to select the parameters such as PER and SNR according to the specific application and scenario rather than using a very wide range of values, which will lead to unnecessary computational (and storage) overhead.

4.7.4 Confidence Interval

Confidence interval refers to the probability that an estimate falls within a given range. This is also applicable to the simulation models that are used to represent different blocks in the communication system. The confidence interval gives a quantitative measure of how well this model compares with the desired model. Given below is the general expression used for confidence interval calculation:

$$Pr\{I - \beta \leq \hat{I} \leq I + \beta\} = 1 - \alpha, \tag{4.4}$$

where $\hat{I}$ refers to the estimated value of I, and $(1 - \alpha)$ gives the probability that the estimate falls between $I \pm \beta$. Typical values for α are 0.01 and 0.05, which correspond to 99 and 95% confidence intervals, respectively.

4.7.5 Convergence/Stopping Criterion

The stopping criterion varies with the specific application/scenario being simulated. However, the basic principle is to ensure the availability of enough statistics to represent the real behavior of any system under given constraints. Both law of large numbers and CLT give credence to the use of a sufficiently high number of iterations to get a good approximation of the underlying phenomena. However, a designer needs to keep in mind the performance vs. complexity vs. time trade-off while deciding on this. This can be illustrated with the SNR vs BER simulation example provided in Chapter 2 (Section 2.1.3) where the simulation runs for a fixed number of samples for different values of SNR. However, in order to optimize the simulation, it is advisable to run it for a sufficient number of bit-errors instead of a fixed number since the number of iterations required to get the same number of bit-errors would vary significantly for different SNR values. Some cases might even necessitate a hybrid approach, i.e. the simulation may be stopped if an extremely large number of iterations is required to acquire the desired number of bit-errors.

4.8 Issues/Limitations of Simulations

The usefulness of any simulation is limited by its accuracy where accuracy refers to the similarity or closeness of the results produced by the simulation to the actual underlying behavior of the system under test. This lack of accuracy generates from two sources, i.e. inaccuracy in modeling or errors in processing.

4.8.1 Modeling Errors

4.8.1.1 Errors in System Model

If the system model does not have a one-to-one mapping of the subsystems present in the actual system, some errors may arise. An approach used in this regard might be to combine cascaded linear devices together (filters, for instance) to streamline the system. On the other hand, there are cases where some blocks which have little impact on the simulation might be removed to reduce the complexity and runtime of the simulation. Antennas are a common example, they are rarely modeled unless the simulation is specifically dealing with them. Another example is the RF front-end, the constituents of which (and their impairments) are not modeled by assuming the presence of some carrier-tracking loops in the transceiver.

4.8.1.2 Errors in Subsystem Model

It is unfair to expect any model to represent its physical counterpart with 100% fidelity. However, it is possible to have an almost equivalent representation of the idealized model of the said component/block/subsystem. Moreover, it is not necessary to replicate the finest details of the behavior of the device as long as its effect on the system can be reproduced.

One thing that should be kept in mind during the subsystem modeling is that estimating the cumulative effect of modeling errors for different subsystems is very difficult, therefore, validation of the models is imperative.

4.8.1.3 Errors in Random Process Modeling

This error originates due to the difference between the models of the random process and the real underlying phenomena. Signal and noise sources are primarily susceptible to this error.

Two kinds of signals can be used in a simulation. One is a known test sequence while the other is an actual source. The former can be exactly generated in an error-free manner. In the latter case, the model needs to faithfully replicate the concerned source. If the source generates samples with some dependencies, they need to be incorporated into the model.

In terms of noise, three possible sources are usually considered, namely, thermal, phase, and impulsive (shot) noise. Thermal noise is modeled as AWGN and can be assumed to be largely error-free. Phase noise, which results from VCO's imperfections, is not completely understood or explicitly modeled. Its effect, however, can be modeled as Gaussian. Impulsive or shot noise represents human activity and noise in optical detectors. Random number generators with proper parameters can emulate this type of noise quite well.

4.8.2 Processing Errors

Processing errors generally occur due to computing limitations such as memory, speed, processing, and precision. Discrete-time representation of continuous signals is an example, where the perfect capturing of the signal would need infinite resources. The numerical approaches used, therefore, introduce some uncertainty into the system.

The sampling of continuous signals leads to aliasing error; however, this error is negligible as compared to the actual noise present for the typical error rates (around 10^{-6}) in digital communication systems. Sampling rates between 8 and 16 samples per symbol are generally adequate to address this. The truncation of the filter's impulse response is another source of error for the simulation. The amount of error is related to the truncated length. It is recommended that the designer investigates this trade-off since it is possible to find a suitable compromise between both.

The processing and speed limitations, on the other hand, can inhibit the ability of Monte Carlo simulations to be carried out long enough for statistical stability. As mentioned earlier (in Section 4.4.2), this is more evident in the case of low BER scenarios where a higher number of simulation iterations are required to get a sufficient number of errors.

4.9 Conclusions

This chapter aims to introduce readers to basic knowledge of simulation regarding wireless systems. Apart from describing different blocks of a wireless link, the general strategies regarding simulation are discussed. Moreover, the readers are given a flavor of network-level simulation from the perspective of the network layout. In the end, however, the authors would like to reiterate that the simulation is heavily dependent on the models being used both for the system blocks and random processes. Therefore, the proper selection of these, keeping in mind the time/computational limitation is extremely important in developing a suitable communication system simulation.

References

1 M. C. Jeruchim, P. Balaban, and K. S. Shanmugan, *Simulation of Communication Systems: Modeling, Methodology and Techniques*. Springer Science & Business Media, 2006.

2 J. E. Hopcroft, R. Motwani, and J. D. Ullman, "Introduction to automata theory, languages, and computation," *ACM Sigact News*, vol. 32, no. 1, pp. 60–65, 2001.

3 W. H. Tranter, T. S. Rappaport, K. L. Kosbar, and K. S. Shanmugan, *Principles of Communication Systems Simulation with Wireless Applications*. Prentice Hall New Jersey, 2004, vol. 1.

4 A.-S. K. Pathan, M. M. Monowar, and S. Khan, *Simulation Technologies in Networking and Communications: Selecting the Best Tool for the Test*. CRC Press, 2014.

5 M. K. Müller, F. Ademaj, T. Dittrich, A. Fastenbauer, B. R. Elbal, A. Nabavi, L. Nagel, S. Schwarz, and M. Rupp, "Flexible multi-node simulation of cellular mobile communications: the Vienna 5G System Level Simulator," *EURASIP Journal on Wireless Communications and Networking*, vol. 2018, no. 1, p. 17, Sept. 2018.

6 M. Rupp, S. Schwarz, and M. Taranetz, *The Vienna LTE-Advanced Simulators*. Springer, 2016.

7 G. Casella, C. P. Robert, M. T. Wells *et al.*, "Generalized accept-reject sampling schemes," in *A Festschrift for Herman Rubin*. Institute of Mathematical Statistics, 2004, pp. 342–347.

8 M. Haenggi, *Stochastic Geometry for Wireless Networks*. Cambridge University Press, 2012.

9 J. G. Andrews, R. K. Ganti, M. Haenggi, N. Jindal, and S. Weber, "A primer on spatial modeling and analysis in wireless networks," *IEEE Communications Magazine*, vol. 48, no. 11, pp. 156–163, 2010.

10 S. Bassoy, M. Jaber, M. A. Imran, and P. Xiao, "Load aware self-organising user-centric dynamic CoMP clustering for 5G networks," *IEEE Access*, vol. 4, pp. 2895–2906, 2016.

11 M. T. Ivrlac, W. Utschick, and J. A. Nossek, "Fading correlations in wireless MIMO communication systems," *IEEE Journal on Selected Areas in Communications*, vol. 21, no. 5, pp. 819–828, 2003.

12 3rd Generation Partnership Project (3GPP), "System architecture for the 5G System (5GS); Stage 2 (Rel-16)," Technical Specification 23.501, ver 16.5.1, Aug. 2020.

5

RF Impairments

Hüseyin Arslan

Department of Electrical Engineering, University of South Florida, Tampa, FL, USA
Department of Electrical and Electronics Engineering, Istanbul Medipol University, Istanbul, Turkey

Radio frequency (RF) front-end, which is an important part of wireless transceivers, greatly influences the system performance. Often, digital algorithm designers model the RF impairments as an additive white Gaussian noise (AWGN). However, understanding, modeling, and accurately simulating the impact of RF impairments are necessary for proper design of wireless systems. The baseband equivalent impairment models characterize the influence of RF front-end effects on the digital baseband signal, which can be integrated into the system simulation setup and analytical studies. In this chapter, baseband equivalent models for various RF impairments are discussed. Also, the impacts of these impairments on the system performance are explained through various tools and measurements.

5.1 Radio Impairment Sources

An overly simplified typical radio transmitter and receiver is shown in Figure 5.1. In this figure, a heterodyne transmitter and receiver is shown. A heterodyne receiver translates the desired RF frequency to one or more intermediate frequencies before digital conversion [1]. However, direct conversion (zero IF or homodyne) transmission and receptions are also possible. A homodyne (no intermediate frequency) receiver translates the desired RF frequency directly to baseband for processing. Baseband signals are typically at frequencies much lower than the carrier frequencies. The range of frequencies for a basedband signal can include DC (zero Hz) all the way to the upper frequencies which depends on the data rate, pulse shape, and other signal and waveform design criteria.

In the transmission side, the I (in-phase) and Q (quadrature-phase) channels are generated in the digital baseband processor depending on the desired waveform and standard requirements. Typically, the waveform generation includes source coding to represent the source with as minimum number of bits as possible (e.g. vocoder are used for speech coding); channel coding for protecting the transmitted bits against the harsh radio channel (e.g. forward error correction encoding and cyclic redundancy check encoding); interleaving to protect the bits against the deep fades; bits-to-symbol mapping, which is the mapping of the bits to in-phase and quadrature-phase (I/Q) plane; upsampling to satisfy the Nyquist criterion (often the bandwidth of the signal is greater than the symbol rate, therefore, upsampling is necessary), and pulse shaping to contain the transmission to a minimal possible bandwidth (increases spectral efficiency). The digital I/Q signal is then passed through I/Q modulator to modulate the I and Q carriers at the intermediate frequencies (IF) frequencies. In this specific figure, IF is analog (I/Q modulation implemented with analog hardware is very common in digital radios). The current trend, with software defined radio (SDR) and

Wireless Communication Signals: A Laboratory-based Approach, First Edition. Hüseyin Arslan.

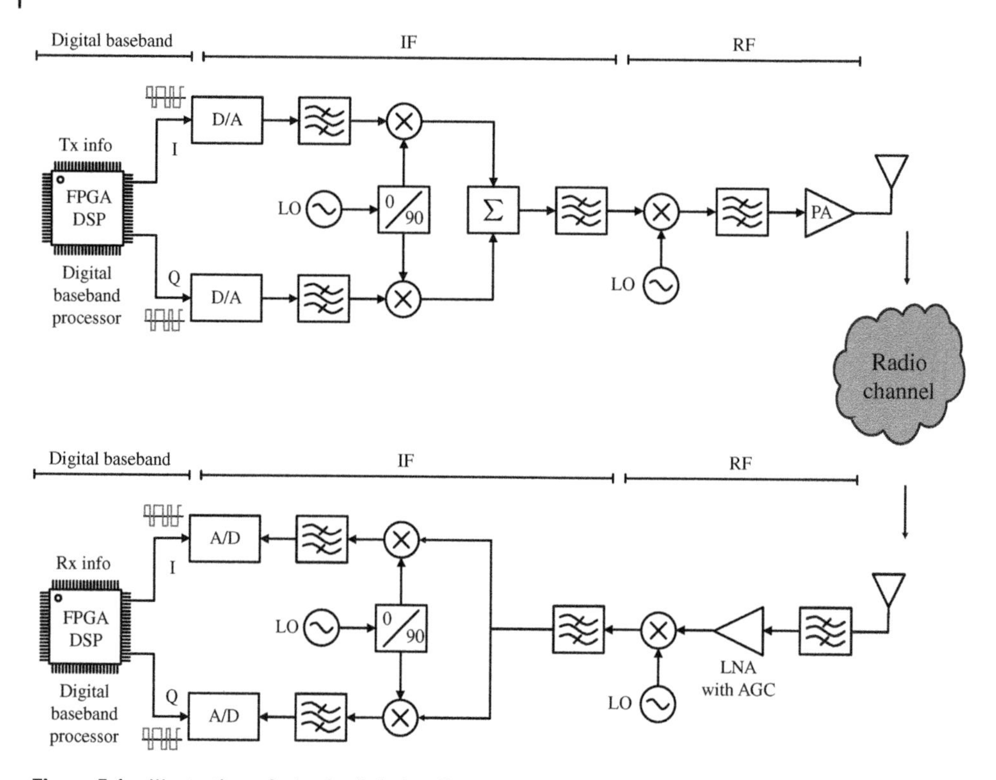

Figure 5.1 Illustration of a basic digital radio system components.

cognitive radio-based transceivers, is to shift the IF to digital domain where the functionalities can be implemented using digital signal processing (DSP) techniques. If IF is done in DSP, then, the digital-to-analog converter (DAC) will be after the I/Q modulator. The IF signal is further mixed to RF frequencies in the RF part. The RF signal is then amplified and transmitted through the transmitter antenna over the wireless medium [2, 3].

Wireless medium is a very harsh environment for radio signals. Transmitted signals are typically reflected and scattered, arriving at the receiver along multiple paths. When these paths have similar delays, they add either constructively or destructively, giving rise to fading [4]. When these paths have very different delays, they appear as signal echoes. Due to the mobility of the transmitter, the receiver, or the scattering objects, the channel changes over time. In addition to all these, interference from other transmitters, and noise will be added to the transmitted signal. The receiver receives corrupted transmitted signal along with other interference and noise signals. Due to the path loss and fading, the received signal is often not powerful enough for processing. Low-Noise Amplifier is used to enhance the power of the received signal. Due to the random nature of fading, the received signal will have a wide dynamic range. Automatic gain controller (AGC) adjusts the received signal level so that roughly constant signal input is provided to the rest of the receiver front-end. Since the multipath and mobility introduce rapid variations to the signal amplitude levels, the AGC should respond these variation quickly. The rest of the process is reciprocal of what is done at the transmitter, where the RF signal is filtered, down-converted from RF frequencies to IF frequency and passed through I/Q demodulator to extract the baseband analog I and Q waveforms, then, these baseband analog waveforms are digitized to be processed using digital baseband signal processing techniques in a DSP or field programmable gate array (FPGA) platform.

The performance of the overall system heavily depends on the transmitter, receiver, and operating channel conditions [5, 6]. Every stage in the transmitter and receiver chain adds noise and impairments to the signal. Some of the critical impairment sources can be listed as:

- **I/Q modulators and demodulators** – modulate and combine two quadrature carriers, 90° out of phase, with signals in the I and Q branch; extract the I and Q branches of a combined signal using quadrature carriers [7]. A pair of periodic signals are said to be in "quadrature" when they differ in phase by 90°. The "in-phase" signal is referred to as "I," and the signal that is shifted by 90° (the signal in quadrature) is called "Q." The I/Q modulator feature a quadrature modulator that accepts *I(t)* and *Q(t)* signals which it then uses to amplitude modulate a pair of quadrature sinusoids followed by a combiner (that adds the outputs of I and Q branch) to create the modulated higher frequency output. Similarly, I/Q demodulator converts the incoming high-frequency signal into their I and Q components.
- **Local oscillators (LO)** – generate periodic sinusoidal waves. The local oscillator is a reference signal required for mixer to enable frequency translation. Frequency accuracy and stability are determined by the local oscillators of the transceivers. Imperfect carrier synchronization in the receiver produces constant or slowly time-varying phase error, which usually can be corrected by baseband processing afterward. The short-term instability of the oscillators appear as phase noise, and it is very critical for the RF performance of both the transmitter and receiver.
- **Mixers** – perform frequency translation by multiplying two signals; frequency conversion or heterodyning. The translation uses a local oscillator (LO) as described above. A follow-up filter selects either the higher or lower (sum or difference) output frequency. One frequency is passed while the other is rejected. Selecting the higher frequency is up-conversion; selecting the lower frequency is down-conversion.
- **Filters** – clean undesired signals from the desired ones by allowing only the desired signal pass through it. The filter blocks signals outside of the passband. The blocked area is known as the stopband. The perfect filter passes desired signals with a flat response and infinitely attenuates signals in the stopband. Practical filters have a gradual roll-off in the transition region from the passband to the stopband, some ripples in the passband and leakage in the stopband.
- **Antennas** – convert electrical signals into electromagnetic waves and vice versa. Antennas are characterized by a number of performance parameters, such as bandwidth, gain, radiation efficiency, beamwidth, beam efficiency, radiation loss, and resistive loss. Antenna gain is often defined relative to a theoretical ideal isotropic antenna that radiates energy equally in all directions. A directional antenna with 10 dBi of antenna gain produces an RF signal with 10 times the power density compared to an isotropic antenna in the direction of maximum transmission. Antennas with gains can be used to utilize the power efficiently, extend the coverage, and minimize interference for other users. Directional antennas and beamforming are especially more important in high frequencies where path loss is severe. Radiation pattern is the geographical distribution of power radiated by the antenna and bandwidth of antenna is range of frequencies (on either side of center frequency) where antenna characteristics (like beamwidth, gain, polarization, etc.) are within acceptable values. Antenna beamwidth is the angle formed by the radiating field where the electric field strength or radiation pattern is 3 dB less than its maximum value. Antenna main lobe is the energy traveling in the primary direction of propagation. Energy traveling to the sides of the primary propagation direction is called side

lobe energy. Antenna matching circuit matches the antenna output impedance to the receiver to minimize the loss and improve the noise figure.

- **Amplifiers** – power amplifier (PA) and low-noise amplifier (LNA) – boost the input power of a signal. LNA is designed to take a very low-level signal at the receiver (after a very lossy medium including wireless channel) and amplify it with minimal additive noise. A low noise amplifier cannot increase signal-to-noise ratio (SNR), it can only raise the power level of both signal and noise. The signal at the output of the LNA should be sufficiently large to enable a detectable voltage range. A power amplifier, on the other hand, takes a signal that is already at a relatively high level and boosts it for transmission over a lossy medium such as wireless channel. The nonlinear behavior of the amplifier is described in many ways and ultimately manifests as both degraded signal quality and legally restricted signal leakage into adjacent channels, as described by the Adjacent Channel Power Ratio (ACPR).
- **ADC and DAC** – provide an interface between the analog and digital world and converts analog signals to digital or vice versa. An ADC uses sampling to make instantaneous measurements of an analog input signal over its voltage range, converting those measurements into digital words with resolution equal to the converter's number of bits. The sampling rate takes place at the frequency (or multiple of that frequency) of the clock oscillator used for timing the ADC. A DAC essentially does the opposite, converting digital input code into analog output signals. The resolution of a converter is expressed in the number of bits. For an ADC, the resolution states the number of intervals or levels which can be divided from a certain analog input range. An n-bit ADC has the resolution of $1/2n$. The speed of a converter is expressed by the sampling frequency. It is the number of times that the converter samples the analog signal, its unit is Hertz (Hz).
- **Signal processors - DSP or FPGA** – are hardware platforms for implementing radio functionalitie using digital signal processing techniques. DSP functions are commonly implemented on two types of programmable platforms: digital signal processors (DSPs) and field programmable gate arrays (FPGAs). DSPs are specialized processors which are optimized for implementation of DSP algorithms, while FPGAs are highly configurable hardware that offer a range of logic resources such as configurable logic cells. FPGAs are programmed using designs which configure and connect the logic resources to implement the desired algorithm.

Proper design of the transmitters and receivers require a good understanding of the impairment sources and their models. Often, optimizing the design of a communications transceivers is inherently a process of compromise. Noise limits the smallest signals that a receiver system is capable of processing. The largest signal is limited by distortions due to nonlinearity of transceiver circuits. The smallest and largest signals define the dynamic range of the receiver system.

5.2 IQ Modulation Impairments

The transmitted signal, that goes through the I/Q vector modulator, experiences several levels of signal distortion due to imperfection in the modulator. These distortions can greatly affect the performance of the received signal and the overall system performance.

The major I/Q impairments can be classified as I/Q offset, I/Q gain imbalance, and I/Q quadrature-error. A model for I/Q impairment in I/Q modulator is given in Figure 5.2. Note that the I/Q impairments in the received signal will have quite different impact depending on the modulation format. For example, the impact on orthogonal frequency division multiplexing (OFDM)-based systems will be different compared to the conventional single-carrier systems [8].

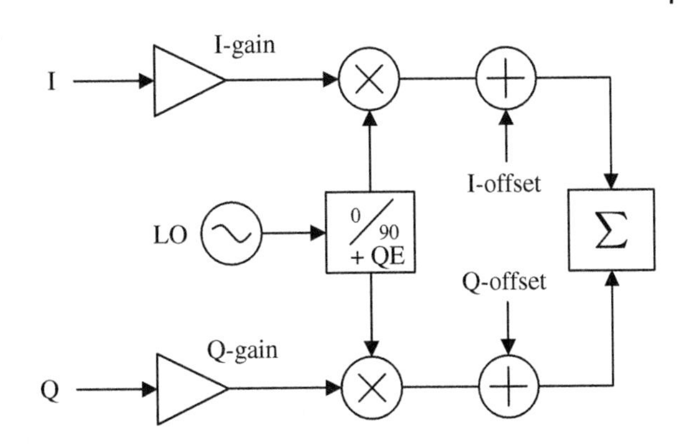

Figure 5.2 Block diagram of I/Q modulator impairment models. The imbalance can be measured as percentage, $IQ_{imbalance} = \{\frac{I_s}{Q_s} - 1\} * 100\%$, or as ratio, $20\log(\frac{I_s}{Q_s})$ dB. The quadrature error is measured in degree. The offset values can also be measured as percentage or in ratio.

I/Q Offset, also called I/Q origin offset or carrier leakage, indicates the magnitude of the carrier feedthrough. I/Q Offset can be observed as an offset (shift) in the constellation. The offset is mainly caused by the mixers and by DC signals introduced through the additional IF amplifiers. Gain mismatch or gain imbalance will result in the amplitude of one channel being smaller than the other. Differences in the parasitic capacitances and inductances along the printed circuit board traces for the I and Q signal paths can cause such I/Q imbalances. Component and design variations in the baseband and front-end integrated circuits, amplitude errors in the DACs, and inconsistencies between each of the analog mixers can also cause I/Q imbalances. By comparing the gain of the I signal with the gain of the Q signal, $20\log(I_s/Q_s)$, the I/Q gain imbalance can be obtained. Baseband representation of the transmitter I/Q origin offset and I/Q gain imbalance that are introduced into time-domain signal can be given as:

$$y_m(n) = I_s\Re\{x_m(n)\} + jQ_s\Im\{x_m(n)\} + I_o + jQ_o, \tag{5.1}$$

where $\Re\{\cdot\}$ is the real part and $\Im\{\cdot\}$ is the imaginary part of a signal. When $I_s = Q_s$, the I/Q imbalance will be zero, and when both I_o and Q_o are zero, there will not be any I/Q offset.

Quadrature Skew Error indicates the orthogonal error between the I and Q signals. Ideally, I and Q channels should be exactly orthogonal (90° apart). When the orthogonality is not ideal (less or more than 90° apart, $90 \pm \psi$), then a quadrature error is observed, resulting in significant problems at the receiver. The source of the quadrature error is typically due to LO splitter, which divides the LO into an in-phase and a quadrature-phase signals.

The time-domain signal at the output of I/Q modulator with quadrature error can be given as:

$$\tilde{x}(n) = \cos(2\pi f_c t)\Re\{x(n)\} + \sin(2\pi f_c t + \psi)\Im\{x(n)\}, \tag{5.2}$$

where ψ is the deviation amount from the perfect quadrature (i.e. the quadrature error in angle). Note that in this analysis, we assume the error is in Q branch for simplicity. However, it can be in I branch or in both. But, at the end, as will be shown later, what we care is the amount of deviation from the perfect quadrature.

Let's assume that the I/Q demodulator at the receiver has perfect quadrature, and the goal is to measure the error caused by the I/Q modulator at the transmitter [9]. Ignoring the other impairments and channel effects initially, the received digital baseband signal after the I/Q demodulator, baseband receiver filtering, analog-to-digital conversion, and synchronization can be written as:

$$\hat{x}(n) = \Re\{x(n)\} + \Im\{x(n)\}\sin(\psi) + j\Im\{x(n)\}\cos(\psi), \tag{5.3}$$

where in a special case of $\psi = 0$, then, $\hat{x}(n) = x(n)$.

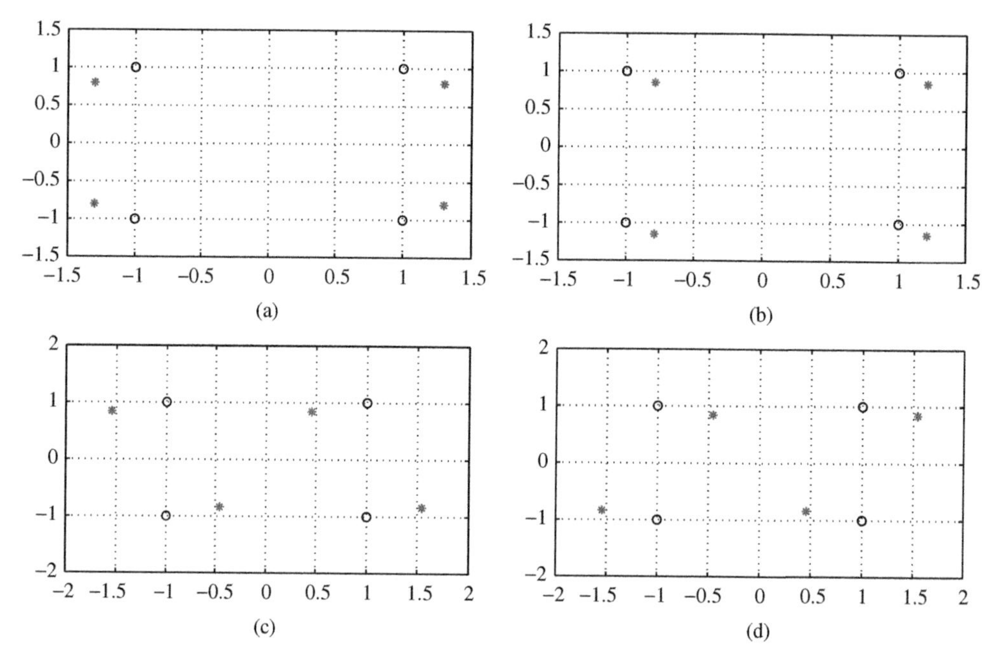

Figure 5.3 Representation of I/Q impairments in single-carrier communication systems with QPSK modulation. Circles represent ideal constellations, stars represent the impaired constellations. (a) I/Q gain imbalance. (b) I/Q offset. (c) Quadrature offset. (d) Quadrature offset.

Figure 5.3 shows the impact of I/Q impairments in single-carrier systems. The corresponding Matlab script is given below. As mentioned before, the impact of the I/Q impairments on multi-carrier systems (like OFDM) will be different. Figures 5.4 and 5.5 show the impact of I/Q gain imbalance and quadrature error on OFDM-based multi-carrier systems. The details of OFDM analysis will be given later in a separate chapter.

```
1  % IQ impairments and their visualization
2  % Observe the effect of:  IQ impairment values on constellation
3  N=1e3; % number of symbols
4  M=4; % modulation order
5  symb=qammod(floor(rand(1,N)*M),M); % ideal symbols
6
7  I_gain=1.3;Q_gain=0.8; % IQ gain imbalances
8  symb_g_imb = I_gain*real(symb) + j*imag(symb)*Q_gain ;
9  subplot(2,2,1)
10 plot(real(symb_g_imb),imag(symb_g_imb),'*')
11 title('(a) I/Q Gain Imbalance')
12 hold on
13 plot(real(symb),imag(symb),'o')
14
15 I_o=0.21;Q_o=-0.15; % DC offset
16 symb_DC_of = real(symb) + j*imag(symb)+I_o+j*Q_o;
17 subplot(2,2,2)
18 plot(real(symb_DC_of),imag(symb_DC_of),'*')
19 title('(b) I/Q offset')
20 hold on
21 plot(real(symb),imag(symb),'o')
22
23 alp=-10; % Quadrature offset
24 symb_DC_qo=real(symb)+imag(symb)*(sin(alp)+j*cos(alp));
25 subplot(2,2,3)
26 plot(real(symb_DC_qo),imag(symb_DC_qo),'*')
27 title('(c) Quadrature offset')
28 hold on
29 plot(real(symb),imag(symb),'o')
```

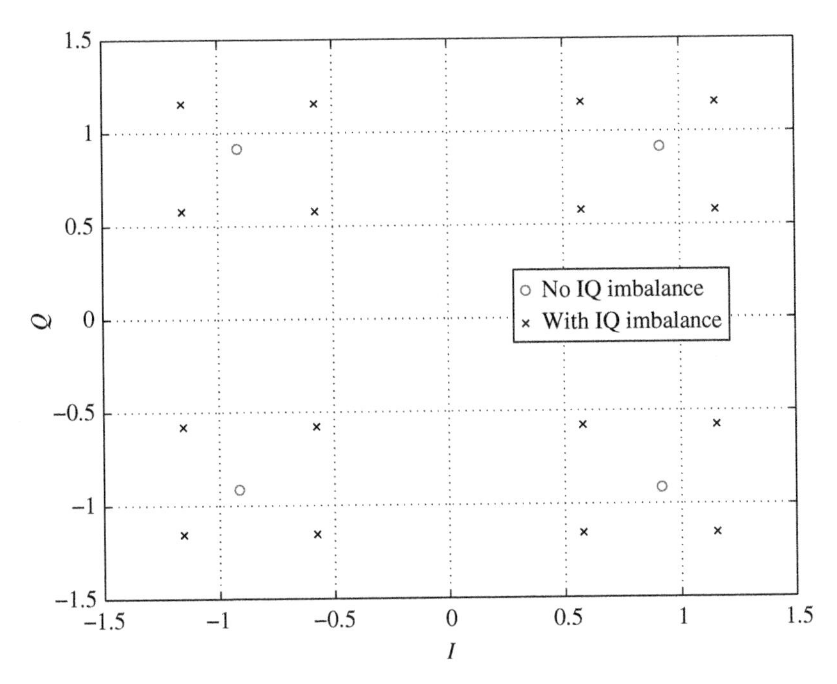

Figure 5.4 Illustration of the effect I/Q gain imbalance in the constellation of the received symbols in OFDM-based WiMAX system.

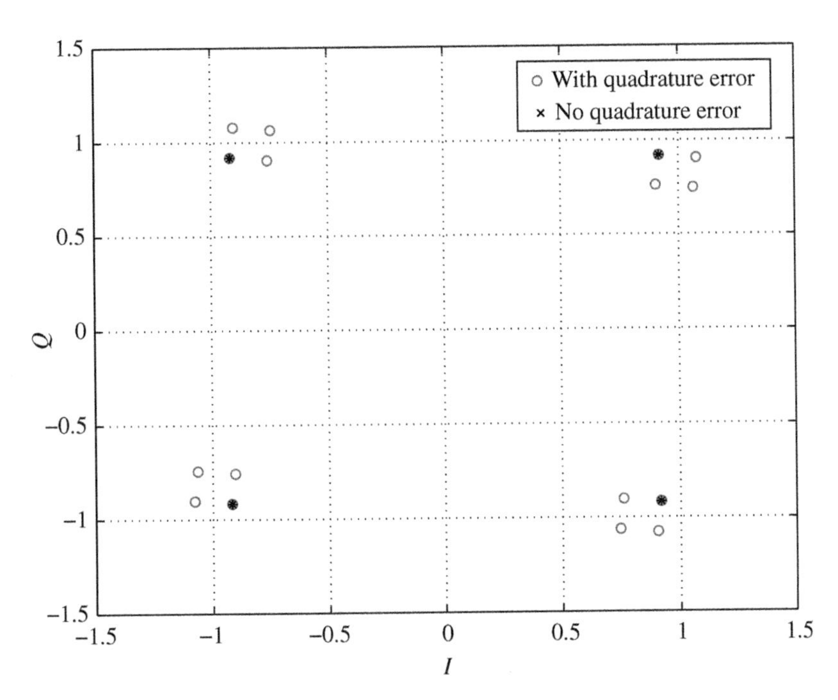

Figure 5.5 Illustration of the effect of quadrature error in the constellation of the received symbols in OFDM-based WiMAX system.

5.3 PA Nonlinearities

Power amplifiers (PAs) and LNAs are two of the essential components of wireless communication systems. PAs are used to amplify the signal before transmission so that the signal can reach to distant receivers with a desired SNR. LNAs usually amplify the weak received signal (at the receiver) with noise. PAs can be divided into two major groups: linear and nonlinear PAs. Linear PAs have the advantage of high linearity which is very important for signals with a wide range of amplitude values. However, they suffer from poor power efficiency, limiting their applications in wireless communication systems. On the other hand, nonlinear PAs can achieve better efficiencies with poor linearity in their responses. The nonlinearity causes several problems like amplitude and phase distortion (AM-to-AM and AM-to-PM effects) and adjacent channel regrowth (i.e. broadening the signal bandwidth). Power back-off technique is widely used in current wireless communication systems (e.g. WLAN) to remedy the problems due to wide dynamic signal ranges [10, 11]. However, this technique sacrifices the efficiency and increases the power consumption. On the other hand, baseband linearization techniques are utilized to pre-distort the signal and hence to compensate the nonlinear effects [12–14]. The efficiency and adaptation of linearization algorithms are, therefore, very important. Depending on the mode of operation and transmitted waveform characteristics, the linearization algorithms need to be adapted. Digital linearization techniques are often based on a feedback scheme and therefore able to react to drifts of the nonlinear power PAs. Since PAs are the major source of power consumption and heat generation in radios, PAs can operate in different power level modes such as high- and low-power modes to save power.

PAs typically operate as a linear device for a small signal (over its linear gain region). However, when the input power is increased, PA operates as a nonlinear device and starts to distort the transmitted signal. The linear and nonlinear operating region of a PA can be defined by a saturation point of an amplifier, where at the saturation point the gain of output signal is reduced by 1 dB, which is also known as 1-dB compression point (see Figure 5.6).

The impact of PAs and related issues have gained a lot of attention recently, especially in multi-carrier systems [15, 16]. One of the major problems with OFDM modulation technique is that it has high peak-to-average power ratio (PAPR) that requires the PAs used in the system to have a large linear operation range to keep the distortion at minimum.

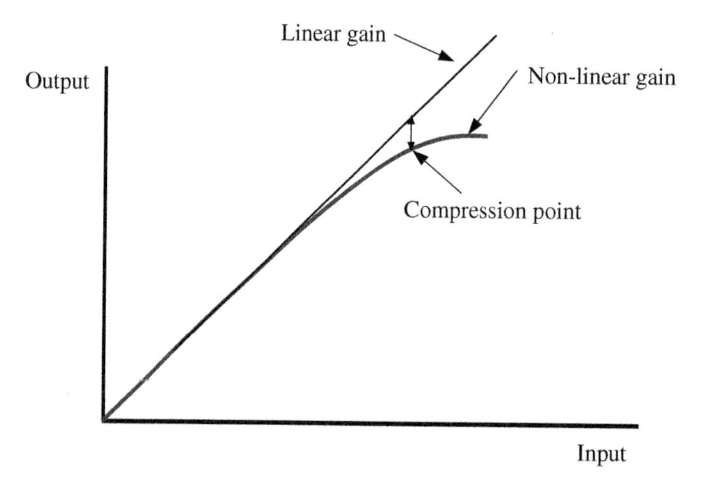

Figure 5.6 Illustration of the PA nonlinearity. In order to minimize power consumption and operate at the highest efficiency, a PA is ideally operated close to its saturation point. The nonlinearity associated with amplifier saturation can then lead to harmonic distortion, intermodulation distortion and spectral regrowth, cross-modulation, SNR degradation, and modulation inaccuracy.

The impairment caused by PAs is related to the PAPR measurement. Discrete-time PAPR of mth OFDM symbol x_m is defined as [15]

$$PAPR_m = \max_{0\leq n\leq N-1} |x_m(n)|^2 / E\{|x_m(n)|^2\}, \tag{5.4}$$

where $E\{\cdot\}$ represents the expected value operation. Although the PAPR is very large for OFDM, high magnitude peaks occur relatively rarely and most of the transmitted power is concentrated in signals of low amplitude, e.g. maximum PAPR for an OFDM system with 32 carriers and QPSK modulation will be observed statistically only once in 3.7 million years if the duration of an OFDM symbol is 100 μs [16]. Therefore, the statistical distribution of the PAPR should be taken into account.

One of the classical and most often used nonlinear model of PA is Saleh's model [17]. It is a simple nonlinear model without memory and it is defined by only two parameters α and β. The model uses two functions to model the AM-to-PM and AM-to-AM characteristics of nonlinear amplifiers and for each function, the parameters are different. For a specific PA and for each function, the two parameters (α and β) can be extracted using a least squares (LS) approximation to minimize the relative error between the measurements of the target PA and the values predicted by the model. The amplitude and phase equations determining the distortions can be written as:

$$A(r) = \frac{\alpha_A r}{1 + \beta_A r^2}, \tag{5.5}$$

and

$$P(r) = \frac{\alpha_P r^2}{1 + \beta_P r^2}, \tag{5.6}$$

where $r(t)$ represents the envelope of the input signal $S_{in}(t)$

$$S_{in}(t) = r(t)\cos(2\pi f_c t + \psi(t)), \tag{5.7}$$

f_c is the carrier frequency and $\psi(t)$ is the phase of the input signal. The output of PA, $S_{out}(t)$, can be written as:

$$S_{out} = A\left(r\left(t\right)\right)\cos\left(2\pi f_c t + \psi(t) + P\left(r\left(t\right)\right)\right). \tag{5.8}$$

The following Matlab code provides the responses of the Saleh's PA model with respect to input voltage level with the α and β parameters corresponding to a traveling-wave tube (TWT) amplifier discussed in [17]. The corresponding plots are given in Figure 5.7.

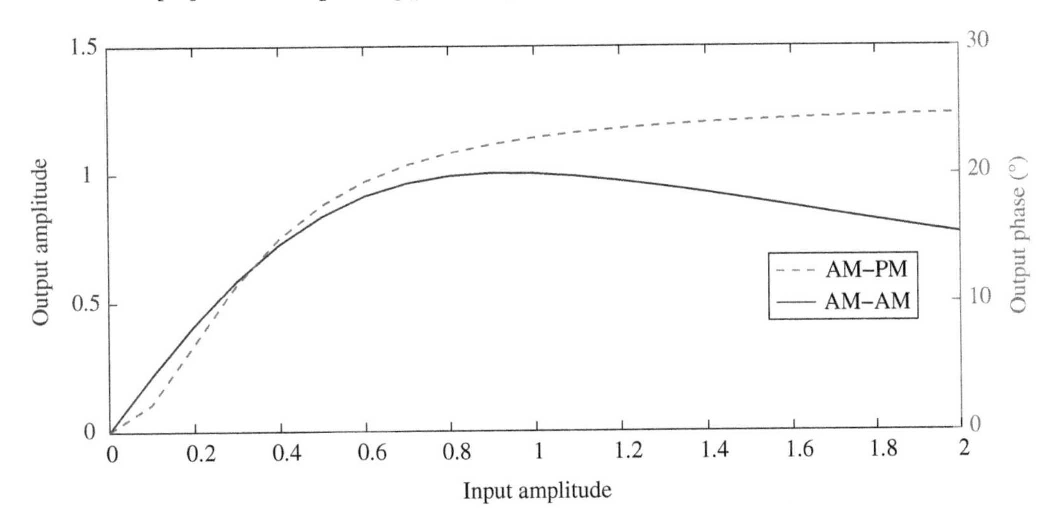

Figure 5.7 AM–AM and AM–PM response of TWT amplifier.

```
1  alpha_A=2.1587;
2  beta_A= 1.1517;
3  alpha_P=4.033;
4  beta_P=9.1040;
5  ind=0;
6  for kk=0:.1:2
7  frame=kk;
8  ind=ind+1;
9  amp_f(ind)=(alpha_A*abs(frame))./(1+beta_A*abs(frame).^2);
10 angle_f(ind)=(alpha_P*abs(frame).^2)./(1+beta_P*abs(frame).^2);
11 end
12 [a,b,c]=plotyy(0:.1:2,amp_f,0:.1:2,angle_f*180/pi)
13 xlabel('Input Amplitude')
14 ylabel('Output Amplitude')
15 set(get(a(2),'Ylabel'),'String','Output Phase (Degree)')
```

In the following Matlab code, a simple OFDM waveform is generated to test the impact of PA on the modulated signal. The OFDM waveform input parameters, like the fast Fourier transform (FFT) size, the modulation type, input power, etc., can be changed to see the impact with various design parameters. The generated baseband signal is passed through the PA. The output of PA can be inspected to see the impact of PA on the waveform. In Figure 5.8, the effect of PA nonlinearity is shown in frequency domain, time domain, and modulation domain. In the spectral plot, you can see the spectral leakage of the input and output signal. If we normalize the adjacent channel power

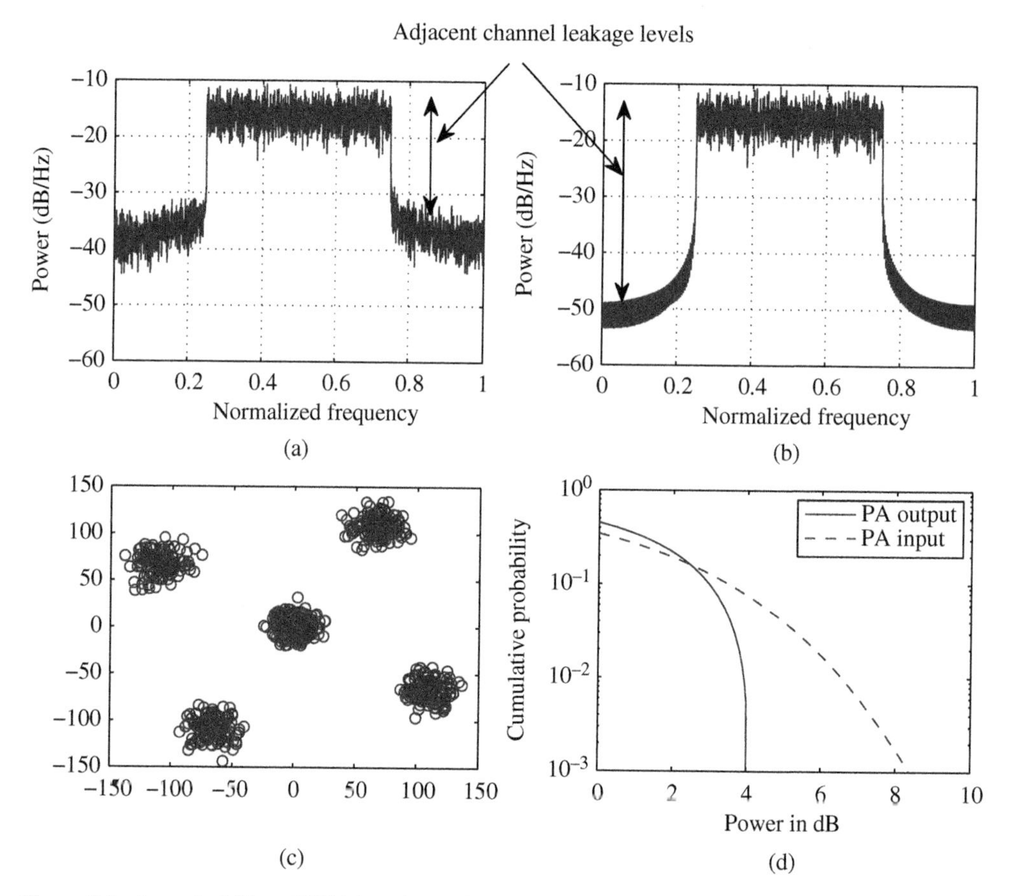

Figure 5.8 Impact of PA on OFDM-based waveform. (a) Impact of PA on spectrum. (b) Spectrum of input signal. (c) Impact of PA on in-band interference. (d) CCDF.

with in-band power, you can see that the PA output has much larger adjacent channel power. In the constellation diagram, the impact of PA on the symbols can be seen clearly. Instead of seeing a clean constellation, due to the in-band interference caused by PA, the constellation looks very noisy (the points around the origin is due to the unused carrier which needs to be ignored at this point). Finally, in complementary cumulative distribution function (CCDF) graph, you can see the power saturation effect of the output power when compared to input power. If the PA were to operate in linear region (without saturation), similar CCDF plots could have been observed. Due to the saturation of the output power, it can be seen from the CCDF curve clearly that the output power does not exceed 4 dB above the average power. On the other hand, in the input power CCDF curve, the power levels exceeding even 8 dB above the average power can be observed.

```
1  % Effect of PA in OFDM signal
2  % Observe the effect of input power on constellation and spectrum
3  MM=1e3; % number of OFDM symbols to run
4  frame=[];
5  N=1024; % FFT size
6  CP_size=1/4;
7  for ii=1:MM   % OFDM signal generation (more details in a later chapter)
8       symb= qammod( floor(rand(1,N)*4)  ,4);
9       used_subc_indices=256:1:N-256;
10      to_ifft=zeros(1,N);
11      to_ifft(used_subc_indices)=symb(used_subc_indices);
12      time=ifft(to_ifft);
13      time_cp=[time(end-N*CP_size+1:end) time];
14      frame=[frame time_cp];
15 end
16 frame=frame*10;
17 % can change input power before feeding to PA
18 % PA parameters - Saleh's nonlinear PA model
19 alpha_A=11.534;
20 beta_A= 1.6242;
21 alpha_P=11.431;
22 beta_P=39.071;
23 amp_f=(alpha_A*abs(frame))./(1+beta_A*abs(frame).^2);
24 angle_f=(alpha_P*abs(frame).^2)./(1+beta_P*abs(frame).^2);
25 f_dist=(amp_f./abs(frame)).*frame.*exp(j*angle_f);
26 subplot(2,2,1)
27 [Pxx1,F]=pwelch(f_dist,[],[],2048,1,'twosided');
28 Pxx1=Pxx1./ sqrt(sum(abs(Pxx1).^2));
29 plot(F,10*log10(Pxx1))
30 xlabel('Normalized Frequency')
31 ylabel('Power (dB/Hz)')
32 title('(a) Impact of PA on spectrum')
33 subplot(2,2,2)
34 [Pxx2,F]=pwelch(frame,[],[],2048,1,'twosided');
35 Pxx2=Pxx2./ sqrt(sum(abs(Pxx2).^2));
36 plot(F,10*log10(Pxx2))
37 xlabel('Normalized Frequency')
38 ylabel('Power (dB/Hz)')
39 title('(b) Spectrum of Input Signal')
40 subplot(2,2,3)
41 rx=f_dist(N*CP_size+1:N*CP_size+N);
42 rx_f=fft(rx);
43 plot(real(rx_f),imag(rx_f),'o') %
44 title('(c) Impact of PA on in-band interference')
45 subplot(2,2,4)
46 [x,y]=hist(abs(f_dist).^2/mean(abs(f_dist).^2),100);
47 semilogy(10*log10(y),1-cumsum(x/sum(x)))
48 hold
49 [x,y]=hist(abs(frame).^2/mean(abs(frame).^2),100);
50 semilogy(10*log10(y),1-cumsum(x/sum(x)),'--')
51 xlabel('power in dB')
52 ylabel('cumulative probability')
53 legend('PA output','PA input')
54 title('(d) CCDF')
```

5.4 Phase Noise and Time Jitter

Phase noise is introduced by local oscillator in any transceiver and can be interpreted as a parasitic phase modulation in the oscillator's signal. The phase noise is popularly modeled as Wiener–Levy (random walk) process (from one symbol to another) [18–20], and the process can be given as:

$$\phi_m = \phi_{m-1} + w_m, \tag{5.9}$$

where w_m is referred as the step size of the walk and modeled with zero mean Gaussian random variable with a variance of $2\pi\beta T$, where βT represents the relative (with respect to symbol rate) two-sided 3-dB bandwidth (i.e. the frequency spacing between 3 dB points of its Lorentzian power spectral density [18–21]). The single-sided power spectrum of the phase noise which can be modeled with a first-order low-pass filter can be given as:

$$S(f) = \frac{\frac{2}{\pi\beta}}{1 + \frac{f^2}{\beta^2}}. \tag{5.10}$$

The phase noise can be assumed to be constant within a symbol duration if the oscillators bandwidth is much smaller than the symbol rate (i.e. $T << 1$).

For the sake of simplicity, if we assume the channel is flat (with a gain of one and phase of zero) and the signal is only effected by phase noise $\phi(n)$ at the receiver, the received time domain signal can be written as:

$$y_m(n) = x_m(n)e^{j\phi_m(n)}. \tag{5.11}$$

The following Matlab code generates phase noise and integrates it to the signal.

```
1  % Simulation of phase noise
2  % observe the effect various phase noise values in eye diagram
3  N=1e3; % Number of symbols
4  sps=8;
5  RMS_deg=5; % desired RMS phase error in degrees.
6  fil1=rcosine(1,sps,'normal',0.3);
7  M=4;
8  symb= qammod(floor(rand(1,N)*M),M);
9  u_frame=upsample(symb, sps);
10 sgnl_tx = conv(fil1,u_frame);
11
12 Wn=1/(5000);
13 % normalized (wrt half sampling rate fs/2) 3 dB bandwidth of filter.
14 [B,A] = butter(1, Wn);
15 % One pole Butterworth filter. Gives 20 dB/dec phase-noise roll-off.
16
17 N=length(sgnl_tx);
18 noise=randn(1,N);
19 % generate AWGN noise samples with variance of one
20 dummy= filter(B,A,[1 zeros([1,N])]);
21 norm_fac=sqrt(sum(abs(dummy).^2));
22 % To make sure that the filter has gain of one
23 Pn=filter(B,A,noise);
24 Pn=Pn/norm_fac;
25 Phase_noise=exp(j*Pn*RMS_deg*(pi/180));
26
27 % Apply the phase noise to the input signal
28 Noisy_signal = sgnl_tx .* Phase_noise;
29 eyediagram(Noisy_signal,sps*2)
```

The power spectrum of an ideal oscillator would have a sharp impulse-like shape at the desired frequency. However, when there is phase noise in the system, the spectrum will be broader and has Lorentzian shape [21]. The ratio of the peak carrier signal to the noise at a specific offset to the carrier characterizes the phase noise (see Figure 5.9). The performance of local oscillator is

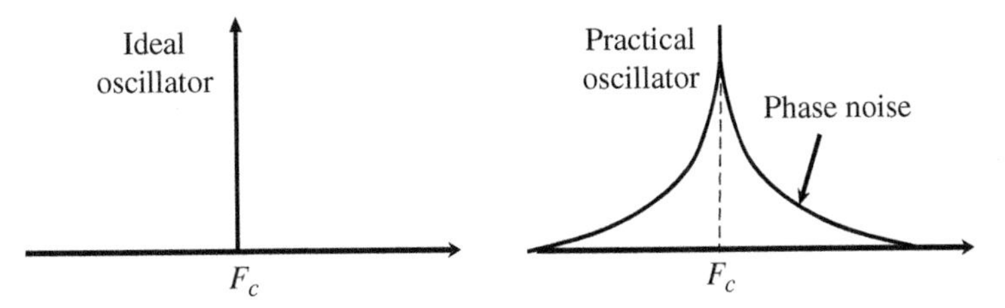

Figure 5.9 Ideal and practical oscillator's spectrum and illustration of phase noise effect in the practical oscillator's spectrum.

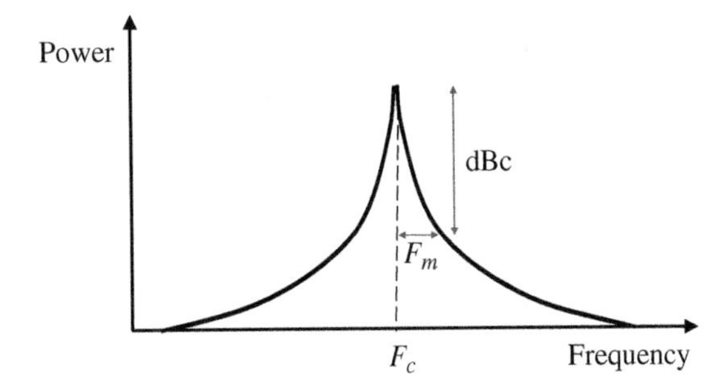

Figure 5.10 Characterizing the phase noise response at a specific frequency around the central frequency. Double-sideband (DSB) phase noise plot of the source is shown in the figure.

usually characterized by measuring the sideband noise of the output signal center frequency. The phase noise is typically specified in terms of dBc/Hz at a specific frequency offset from the carrier frequency as shown in Figure 5.10. The single-sideband phase noise expression can be written as:

$$L(F_m) = 10 * \log_{10} \frac{P_s(f_c + f_m)_{1\text{Hz}}}{P_c(f_c)} \text{ dBc/Hz}, \tag{5.12}$$

where f_m is the sideband frequency and f_c is the frequency of the carrier signal. As mentioned above, the expression is normalized to a 1 Hz bandwidth, where dBc indicates the power level referenced to the carrier power level in dBm.

Effect of random phase noise on the constellation diagram for single-carrier systems is shown in Figure 5.11. If the phase noise is constant over a block of data (or burst), then the effect will be a phase rotation of the symbols. Chanel estimation can estimate this phase rotation and it can be corrected easily by derotating the received samples by the estimated phase offset. However, if the phase noise is randomly changing very fast where the channel estimators cannot track the variation, then it will cause an arc around the true constellation points, leading to modulation errors [22, 23].

Phase noise in OFDM systems is very critical. If it is a constant phase noise over an OFDM symbol, then the problem is easy to handle as it only causes a fixed phase rotation on the symbols. By inserting pilot carriers, this constant phase rotation can be estimated and corrected. On the other hand, if the phase noise is randomly changing over an OFDM symbol, then it will cause spreading the carriers, leading to loss of orthogonality and hence inter-carrier interference (ICI). ICI will cause modulation errors. The impact of phase noise on the OFDM carriers is shown in Figure 5.12.

Jitter is defined as the short-term variations of a digital signal's significant instants from their ideal positions in time. For example, in digital communication receiver, the significant instants could be the optimum sampling instants. The deviation of the sampling timing due to jitter causes degradation in error vector magnitude (EVM) and bit-error-rate (BER) performance of the system. Eye diagram is an excellent way of visualizing the time jitter. Jitter can be expressed in absolute time or normalized to a unit interval (UI), like the symbol duration. It can be quantified with its average, peak, and peak-to-peak values.

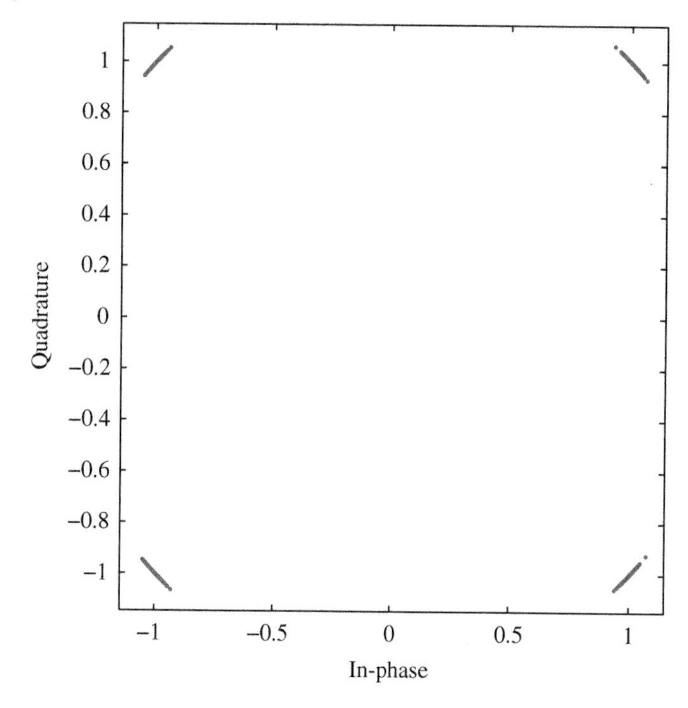

Figure 5.11 Impact of random phase noise on the constellation diagram of QPSK-modulated single-carrier signal.

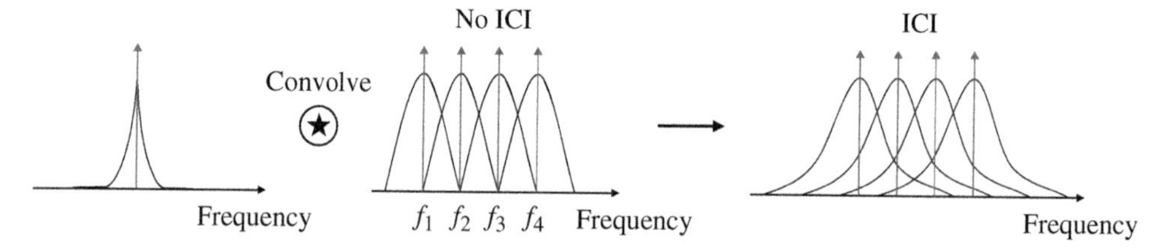

Figure 5.12 Simplified illustration of the impact of random phase noise on the OFDM carriers.

Jitter can be divided into two fundamental types: random jitter (which is often caused by thermal noise) and deterministic jitter (which is caused by process or component interactions in the system). Random jitter is unpredictable and has a Gaussian probability density function. On the other hand, deterministic jitter is predictable and bounded.

Jitter and phase noise are different interpretations of the same underlying phenomenon, one is the time-domain and the other is the frequency-domain interpretation. Jitter can be characterized in the time domain by its distribution histogram, and phase noise can be characterized by its power spectrum.

5.5 Frequency Offset

Frequency offset is caused by the local oscillator mismatches between the transmitter and the receiver and wrong carrier synchronization. There are no two perfect oscillators in the transmitter and receiver which will not cause any frequency offset. Therefore, in each wireless communication system, there will be a frequency offset to deal with. The frequency offset to a baseband model of a signal $x(n)$ can be introduced as:

$$\hat{x}(n) = x(n)e^{\frac{j2\pi f_o n}{f_s}}, \tag{5.13}$$

where f_s is the sampling frequency, and f_o is the frequency offset value.

Frequency offset causes a phase rotation in the constellation diagram in single-carrier systems. The amount of rotation introduced to each constellation point is dependent on the frequency offset value and the sample index, and it keeps growing. Due to the growing phase rotations, the constellation diagram forms a circular shape. When the signal is downsampled, with a good SNR value and in a flat-fading channel, the constellation diagram can be interpreted as rings. The amount of phase rotation between samples of the same distance is the same since the frequency offset value is constant for the entire received signal. Frequency offset is a very critical factor in OFDM system design. It results in ICI and destroys the orthogonality of subcarriers [24, 25].

5.6 ADC/DAC Impairments

Analog-to-digital converter (ADC) and DAC constitute the interface (boundary) between the analog and digital world. A typical ADC consists of a sampler followed by a quantizer. Sampling rate is one of the important features of ADCs. One of the limitations toward ideal SDR is the sampling rate requirements as the bandwidth of the signal before ADC must be smaller than half of the sampling rate according to the Nyquist theorem. If we are working with a signal-containing frequency components from DC to some maximum frequency, f_{max}, then the sampling rate, f_s, must be equal to or greater than twice f_{max}, i.e. $f_s \geq 2\, f_{max}$. Otherwise, a phenomenon called aliasing will occur. Often, the signals might contain components at high frequencies that are not essential for the integrity of it. Therefore, to avoid aliasing, the signal is low-pass filtered (sometimes also referred as anti-aliasing filtering) to remove high-frequency components and hence band-limiting the signal before sampling.

Once the analog signal is properly sampled, the quantizer converts these discrete samples (with continuous amplitude levels) into bits with a word length. The word length determines the resolution of ADC and also determines the quantization error. Each additional bit introduces 6 dB more dynamic range to the receiver (note that a factor of two means 6 dB for voltage quantity, 20 $\log_{10}(2)$ dB). In addition to the quantization noise, distortions due to the static and dynamic nonlinearity features of ADCs also affect the performance of ADC. Signal-to-noise-and-distortion (SINAD) is the ratio of the root-mean-square (rms) signal amplitude to the mean value of the root-sum-square of all other spectral components, including harmonics, but excluding DC. SINAD is a good indication of the overall dynamic performance of an ADC, because it includes all components which make up noise and distortion. Due to the distortions, the effective number of bits (ENOB) that specifies the dynamic performance of an ADC at a specific frequency, amplitude, and sampling rate will be different than what is expected from an ideal ADC that only includes the quantization noise.

A generalized DAC structure consists of DAC register, resistor string and followed by output buffer amplifier blocks is given in [26]. We refer to [26, 27] for the basic operation of DAC. In a nutshell, a typical DAC converts the bits into a sequence of impulses that are then processed by a reconstruction filter using some form of interpolation to fill in data between the impulses. Similar to ADC, the main characteristics of DACs are resolution, maximum sampling rate, monotonicity, dynamic range, and phase distortion. If we consider a system that converts an N-bit digital number into an analog signal that will range in value from V_{min} to V_{max}, the resolution ϑ can be given as

$$\vartheta = \frac{V_{max} - V_{min}}{2^N - 1}, \tag{5.14}$$

where 2^N is the number of levels in the system. Increasing the number of bits N used in each sample improves the resolution.

5.7 Thermal Noise

Thermal is common to most electrical components such as resistors, capacitors, inductors, and transistors. It is caused by the thermal excitement of electrons in electrical conductors and the noise power depends on the temperature and bandwidth, $P = kTB$, where k is Boltzmanzz's constant $1.38 * 10^{-23}$ Joules/Kelvin. At a room temperature (68°F), the per hertz power will be -173.93 dBm/Hz. For 20 MHz of BW, the total noise BW will be $P = -174 + 10 * \log_{10}(20 * 10^6) \cong -101$ dBm. The thermal noise distribution is Gaussian in nature with a frequency-independent spectrum (often referred as white noise).

5.8 RF Impairments and Interference

Wireless communication systems are known to be limited by interference. There are several reasons for interference. Wireless channel is one of the main sources of interference, which will be discussed in detail in a later chapter. Another significant source of interference is RF impairments. The interference caused by RF impairments can be linear or nonlinear. Also, the impairments can cause self-interference, e.g. in the form of inter-symbol interference (ISI) and ICI, multiuser interference in the form of adjacent channel interference (ACI). The interference can also depend on the type of waveform used in communication. In the following subsections, some of these interferences caused by RF impairments are discussed.

5.8.1 Harmonics and Intermodulation Products

Practical radio front-ends consist of several nonlinear components. Nonlinearity in RF and IF circuits leads to two undesirable outcomes: harmonics and intermodulation distortion. If a sinusoidal signal with a frequency f_1 is applied to a nonlinear system, the output of the system includes frequency components at the integer multiples of the input frequency which are called harmonics, i.e. $2f_1, 3f_1, 4f_1, 5f_1$, etc., as illustrated in Figure 5.13. On the other hand, when two signals with different

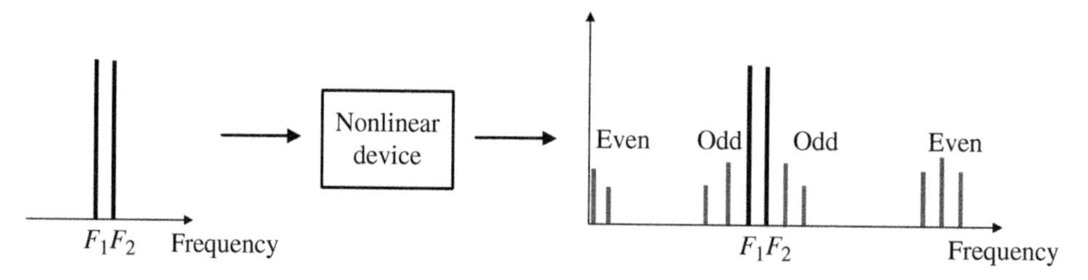

Figure 5.13 Intermodulation products.

frequencies are applied to a nonlinear system, then the output of the system not only includes the harmonics of the input frequencies but also includes components at frequencies to the sums and differences of integer multiples of the input frequencies ($\pm mf_1 \mp nf_2$), i.e. $f_1 + f_2$, $2f_1 - f_2$, $2f_2 - f_1$, $3f_1 - 2f_2$, etc. To define the order, we add the harmonic multiplying constants of the two frequencies producing the intermodulation product ($IMD = \pm mf_1 \mp nf_2$, the order is $m + n$). For example, $f_1 - f_2$ is second order, $2f_1 - f_2$ is third order, $3f_1 - 2f_2$ is fifth order, etc. Intermodulation distortion is often characterized by the third-order intermodulation and the performance is measured as third-order intercept point, IP_3. IP_3 is defined as the intersection between the linear gain curve of the input signal and the third-order product curve of the input signal, as shown in Figure 5.14. Note that the slope of the fundamental tone and third-order intermodulation are not the same. The third-order intermodulation has three times bigger slope (in the linear region). If we extrapolated the lines from the linear region, the point at which they intersect will be the IP3.

The interference to a signal due to third-order intermodulation of two nearby interferers is very common; therefore, third-order intermodulation and its characterization is very critical. IP3 determines the maximum power level that the front-end can accept at the input before the third-order products become problematic (critically interfere with the desired signal).

The following Matlab script explains the interference caused by the harmonics and intermodulation products clearly. Let's assume a nonlinear device defined with the following function

$$y(t) = 7x(t) - 0.095x^3(t). \tag{5.15}$$

We feed the input of this device with a two tone bandpass signal as

$$x(t) = \cos(2 * \pi f_1 t) + 0.63\ \cos(2\pi f_2 t). \tag{5.16}$$

By increasing the input power of the signal, third-order harmonics and third-order intermodulations can be seen as shown in Figure 5.15. Figure 5.15b is the spectrum of the input signal with two

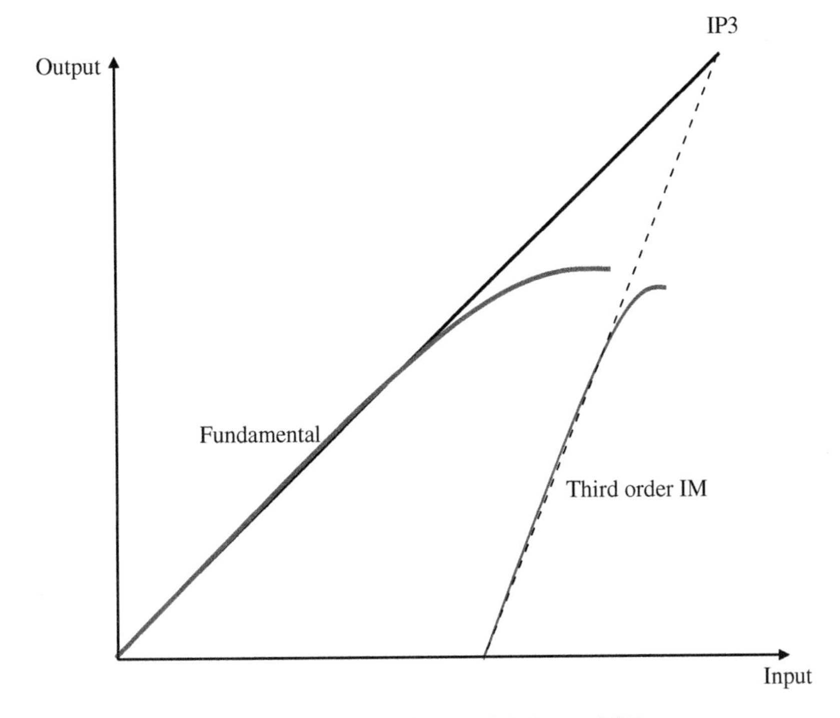

Figure 5.14 Illustration of third order intermodulation and IP3.

```
1  % Intermodulation products
2  f1=1000;
3  f2=900;
4  time=[0:1/8000:150];
5  x=0:0.01:6;
6  y=7*x-0.095*x.^3;
7  subplot(3,2,1)
8  plot(x,y)
9  xlabel('Input')
10 ylabel('Output')
11 x=cos(2*pi*f1*time)+0.63*cos(2*pi*f2*time);
12 xg=x*0.1;
13 y=7*xg-0.095*xg.^3;
14 subplot(3,2,2)
15 pwelch(x)
16 subplot(3,2,3)
17 pwelch(y)
18 xg=x*0.5;
19 y=7*xg-0.095*xg.^3;
20 subplot(3,2,4)
21 pwelch(y)
22 xg=x*1.5;
23 y=7*xg-0.095*xg.^3;
24 subplot(3,2,5)
25 pwelch(y)
26 xg=x*15;
27 y=7*xg-0.095*xg.^3;
28 subplot(3,2,6)
29 pwelch(y)
```

tones. Figure 5.15c through f are the outputs of the nonlinear device with the input power increasing. As can be seen, as the input power increases, the harmonics and intermodulation products increase as well.

5.8.2 Multiple Access Interference

In wireless communication, spectrum is a very valuable resource and needs to be shared among multiple users needing communication. Signal separation could be achieved by exploring the property of orthogonality in at least one of the orthogonal domains in electrospace: frequency, time, code, space, modulation, etc., [28–30]. In the past, **spectrum** using frequency division multiple accessing (FDMA), **time** using time division multiple accessing (TDMA), **code** using code division multiple accessing (CDMA), **angle** using space division multiple accessing (SDMA), **modulation** using orthogonal frequency division multiple accessing (OFDMA), and **geographical spacing (re-use)** using cellular type of structure are used as orthogonal dimensions of electrospace to provide multiple users the ability to access the spectrum. For example, in FDMA, the spectrum is splitted into non-overlapping bands and users are assigned to these bands to achieve multiple accessing. Similarly, reuse of frequencies in different geographical locations allows multiple users to use the same spectrum over and over without interfering with each other.

In a perfect world, at least in one of the dimensions of the electrospace, orthogonality is satisfied. Hence, interference will not be an issue. However, due to the practical channel and impairment conditions discussed in previous sections, maintaining the orthogonality is very difficult and sometimes there is a trade of in maintaining the orthogonality with spectral efficiency [31, 32]. Guaranteeing the orthogonality for all types of channel and impairment conditions is not always a good idea as it reduces the spectral efficiency significantly. For example, in FDMA, by allowing a significantly large spacing between adjacent channels, we can control the adjacent channel interference (ACI). But, this comes at the expense of allocating more spectrum to each user. By allowing

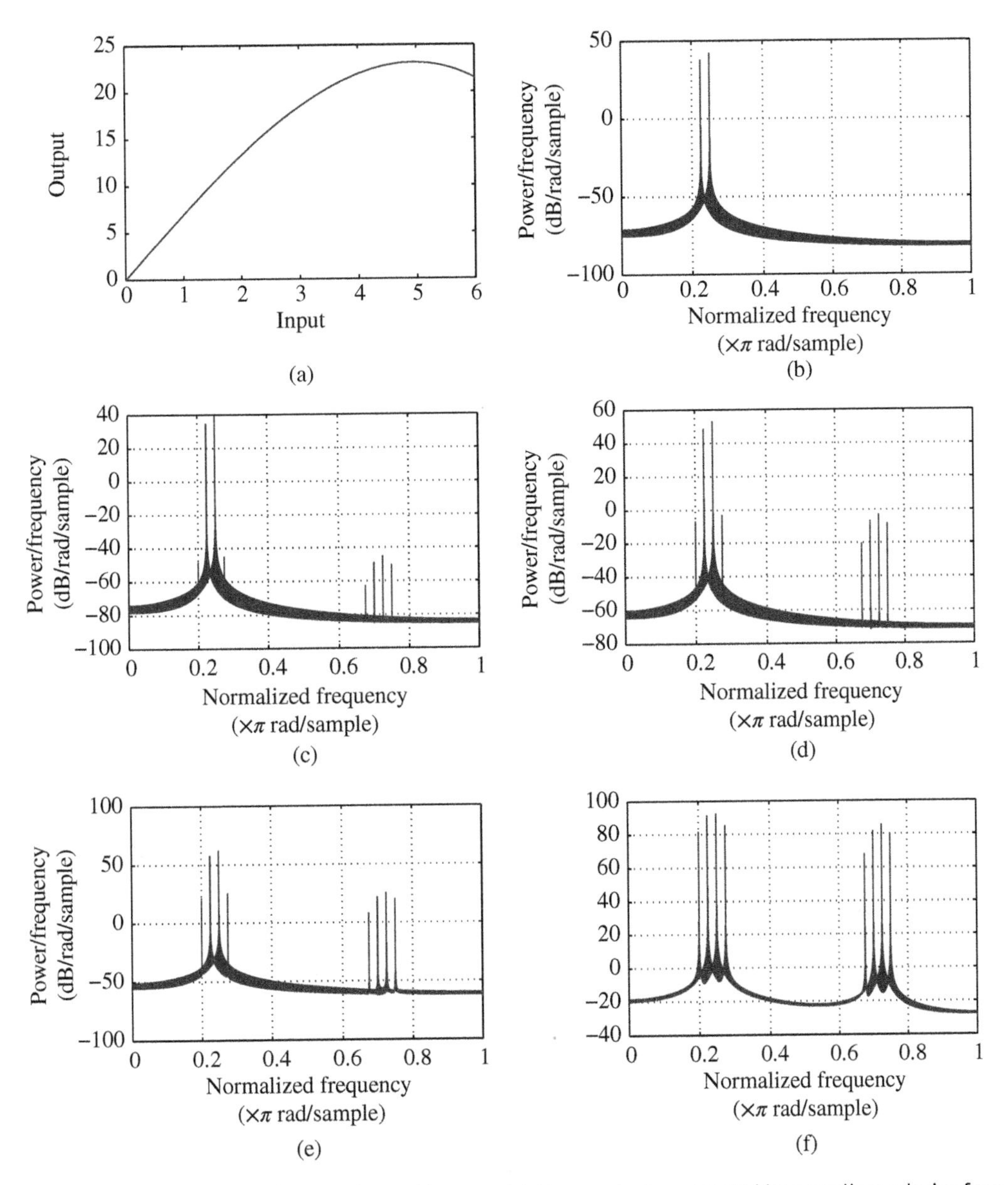

Figure 5.15 Illustration of harmonics and intermodulation products generated by a nonlinear device for a two-tone input signal.

some spectral overlapping, spectral efficiency can be increased significantly. However, this will introduce ACI. Also, spectral regrowth (adjacent channel regrowth) due to the PA nonlinearity causes significant amount of ACI if sufficient guards are not allocated in between the adjacent channels. In general, any RF nonlinearity can potentially cause ACI due to the difficulty of band-limiting the transmitted signal and leakage of the side lobes to the adjacent channels.

Understanding the relation between RF impairments, waveform, and multiple accessing is critical for the design of proper radio-access technologies [33, 34]. Within this context, single-carrier and multi-carrier waveform designs have different effects on the RF impairments.

In Nyquist-shaped overlapped orthogonal systems, like raised cosine pulse-shaped single-carrier system or Sinc-shaped OFDM systems, the orthogonality in the overlapped domain can be lost easily with any shift or dispersion in that domain. For example, frequency offset and phase noise impairments can destroy the orthogonality in OFDM leading to ICI. Similarly, timing jitter and sample clock error can destroy the orthogonality in raised cosine pulse-shaped single-carrier systems, causing ISI. Both ISI and ICI can lead to self-interference and/or multi-user interference depending on the type of used multiple accessing scheme. I/Q impairments as discussed earlier can also cause both self- and multiuser interference again depending on the multiple accessing scheme. Further details of the impairments on various multiple accessing schemes and waveforms will be discussed in more detail in later chapters.

5.9 Conclusions

Understanding RF impairments is important not only for accurate evaluation of the system performance but also in developing proper techniques to minimize their effects. In order to avoid complication and risks during the deployment stage, it is necessary to take RF impairments into account during the air interface design stage. The impact of certain imperfection depends on the applied digital transmission technology. For example, while single-carrier systems are more robust to some impairments, multi-carrier communication can significantly suffer from these (or the other way around). Even though the basic models of some of the common impairments are discussed in this chapter, it is important to investigate these in more detail for each specific radio-access technology.

References

1 B. Razavi, *RF Microelectronics*. Prentice Hall PTR, Upper Saddle River, NJ, 1998.

2 K. Chang, *RF and Microwave Wireless Systems*. Wiley, New York, 2000.

3 A. F. Demir, B. Pekoz, S. Kose, and H. Arslan, "Innovative telecommunications training through flexible radio platforms," *IEEE Commun. Mag.*, vol. 57, no. 11, pp. 27–33, 2019.

4 P. Rykaczewski, D. Pienkowski, R. Circa, and B. Steinke, "Signal path optimization in software defined radio systems," *IEEE Trans. Microwave Theory Techn.*, vol. 53, no. 3, pp. 1056–1064, 2005.

5 A. Scott and R. Frobenius, *RF Measurements for Cellular Phones and Wireless Data Systems*. Wiley, Inc., Hoboken, NJ, 2008.

6 M. Karabacak, A. H. Mohammed, M. K. Özdemir, and H. Arslan, "RF circuit implementation of a real-time frequency spread emulator," *IEEE Trans. Instrum. Meas.*, vol. 67, no. 1, pp. 241–243, 2017.

7 H. Arslan, "IQ gain imbalance measurement for OFDM based wireless communication systems," in *MILCOM 2006 – 2006 IEEE Military Communications Conference*, Washington, DC, 23–25 Oct. 2006, pp. 1–5.

8 A. Tusha, S. Doğan, and H. Arslan, "Performance analysis of frequency domain IM schemes under CFO and IQ imbalance," in *2019 IEEE 30th Annual International Symposium on Personal, Indoor and Mobile Radio Communications (PIMRC)*. IEEE, Istanbul, 8–11 Sept. 2019, pp. 1–5.

9 A. Tarighat, R. Bagheri, and A. H. Sayed, "Compensation schemes and performance analysis of IQ imbalances in OFDM receivers," *IEEE Trans. Signal Process.*, vol. 53, no. 8, pp. 3257–3268, 2005.

10 J. Li and M. Kavehrad, "OFDM-CDMA systems with nonlinear power amplifier," in *Proc. IEEE Wireless Commun. and Networking Conf.*, vol. 3, New Orleans, LA, 21–24 Sept. 1999, pp. 1167–1171.

11 H. J. R. J. Park, S. R. Park, and K. H. Koo, "Power amplifier back-off analysis with AM-to-PM for millimeter-wave OFDM wireless LAN," in *Proc. IEEE Radio and Wireless Conference*, Waltham, MA, 19–22 Aug. 2001, pp. 189–192.

12 R. Koch, "Linearization: reducing distortion in power amplifiers," *IEEE Microwave*, vol. 2, pp. 37–49, Dec. 2001.

13 S. Chang and E. J. Powers, "A simplified predistorter for compensation of nonlinear distortion in OFDM systems," in *Proc. IEEE Global Telecommunications Conf.*, vol. 5, San Antonio, TX, 25–29 Nov. 2001, pp. 3080–3084.

14 J. Y. Hassani and M. Kamarei, "A flexible method of LUT indexing in digital predistortion linearization of RF power amplifiers," in *Proc. IEEE International Symposium on Circuits and Systems*, vol. 1, Sydney, 6–9 May 2001, pp. 53–56.

15 S. Muller and J. Huber, "A comparison of peak power reduction schemes for OFDM," *GLOBE-COM 97. IEEE Global Telecommunications Conference*, vol. 1, Phoenix, AZ, 3–8 Nov. 1997, pp. 1–5.

16 H. Ochiai and H. Imai, "On the distribution of the peak-to-average power ratio in OFDM signals," *IEEE Trans. Commun.*, vol. 49, no. 2, Feb. 2001, pp. 282–289.

17 A. Saleh, "Frequency-independent and frequency-dependent nonlinear models of TWT amplifiers," *IEEE Trans. Commun.*, vol. COM-29, pp. 1715–1720, Nov. 1981.

18 S. Wu and Y. Bar-Ness, "OFDM systems in the presence of phase noise: consequences and solutions," *IEEE Trans. Commun.*, vol. 52, no. 11, pp. 1988–1996, Nov. 2004.

19 T. Lee and A. Hajimiri, "Oscillator phase noise: a tutorial," *IEEE Journal of Solid-State Circuits*, vol. 35, no. 3, pp. 326–336, Mar. 2000.

20 A. G. Armada, "Understanding the effects of phase noise in orthogonal frequency division multiplexing (OFDM)," *IEEE Trans. Broadcast.*, vol. 47, no. 2, pp. 153–159, June 2002.

21 J. Roychowdhury, A. Demir, and A. Mehrotra, "Phase noise in oscillators: a unifying theory and numerical methods for characterization," *IEEE Trans. Circ. Syst.*, vol. 47, pp. 655–674, May 2000.

22 H. Arslan and D. Singh, "The role of channel frequency response estimation in the measurement of RF impairments in OFDM systems," in *2006 67th ARFTG Conference*, San Francisco, CA, 16 June 2006, pp. 241–245.

23 H. Arslan, "Channel frequency response estimation under the effect of RF impairements in OFDM based wireless systems," in *IEEE Vehicular Technology Conference*, Montreal, QC, 25–28 Sept. 2006, pp. 1–5.

24 P. H. Moose, "A technique for orthogonal frequency division multiplexing frequency offset correction," *IEEE Trans. Commun.*, vol. 42, no. 10, pp. 2908–2914, Oct. 1994.

25 K. Sathananthan and C. Tellambura, "Performance analysis of an OFDM system with carrier frequency offset and phase noise," *IEEE 54th Vehicular Technology Conference. VTC Fall 2001. Proceedings (Cat. No.01CH37211)*, vol. 4, Atlantic City, NJ, 7–11 Oct. 2001, pp. 2329–2332.

26 Texas Instruments, "12-BIT, quad, ultralow glitch, voltage output digital-to-analog converter." *12-Bit, Quad Channel, Ultra-Low Glitch, Voltage Output Digital-to-Analog Converter with 2.5V, 2ppm/°C Internal Reference*, Technical note, Feb. 2008 [Online]. Available: http://focus.ti.com/

27 D. Comer, "A monolithic 12-bit DAC," *IEEE Trans. Circ. Syst.*, vol. 25, no. 7, pp. 504–509, Jul. 1978.

28 A. Gorcin and H. Arslan,"An OFDM signal identification method for wireless communications systems," *IEEE Trans. Vehicul. Technol.*, vol. 64, no. 12, pp. 5688–5700, 2015.

29 ——, "Signal identification for adaptive spectrum hyperspace access in wireless communications systems," *IEEE Commun. Mag.*, vol. 52, no. 10, pp. 134–145, 2014.

30 A. M. Hisham and H. Arslan, "Multidimensional signal analysis and measurements for cognitive radio systems," in *2008 IEEE Radio and Wireless Symposium*, Orlando, FL, 22–24 Jan. 2008, pp. 639–642.

31 S. Guzelgoz and H. Arslan, "Modeling, simulation, testing, and measurements of wireless communication systems: a laboratory based approach," in *2009 IEEE 10th Annual Wireless and Microwave Technology Conference*, Clearwater, FL, 20–21 Apr. 2009, pp. 1–5.

32 S. Ahmed and H. Arslan, "Cognitive intelligence in UAC channel parameter identification, measurement, estimation, and environment mapping," in *OCEANS 2009-EUROPE*, Bremen, 11–14 May 2009, pp. 1–14.

33 M. E. Sahin and H. Arslan, "MIMO-OFDMA measurements; reception, testing, and evaluation of wimax MIMO signals with a single channel receiver," *IEEE Trans. Instrum. Meas.*, vol. 58, no. 3, pp. 713–721, 2009.

34 M. Sahin, H. Arslan, and D. Singh, "Reception and measurement of MIMO-OFDM signals with a single receiver," in *2007 IEEE 66th Vehicular Technology Conference*, Baltimore, MD, 30 Sept.–3 Oct. 2007, pp. 666–670.

6

Digital Modulation and Pulse Shaping

Hüseyin Arslan

Department of Electrical Engineering, University of South Florida, Tampa, FL, USA
Department of Electrical and Electronics Engineering, Istanbul Medipol University, Istanbul, Turkey

Fundamental to all communication systems is the process of modulation, which maps the information to physical signals. In digital modulation, the information bits are mapped to a finite set of distinct physical signals. The process involves mapping the bits to constellations followed by pulse shaping, which are done in digital baseband. Carrier modulation is employed often in analog domain after digital to analog conversion. The choice of digital modulation and pulse-shaping techniques impacts the performance of communication systems. There is no solution which can be considered as the best modulation and pulse-shaping techniques for all situations. Depending on the channel conditions, radio frequency (RF) hardware cost and properties, allowed complexity, required data rate, available bandwidth, and desired levels of performance some will provide a better fit than others. In this chapter, some of these popularly used digital modulation and pulse-shaping techniques will be discussed.

6.1 Digital Modulation Basics

Through the recent decades, wireless communications have observed a tremendous shift from analog communications to digital communications. The trend to digital systems was motivated by achieving more information capacity, better quality communications, more secure transmission and reception of the information, and capability to interchange information with other digital data services. These advantages helped digital communication techniques replace the analog devices in today's world. This resulted in the analog communications, which are commonly implemented by means of amplitude modulation (AM) and frequency modulation (FM) techniques, typically being constrained to few areas such as radio and television broadcasting.

The term modulation is often used to imply the operation of mixing the information bearing baseband signal with an analog sinusoidal carrier to upconvert them into intermediate frequencies (IF) and/or radio frequencies (RF). This is for the purpose of sharing the available spectrum with multiple applications and users and is also commonly referred as modulation of the signal using a carrier (or carrier modulation) [1]. In digital modulation, the analog sinusoidal carrier ($A_c \cos(2\pi f_c t + \theta)$, where A_c is the amplitude, f_c is the frequency, and θ is the phase of the carrier) is modulated with a digital baseband signal, i.e. the source is digital and a finite number of distinct signals are used to represent the digital data. Similar to classical analog AM, FM, and PM, the amplitude, frequency, and phase of the carrier are modulated with the baseband signal (see Figure 6.1). Unlike the classical analog modulation, in digital modulation, the modulating symbols are chosen from a finite set. Therefore, shift keying term is used in digital modulations, like amplitude-shift

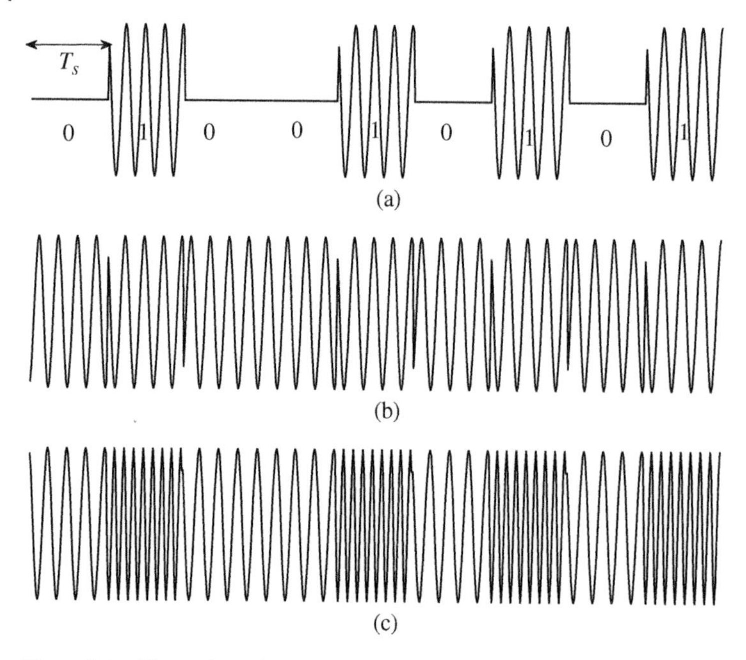

Figure 6.1 Illustration of carrier modulation options: on–off keying (special case of amplitude modulation), binary PSK, and binary FSK. Here, T represents the symbol duration. (a) ASK (special case OOK). (b) PSK. (c) FSK.

keying (ASK), frequency-shift keying (FSK), and phase-shift keying (PSK). Even though modulation operation described above requires carrier, in impulse radio (IR)-based ultra-wideband systems and other impulse radio communication systems' carrier is not needed to modulate the bits.

Digital communication requires *mapping* of the stream of bits into waveforms at the transmitter, and *demapping* is achieved at the receiver through various reception algorithms. These mapping and demapping operations are also commonly referred as *modulation* and *demodulation*. In a more formal expression, Proakis in his well-known digital communications book defines modulation as "the mapping of a sequence of binary digits into a set of corresponding waveforms" [2]. The mapped waveform can correspond to a single bit, or a sequence of bits (a symbol), and can change in amplitude, phase (delay), or frequency, or in their combination. Another common definition of modulation that implies a similar meaning is "the process where the message information is embedded into the radio carrier" [1].

The relation between bit and symbol is defined based on how the bits are mapped to symbols. Bit rate indicates the number of bits transmitted per second. On the other hand, symbol rate (sometimes also referred as baud rate) indicates the number of symbols transmitted per second. Depending on the mapping, symbol rate is less than or equal to the bit rate, more specifically, the symbol rate is the bit rate divided by the number of bits that represents each symbol. For example, if one bit is mapped to a symbol (like binary modulations, BPSK, BFSK), then the symbol rate is equivalent to the bit rate. If three bits are mapped to a symbol, like in 8PSK used in enhanced data rates for GSM evolution (EDGE), then the symbol rate is one-third of bit rate.

As discussed in earlier chapters, in-phase and quadrature-phase (I/Q) modulation is popularly used in communication systems. In I/Q modulation, the I and Q channels are modulated with two orthogonal quadrature carriers, 90° out of phase of each other. The I and Q channels are obtained by mapping the bits to I/Q plane, where in the I/Q plane, depending on the modulation type, discrete points represent each symbol.

When selecting a suitable modulation scheme, one needs to understand the requirements of the system as the choice of the modulation impacts the performance and complexity of the system significantly. There is not really a single perfect modulation that satisfies all the requirements. Therefore, a system designer needs to understand the trade-offs between various modulation options given the physical characteristics of the channel that the system needs to be operated, the target performance specifications, desired data rate, and hardware limitations. In an ideal world, one wishes to obtain the following characteristics from a modulation option:

- Low bit-error-rate (BER) at low signal-to-noise ratio (SNR) (high power efficiency)
- Occupies minimal bandwidth (high spectral efficiency)
- Performs well in multipath fading (robust to multipath effects like Doppler spread, delay spread, and fading)
- Performs well in time varying channels (symbol timing jitter)
- Low out of band radiation
- Low cost and easy to implement
- Constant or near-constant envelope (performs well under nonlinear distortion)

Especially, power efficiency and spectrum efficiency of various digital modulation techniques have been studied and evaluated extensively. Power efficiency is important as it allows the signal to be received with a target BER performance with less power. Alternatively, for the same transmitted power, a power efficient modulation allows better BER performance, or for the same power and BER performance it allows better coverage and immunity against fading and other impairments. On the other hand, spectral efficient modulation allows the information transfer with minimal amount of bandwidth, or for the same bandwidth more information can be transferred with spectrally efficient modulation. Spectrum efficiency is often measured as the data-rate-per-hertz, i.e. R/B bits-per-second-per-Hz.

6.2 Popularly Used Digital Modulation Schemes

6.2.1 PSK

As described above, in PSK, the phase of the carrier is changed depending on the information to be transmitted. For example, in the simple binary PSK (BPSK) which is used popularly in various wireless communication systems, the phase shifts can be 0° or 180° for representing a bit value of 0 or 1, respectively. As can be seen in BPSK, a bit is represented with one symbol. Since there are two possible bit values (0 or 1), then the symbol alphabet in BPSK will have two possible symbols.

Quadrature phase-shift keying (QPSK) and various derivatives (offset QPSK, $\frac{\pi}{4}$-QPSK, $\frac{\pi}{4}$-DQPSK) are also very popular PSK modulations used in wireless communication systems. In QPSK, the number of phase shifts is four, i.e. four possible symbol values corresponding to each phase shift. Since the number of alphabet is increased to four, we can map two bits to each symbol. Hence, compared to BPSK, QPSK is more spectrally efficient modulation. In general, the number of bits that can be represented with each symbol can be calculated as:

$$M = 2^B, \tag{6.1}$$

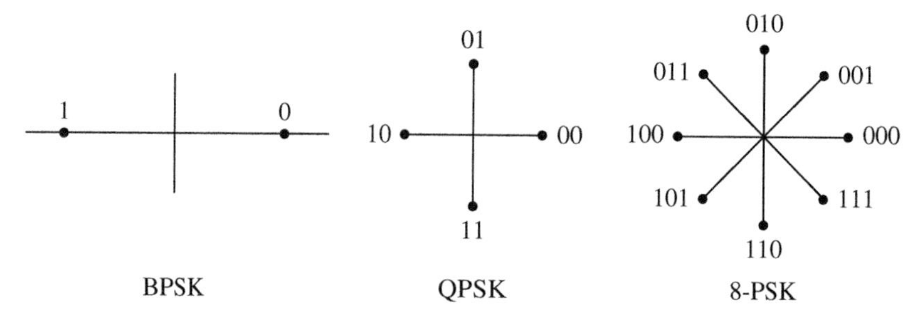

Figure 6.2 Constellation plots for various PSK (PSK, QPSK, 8PSK).

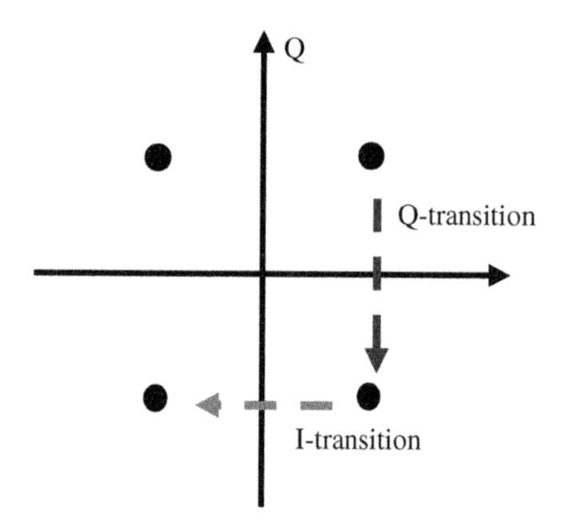

Figure 6.3 Illustration of OQPSK. Notice that the transition from one state to the exact opposite happens in two stages. In other words, the I and Q channels do not change simultaneously. First, one of the channels switches the states, then half-symbol duration later, the other channel starts transitioning.

where M represents the number of symbols in the symbol alphabet and B represents the number of bits that can be represented with each symbol. In QPSK, $B = 2$ and $M = 4$. Hence, if we generalize the BPSK and QPSK modulations to MPSK (M-level PSK modulation, or sometimes referred as Mary PSK), there will be M possible phase shifts that can be applied to the carrier. Figure 6.1b shows the modulated carrier for BPSK and Figure 6.2 shows the constellations for various PSK (PSK, QPSK, and 8PSK).

In QPSK, the transitions from one symbol to another might go through the origin. The origin here means the point where the carrier amplitude is zero, i.e. both in-phase (I) and quadrature-phase (Q) values are zero. This happens when both I and Q signals change their value (i.e. $1 + j$ symbol value in one symbol interval will be $-1 - j$ in the next symbol interval) at the same time. This change also corresponds to a 180° phase shift from one state to another. Going from one state in the I/Q plane to the exact opposite results in large amplitude variation. The large signal amplitude variation puts a lot of constraints on power amplifier (PA) designs as the nonlinear properties of PAs cause distortions and out-of-band radiation. If we were to limit the simultaneous transition of I and Q signals to the exact opposite state, we can avoid the transition through the origin. Offset quadrature phase-shift keying (OQPSK) can be developed from QPSK by delaying Q channel half a symbol with respect to I channel (see Figure 6.3). By introducing this delay, the signal is made closer to constant envelope, and hence the signal transition through the origin is avoided.

Another way of avoiding signal transition through the origin is the use of $\frac{\pi}{4}$-shifted-QPSK (or simply $\frac{\pi}{4}$QPSK). Instead of using a single QPSK constellation, in $\frac{\pi}{4}$QPSK, two constellations are used, one shifted by 45° compared to the other. If you look at the constellation diagram (see Figure 6.4), the constellation looks like 8PSK, but actually it is broken into two QPSK constellations. The even and odd indexed symbols (or symbols that are immediately following each other) use different

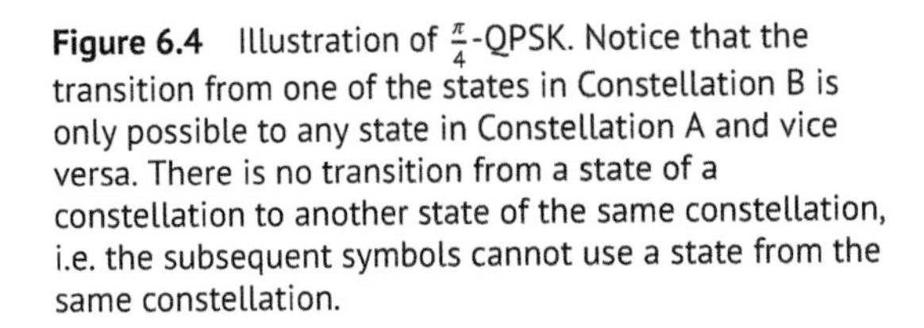

Figure 6.4 Illustration of $\frac{\pi}{4}$-QPSK. Notice that the transition from one of the states in Constellation B is only possible to any state in Constellation A and vice versa. There is no transition from a state of a constellation to another state of the same constellation, i.e. the subsequent symbols cannot use a state from the same constellation.

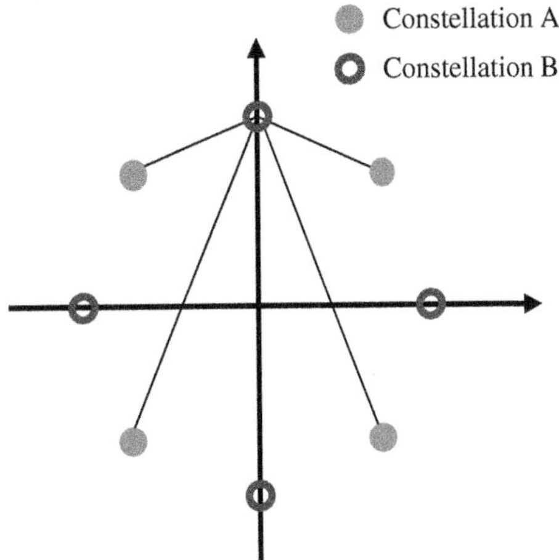

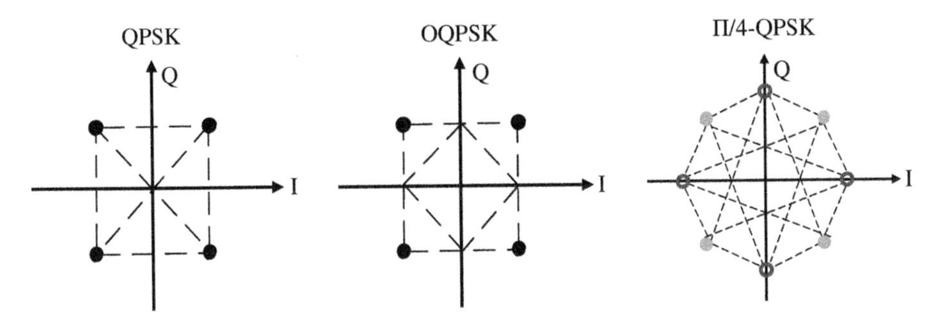

Figure 6.5 Constellation and transitions of QPSK, OQPSK, and $\frac{\pi}{4}$QPSK modulations. In $\frac{\pi}{4}$QPSK, the maximum phase change is limited to 135° compared to 180° for QPSK and 90° for OQPSK.

constellations for mapping. In other words, if the symbol at time n is using one of these two constellations, the symbol at time $n+1$ should use the other constellation, and so on. Using the two constellations in alternating manner, the transition through the origin is eliminated (i.e. 180° phase shift is completely eliminated). Figure 6.5 shows constellation and transitions of QPSK, OQPSK, and $\frac{\pi}{4}$QPSK modulations. The I/Q vector plots of QPSK and its variants are shown in Figure 6.6.

Figures 6.7 and 6.8 show the I-eye and Q-eye plots of OQPSK and $\frac{\pi}{4}$QPSK modulations. In OQPSK modulation, half-symbol delay between I-eye and Q-eye plots can be observed clearly. In the case of $\frac{\pi}{4}$QPSK, unlike the regular QPSK or OQPSK modulation, the eye diagrams do not show a single eye formation. Since the consecutive symbols are using different constellation, the eye formation is not the same in alternating symbol intervals. This feature can be used to differentiate a $\frac{\pi}{4}$QPSK modulation from 8PSK modulation.

6.2.2 FSK

In FSK, the frequency of the carrier is changed with the modulating signal, while the amplitude remains constant. Constant envelope of the carrier is a desirable characteristic for improving the power efficiency of transmitters. On the other hand, due to the spreading of the signal over a wider bandwidth, FSK modulation has poor spectral efficiency. In binary FSK (BFSK or 2FSK), 1 is represented by one frequency (let's say f_1) and 0 is represented by another frequency (let's say

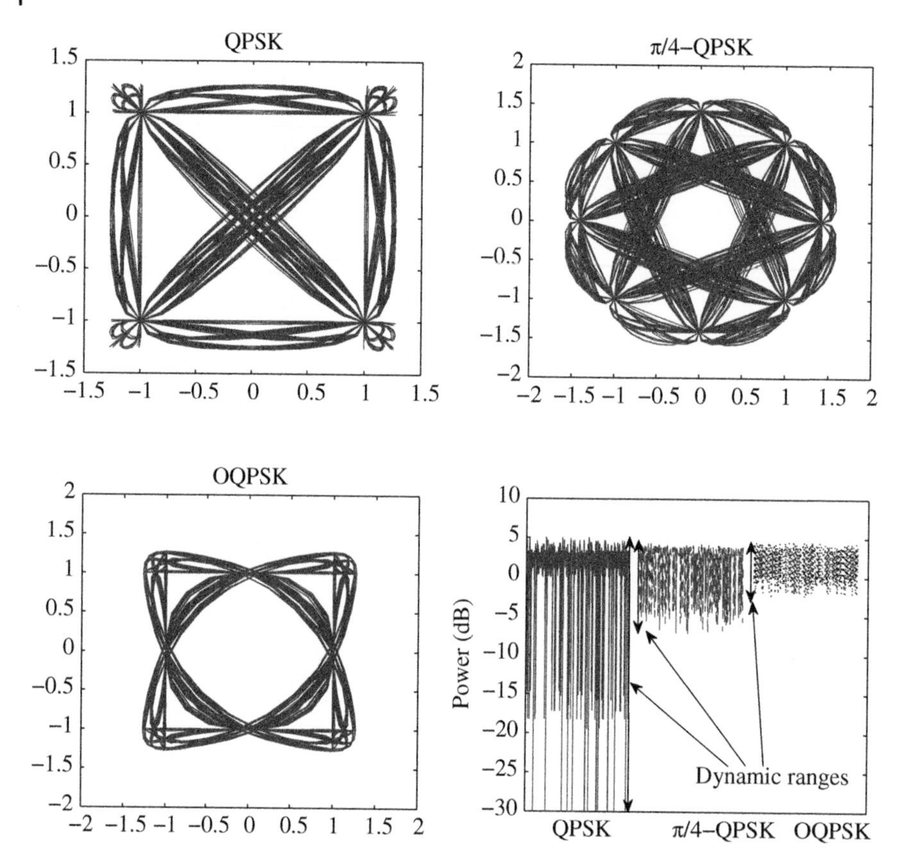

Figure 6.6 Simulation of various QPSK constellations. The dynamic ranges of QPSK, $\frac{\pi}{4}$QPSK and OQPSK is also given over a short frame of data duration. A raised cosine filter with a roll-off factor of 0.7 is used in this figure.

$f_2 = f_1 + \Delta f$, where Δf represents the frequency separation or frequency deviation). Usually, Δf is chosen in such a way that the modulated symbols are orthogonal to each other. When Δf is equal to half the bit rate (which corresponds to modulation index of 0.5), a special case of FSK which is referred as minimum-shift keying (MSK) is obtained. This frequency deviation is also the minimum frequency deviation that yields orthogonality of I and Q, resulting in a phase shift of $\pm\pi/2$ radians per symbol. Due to this exact phase relationship, MSK can be considered as either phase or frequency modulation. For example, MSK can be obtained from OQPSK by simply changing the pulse shape to a half-cycle sinusoid. MSK is also called as continuous phase FSK (CPFSK) where the frequency changes occur at the carrier zero crossings. Normalized spectrum plots of BFSK and MSK signals are shown in Figure 6.9.

Generalizing 2FSK to MFSK leads to M separate frequencies that the carrier can be shifted depending on the information bits. Again, the frequency separation can be defined as the difference between two successive frequencies $\Delta f = f_k - f_{k-1}$. Note that Δf will determine the degree of separability of M possible transmitted waveforms from each other. The minimum frequency separation between successive frequencies that satisfies orthogonality is equal to half the symbol rate.

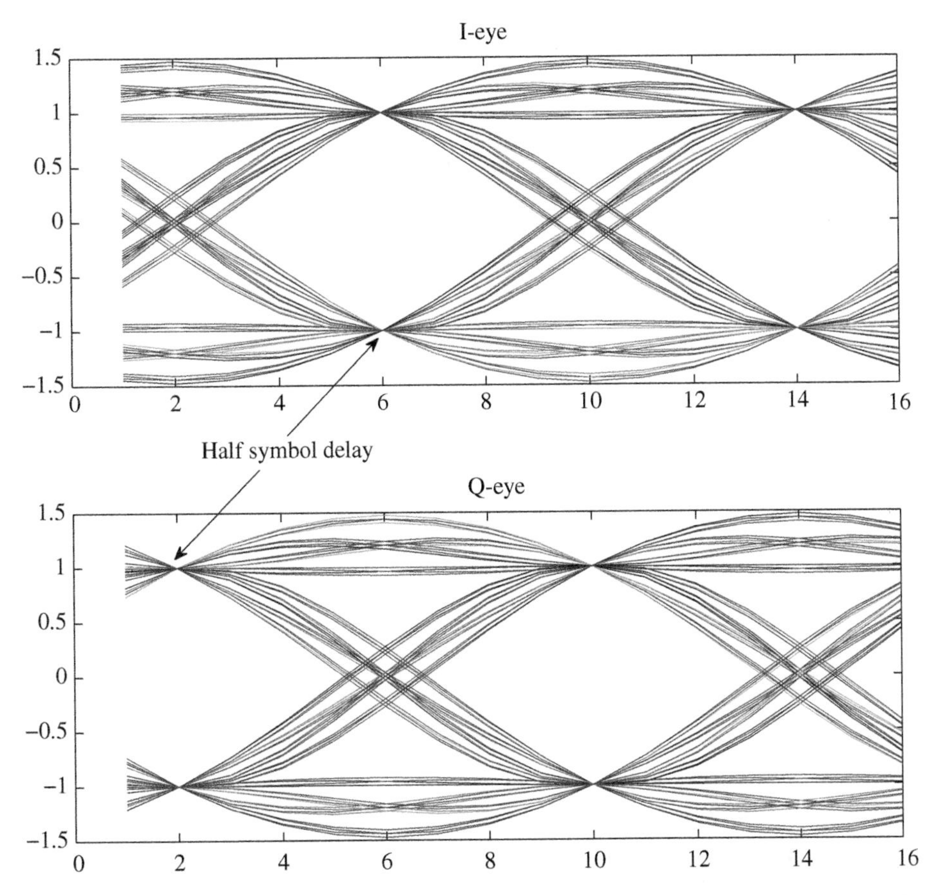

Figure 6.7 Eye diagram of OQPSK-modulated signal using raised cosine filtering with roll-off of 0.5. Observe that the I-eye and Q-eye are delayed by half a symbol with respect to each other.

6.2.2.1 GMSK and Approximate Representation of GSM GMSK Signal

In order to simplify the receiver structures, the constant envelope binary phase modulation signal can be expressed as the superposition of amplitude-modulated pulses. A continuous phase-modulated (CPM) baseband signal is represented as:

$$s(t) = e^{j\phi(t,\underline{\alpha})}, \tag{6.2}$$

where $j = \sqrt{-1}$, $\underline{\alpha} = (..., \alpha_k, ...)$ is differentially encoded transmitted data sequence, $\phi(t, \underline{\alpha})$ is the modulating phase function which can be written as:

$$\phi(t, \underline{\alpha}) = \sum_i \alpha_i q(t - iT), \tag{6.3}$$

where $\alpha_i = b_i b_{i-1}$, $b_i = [-1,1]$, T is the symbol duration, and $q(t)$ is the phase pulse which is obtained by integrating the frequency pulse $g(t)$ as follows:

$$q(t) = \int_{-\infty}^{t} g(\tau)d\tau. \tag{6.4}$$

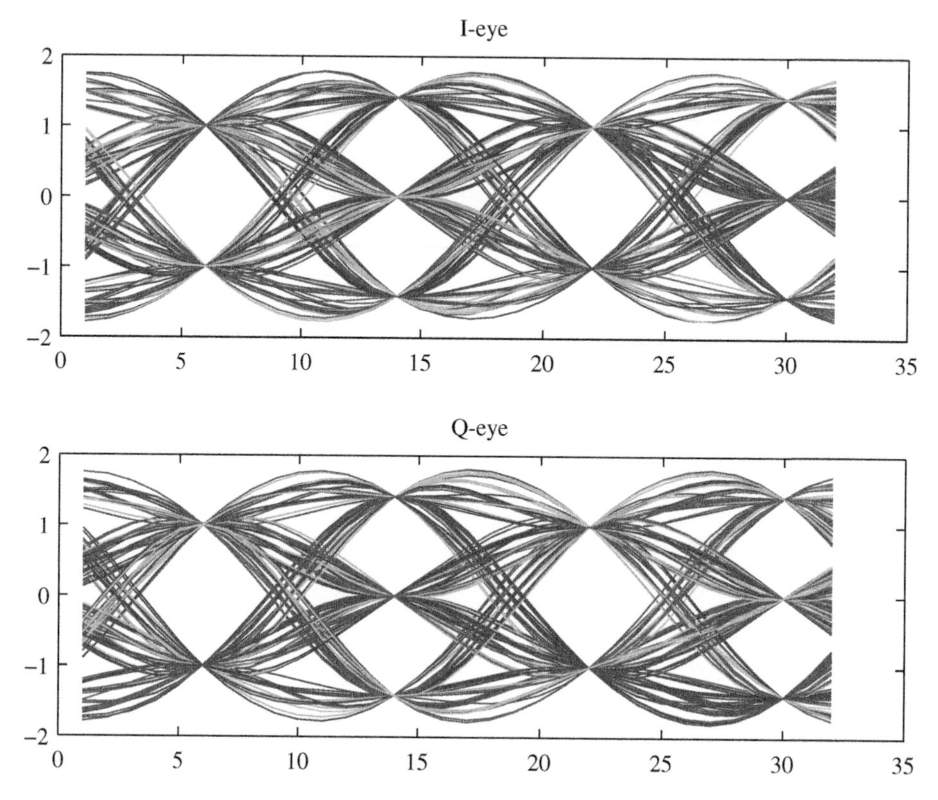

Figure 6.8 Eye diagram of $\frac{\pi}{4}$QPSK-modulated signal using raised cosine filtering with roll-off of 0.5. Observe that the eye formation is not the same in alternating symbol intervals. This is due to the fact that in $\frac{\pi}{4}$QPSK the subsequent symbols are not using the same constellation. By looking at the eye diagram of $\frac{\pi}{4}$QPSK, one can clearly differentiate it from an 8PSK modulation, which was not the case in constellation plot.

For the case of global system for mobile communications (GSM), Gaussian minimum shift keying (GMSK), $g(t)$ is Gaussian-shaped and time limited to an interval equal to $3T$ ($BT = 0.3$, where BT represents time-bandwidth product which will be described later). The frequency and phase pulse of GSM GMSK signal are shown in Figure 6.10.

This GSM GMSK signal can be approximated as linear amplitude-modulated signal. Such an approximation simplifies the receiver design without much sacrifice of the performance. In the transmission, the exact GMSK-modulated signal is generated and send through radio channel. However, at the receiver, this signal is approximated with linear amplitude-modulated signal as follows:

$$s(t) = \sum_{k} b_k j^k p(t - kT), \tag{6.5}$$

where $p(t)$ is the effective pulse shape due to the linear approximation. The error because of the linear model approximation of GMSK signal is very small as long as we chose a reasonable BT value. For the GSM case ($BT = 0.3$), this approximation is acceptable. More detailed analysis concerning the approximation of continuous phase-modulated signals by superposition of amplitude-modulated pulses can be found in [3, 4].

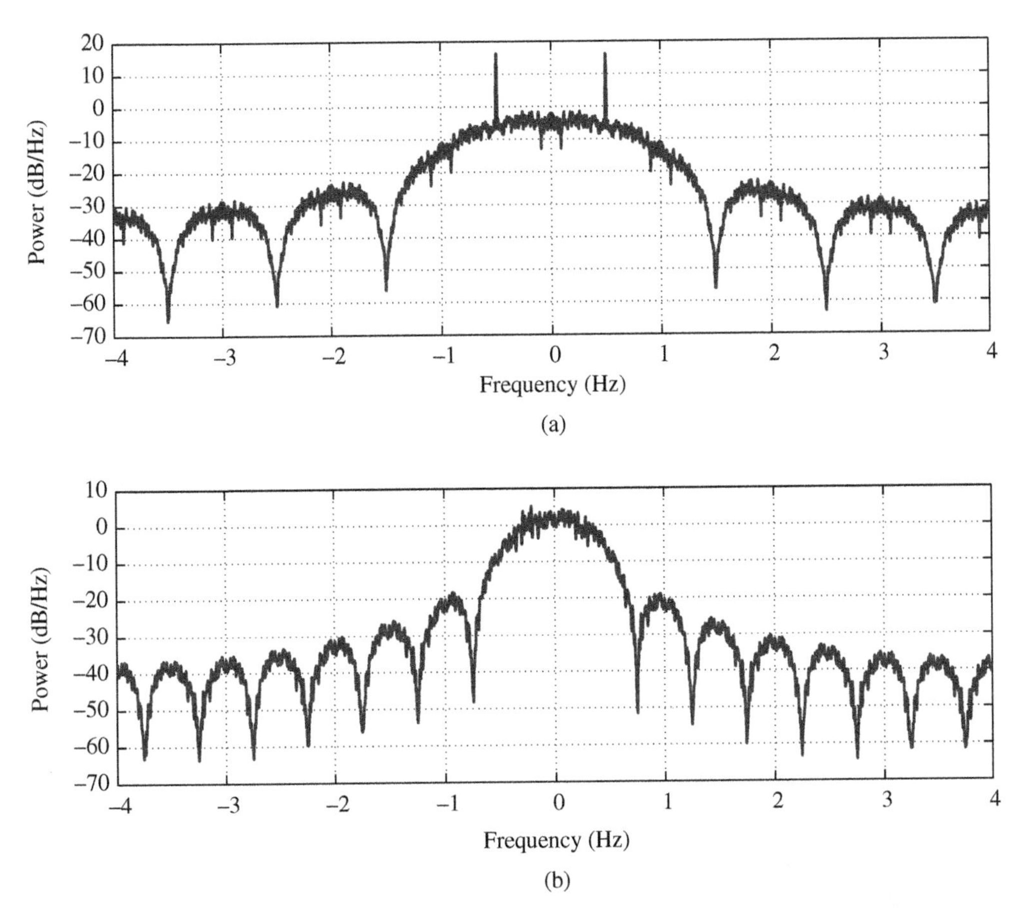

Figure 6.9 Normalized (with symbol rate) spectrum of the BFSK and MSK signals. (a) Normalized spectrum of unfiltered BFSK with modulation index of 1. (b) Normalized spectrum of unfiltered MSK (modulation index = 0.5).

The differential precoding is specified by the GSM recommendations [5]. At the modulator, the differential encoder needs the first bit value which can be set to "1" as stated in the standard [5]. The following Matlab code generates GMSK-modulated GSM signal and its approximate linear modulation representation. Figure 6.11 shows the corresponding modulated signals.

6.2.3 QAM

Quadrature amplitude modulation (QAM) is another family of digital modulation techniques where both phase and amplitude of the carrier are varied. Use of the family of QAM has gained significant attention in recent wireless communication systems and is widely adopted in various communication standards and commercial applications. There are various types and levels of QAM modulations including 16-state quadrature amplitude modulation (16-QAM), 32-state quadrature amplitude modulation (32-QAM), 64-state quadrature amplitude modulation (64-QAM), 128-state quadrature amplitude modulation (128-QAM), and so on [6]. Here, M-QAM stands for M state QAM modulation, where M defines the number of constellation points. For example, in 16-QAM, four bits are transmitted with a symbol, and hence, there are $2^4 = 16$ constellation points in the

```
1  % PAM representation of GSM GMSK signal
2  sps=8;
3  tap=5;
4  n=sps*(tap-1)/2;
5  BT=0.3; % time-bandwidth product
6  % Gaussian pulse shape generation
7  i=-n:n ;
8  q1=pi*sqrt(2/log(2))*BT*(i/sps+.5);
9  q2=pi*sqrt(2/log(2))*BT*(i/sps-.5);
10 ps(i+n+1)=1/(4*sps)*(erf(q1)-erf(q2));
11
12 It=[1,1,-1,1,1,-1,1,-1,-1,-1,1,1,1,1,-1,1,1,1,-1,1,1,-1,1,-1,-1,-1];
13 % the first training sequence used in GSM
14 Id=[-ones(1,8),(randn(1,54) > 0)*2-1,It,(randn(1,54) > 0)*2-1,-ones(1,8)];
15 m=length(Id);
16 dif_en = Id.* [1 Id(1:m-1)];
17
18 II=(dif_en'*[1 zeros(1,sps-1)])';
19 up_sampled=reshape(II,1,m*sps);
20
21 c11=conv(ps,up_sampled);
22 %integration of the BPSK generated signal to obtain GMSK phase
23 c112=cumsum(c11)*j*pi;
24 %%%%%%%%%%%%%%%%%% Generation of the GMSK signal
25 s=exp(c112); % desired GMSK signal, constant envelope
26
27 % Approximate pulse shape
28 q=cumsum(ps);
29 q1=q(1:length(q));
30 q2=1/2-q1;
31 z=zeros(1,length(q));
32 pay=[q1 q2(2:length(q)) z];
33 S0=sin(pay);
34 S1=sin(pay(sps:length(pay)));
35 S2=sin(pay(2*sps:length(pay)));
36 S3=sin(pay(3*sps:length(pay)));
37 %%%%%%%%%%%%%%%%%%%%%%%%%%%%%
38 L=length(ps)+sps;
39 L0=S0(1:L).*S1(1:L).*S2(1:L).*S3(1:L);
40 gp=L0(1:length(L0));
41
42 Idj=j.^(1:length(dif_en)).*dif_en;
43 II=(Idj'*[1 zeros(1,sps-1)])';
44 up_sampled=[reshape(II,1,m*sps)];
45 approx=conv(gp,up_sampled);
46
47 subplot(1,2,1)
48 plot(real(s),imag(s))
49 title('True GMSK signal')
50 subplot(1,2,2)
51 lgp=length(gp);
52 plot(real(approx(lgp:end-lgp))), imag(approx(lgp:end-lgp))
53 title('Approximate GMSK signal')
```

I/Q plane. There are several implementations of QAM modulations in the constellation diagram. For instance, in Figure 6.12, three different implementations of 16-QAM modulation can be seen.

For a given average power, more bits per symbol can be encoded in QAM which makes it more spectrally efficient compared to other modulation options. Using higher-order modulation formats, i.e. more points on the constellation, it is possible to transmit more bits per symbol. However, the points are closer together and they are therefore more susceptible to noise and data errors. When using QAM, the constellation points are normally arranged in a square grid with equal vertical and horizontal spacing and as a result the most common forms of QAM use a constellation with the number of points equal to a power of 2, i.e. 2, 4, 8, 16, and so on. Even though high-order QAM provides spectral efficiency advantage, this comes at the price of larger dynamic range and with large the peak-to-average power ratio (PAPR) and so, with a reduced the energy efficiency.

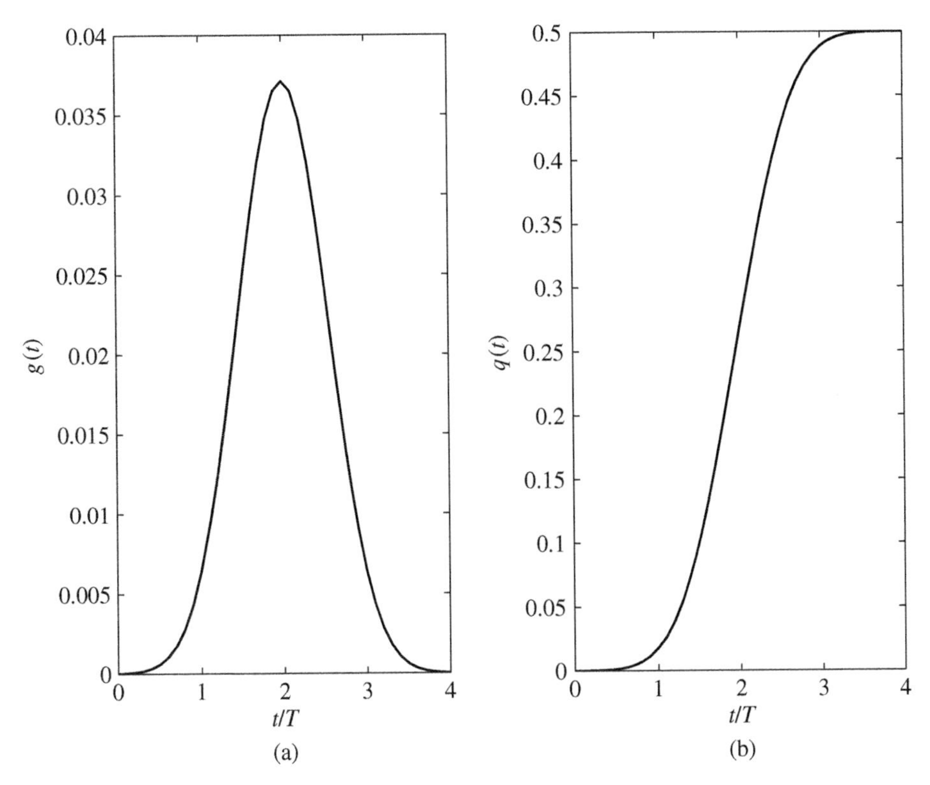

Figure 6.10 The frequency and phase pulse of GSM GMSK signal. (a) The frequency pulse. (b) The phase pulse.

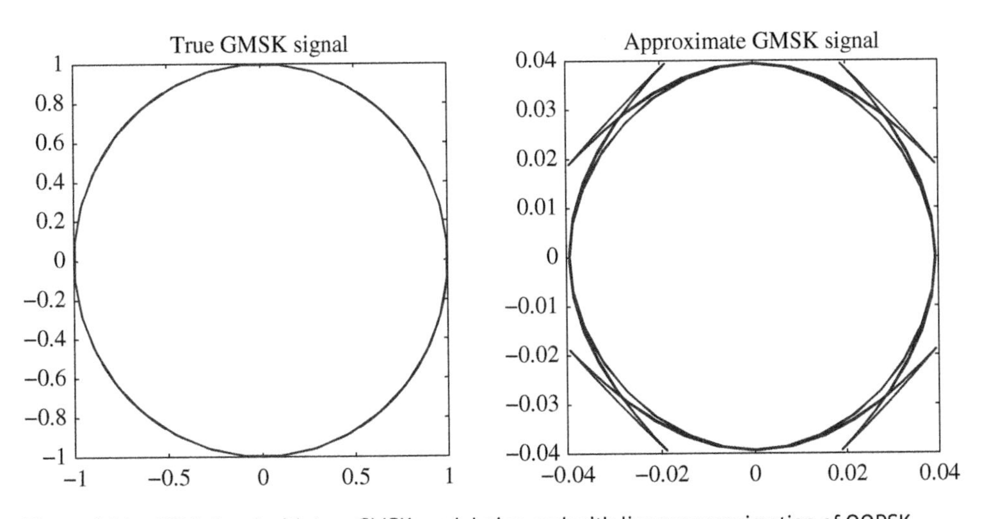

Figure 6.11 GSM signal with true GMSK modulation and with linear approximation of OQPSK.

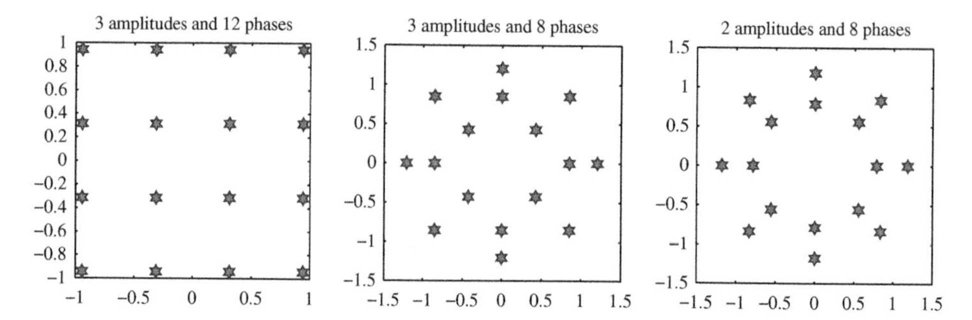

Figure 6.12 Different implementations of 16-QAM modulation.

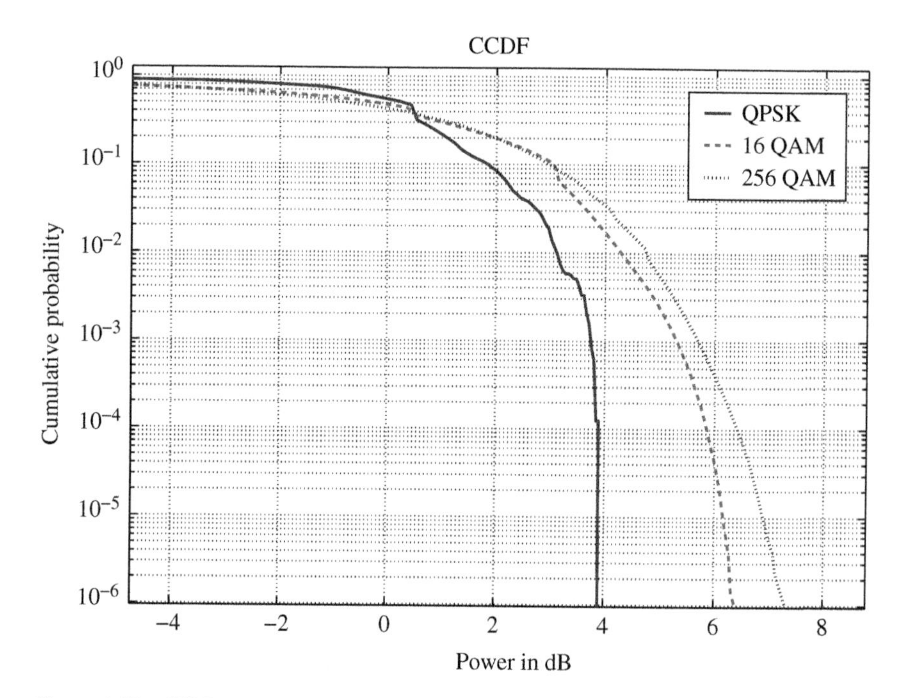

Figure 6.13 CCDF plots of various QAM modulations.

Figure 6.13 shows the complementary cumulative distribution function (CCDF) plots for different QAM modulations. As can be seen, as the modulation order increases, the dynamic range also increases.

6.2.4 Differential Modulation

Differential modulation allows noncoherent detection of the transmitted signal and hence relaxes the need for sending separate reference symbols along with data. Also, they allow the design of simpler and low-cost receivers [7]. Differential modulations are popularly applied to PSK-type mappings like differential PSK (DPSK) and differential QPSK (DQPSK). For coherent detection of PSK signals, the carrier phase needs to be estimated with no phase ambiguity using training (or pilot) symbols [8]. In DPSK, the information is not in the absolute phase, but it is in the transition (difference) of the phases (i.e. information bits encoded as the differential phase between current and previous symbol). For example, in DPSK, a 0 bit can be encoded by no change in phase between previous and current symbols, whereas a 1 bit can be encoded as a phase change of π between previous and current symbols. At the receiver, the phase difference between the signals received at two subsequent symbol samples is used for detection. Therefore, a phase shift by the channel which is the same for both subsequent symbols will not effect the decision.

Differentially modulated signal can be demodulated using both noncoherent and coherent demodulation techniques. If the data are demodulated using noncoherent demodulation, approximately there is 3-dB performance degradation.

6.3 Adaptive Modulation

One of the main objectives for the recent and future generations of wireless communications systems is to extend the services provided by the current systems with higher-rate data capabilities. Higher-order modulation techniques allow more data rate. However, higher-order modulations require a better link quality to provide the same performance for the modem bits. In wireless mobile communication systems, due to multipath and location of the receiver with respect to the transmitter, the link quality changes dramatically. In general, adaptive design methodologies typically identify the user's requirements and then allocate just enough resources, thus enabling more efficient utilization of system resources and consequently increasing capacity.

A reliable link must ensure that the receiver is able to capture and reproduce the transmitted information bits. Therefore, the target link quality must be maintained all the time in spite of the changes in the channel and interference conditions. Higher-order modulations (HOM) allow more bits to be transmitted for a given symbol rate. On the other hand, HOM is less power efficient, requiring higher energy per bit for a given BER. Therefore, HOM should be used only when the link quality is high, as they are less robust to channel impairments. High-quality link can be obtained by transmitting the signals with spectrally less efficient modulation schemes, like BPSK and QPSK. On the other hand, new generation systems aim for higher data rates made possible through spectrally efficient higher-order modulations. Therefore, a reliable link with higher information rates can be accomplished by continuously controlling the modulation levels [9–11]. Higher modulation orders are assigned to users that experience good link qualities, so that the excess signal quality can be used to obtain higher data rates. Recent designs have exploited this with adaptive modulation techniques that change the order of the modulation [12–14]. For example, EDGE standard introduces both Gaussian minimum-shift keying (GMSK) and 8PSK modulations with different coding rates through link adaptation and hybrid automatic repeat request (HARQ). The channel quality is estimated at the receiver, and the information is passed to the transmitter through appropriately defined messages. The transmitter adapts the coding and modulation based on this channel quality feedback. Similarly, IEEE 802.16-based wireless metropolitan area network standard (commonly known as WiMAX), 802.11a/g/n/ac-based wireless local area network standard (commonly known as WiFi), and Long Term Evolution (LTE) all uses adaptive modulation and coding (BPSK, QPSK, 16-QAM, and 64-QAM). For example, in 802.11a/g, through the adaptive modulation and coding, a data rate ranging from 6 to 54 Mbit/s can be obtained.

Figures 6.14 and 6.15 illustrates the capacity gain that can be achieved by employing adaptive modulation. First, the BER performances of different modulations as a function of SNR are given in Figure 6.14. As can be seen, a desired BER can be achieved with low-order modulations for lower SNR values. Higher-order modulations need better link quality (higher SNR) in order to obtain the

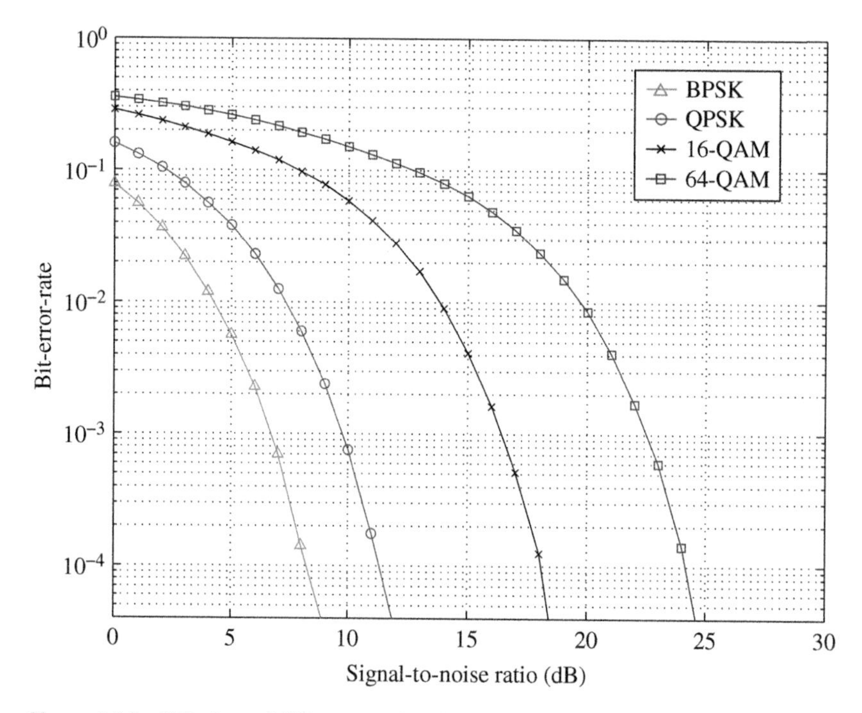

Figure 6.14 BER plots of different modulations as a function of SNR.

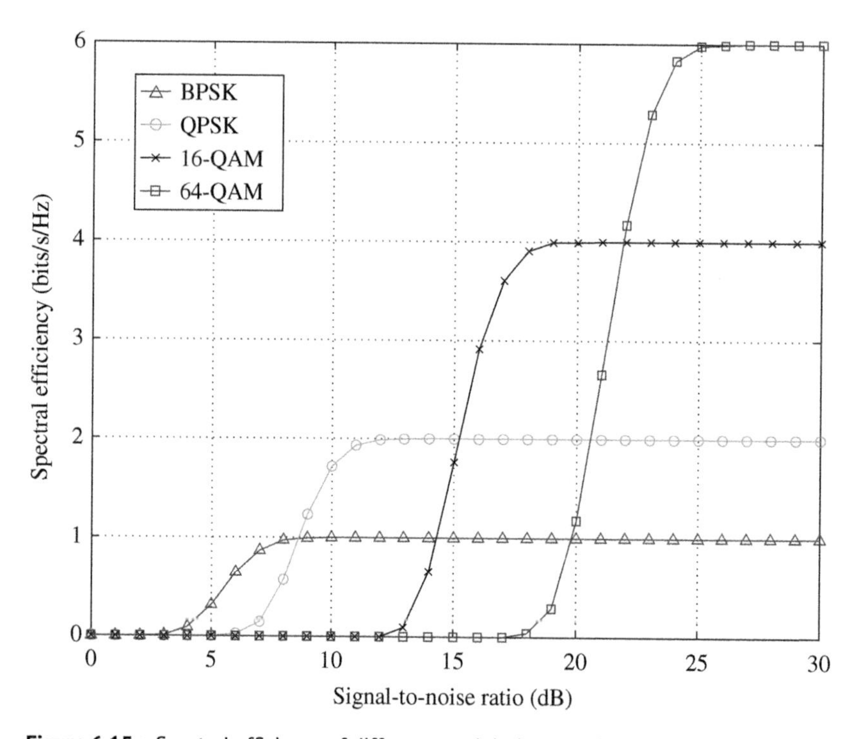

Figure 6.15 Spectral efficiency of different modulation as a function of SNR.

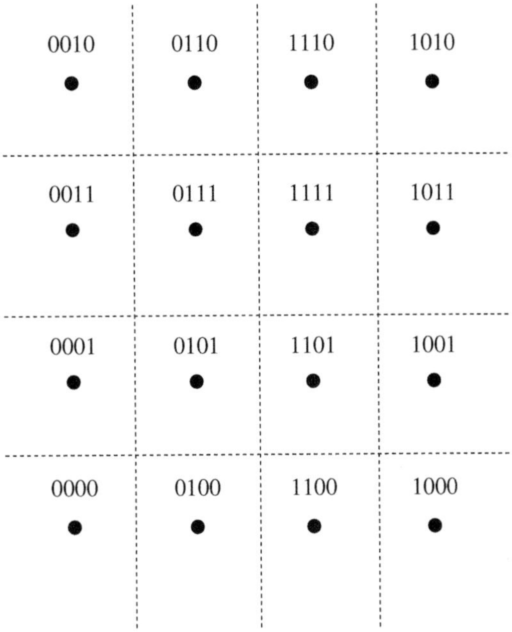

Figure 6.16 Illustration of Gray mapping on 16-QAM modulation. Decision boundaries are also shown in this figure.

0010	0110	1110	1010
0011	0111	1111	1011
0001	0101	1101	1001
0000	0100	1100	1000

same BER performance. Figure 6.15 shows the spectral efficiencies of different uncoded modulations, where an arbitrary packet size of 200 bits is used. Notice that the optimal spectral efficiency for different SNR regions can be obtained through the use of different modulation depending on the value of SNR.

6.3.1 Gray Mapping

The bits can be mapped to the constellation points depending on various criteria. One of the most important aspects that is considered in mapping is to minimize the BER performance. Gray mapping is popularly used for minimizing the BER. Gray codes are binary sequences where two neighboring symbol values differ in only one digit. Figure 6.16 illustrates an example of Gray mapping applied to 16-QAM modulation.

6.3.2 Calculation of Error

BER performance has been used popularly to characterize the performance of various modulations. Depending on the used modulation and the type of the receiver, the BER performance can change. In additive white Gaussian noise channel, the baseband received signal can be modeled as:

$$y(t) = s(t) + w(t), \tag{6.6}$$

where $s(t)$ is the transmitted symbol and $w(t)$ is the AWGN noise. Let's assume a simple binary orthogonal modulation where a bit value of "0" is mapped to $s_0(t)$ and a bit value of "1" is mapped to $s_1(t)$, for the duration T. At the receiver, after the matched filtering and sampling, the received samples at the two matched filter outputs at each symbol interval can be represented as:

$$r_i = V_i + z_i, \tag{6.7}$$

where the subscript i represents two different values corresponding to matched filter outputs (r_0 is the first matched filter output, r_1 is the second matched filter output; V_0 is the decision statistics for the signal $s_0(t)$ and V_1 is the decision statistics of the signal $s_1(t)$; z_0 is noise at the first matched filter output and z_1 is the noise at the second matched filter output). If $s_0(t)$ is transmitted, the first matched filter output will be $r_0 = V_0 + z_0$ and the second matched filter output will be $r_1 = z_1$. On the other hand, if $s_1(t)$ is transmitted, the first matched filter output will be $r_0 = z_0$ and the second matched filter output will be $r_1 = V_1 + z_1$. The detector compares the matched filter outputs r_1 and r_0 and decides whether a "0" or "1" is transmitted. If $s_0(t)$ is transmitted, and assuming the noise at both matched filter outputs has the same statistics and Gaussian distributed with mean zero and variance of σ^2, the probability density functions of the received samples at both correlator outputs can be represented as:

$$p(r_0|s_0(t)) = \frac{1}{\sqrt{2\pi}\sigma} e^{-(r-V_0)^2/2\sigma^2} \tag{6.8}$$

and,

$$p(r_1|s_0(t)) = \frac{1}{\sqrt{2\pi}\sigma} e^{-r^2/2\sigma^2}. \tag{6.9}$$

If $s_0(t)$ is transmitted, an error occurs if the received signal value at the second matched filter output exceeds the value of the first matched filter output, i.e. $P_{1|0} = P(r_1 > r_0) = P(z_1 > V_0 + z_0)$. Similarly, if $s_1(t)$ is transmitted, an error occurs if the received signal value at the first matched filter output exceeds the value of the second matched filter output, i.e. $P_{0|1} = P(r_0 > r_1) = P(z_0 > V_1 + z_1)$. For equal probable symbol transmission and assuming that $V_1 = V_2 = V$, the combined probability of error will be calculated as:

$$P_e = 0.5(P_{1|0} + P_{0|1}) = Q\left(\sqrt{\frac{V^2}{2\sigma^2}}\right) = Q\left(\sqrt{\frac{1}{2}SNR}\right) = Q\left(\sqrt{\frac{E_b}{N_o}}\right), \tag{6.10}$$

where $Q(\cdot)$ represents the Gaussian Q-function[1]. $\frac{E_b}{N_o}$ represents the energy per bit to noise power spectral density ratio. Noise variance is related to the noise power spectrum density as $\sigma^2 = \frac{N_o}{2}$. Also, $\frac{E_b}{N_o}$ is related to SNR with

$$\frac{E_b}{N_o} = SNR\frac{B}{R}, \tag{6.11}$$

where $\frac{R}{B}$ (bps/Hz) represents the spectral efficiency of the modulation. Here, R represents the bit rate and B represents the bandwidth.

If the modulated symbols are antipodal signals (like in BPSK), where one signal waveform is the negative of the other, one filter will be sufficient at the receiver. The resulting received sample can be represented as:

$$r = \pm V + z. \tag{6.12}$$

Assuming z is Gaussian random variable with zero mean and variance of σ^2, probability density functions of $p(r|1)$ and $p(r|0)$ can be obtained as:

$$p(r|0) = \frac{1}{\sqrt{2\pi}\sigma} e^{-(r-V)^2/2\sigma^2} \tag{6.13}$$

and,

$$p(r|1) = \frac{1}{\sqrt{2\pi}\sigma} e^{-(r+V)^2/2\sigma^2}. \tag{6.14}$$

---

1 The Gaussian Q-function is defined as $Q(x) = \frac{1}{\sqrt{2\pi}} \int_x^{\infty} e^{-\frac{r^2}{2}} dr$.

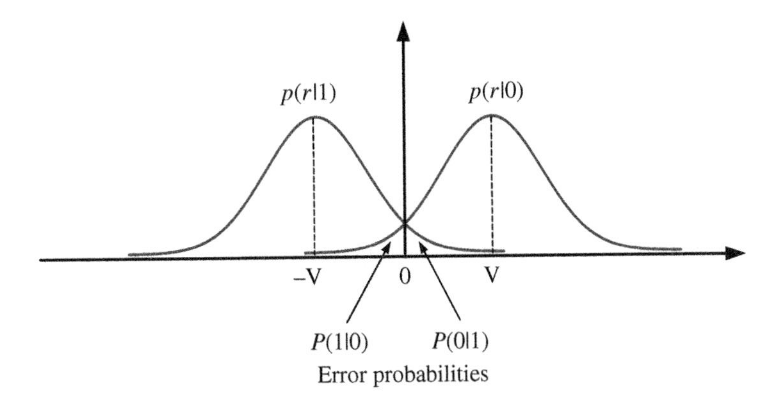

Figure 6.17 PDF of the received samples depending on the transmitted symbols.

These probability density functions are shown in Figure 6.17.

For equal probable bit transmission ($P(0) = P(1)$), the optimal detector for BPSK compares the received sample, r, with a threshold of zero. If the received sample is greater than the threshold, a decision of "zero" bit transmission is made. Similarly, if the received sample is less than the threshold, a decision of "one" bit transmission is made. The noise might push the received signal to the other side of the threshold giving rise to bit errors. The bit error rates can be calculated easily by finding the total probabilities of the PDFs that fall on the other side of the threshold ($P_e = \frac{1}{2}(P_{0|1} + P_{1|0})$, where $P_{0|1}$ is the probability of receiving 0 given that 1 is transmitted, and $P_{1|0}$ is the probability of receiving 1 when 0 is transmitted) and it can be given as:

$$P_e = \frac{1}{\sqrt{2\pi\sigma}} \int_{-\infty}^{0} e^{-(r-V)^2/2\sigma^2} dr, \tag{6.15}$$

which is equal to $Q(\sqrt{\frac{V^2}{\sigma^2}}) = Q(\sqrt{SNR}) = Q(\sqrt{2\frac{E_b}{N_o}})$.

The following Matlab code finds the theoretical BER curve for BPSK modulation and compares it with simulation. The corresponding plots are given in Figure 6.18.

```
1  % Theoretical BER vs SNR for BPSK signal
2  dB = 0: 0.1: 10;
3  SNR = 10.^ (dB ./ 10);
4  Pb = 0.5 * erfc(sqrt(SNR));
5  figure(1);clf
6  semilogy (dB, Pb);
7  grid on;
8  xlabel('SNR (dB)');
9  ylabel('Bit error rate');
10 title('Performance of BPSK modulation over AWGN channel');
11 hold on
12 % Simulation
13 N=1e6;
14 bits= (randn(1,NN) > 0)*2-1;
15 noise =  (randn(1, NN) + j*randn(1, NN))/sqrt(2);
16 for kk=1:length(dB)
17     sigma= sqrt(10.^ (-dB(kk)/ 10));
18     y=bits+noise*sigma;
19     det=(real(y) > 0)*2-1;
20     err(kk)= length(find(det ≠ bits))/ NN ;
21 end
22 semilogy (dB, err ,'ro');
```

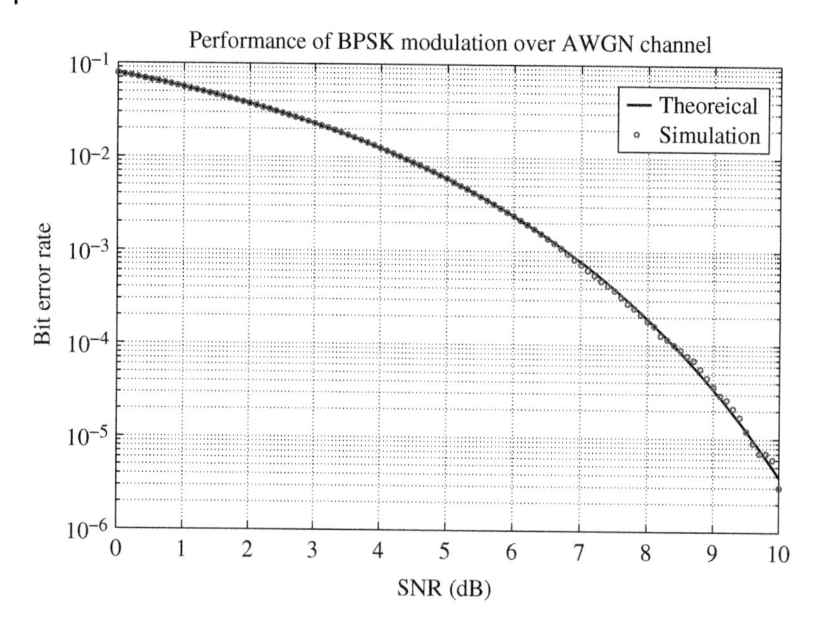

Figure 6.18 Performance of BPSK with both simulation and theoretical calculation.

6.3.3 Relation of $\frac{E_b}{N_o}$ with SNR at the Receiver

In developing computer simulations to plot BER versus $\frac{E_b}{N_o}$ or SNR curves, understanding the relation between them would be useful. These two quantities can be related as:

$$SNR = \frac{E_b}{N_o}\log_2(M)\frac{1}{B_e T}, \tag{6.16}$$

where M is the modulation order, T is the symbol duration, and B_e is the effective bandwidth which can be calculated as:

$$B_e = f_s \sum_{l=0}^{L-1} |h_l|^2, \tag{6.17}$$

where f_s is the sampling rate and h_l is the normalized FIR filter coefficients, i.e. the summation of the absolute filter coefficients is equal to unity. The following Matlab simulation illustrates the relation between $\frac{E_b}{N_o}$ and SNR. For a desired $\frac{E_b}{N_o}$ value, the target SNR is calculated as given in the code. The target (desired) SNR is also compared with the actual SNR value after the receiver filter. Ideally, they should match. But, because of the random noise statistics, there is a slight difference between the target SNR and actual SNR values. The difference decreases when the simulated number of symbols increases. In plotting the SNR versus BER curves, the estimated SNR values (after the receiver filter) should be used.

6.4 Pulse-Shaping Filtering

Baseband pulse-shaping filtering is used in digital modulation to limit bandwidth and reduce inter-symbol interference. Square pulse shapes are not practical to be used in wireless communication systems as they require significant amount of bandwidth. Even though they are Nyquist pulses and very robust to timing errors and inter-symbol interference (ISI), they have terrible spectral

```
1  % Relation between SNR, EbNo, and EsNo when a receiver filtering is used
2  EbNo_target= 10 ; %dB  Desired Eb/No in the simulation
3  N=1e4; % Number of symbols
4  sps=8; % oversampling ratio
5  fltr=rcosine(1,sps,'sqrt',0.3); %RRC filtering
6  M=4; % modulation order
7  symb= qammod(floor(rand(1,N)*M),M);
8  % normalize to get average symbol power of 1.
9  symb=symb./sqrt(mean(abs(symb).^2));
10 u_frame=upsample(symb, sps); % upsample the symbols before filtering
11 sgnl_tx = conv(fltr,u_frame); % transmit filter
12 noise=sqrt(1/2)*(randn(1,length(sgnl_tx))+j*randn(1,length(sgnl_tx)));
13 % set the variance of the complex noise to 1.
14 fltr_normalized=fltr/sum(fltr);
15 % The SNR desired at the output of the matched filter
16 SNR_target = 10*log10(exp(log(10)*EbNo_target/10) * log2(M) * ...
17     (1 /(sum(fltr_normalized .* fltr_normalized) *sps)))
18 % The desired noise standard deviation to get the desired EsNo value
19 std=sqrt( 10.^( -(10*log10((log2(M))) + EbNo_target )/10));
20 rx=sgnl_tx+noise*std;
21 rx_f=conv(sgnl_tx,fltr);
22 rx_n=conv(noise*std,fltr);
23 SNR_before= 10*log10(mean(abs(sgnl_tx).^2) /mean(abs((noise*std).^2)))
24 SNR_after= 10*log10(mean(abs(rx_f).^2) /mean(abs((rx_n).^2)))
```

side lobes (see Figure 6.19). Therefore, various pulse shapes are used to adjust time and spectrum characteristics of the signal. Some commonly used pulse shapes are root-raised-cosine (RRC) pulse shaping which is popularly used in flat fading channels with PSK and QAM modulations like in WiMAX; half-sinusoidal pulse shaping which is used with MSK; and Gaussian filtering which is again used with MSK (GMSK) like in GSM.

Raised cosine filter is a Nyquist filter and it is robust to ISI [15]. If both transmitter and receiver uses RRC filter, then the combined effective filter results in raised cosine filter [16]. Therefore, the signal at the receiver after the RRC filtering (which is matched to the exact transmitter filter response) can lead to ISI free samples (assuming that the channel is frequency flat and the sample timing is perfectly achieved). The Nyquist filter impulse response has its peak amplitude at the symbol instant ($t = 0$) and is zero at all other surrounding symbol instants (see Figure 6.20). That is, it crosses zero at integer multiples of the symbol period. This means that Nyquist-filtered symbols do not interfere with surrounding symbols (zero intersymbol interference). The effects of pulse shaping on QPSK modulation is investigated in the following Matlab code and the corresponding plots are given in Figures 6.21, 6.22, 6.24, and 6.25.

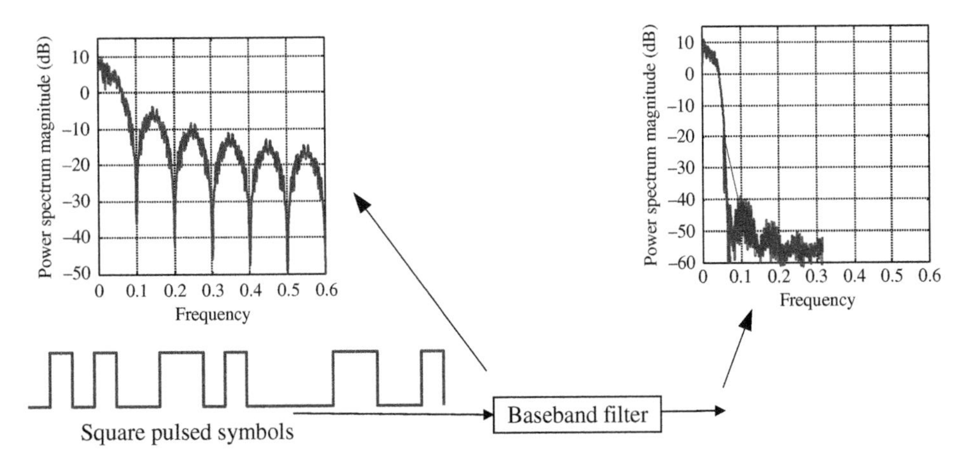

Figure 6.19 Illustration of the effect of baseband filtering on the transmitted signal's spectrum.

```
% Effect of pulse-shaping % QPSK versions
N=1e4; % Number of symbols
sps=8;
%fil=rcosine(1,ovrs,'normal',0.3);
fltr=ones(1,sps);
M=4;
figure(1)
clf
symb= qammod(floor(rand(1,N)*M),M);
u_frame=upsample(symb, sps);
sgnl_tx = conv(fltr,u_frame);
sgnl_tx=sgnl_tx(length(fltr):end-length(fltr));
[Pxx,F]=pwelch(sgnl_tx,[],[],2048,sps);
plot(F-sps/2,fftshift(10*log10(Pxx)),'b-')
hold on
fltr=rcosine(1,sps,'normal',0.3);
symb= qammod(floor(rand(1,N)*M),M);
u_frame=upsample(symb, sps);
sgnl_tx = conv(fltr,u_frame);
tx_4QAM1=sgnl_tx(length(fltr):end-length(fltr));
[Pxx,F]=pwelch(tx_4QAM1,[],[],2048,sps);
plot(F-sps/2,fftshift(10*log10(Pxx)),'r--')
fltr=rcosine(1,sps,'normal',0.9);
symb= qammod(floor(rand(1,N)*M),M);
u_frame=upsample(symb, sps);
sgnl_tx = conv(fltr,u_frame);
tx_4QAM2=sgnl_tx(length(fltr):end-length(fltr));
[Pxx,F]=pwelch(tx_4QAM2,[],[],2048,sps);
plot(F-sps/2,fftshift(10*log10(Pxx)),'k:')
figure(2)
clf
[x1,y1]=hist(abs(tx_4QAM1).^2,100) ;
[x2,y2]=hist(abs(tx_4QAM2).^2,100) ;
semilogy(10*log10(y1),1-cumsum(x1/sum(x1)),'b-')
hold on
semilogy(10*log10(y2),1-cumsum(x2/sum(x2)),'r--')
grid
xlabel('power in dB')
ylabel('cumulative probability')
title('CCDF')
figure(3)
subplot(1,2,1)
plot(real(tx_4QAM1),imag(tx_4QAM1))
title('roll-off=0.3')
subplot(1,2,2)
plot(real(tx_4QAM2),imag(tx_4QAM2))
title('roll-off=0.9')
figure(4)
clf
subplot(2,1,1)
x=vec2mat(tx_4QAM1(100:end-100),sps*2);
plot(real(x(1:end-1,:).'))
title('roll-off=0.3')
subplot(2,1,2)
x=vec2mat(tx_4QAM2(100:end-100),sps*2);
plot(real(x(1:end-1,:).'))
title('roll-off=0.9')
```

Even though Nyquist filtering provides ISI free signaling in flat fading channels, there are applications when ISI is not the most important criteria. Especially if the channel is already frequency-selective, the ISI free filtering might not be the most critical thing anymore. Gaussian filtering, for example, is popularly used with MSK modulation (even though it does not have a good ISI characteristics) because of the superior spectral characteristics.

Gaussian filter impulse response is defined by a Gaussian-shaped pulse. Bandwidth-symbol time product (BT) is related to the filters 3-dB bandwidth and data rate by $BT = B_{3\text{ dB}}T$. As can be seen from Figure 6.23, the filter impulse response spreads to a duration larger than a symbol duration,

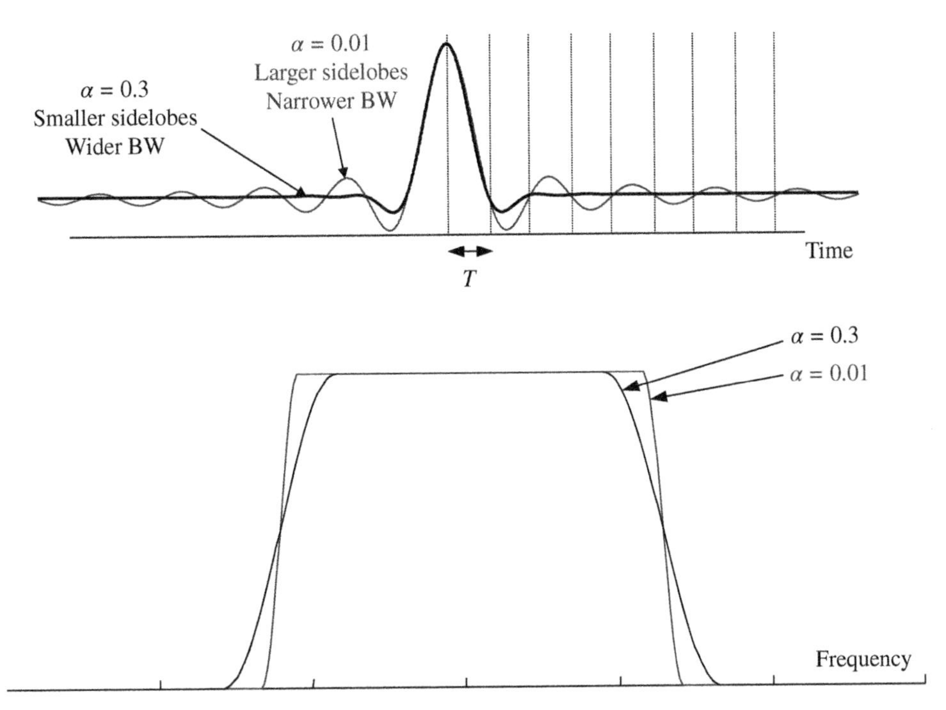

Figure 6.20 Impact of roll-off factor in RC filtering. Roll-off factor provides a direct measure of the occupied bandwidth of the system, $B = (1 + \alpha) * (symbol\ rate)$. Roll-off factor determines the excess bandwidth.

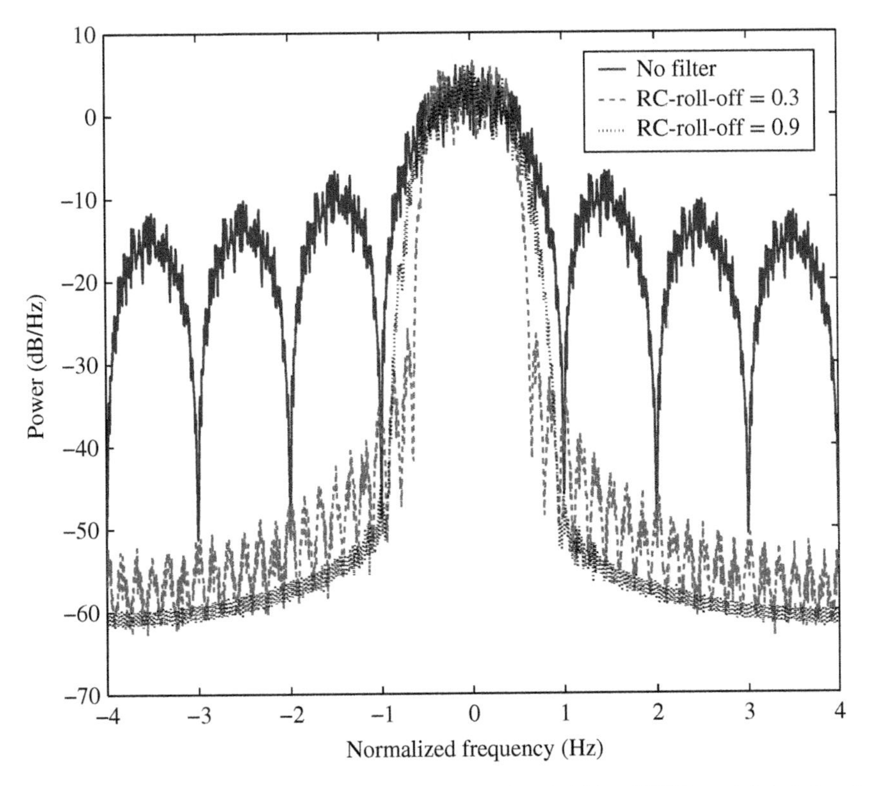

Figure 6.21 Effect of raised cosine filtering in the bandwidth of QPSK-modulated signals.

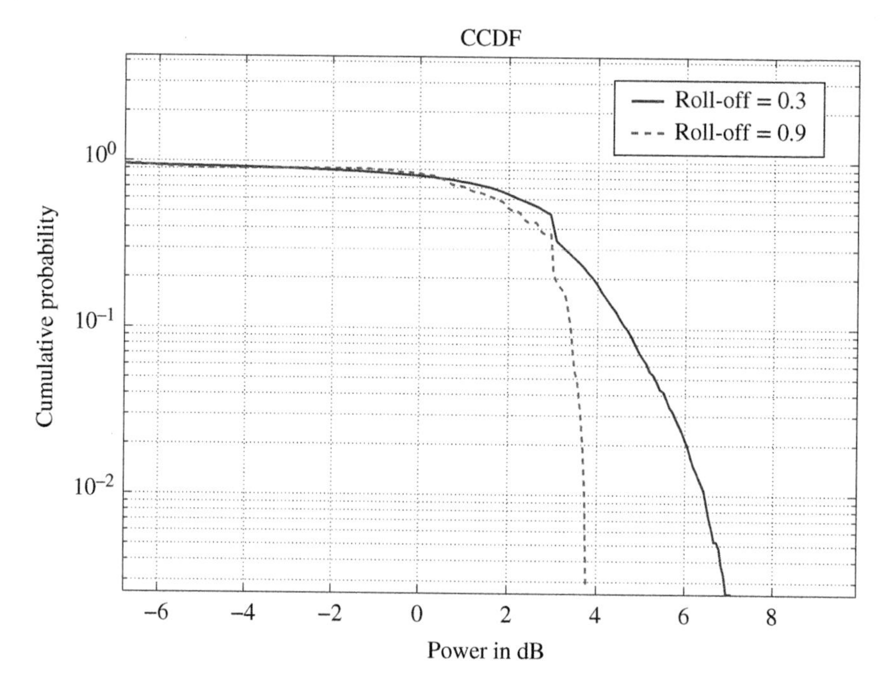

Figure 6.22 Effect of roll-off factor on CCDF in raised cosine filtering. Reducing the roll-off increases the ISI, which is in sense-making each time sample a superposition of multiple data symbols, and hence increases the fluctuation of the time samples. The increase in the dynamic range of the signal can be seen easily in the CCDF plot.

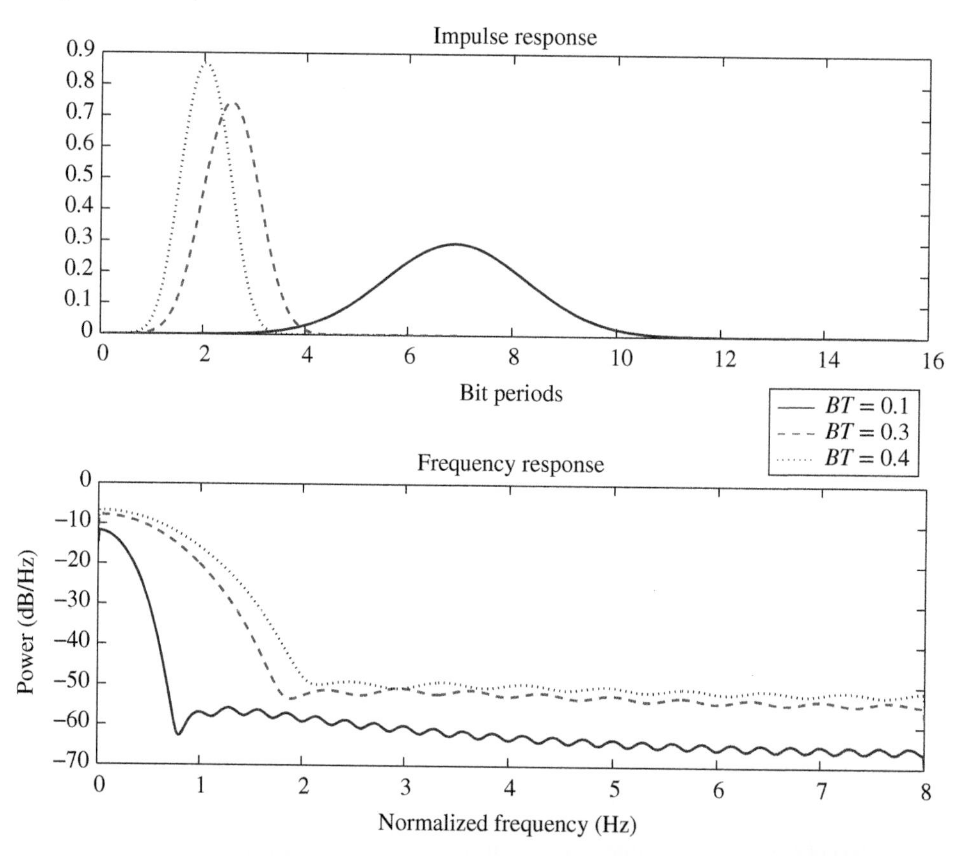

Figure 6.23 Gaussian filter response.

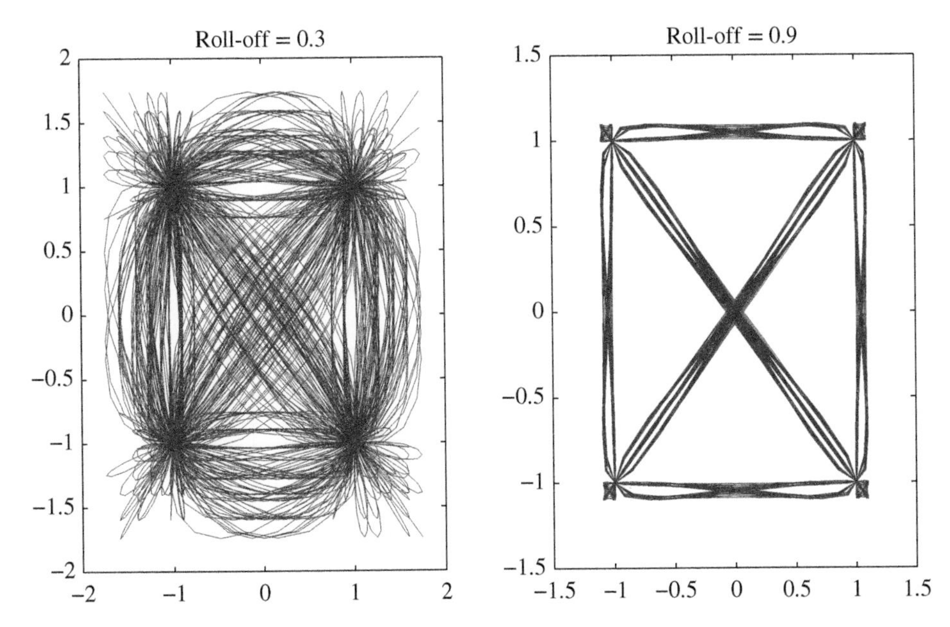

Figure 6.24 Effect roll-off factor on I/Q vector plot in raised cosine filtering. Roll-off factor impacts the transition from one state to another. Increasing roll-off will also increase the variation of this transition.

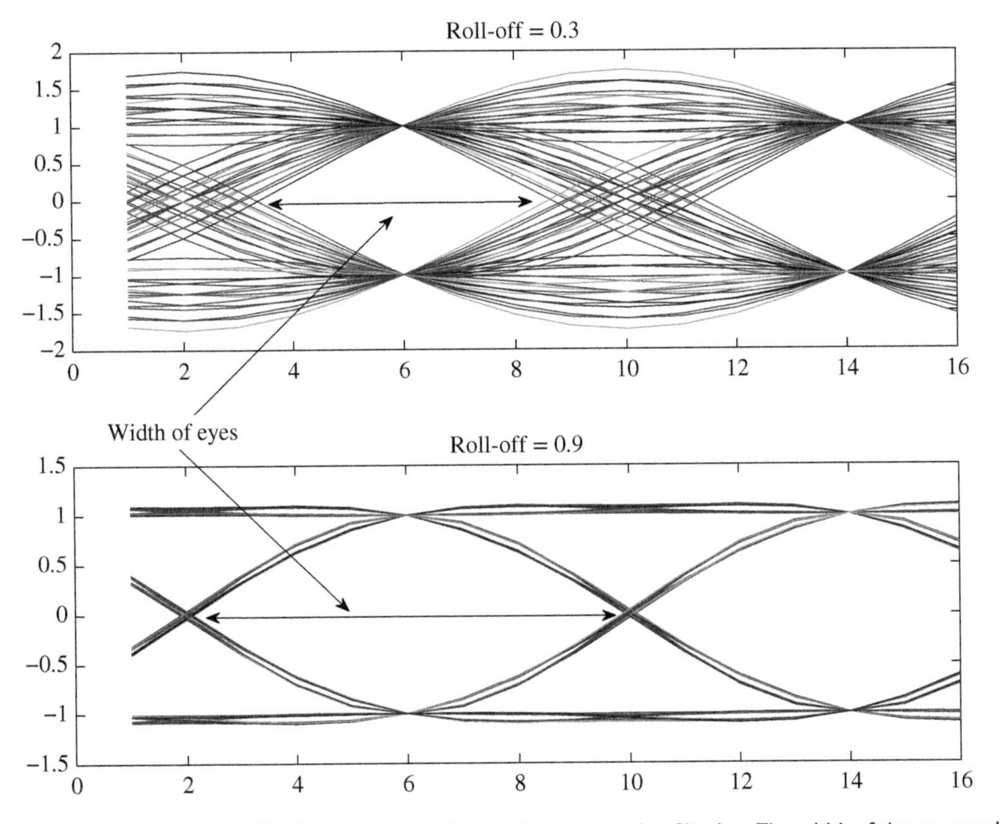

Figure 6.25 Effect of roll-off factor on eye diagram in raised cosine filtering. The width of the eye provides information about tolerance to jitter and sample timing errors. Increasing the roll-off also increases the width of the eye.

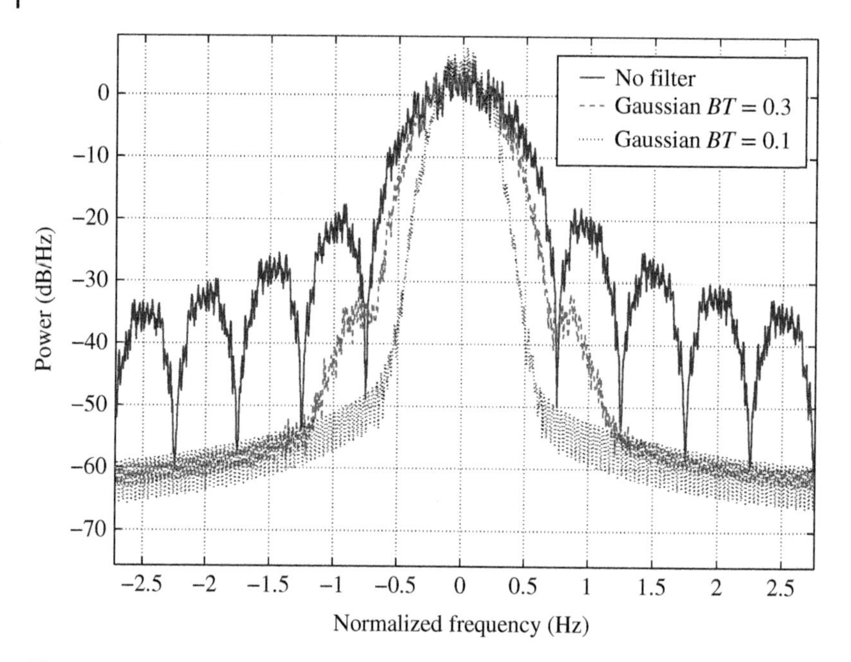

Figure 6.26 Effect of Gaussian filtering in MSK signal's spectrum.

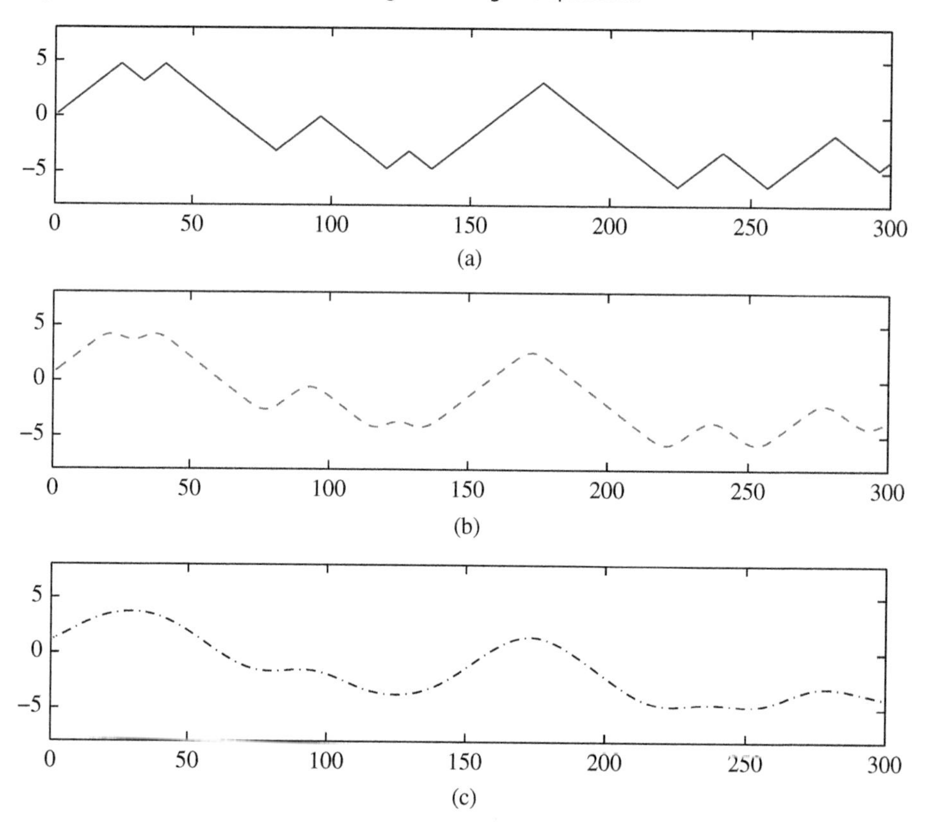

Figure 6.27 Effect of Gaussian filtering in MSK signal's phase. Filtering smooths sudden phase changes. (a) No filter. (b) $BT = 0.3$. (c) $BT = 0.1$.

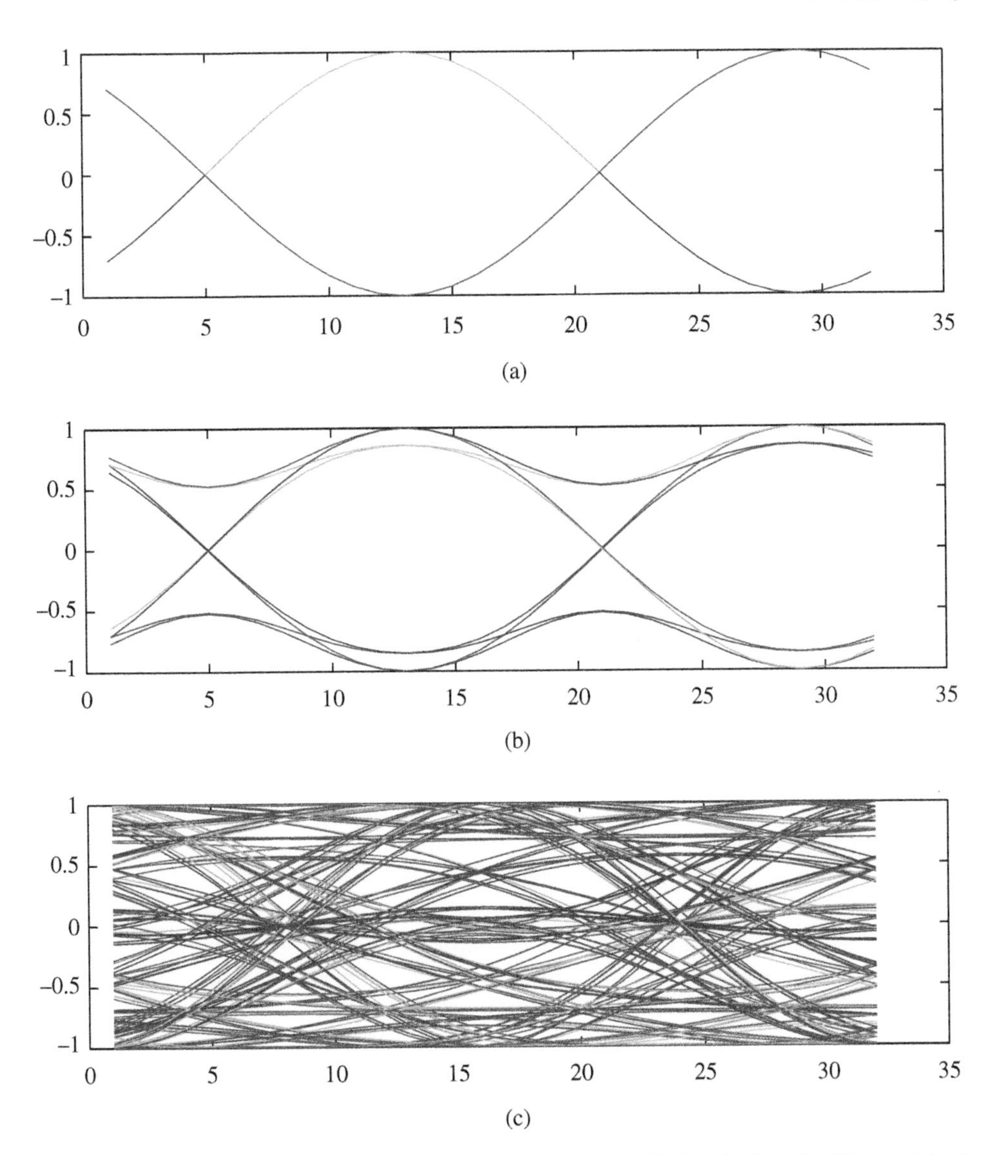

Figure 6.28 Effect of Gaussian filtering in the eye diagram of MSK signal. Gaussian filtering introduces ISI as can be seen from the eye diagram. As the BT decreases, the ISI increases. (a) No filter. (b) $BT = 0.3$. (c) $BT = 0.1$.

leading to ISI. As the BT increases, ISI decreases. On the other hand, lower BT produces a faster power spectrum roll-off, as can be seen in Figure 6.26. Therefore, there is a trade-off between time response and frequency response. For example, in GSM, BT of 0.3 is used to compromise between spectral efficiency and intersymbol interference. In GSM, because of the ISI caused by the pulse-shaping filtering, at the receivers equalizers are needed even if the channel is frequency flat. Figures 6.27 and 6.28 show the effect of Gaussian filtering on the phase and eye diagram of MSK signals, respectively.

6.5 Conclusions

Modulation has always been one of the critical research topics for communication system design as it has direct impact on the performance of the overall system. Digital modulation provides wide variety of options for the system designers depending on the desired metrics to satisfy. In this chapter, basic principles of digital communication and pulse shaping are discussed. There is a lot more than what is discussed in this chapter. There are a lot of great reference materials that cover the theoretical aspects of digital modulation and pulse shaping. This chapter mainly focused on the basics and practical aspects to complement the other chapters in the book.

References

1 G. L. Stuber, *Digital Communications*, 4th ed. McGraw-Hill, 2001.

2 J. G. Proakis, *Principles of Mobile Communication*, 2nd ed. Massachusetts: Kluwer Academic Publishers, 2001.

3 P. Jung, "Laurent's representation of binary digital continuous phase modulation modulated signals with modulation index 1/2 revisited," *IEEE Trans. Commun.*, vol. 42, pp. 221–224, Feb./Mar./Apr. 1994.

4 P.A. Laurent, "Exact and approximate construction of digital phase modulations by superposition of amplitude modulated pulses (AMP)," *IEEE Trans. Commun.*, vol. 34, pp. 150–160, Feb. 1986.

5 European Telecommunications Standards Institute, "European digital cellular telecommunication system (Phase 2)," *Modulation. GSM 05.04. vers. 4.0.3*, Valbonne, Sept. 1994.

6 A. M. Jaradat, J. M. Hamamreh, and H. Arslan, "Modulation options for OFDM-based waveforms: classification, comparison, and future directions," *IEEE Access*, vol. 7, pp. 17263–17278, 2019.

7 L. W. Couch, *Digital and Analog Communications.* Upper Saddle River, NJ: Prentice-Hall, 1997.

8 H. B. Çelebi and H. Arslan, "A joint blind carrier frequency and phase offset detector and modulation order identifier for MPSK signals," in *2010 IEEE Radio and Wireless Symposium (RWS)*, New Orleans, LA, 10–14 Jan. 2010, pp. 348–351.

9 J. Jacobsmeyer, "Adaptive data rate modem," July 1996, U.S. Patent 5541955.

10 S. Sampei, *Applications of Digital Wireless Technologies to Global Wireless Communications.* New Jersey: Prentice Hall, 1997.

11 K. Balachandran, S. Kabada, and S. Nanda, "Rate adaptation over mobile radio channels using channel quality information," in *Proc. IEEE Globecom'98 Commun. Theory Mini Conf. Record*, Sydney, 8–12 Nov. 1998, pp. 46–52.

12 J. M. Hamamreh, M. Yusuf, T. Baykas, and H. Arslan, "Cross MAC/PHY layer security design using ARQ with MRC and adaptive modulation," in *2016 IEEE Wireless Communications and Networking Conference.* IEEE, Doha, 3–6 Apr. 2016, pp. 1–7.

13 M. Karabacak, H. A. Çirpan, and H. Arslan, "An efficient modulation identification algorithm without constellation map knowledge," in *2011 IEEE Radio and Wireless Symposium*, Phoenix, AZ, 16–19 Jan. 2011, pp. 146–149.

14 M. Karabacak, H. A. Çirpan, and H. Arslan, "Adaptive pilot based modulation identification and channel estimation for OFDM systems," in *21st Annual IEEE International Symposium on Personal, Indoor and Mobile Radio Communications*, Istanbul, 26–30 Sept. 2010, pp. 730–735.

15 J. G. Proakis, *Digital Communications.* McGraw Hill, 1995.

16 S. Haykin, *Digital Communications.* Toronto: Wiley, 1988.

7

OFDM Signal Analysis and Performance Evaluation

Hüseyin Arslan

Department of Electrical Engineering, University of South Florida, Tampa, FL, USA
Department of Electrical and Electronics Engineering, Istanbul Medipol University, Istanbul, Turkey

Multi-carrier modulation techniques are widely adopted in various telecommunication standards such as wireless wide area, local area, and personal area networks. Orthogonal frequency division multiplexing (OFDM) is the most popular multi-carrier waveform that is used to achieve high spectral efficiencies. It also has other advantages, such as efficient hardware implementation, low-complexity equalization, and easy multiple-input multiple-output (MIMO) integration. However, OFDM-based systems suffer from high out-of-band emissions (OOBEs), peak-to-average power ratio (PAPR) and strict synchronization requirements. This chapter aims to provide an in-depth understanding of OFDM as well as its performance in the presence of a multipath channel and various radio frequency (RF) impairments.

7.1 Why OFDM?

Multi-carrier modulation techniques are widely adopted in various telecommunication standards. Their primary advantage over single-carrier transmission schemes is their ability to cope with frequency-selective channels for broadband communications. The transmission bandwidth is divided into several narrow subchannels, and data are transmitted in parallel over these subchannels with a set of narrow carriers. If the bandwidth of each carrier is set to be less than the coherence bandwidth of the channel, each carrier experiences a single-tap flat fading channel. As a result, the complex equalizers that are required for single-carrier communications to combat inter-symbol interference (ISI) can be avoided with properly designed multi-carrier systems.

OFDM is the most popular multi-carrier modulation technique that is currently being used in various standards, such as fourth generation (4G) long term evolution (LTE) and the IEEE 802.11 family [1]. It provides several tempting features such as efficient hardware implementation, low-complexity equalization, and easy MIMO integration. OFDM also enables high spectral efficiency as it allows subcarriers to overlap while preserving their orthogonality. On the other hand, OFDM severely suffers from its high OOBE, PAPR, and strict synchronization requirements.

OFDM has found applications in many wireless technologies, including digital audio/video broadcasting in Europe and high-speed wireless local area network (WLAN) standards such as IEEE 802.11a/g/n/ac/ax. It has also been applied to wireless metropolitan area networks (WMANs) for fixed and mobile wireless access (e.g. 802.16-2004 for fixed and 802.16-2005 for mobile WiMAX), wireless personal area networks (WPANs) through multiband ultrawideband, and wireless regional area networks (WRANs) through cognitive radio (e.g. IEEE 802.22).

Wireless Communication Signals: A Laboratory-based Approach, First Edition. Hüseyin Arslan.

The system parameters of various wireless communication standards using OFDM technologies are given in Table 7.1.

OFDM has also become a dominant radio access method for cellular standards. Initially, LTE and LTE-advanced (LTE-A) adopted OFDM for the 4G cellular wireless. Later, numerous waveforms had been proposed [2] considering the disadvantages of OFDM, but none of them can address all requirements of a mobile network with diverse applications. Therefore, OFDM remains as the waveform of the fifth generation (5G) new radio (NR) [3–6], and the multi-numerology OFDM concept [7, 8], which allowed a flexible way of using OFDM, is introduced for different applications. The physical layer of 5G is designed to achieve better flexibility in an effort to support diverse requirements. In a broad sense, numerology refers to the waveform selection (or parameterization) to form resource blocks based on the user requirements and channel conditions. Different users within the same frame are allowed to utilize different waveforms or the same base waveform with a different parameterization that best meets their requirements. Based on this definition, the implementation of the multi-numerology frame structure can be principally categorized into two types as hybrid/mixed waveforms-based and single-waveform-based mixed numerologies. 5G standard has opted for the single-waveform-based numerology structure whereby OFDM has been selected as the parent waveform whose numerologies are defined by their subcarrier spacing values. The differences between 4G LTE and 5G NR are summarized in Table 7.2.

Table 7.1 Exemplary OFDM-based wireless standards and their basic parameters.

Standard	IEEE 802.11(a/g)	IEEE 802.16(d/e)	IEEE 802.22	DVB-T
FFT size	64	128, 256, 512, 1024, 2048	1024, 2048, 4096	2048, 8192
CP ratio	1/4	1/4, 1/8, 1/16, 1/32	Variable	1/4, 1/8, 1/16, 1/32
Bits per symbol	1, 2, 4, 6	1, 2, 4, 6	2, 4, 6	2, 4, 6
Pilots	4	Variable	96, 192, 384	62, 245
Bandwidth (MHz)	20	1.75–20	6, 7, 8	8
Multiple accessing	CSMA	OFDMA/TDMA	OFDMA/TDMA	N/A

Table 7.2 4G LTE and 5G NR comparison.

Criterion	4G LTE	5G NR
Downlink waveform	CP-OFDM	CP-OFDM
Uplink waveform	SC-FDM	CP-OFDM and SC-FDM
Numerology structure	Fixed	Flexible
Resource element	Fixed	Variable
Subcarrier spacing (Δf)	15 kHz	15, 30, 60, and 120 kHz
Max. transmission BW	Up to 20 MHz	Up to 400 MHz
Max. # of RBs	Up to 100	Up to 275
FFT size for max. BW	2048	4096
Normal CP length	4.69 or 5.21 μs	0.60, 1.19, 2.38, and 4.76 μs
Extended CP length	16.67 μs	4.17 μs for 60 kHzΔf

Table 7.3 OFDM for current and future wireless communication systems.

Requirements	OFDM's strength
Broadcasting	OFDM enables the same radio/TV channel frequency usage throughout a country. Large delay spread created by multiple geographically separated stations can be handled easily. Signals received from multiple stations can be combined simply.
Efficient spectrum utilization	The OFDM waveform can easily be shaped by turning off some subcarriers where interference or other users exist. It is a compelling feature, especially for cognitive radio. Variable bandwidth can also be allocated to different users. Scheduling in the frequency domain can be done flexibly and efficiently.
Adaptation/scalability	OFDM systems can be adapted to different transmission environments and available resources easily, due to a large number of parameters suitable for adaption and flexibility. For example, the cyclic prefix (CP) size can be adapted depending on the time dispersion of the channel.
Advanced antenna techniques	MIMO techniques are commonly used with OFDM, mainly because of the reduced equalizer complexity. OFDM also supports smart antennas.
Interoperability	Interoperability can be defined as the ability of two or more systems to exchange information and utilize it. WLAN (e.g. IEEE 802.11), WMAN (e.g. IEEE 802.16), WRAN (e.g. IEEE 802.22), WPAN (e.g. IEEE 802.15.3a), 4G LTE, and 5G NR are using OFDM as their physical layer techniques. Therefore, interoperability becomes easier compared to other technologies.
Multiple accessing and spectral allocation	Support for multiuser access are already inherited in the system design by assigning groups of subcarriers to different users like in orthogonal frequency division multiple access (OFDMA). Interleaved, randomized, or clustered assignment schemes can be used for this purpose. Hence, it offers very flexible multiple accessing and spectral allocation capability for future radios without any extra complexity or hardware.
NBI immunity	Narrow-band interference (NBI) affects only some subcarriers in OFDM systems. These subcarriers can be simply turned off.
Immunity to spectral fading	Poor (faded) subcarriers can be loaded with more robust modulations or can be avoided completely. Subchannel-based adaptation is possible, especially for low-mobility applications.
Opportunistic scheduling	Carriers or a group of carriers can be assigned adaptively to the users experiencing the best conditions in these channels.

The flexibility and adaptiveness of OFDM, along with several other great features, made it a desirable access technology of many current and future wireless communication systems. In Table 7.3, a summary of the requirements for current/future wireless communication systems and how OFDM can fulfill these requirements are described.

7.2 Generic OFDM System Design and Its Evaluation

In this section, a generic OFDM system is described in detail. First, a simple OFDM system in a multipath channel with additive white Gaussian noise (AWGN) is modeled and analyzed, then the effect of the multipath channel and RF impairments on the OFDM signal are investigated. Finally, coherent and differential OFDM signaling is discussed.

7.2.1 Basic CP-OFDM Transceiver Design

A simplified block diagram of a basic OFDM transceiver is given in Figure 7.1. In a multipath fading channel, each subcarrier might experience different attenuation due to the frequency selectivity. Furthermore, the power level on some subcarriers can be significantly less than the average power level because of the deep fades. As a result, the overall bit-error-rate (BER) might be dominated by a few subcarriers with low power levels. Channel coding is used to reduce the degradation of system performance due to this problem. In addition, interleaving is applied to randomize the occurrence of bit errors and provide immunity to burst errors. Initially, coded and interleaved data bits are mapped to the constellation points in order to obtain data symbols. Afterward, the serial data symbols are converted to parallel data symbols, which are fed to the inverse fast Fourier transform (IFFT) block to obtain the time-domain OFDM symbol as follows:

$$x_m(t) = \frac{1}{\sqrt{N}} \sum_{k=0}^{N-1} X_m(k) e^{j2\pi k\Delta f t} \quad 0 \leq t \leq T_s, \tag{7.1}$$

where $X_m(k)$ is the transmitted data symbol at the kth subcarrier of the mth OFDM symbol, N is the number of subcarriers, T_s is the OFDM symbol duration, and Δf is the subcarrier spacing. The discrete version of the time-domain OFDM symbol can be given with a sampling rate of T_s/N as follows:

$$x_m(n) = \frac{1}{\sqrt{N}} \sum_{k=0}^{N-1} X_m(k) e^{j2\pi k\Delta f(nT_s/N)} \quad 0 \leq n \leq N-1. \tag{7.2}$$

The orthogonality among the subcarriers is achieved by ensuring $T_s = \frac{1}{\Delta f}$. Accordingly, the OFDM symbol is expressed as follows:

$$x_m(n) = \frac{1}{\sqrt{N}} \sum_{k=0}^{N-1} X_m(k) e^{j2\pi kn/N} \quad 0 \leq n \leq N-1. \tag{7.3}$$

In the following step, the CP is added by copying the last part of the IFFT sequence and appending it to the beginning as a guard interval. The CP ensures the circularity of the channel and its length is determined based on the maximum excess delay of the channel. The CP-OFDM symbol formation

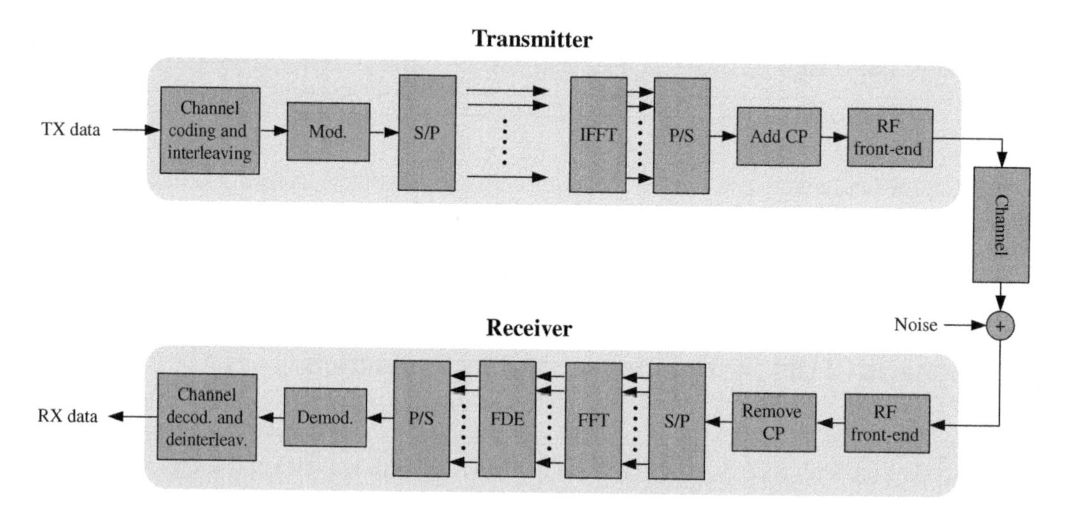

Figure 7.1 Block diagram of a basic CP-OFDM transceiver.

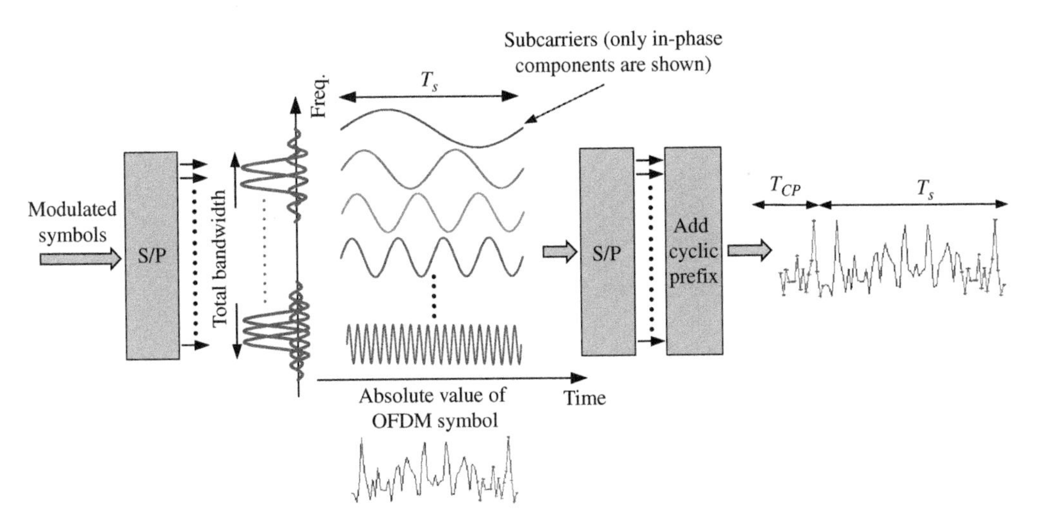

Figure 7.2 Illustration of a CP-OFDM symbol formation at the transmitter.

at the transmitter is shown in Figure 7.2. Subsequently, the time-domain baseband signal passes through the RF front-end components before transmission.

The signal is received along with noise after passing through the multipath radio channel. Initially, a band-pass filter is applied to the received analog signal and then it is down-converted to baseband and digitized by the RF front-end components. Afterward, the guard interval, CP, is removed in the following step. Assuming perfect time and frequency synchronization, the baseband OFDM signal at the receiver can be formulated as follows:

$$y_m(n) = \sum_{l=0}^{L-1} x_m(n-l)h_m(l) + w_m(n), \tag{7.4}$$

where L is the number of sample-spaced channel taps, $w_m(n)$ is complex-valued AWGN samples, and $h_m(l)$ shows the time-domain channel impulse response (CIR) for mth OFDM symbol, and it is assumed as a time-invariant linear filter, which is expressed as $h_m(l) = \sum_{i=0}^{L-1} h_m(i)\delta(l-i)$. In the following step, a fast Fourier transform (FFT) is performed, and the frequency-domain OFDM signal at the receiver is formulated as follows:

$$Y_m(k) = FFT\{y_m(n)\} = X_m(k)H_m(k) + W_m(k) \quad 0 \le k \le N-1, \tag{7.5}$$

where $H_m(k) = \sum_{l=0}^{L-1} h_m(l)e^{-j2\pi kl/N}$ is the channel frequency response, and $W_m(k)$ is the component due to the noise contribution. It should be noted that CP-OFDM converts the convolution operation in the time domain into the multiplication operation in the frequency domain. Hence, simple one-tap frequency domain equalization (FDE) can be used to recover the transmitted symbols. After the FDE operation, symbols are demodulated, deinterleaved, and decoded to obtain the transmitted data bits.

7.2.2 Spectrum of the OFDM Signal

OFDM signal can be considered as a set of orthogonal carriers, which are windowed in the time domain. The inherent rectangular window in the time domain leads to sinc-shaped subcarriers in the frequency domain. The superposition of orthogonally overlapped subcarriers are shown in Figure 7.3. Assuming that the data at each subcarrier are statistically independent and mutually

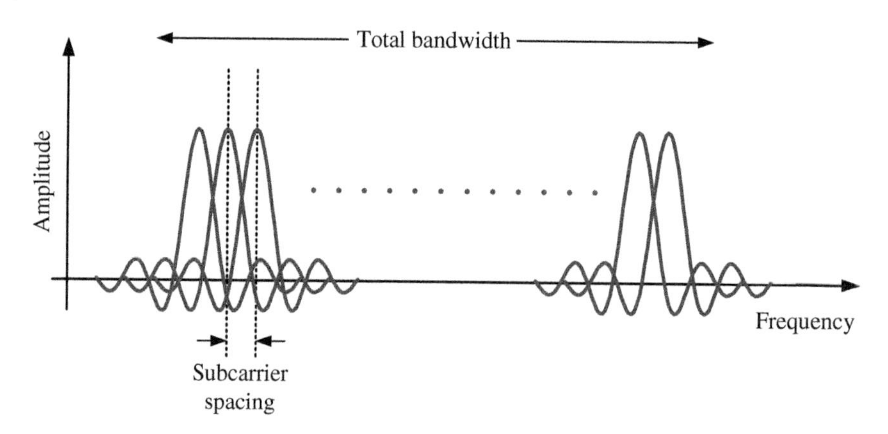

Figure 7.3 Orthogonally overlapped OFDM subcarriers.

orthogonal, the power spectral density (PSD) of an OFDM signal is obtained by summing the power spectra of individual subcarriers, and it is expressed by the following equation [9]:

$$PSD = \frac{\sigma_d^2}{T_s} \sum_{k=0}^{N-1} \left(\operatorname{sin c}\left[(f - k\Delta f)\, T_s\right]\right)^2, \tag{7.6}$$

where σ_d^2 represents the variance of the data symbols and k stands for the subcarrier index.

The following Matlab script provides a simple example of generating multiple OFDM symbols. The time- and frequency-domain plots of the OFDM signal are shown in Figure 7.4. It should be

```
1  % Generation of a simple OFDM signal
2  N=256; % FFT size
3  CP_size=1/4; % CP ratio
4  used_subc_indices=35:1:N-35; % used subcarriers
5  symbols=2*(randn(1,N)>0)-1; % BPSK symbol generation
6  to_ifft=zeros(1,N);
7  to_ifft(used_subc_indices)=symbols(used_subc_indices);
8  time=ifft(to_ifft);
9  time_cp=[time(end-N*CP_size+1:end) time]; % adding CP
10
11 clf; subplot(2,2,[1,2])
12 plot(10*log10(abs(time_cp).^2))
13 hold on
14 plot(10*log10(abs(time_cp(1:64)).^2),'r')
15 title('(a) An OFDM time signal')
16 xlabel('Time sample index')
17 ylabel('Power (dB)')
18 subplot(2,2,3)
19 pwelch(time_cp)
20 title('(b) OFDM spectrum (only one symbol)')
21
22 % -------- Multiple Symbols --------------
23 num_OFDM_symbols=1000; % Number of OFDM symbols to be generated
24 frame=[];
25 N=256;
26 CP_size=1/4;
27 for ii=1:num_OFDM_symbols
28     symbols=2*(randn(1,N)>0)-1;
29     used_subc_indices=25:1:256-25;
30     to_ifft=zeros(1,N);
31     to_ifft(used_subc_indices)=symbols(used_subc_indices);
32     time=ifft(to_ifft);
33     time_cp=[time(end-N*CP_size+1:end) time]; % CP removal
34     frame=[frame time_cp]; % Concatenation of generated symbols
35 end
36 subplot(2,2,4)
37 pwelch(frame)
38 title('(c) OFDM spectrum (multiple symbols)')
```

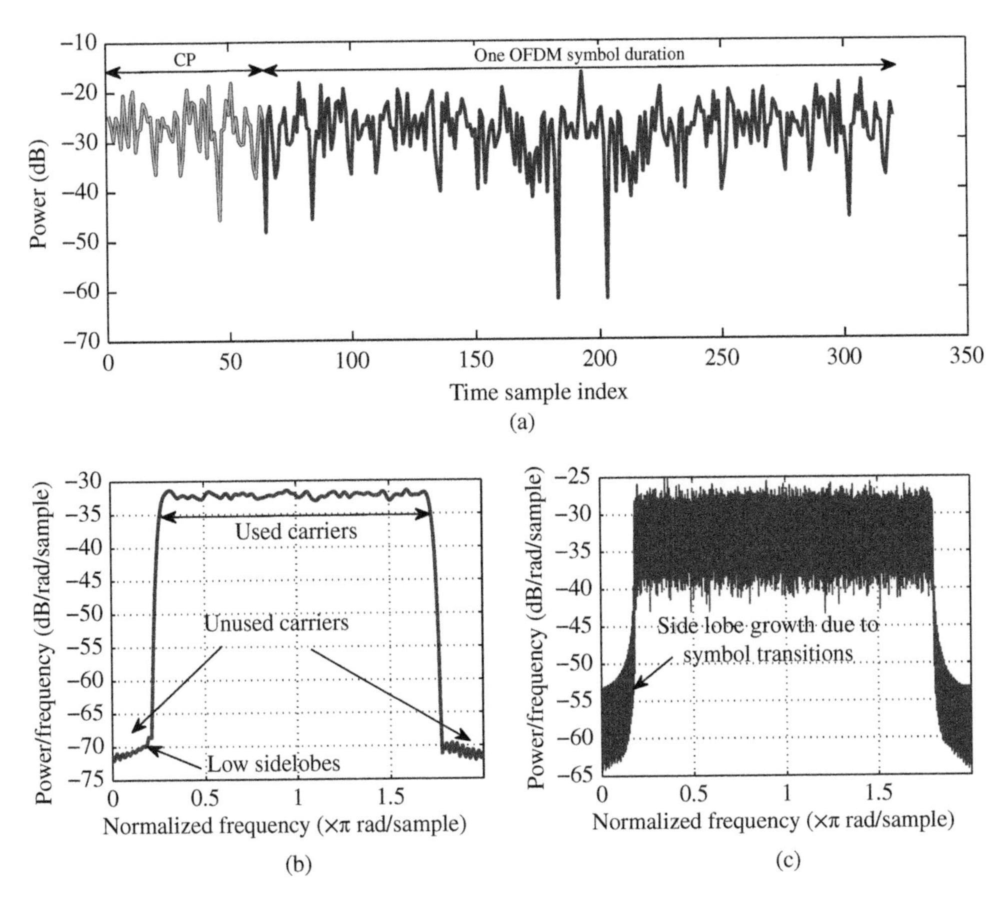

Figure 7.4 OFDM signal in the time and frequency domains. (a) An OFDM time signal. (b) OFDM spectrum of only one symbol (i.e. no symbol transition). (c) OFDM spectrum of multiple symbols (i.e. with symbol transitions).

noted that when the OFDM symbols are concatenated to each other, the sharp symbol transition between symbols creates spectral regrowth in the adjacent channels.

The sidelobes of the sinc-shaped subcarriers at the edges and the discontinuity at the boundaries of subsequent OFDM symbols cause significant interference to the neighboring bands. Typically, the interference is managed by various windowing/filtering approaches along with the guard band allocation [10–12] to meet the spectral mask requirements of various standards. The windowing/-filtering operations reduce the unwanted emissions, but they need an extra period, which extends the guard duration between the consecutive OFDM symbols. Also, additional guard bands might still be required between adjacent channels to deal with the remaining interference. In other words, better interference mitigation is realized with the cost of reduced spectral efficiency. Accordingly, the future communication systems have to optimize the guards in both time and frequency domains to improve the spectral efficiency [13].

A time-domain windowing operation is implemented to demonstrate how to control the OOBEs of OFDM in this part. The windowing operation can smooth the sharp symbol transitions to improve the spectral confinement with low complexity. Various windowing functions have been compared thoroughly [14] with different trade-offs between the main lobe width and the sidelobe suppression. For example, raised cosine (RC) windowing function is one of the commonly employed windowing functions in the literature [15–17]. A significant interference reduction can be achieved by changing the roll-off factor of the RC window. However, the OOBE improvement comes at the cost of longer OFDM symbol duration and lower spectral efficiency, as mentioned earlier. The RC window function can be defined as follows:

$$g_n = \begin{cases} \frac{1}{2} + \frac{1}{2}\cos\left(\pi + \frac{\pi n}{\alpha N_s}\right), & \text{for } 0 \leq n < \alpha N_s \\ 1, & \text{for } \alpha N_s \leq n < N_s \\ \frac{1}{2} + \frac{1}{2}\cos\left(\pi \frac{n - N_s}{\alpha N_s}\right), & \text{for } N_s \leq n < (1+\alpha)N_s, \end{cases} \tag{7.7}$$

where $N_s = N + N_g$ is the symbol length in samples and α is the roll-off factor. Here, N_g represents the symbol extension. The total symbol length is $(1 + \alpha)N_s$. However, adjacent symbols partially overlap over a length of αN_s from each side, causing the actual symbol time to be N_s. To maintain the orthogonality between the OFDM subcarriers and the system resistance to ISI, the symbols are extended using both prefix and postfix. These durations for both prefix and postfix extensions are chosen in such a way that they cover the overlapping period of the symbols. The following Matlab script demonstrates the effect of raised cosine windowing and the results are plotted in Figure 7.5 for a windowing length of 1/16th of the OFDM symbol duration.

```
1  %Effect of the windowing operation on spectrum of an OFDM signal
2  %Please rerun the script with different roll-off factors
3  N = 256;
4  CP = 1/4;
5  G = CP*N;
6  %RC windowing CP
7  CP2 = 1/16; % the ratio of the additional guard period
8  G2 = CP2*N;
9  num_OFDM_symbols=100;
10 frame=[]; frame_RC=[]; x0 = zeros(1, N+G);
11 %------Raised-cosine (comparison)-------------
12 RRC = 0.5 + 0.5.*cos(pi + pi.*[0:G2-1]./G2);
13 LRC = 0.5 + 0.5.*cos(pi.*[0:G2-1]./G2);
14 %--------------------------------------------
15
16 for ii=1:num_OFDM_symbols
17     %symbols=2*(randn(1,N)>0)-1;
18     symbols= qammod( floor(rand(1,N)*4) ,4);
19     used_subc_indices=[floor(N/4):1:N/2-1 N/2+1:1:N-floor(N/4)];
20     to_ifft=zeros(1,N);
21     to_ifft(used_subc_indices)=symbols(used_subc_indices);
22     time=ifft(to_ifft);
23     time_cp=[time(end-G+1:end) time];
24     frame=[frame time_cp];
25
26 %------Raised-cosine (comparison)-------------
27     Ccp = LRC.*x0(G+[1:G2]) + RRC.*time_cp(end-G-G2+[1:G2]);
28     frame_RC = [frame_RC, Ccp, time_cp];
29     x0 = time_cp;
30 end
31
32 figurc(1)
33 clf
34 [Pxx,F]=pwelch(frame,[],[],2048,1,'twosided');
35 plot(F,10*log10(Pxx))
36 hold on
37 [Pxx,F]=pwelch(frame_RC,[],[],2048,1,'twosided');
38 plot(F,10*log10(Pxx),'r-')
```

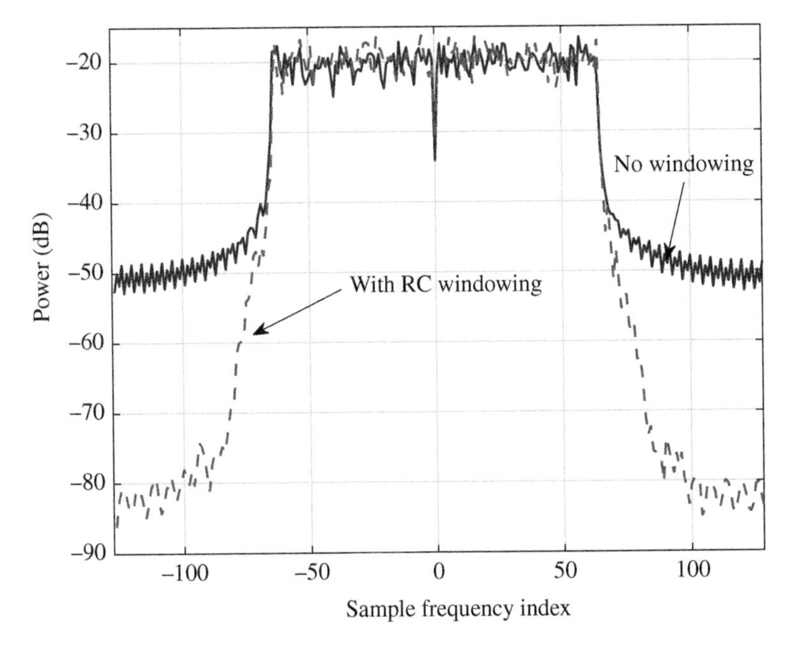

Figure 7.5 Spectrum of the windowed-OFDM signal.

Edge windowing [18] is a more efficient windowing approach to reduce the high OOBE of CP-OFDM. Although the outer subcarriers have a higher contribution to the OOBE compared to the inner subcarriers, conventional windowing techniques apply the same window for all subcarriers within an OFDM symbol. On the other hand, the edge windowing approach applies windows with higher sidelobe suppression capabilities to edge subcarriers while it applies lighter windows to inner ones.

7.2.3 PAPR of the OFDM Signal

PAPR is a critical issue related to CP-OFDM systems, including other multi-carrier modulation schemes, due to the random addition of subcarriers in the time domain [19]. The theoretical upper bound for PAPR of an OFDM signal is 10 $\log_{10}(N)$ dB assuming all N subcarriers are active and modulated with the same phase shift keying (PSK) modulation. Also, the PAPR increases as the modulation order increases when quadrature amplitude modulation (QAM) modulation is considered. However, the theoretical upper bound is only achievable when all subcarriers are equally modulated, which is unlikely, especially as N increases. Therefore, the theoretical upper bound does not have practical usage to describe the amplitude statistics of an OFDM signal. Usually, the statistical distribution of the PAPR is used for developing efficient PAPR reduction techniques.

Assume that the modulated symbols at each subcarrier are statistically independent and mutually orthogonal. These symbols are mapped to the time domain through an IFFT operation. As a result, each OFDM time sample consists of a combination of N independent and identically distributed random variables. When N is sufficiently large, time samples of an OFDM signal can be modeled with Gaussian distribution according to the central limit theorem. If both the real and imaginary parts of the signal are Gaussian distributed with zero mean and variances of $\sigma_r^2 = \sigma_i^2 = \sigma^2/2$, the amplitude of the time-domain signal becomes Rayleigh distributed, $f(r) = \frac{2r}{\sigma_r^2} e^{r^2/\sigma_r^2}$, where σ_r^2 is the average power [20].

The probability of the power that exceeds a threshold for a given OFDM sample can be expressed as follows:

$$P(f(r) > r_o) = \int_{r_o}^{\infty} f(r)dr = e^{r_o^2/\sigma_r^2} = e^{PAPR_o}. \tag{7.8}$$

Assuming that all N samples are independent in an OFDM symbol, the complementary cumulative distribution function (CCDF) of the PAPR can be calculated as follows:

$$CCDF[PAPR] = 1 - (1 - e^{r_o^2/\sigma_r^2})^N = 1 - (1 - e^{PAPR_o})^N. \tag{7.9}$$

Figure 7.6 shows the CCDF of the PAPR for various N using the analytical expression above. The PAPR increases with an increase in N. Similar curves can be obtained through simulations using the following Matlab script. It should be noted that the OFDM signal is oversampled before passing through the digital-to-analog converter (DAC) in practical communication systems and the analog signal after the DAC usually has a higher PAPR value compared to the given PAPR expression in Eq. (7.9).

```
1  % Theoretical and numerical evaluation for CCDF of the PAPR
2  % Please rerun the script for various FFT sizes
3  N=256;
4  Pr=1-(1-exp(-[1:15])).^N;
5  semilogy(10*log10(1:15),Pr,'k-') % Theoretical plot
6
7  % Simulated plot for the CCDF of the PAPR
8  for kk=1:300000 %
9      symbols=2*(randn(1,N)>0)-1; % random BPSK symbols
10     time=ifft(symbols); %IFFT
11     time_cp1=[time(end-N*CP_size+1:end) time];
12     meanP = mean(abs(time_cp1).^2);
13     peakP = max(abs(time_cp1).^2);
14     PAPR2(kk)=peakP/meanP;
15 end
16
17 [x,y]=hist(10*log10(PAPR2),[4:0.2:13]);
18 semilogy(y,1-cumsum(x/sum(x)),'r—')
19 grid
```

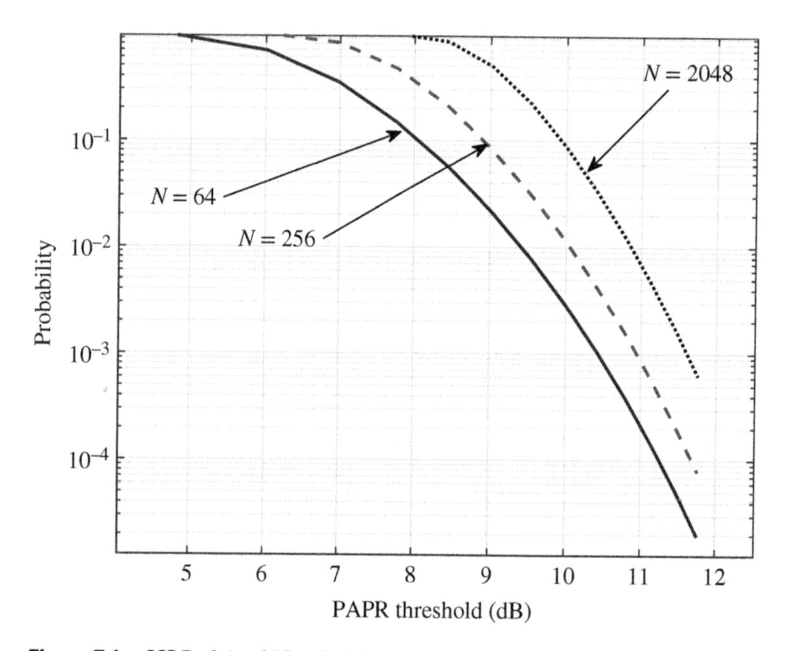

Figure 7.6 CCDF of the OFDM PAPR for various number of subcarriers.

7.2.4 Performance in Multipath Channel

A transmitted signal may arrive at a receiver either directly (i.e. line-of-sight [LOS]) or after being reflected from various objects in the environment (i.e. non-LOS [NLOS]). These reflected signals from different surfaces travel through different paths and accordingly reach the transmitter with different delays and gains. This propagation environment is usually referred to as a multipath channel and illustrated in Figure 7.7. Multipath propagation creates small-scale (a.k.a. Rayleigh) fading effects on the received signal, as shown in Figure 7.8. The performance of OFDM systems in time and frequency-dispersive multipath channels is described in this section.

7.2.4.1 Time-Dispersive Multipath Channel

The multipath channel causes dispersion in the time domain and produces ISI. The dispersion in the time domain might lead to a frequency-selective fading, depending on the transmission bandwidth of the signal. The coherence bandwidth of the channel (B_c) is defined as the bandwidth, in which the channel frequency response can be considered as flat (i.e. highly correlated). It is inversely proportional to the delay spread in the propagation environment. When the transmission bandwidth exceeds the coherence bandwidth of the channel, the signal experiences a frequency-selective fading.

The frequency-selective fading and ISI result in significant communication performance degradation. Channel equalizers are used to compensate for the ISI effect of the multipath channel. The complexity of these equalizers depends on the number of resolvable channel taps. Single-carrier systems transceive signals with shorter symbol duration compared to multi-carrier systems, which utilize the same transmission bandwidth, and they resolve more channel taps. As a result, sophisticated equalizers are required for broadband single-carrier systems.

OFDM has been promoted for broadband communications due to its high performance in time-dispersive channels. The bandwidth of each subcarrier is set to be less than the coherence bandwidth of the channel. Hence, each subcarrier experiences a single-tap flat fading channel, and no complex multi-tap channel equalizer is needed. To avoid the multipath components that leak from the tail of the previous symbol to the head of the following symbol, OFDM symbol duration is extended by adding a guard interval with a period of T_g to the beginning of each symbol either with zero-padding (ZP) or CP. Figure 7.9 illustrates the effect of time-dispersive multipath channel on OFDM symbols and the use of the CP to avoid ISI. The guard interval should be longer than the maximum excess delay of the channel, which is defined as the delay between the first and last received paths over the channel.

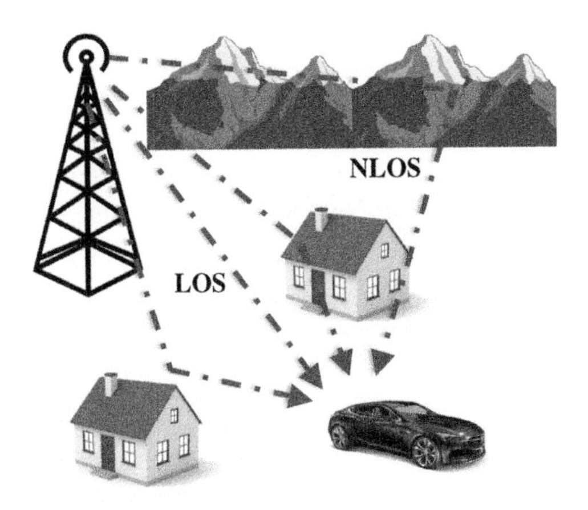

Figure 7.7 Wireless communications in the presence of a multipath propagation.

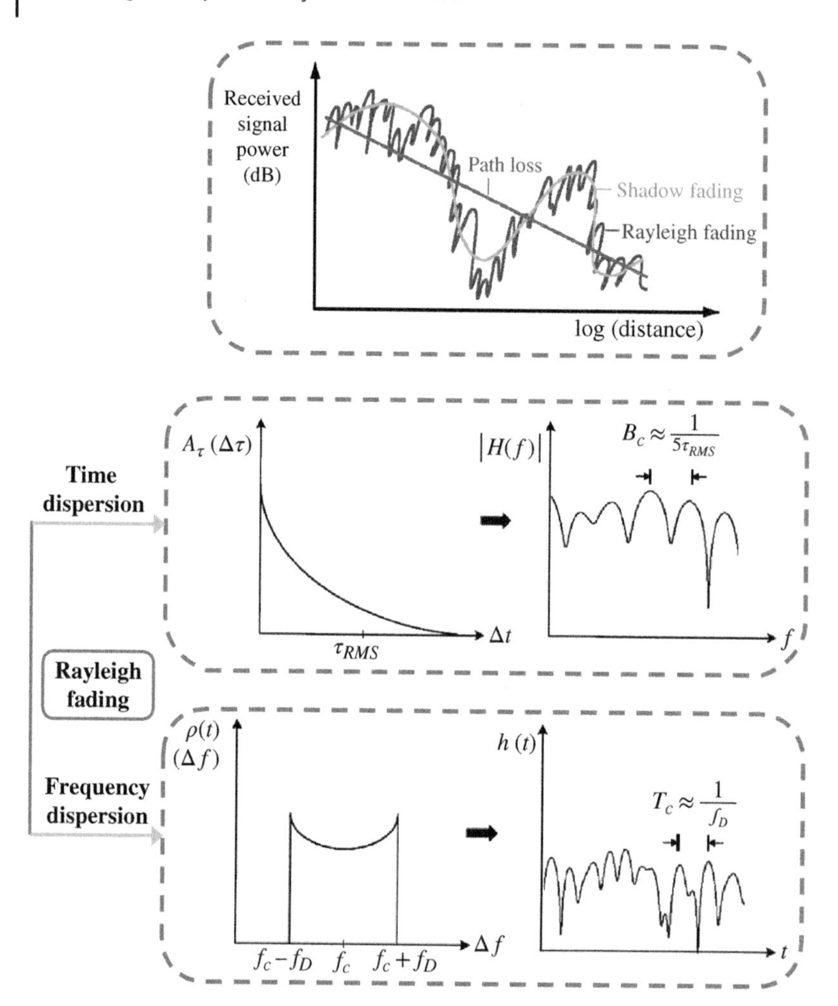

Figure 7.8 Time- and frequency-dispersive multipath channel.

The utilization of CP is more common than ZP due to its advantages against various impairments, as discussed in the following section. It should be noted that the CP constitutes redundant information, and hence, it reduces the spectral efficiency. Also, a portion of the transmission power is wasted for CP, and it reduces the power efficiency as well. The CP duration is hard-coded in 4G LTE and does not take into account the individual user's channel delay spread. As a result, the fixed guard interval leads to a degradation in the spectral and power efficiencies.

The channel can be considered as a filter, and a transmitted signal arrives at a receiver after convolving with the channel. This convolution operation in the time domain corresponds to a multiplication operation in the frequency domain if the channel is circular. The CP part of the OFDM signal ensures the circularity of the channel and enables easy FDE with a simple multiplication operation. Assuming the channel is slowly varying in time, the received signal at each subcarrier can be expressed as follows:

$$Y_m(k) = X_m(k)H(k), \tag{7.10}$$

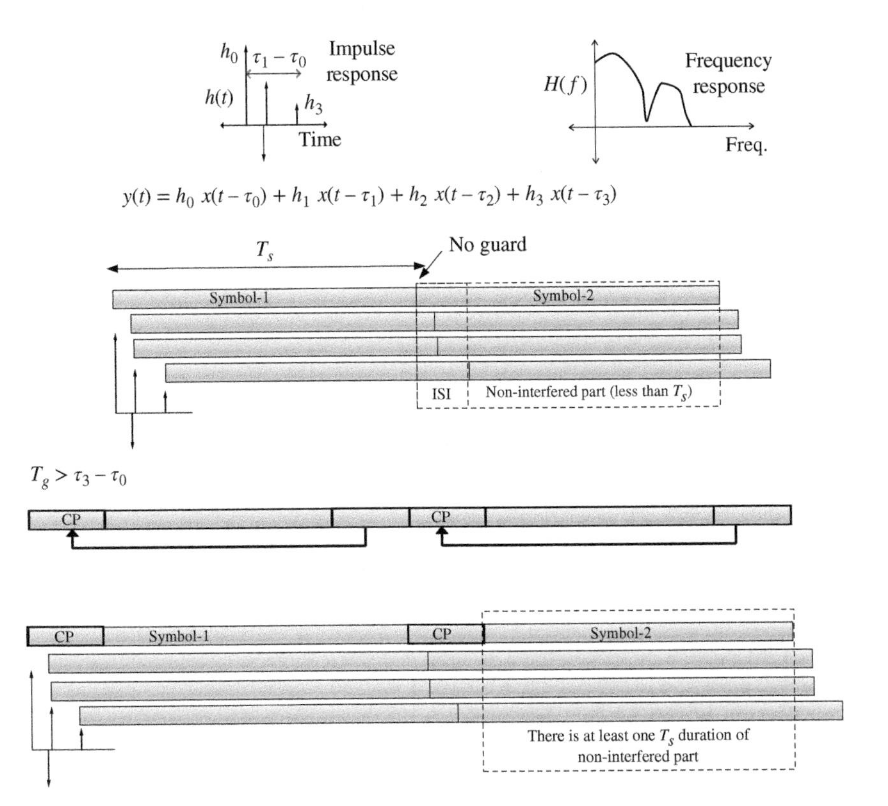

Figure 7.9 Effect of time-dispersive multipath channel on OFDM symbols and the use of the CP to avoid ISI.

where $H(k)$ represents the complex channel frequency response, which is assumed to be constant within a subframe. It should be noted that the channel frequency response is subcarrier-dependent, as shown in Figure 7.10. However, the variation across the subcarriers is smooth. In other words, the channel frequency response of closely spaced subcarriers is correlated. Once the channel frequency response is estimated, the equalization can be performed as shown below:

$$Y_{equalized}(k) = \frac{X_m(k)H(k)H^*(k)}{|H(k)|^2}. \tag{7.11}$$

The BER performance of OFDM in time-dispersive/frequency-selective multipath fading channel could be poor due to the deep fadings in some carriers. In single-carrier systems with perfect channel equalization capability, multipath time dispersion allows path diversity, and in theory, it provides a BER performance close to the AWGN channel. However, in OFDM, the path diversity is lost as OFDM converts the frequency-selective channel to frequency flat channel for each subcarrier. Each subcarrier experiences a random fading with an envelope that has a Rayleigh distribution. The error floor and poor performance in fading channels can be handled with the forward error correction (FEC) and frequency interleaving mechanisms. The following Matlab script simulates an OFDM system in both AWGN and multipath fading channel. Figure 7.11 presents the corresponding BER performance of OFDM system.

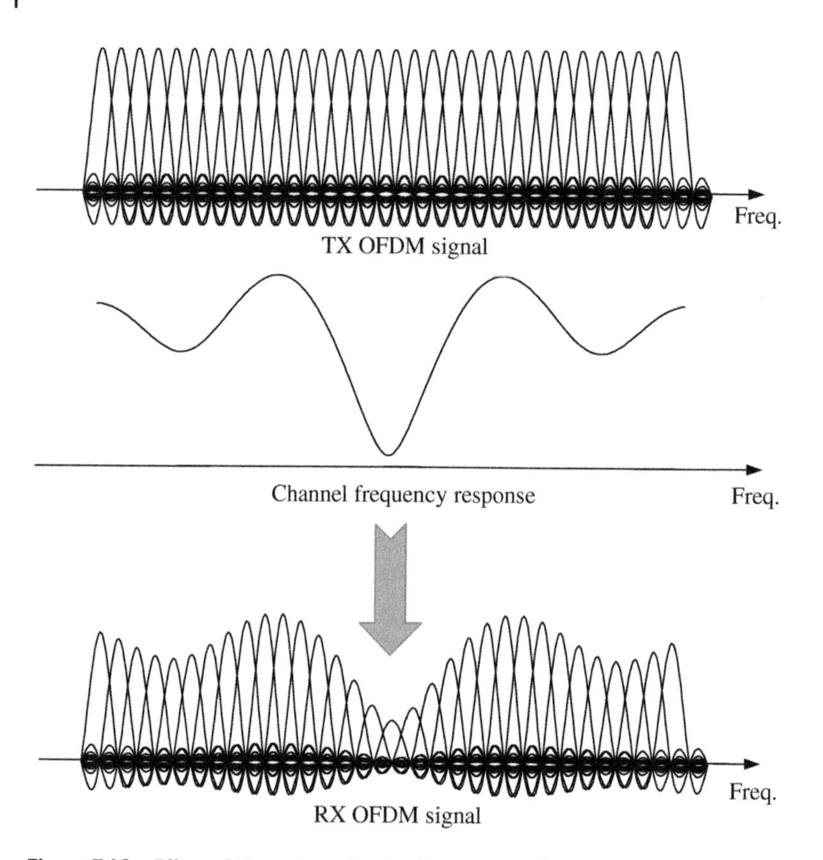

Figure 7.10 Effect of time-dispersive (i.e. frequency selective) channel on OFDM subcarriers.

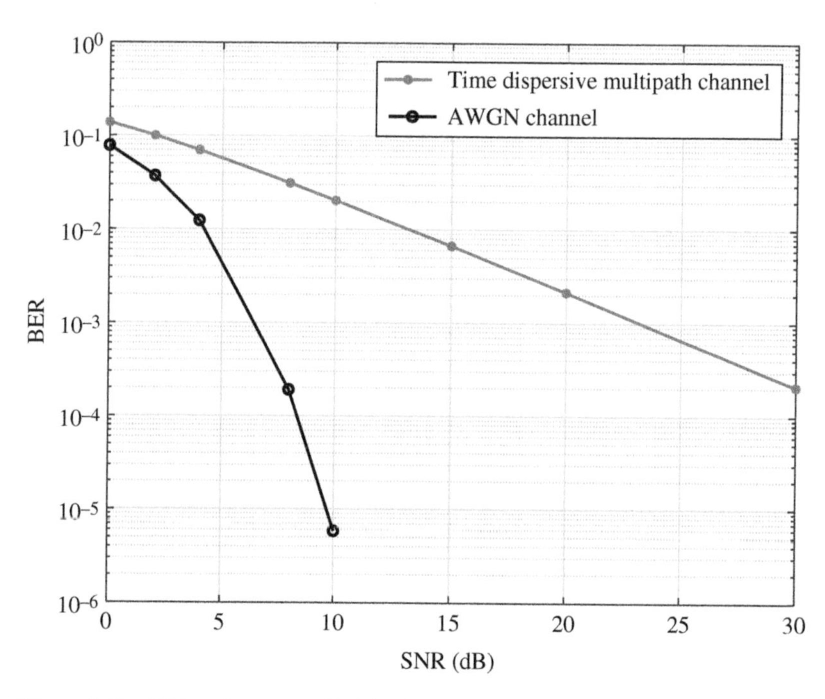

Figure 7.11 BER performance of OFDM in time-dispersive multipath channel and AWGN channel.

```
1  % BER performance of OFDM in frequency-selective and AWGN channel
2  SNR=[0 2 4 6 8 10 15 20 30];% dB
3  N=256; % FFT size
4  CP_size=1/4; % CP length as a ratio of the OFDM symbol duration
5  G=N*CP_size;
6
7  for num=1:1000
8   symbols=2*(randn(1,N)>0)-1; % Random BPSK symbols in each carrier
9   time=ifft(symbols)*sqrt(N); % IFFT (using all the subcarriers)
10  time_cp=[time(end-N*CP_size+1:end) time];
11  L=7;% Number of channel taps
12  cir=(randn(1,L)+j*randn(1,L))/sqrt(2)/sqrt(L);
13  cfr=fft(cir,N);
14  noise = 1/sqrt(2)*(randn(1,N+G) + j*randn(1,N+G));
15  rec_fading=filter(cir,1,time_cp);
16  rec_fading=rec_fading;
17  rec_awgn=time_cp;
18
19  for kk=1:length(SNR)
20   rec_fading_noisy = rec_fading + noise*10^(-SNR(kk)/20);
21   rec_awgn_noisy = rec_awgn + noise*10^(-SNR(kk)/20);
22   rec_fading_freq= fft(rec_fading_noisy(G+1:end))/sqrt(N);
23   rec_fading_eqz=sign(real(rec_fading_freq.* conj(cfr)./(abs(cfr).^2)));
24   rec_awgn_bits= sign(real(fft(rec_awgn_noisy(G+1:end))/sqrt(N)));
25   BER_fading(num,kk)=length(find(rec_fading_eqz≠symbols))/N;
26   BER_awgn(num,kk)= length(find(rec_awgn_bits≠symbols))/ N;
27  end
28 end
29
30 figure(1);clf
31 semilogy(SNR,mean(BER_fading),'r-*','LineWidth',2)
32 hold
33 semilogy(SNR,mean(BER_awgn),'k-o','LineWidth',2)
34 xlabel('SNR (dB)'); ylabel('BER'); grid;
35 legend('Multipath dispersive channel','AWGN channel')
```

7.2.4.2 Frequency-Dispersive Multipath Channel

Mobility in free space or In the previous sections line-of-sight (LOS) multipath propagation environments, where a single dominant multipath component exists, leads to a Doppler shift issue. Handling the Doppler shift is straightforward, and pilot-based techniques can be used to estimate and compensate the frequency offset resulting from the Doppler shift effect. However, if the number of multipath components is large, and they arrive at a receiver from different angles, Doppler spread occurs. Doppler spread is a combination of different Doppler shifts, and unlike the Doppler shift issue, it is hard to deal with due to its random nature.

Mobility in a multipath channel causes dispersion in the frequency domain and produces inter-carrier interference (ICI) for OFDM systems. The dispersion in the frequency domain might lead to a time-selective fading, depending on the symbol duration of the signal. The coherence time of the channel (T_c) is defined as the duration, in which the channel time response can be considered as flat (i.e. highly correlated). It is inversely proportional to the Doppler spread in the propagation environment. When the symbol duration exceeds the coherence time of the channel, the signal experiences a time-selective fading.

The maximum Doppler shift occurs when a multipath component and the direction of mobility are aligned (i.e. the angle of arrival is 0° or 180° with respect to the direction of travel). The maximum Doppler shift can be calculated as follows:

$$f_{d\,max} = \frac{v}{\lambda} = \frac{v f_c}{c}, \tag{7.12}$$

where v is the speed of the mobile, f_c is the carrier frequency, and c is the speed of the light. The ICI created by Doppler spread can degrade the performance of OFDM systems seriously. A theoretical

ICI power derivation due to Doppler spread issue [21] and a universal upper bound [22] are given respectively in the following equations:

$$P_{ICI} = \frac{(f_{d\,max}T_s)^2}{2} \sum_{\substack{k=1 \\ k \neq i}}^{N} \frac{1}{(k-i)^2}, \tag{7.13}$$

$$P_{ICI} \leq \frac{2\pi f_{d\,max} T_s}{12}. \tag{7.14}$$

Figure 7.12 shows the effect of Doppler spread on ICI power as a function of normalized Doppler spread with respect to the subcarrier spacing. The results are obtained through analytical equations above and numerical evaluation using the below Matlab script. As the mobility (e.g. vehicular speed) increases, the maximum Doppler shift and spectral spreading increases as well. Accordingly, it leads to more ICI and degrades the communication system performance.

```
1  % Simulation of ICI power due to the Doppler spread issue.
2  % To obtain a reliable ICI power over Monte Carlo simulations,
3  % there should be sufficient number of symbols simulated.
4  frame=[];
5  for kk=1:1000
6      N=256; %FFT size
7      CP_size=1/4; %CP length as a ratio of the OFDM symbol duration
8      G=N*CP_size;
9      used_subc_indices=[1:1:N/2-1 N/2+1:1:N];
10     %Leave one of the carriers as zero carrier
11     symbols=2*(randn(1,N)>0)-1; %Random BPSK symbols
12     to_ifft=zeros(1,N);
13     to_ifft(used_subc_indices)=symbols(used_subc_indices);
14     time=ifft(to_ifft); %IFFT
15     time_cp=[time(end-N*CP_size+1:end) time];
16     %Insertion of CP. Has no value for this specific simulation
17     frame=[frame time_cp];
18 end
19
20 for kk=1:length(fd_normalized)
21 % Test for various Doppler spread values
22     f_inst=fd_normalized(kk);
23     CHAN = rayleighchan(1/N, f_inst);
24     % Generating the channel for a given Doppler spread
25     % A single tap Rayleigh fading is considered for simplicity
26     y=filter(CHAN,frame);
27
28     x1=vec2mat(y,N+G);
29     x11=x1(:,G+1:end);
30     xf1=fft(x11.');
31     ICI(kk)=mean(abs(xf1(N/2,:)).^2);
32 end
33 plot(fd_normalized,10*log10(ICI),'k:d')
```

7.2.5 Performance with Impairments

Several impairments degrade the performance of OFDM systems if the system is not properly designed. The integrated design and research for the baseband and RF challenges of OFDM systems require a thorough understanding of these impairments. This section presents critical impairments and their effects on OFDM systems. Also, the measurement of these impairments are investigated using various tools that are discussed in previous chapters.

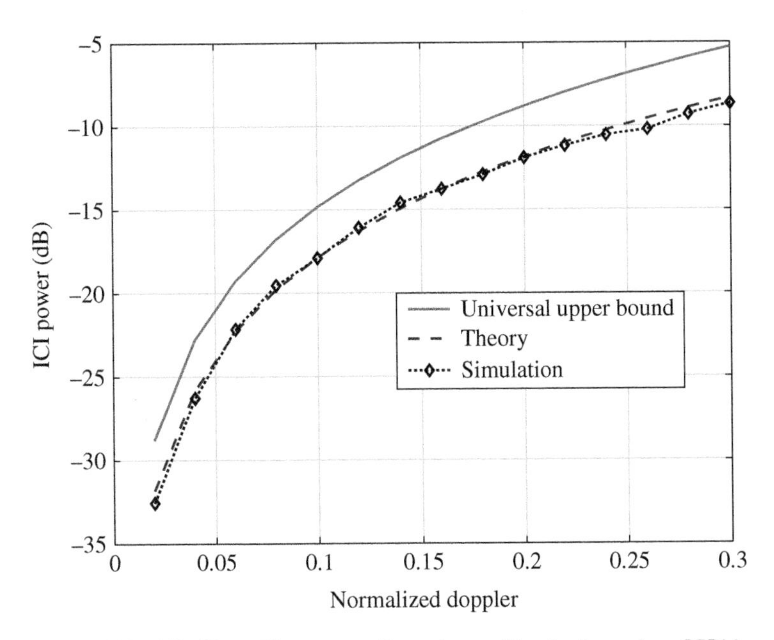

Figure 7.12 ICI effect of frequency-dispersive multipath channel on OFDM.

7.2.5.1 Frequency Offset

The frequency offset occurs when there is a difference between the transmitter local oscillator and receiver local oscillator. It results in ICI and destroys the orthogonality of subcarriers. The frequency offset is usually compensated using adaptive frequency correction (AFC); however, any residual error results in degraded system performance.

Assume a time-domain OFDM signal which encounters a frequency offset issue as given below,

$$X(k) \overset{IFFT}{\rightarrow} x(n) \overset{Frequency\ Offset}{\rightarrow} y(n) \overset{FFT}{\rightarrow} Y(k), \tag{7.15}$$

where $y(n)$ represents the received baseband signal with frequency offset error, and $Y(k)$ denotes the recovered data symbols. After going through detailed derivation, the recovered symbols can be related to the transmitted symbols considering a frequency offset of f_o as follows:

$$Y_m(k) \approx X_m(k)S_m(k,k) + \sum_{l=0,l\neq k}^{N-1} X_m(l)S_m(l,k), \tag{7.16}$$

where

$$S_m(l,k) = \frac{\sin(\pi(l-k+\epsilon))}{\pi(l-k+\epsilon)} e^{j\pi(l-k+\epsilon)} e^{j2\pi\epsilon(m-1)\frac{N_s}{N}} e^{j2\pi\epsilon\frac{N_s-N}{N}}, \tag{7.17}$$

$$\epsilon = \frac{\delta f}{\Delta f}. \tag{7.18}$$

The first term in Eq. (7.16) is equal to the transmitted symbol multiplied with an attenuation and phase rotation term that depends on ϵ and OFDM symbol index (please note that it does not depend on subcarrier index k, and hence the effect of frequency offset on the subcarriers is the same for all carriers). Therefore, $S_m(k,k)$ term introduces a constant phase shift of $2\pi\epsilon mN_s/N$ and an attenuation of $\sin(\pi\epsilon)/\pi\epsilon$ in magnitude. In addition to the attenuation of desired symbols, there is also interference between subcarriers. The second term in Eq. (7.16) represents the interference from other subcarriers, which is often referred to as ICI.

The following Matlab script simulates the effect of frequency offset considering a simple OFDM system. Correspondingly, Figure 7.13 shows the effect of frequency offset on the constellation of received symbols. It should be noted that the frequency offset error introduces both phase rotation and ICI. The phase rotation can be handled by simple pilot-based phase-tracking techniques. However, ICI creates a circular noise-like effect on the constellation and needs and requires more advanced receiver algorithms.

```
1  % Understanding the effect of freq. offset in the constellation
2  % The script can be run for various offset values
3  num_OFDM_symbols=2;
4  frame=[];
5  params.fft_size=256;
6  N=params.fft_size;
7  CP_size=1/4;
8
9  for ii=1:num_OFDM_symbols
10   symbols= qammod( floor(rand(1,N)*4) ,4);
11   tx_symbols(ii,:)=symbols;
12   used_subc_indices=1:1:256;
13   to_ifft=zeros(1,params.fft_size);
14   to_ifft(used_subc_indices)=symbols(used_subc_indices);
15   time=ifft(to_ifft);
16   time_cp=[time(end-N*CP_size+1:end) time];
17   frame=[frame time_cp];
18 end
19
20 f_offset=.1;
21 frame_o=frame.*exp(j*2*pi*f_offset*[1:length(frame)]/N);
22 rx=frame_o(N*CP_size+1:N*CP_size+N);
23 rx_f=fft(rx,params.fft_size);
24 scatterplot(rx_f) % with freq offset
```

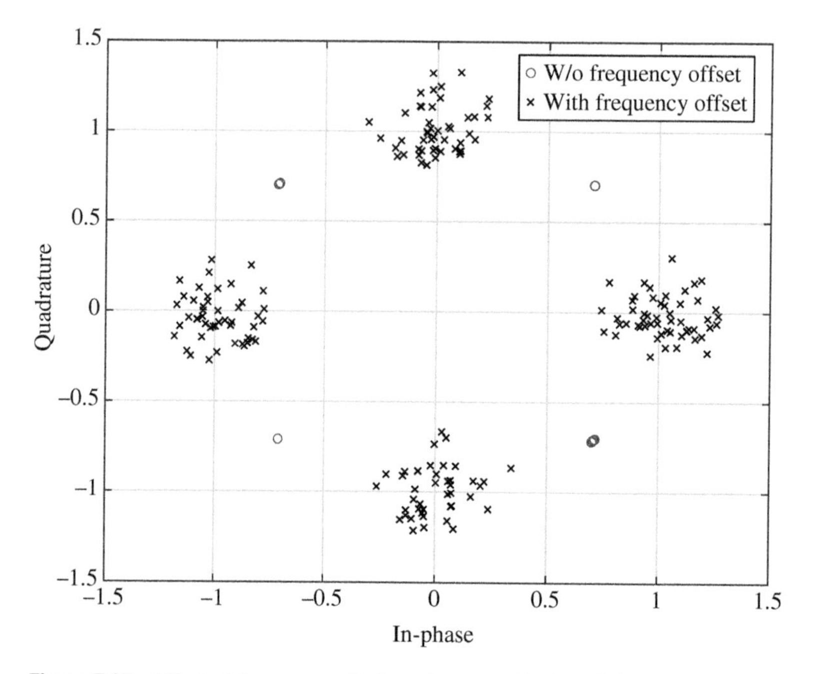

Figure 7.13 Effect of frequency offset on the constellation of the received symbols.

The ICI power due to the frequency offset can be theoretically calculated using the following equation [1]:

$$P_{ICI} = E\left\{\left|\sum_{l=0,l\neq k}^{N-1} X_m(l)S_m(l,k)\right|^2\right\} = \frac{\pi^2}{3}\left(\frac{f_o}{\Delta f}\right)^2, \tag{7.19}$$

where f_o is the fractional frequency offset and Δf is the subcarrier spacing. The following Matlab script calculates the ICI power due to the frequency offset using Eq. (7.19) and compares it with numerical evaluation results as illustrated in Figure 7.14.

```
1   % Calculation of ICI power due to frequnecy offset
2   % Theoretical Evaluation
3   df=0.01:0.01:0.7; % normalized frequency offset
4   ici=(pi.^2/3)*df.^2;
5   plot(df,10*log10(ici))
6   grid
7   hold on
8
9   % Numerical Evaluation
10  df=0.01:0.01:0.7;
11  for of=1:length(df)
12   f_offset=df(of);
13
14   for kk=1:10000
15     %To obtain a reliable ICI power over Monte Carlo sims.,
16     %there should be sufficient number of symbols simulated.
17     N=256; % FFT size
18     CP_size=1/4;
19     %CP length as a ratio of the OFDM symbol duration
20     used_subc_indices=[1:1:N/2-1 N/2+1:1:N];
21     %Leave one of the carriers as zero carrier
22     symbols=2*(randn(1,N)>0)-1; %Random BPSK symbols
23     to_ifft=zeros(1,N);
24     to_ifft(used_subc_indices)=symbols(used_subc_indices);
25     time=ifft(to_ifft); %IFFT
26     time_cp=[time(end-N*CP_size+1:end) time];
27     %Insertion of CP, which has no value for this specific simulation.
28     % Introduce freq. offset
29     frame_o=time_cp.*exp(j*2*pi*f_offset*[1:length(time_cp)]/N);
30     rx=frame_o(N*CP_size+1:N*CP_size+N); % removal of CP
31     rx_f=fft(rx,N); % FFT
32     ici(kk)=abs(rx_f(N/2)).^2;
33     %This is the zero carrier.
34     %Without ICI, it should have zero power
35   end
36   av_ICI(of)=mean(ici);
37  end
38
39  plot(df,10*log10(av_ICI),'r-')
40  legend('theory','simulation')
41  xlabel('Normalized frequency offset')
42  ylabel('ICI Power')
43  grid
```

The following Matlab script demonstrates the impact of an active subcarrier on other subcarriers as a function of frequency offset and spacing between the active subcarrier and the subcarriers that it interferes with. The evaluation results are illustrated in Figure 7.15. As the frequency offset increases, the ICI power that a subcarrier creates on the neighboring subcarriers increases as well. Also, the ICI effect from an active subcarrier to the neighboring subcarriers can be formulated as follows:

$$P_{ICI}(k) = \left|\sin c\left(k - \frac{f_o}{\Delta f}\right)\right|^2. \tag{7.20}$$

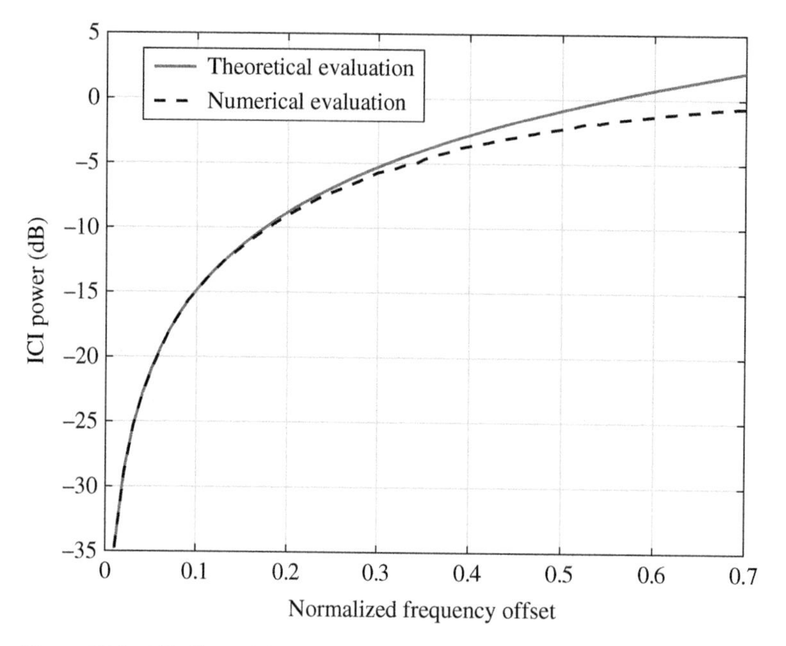

Figure 7.14 ICI effect of frequency offset on OFDM.

```
1  % Understanding the impact of a subcarrier on the neighboring
2  % carriers when there is a frequency offset.
3  N=256; % FFT size
4  CP_size=1/4; % CP length w.r.t. the OFDM symbol duration
5  used_subc_indices=N/2; % use only one active subcarrier
6  symbols=1; %
7  to_ifft=zeros(1,N);
8  to_ifft(used_subc_indices)=symbols;
9  time=ifft(to_ifft); %IFFT
10 time_cp=[time(end-N*CP_size+1:end) time];
11 %Insertion of CP, which has no value for this specific simulation.
12 f_offset=0.01;
13 frame_o=time_cp.*exp(j*2*pi*f_offset*[1:length(time_cp)]/N);
14 %Introduce frequency offset
15 rx=frame_o(N*CP_size+1:N*CP_size+N); %CP removal
16 rx_f=fft(rx,N); %FFT
17 plot(10*log10(abs(rx_f).^2))
18 hold
19 
20 %**********************************
21 % Theoretical evaluation
22 k=-127:1:128;
23 ici=abs(sinc(k-f_offset)).^2;
24 plot(10*log10(ici),'ro—')
25 legend('simulation','theory')
26 xlabel('carrier index')
27 ylabel('ICI power')
```

When the frequency offset amount is large, the impact is not only ICI but also the shift of subcarriers, which leads to the demodulated data at the receiver being in a wrong subcarrier position with respect to the subcarrier mapping deployed at the transmitter. For large values of frequency offset, the offset can be split into integer and fractional parts $f_{total} = f_{int} + f_o = \Gamma\Delta f + \epsilon\Delta f$, where Γ is an integer and $-0.5 \leq \epsilon \leq 0.5$. The fractional part causes ICI, as discussed above in detail. The integer part corresponds to the multiples of the subcarrier spacing Δf, which can cause subcarrier shift to the left or right depending on the sign of the offset. In other words, the symbol that is transmitted on the kth subcarrier appears on the neighboring carriers. Although the integer part of the frequency offset creates a significant problem, estimating it is quite easy, especially when pilots

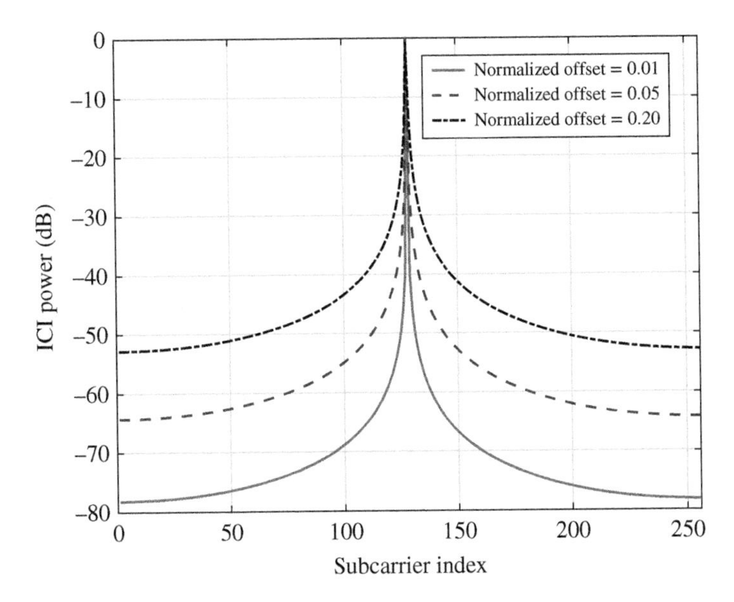

Figure 7.15 ICI effect of frequency offset from an active subcarrier to the neighboring subcarriers.

are transmitted along with data. The offset can be estimated by correlating the known pilots with the received signal, and the frequency offset is compensated accordingly.

7.2.5.2 Symbol Timing Error

In OFDM systems, the symbol timing synchronization errors at the receiver causes two major problems depending on whether the timing error is toward the left of the perfect synchronization point or the right.

When the timing synchronization error is to the left of the perfect synchronization position, the timing error only leads to a rotation of symbols for small offset values. However, if the timing offset is large, it causes ISI as well. Usually, the CP duration T_g is selected larger than the maximum excess delay of the channel τ_{max}. Therefore, we have an error margin of $T_g - \tau_{max}$. If the timing position is shifted to a point within this margin of the CP duration, then, only subcarrier-dependent phase rotation is observed. However, if the shift is larger than this margin, then ISI occurs as well. Rotation of symbols can be folded into the estimated channel and corrected easily if the timing offset is smaller than the unused part of the CP. The unused part of the CP is the part, which does not interfere with the previous symbol.

Assume a time-domain OFDM signal which encounters time offset issue as given below,

$$X(k) \overset{IFFT}{\rightarrow} x(n) \overset{Symbol\ Timing\ Error}{\rightarrow} y(n) \overset{FFT}{\rightarrow} Y(k). \tag{7.21}$$

where transmitted symbols are represented with $X(k)$, and the baseband equivalent of the time-domain signal is represented with $x(n)$. The symbol timing error is a result of an incorrect start position assumption of the OFDM symbol. Therefore, $y(n)$ is nothing but the shifted version of $x(n)$ in the time domain. For example, if there is a timing error of θ_τ, $y(n)$ can be expressed as follows:

$$y(n) = x(n \pm \theta_\tau) \tag{7.22}$$

$$= \sum_{k=0}^{N-1} X(k) e^{j\frac{2\pi k}{N}(n \pm \theta_\tau)}. \tag{7.23}$$

The sign of θ_τ depends on whether the sampling is done before the exact start position or after the exact position. In the following step, $Y(k)$ can be calculated from $y(n)$ using FFT as follows:

$$Y(k) = \frac{1}{N}\sum_{n=0}^{N-1}\left\{\sum_{l=0}^{N-1} X(l)e^{j\frac{2\pi l}{N}(n\pm\theta_\tau)}\right\}e^{-j\frac{2\pi kn}{N}}$$
$$= X(k)e^{\pm j\frac{2\pi k\theta_\tau}{N}}. \tag{7.24}$$

Equation (7.24) shows that the timing offset of θ_τ causes only rotation on the recovered data symbols. Also, the value of the recovered symbol depends only on the transmitted data but not on the neighboring carriers. In other words, the symbol timing error does not destroy the orthogonality of the subcarriers, and the effect of timing error is a phase rotation that linearly changes with the carriers' order.

When the timing synchronization error is to the right of the perfect synchronization position, the timing offset causes ISI, ICI, and phase rotation. This type of timing error is undesirable since it causes loss of subcarrier orthogonality and leakage to the next OFDM symbol, which leads to ICI and ISI, respectively. As a result, the symbol timing is often intentionally slightly shifted toward the CP (i.e. toward the left of the actual estimated timing position), so that any possible error in symbol timing estimation that might create the loss of orthogonality can be avoided. Even though this intentional bias in synchronization prevents the loss of orthogonality of the subcarriers and ICI, it results in the effective channel frequency response to be less correlated due to the additional subcarrier-dependent phase shift. As a result, the channel estimation performance degrades since the noise averaging effect is reduced. However, a well-designed channel estimation algorithm can take care of this problem.

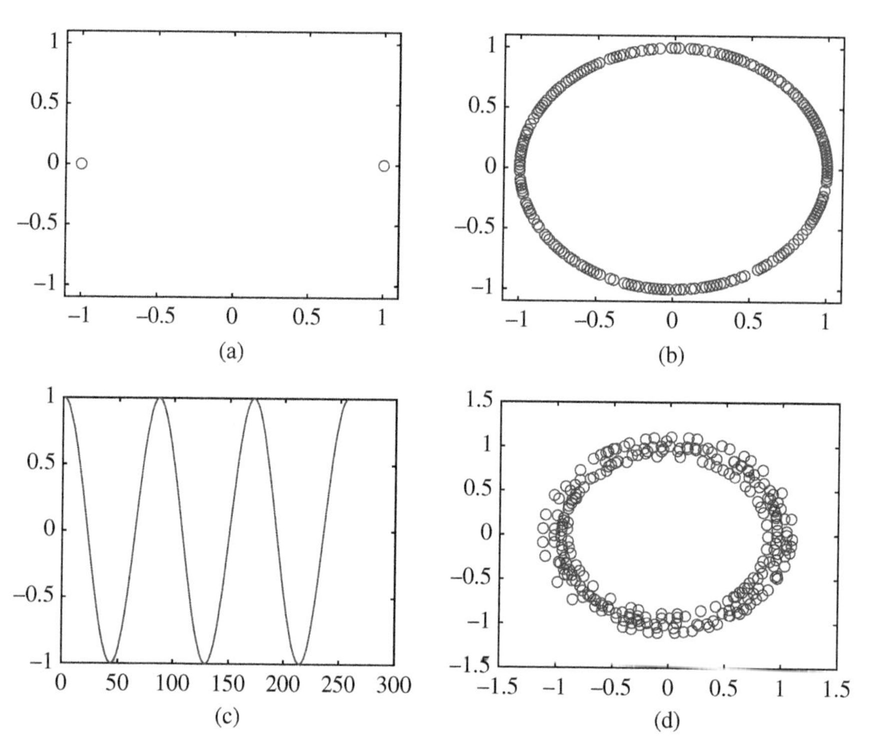

Figure 7.16 Effect of symbol timing error on the constellation of the received symbols and channel frequency response. (a) No timing offset. (b) Timing offset within CP. (c) Rotation of the effective CFR due to timing offset. (d) Timing offset outside CP.

The following Matlab script evaluates the effects of symbol timing offset and the results are illustrated in Figure 7.16.

```
1  % Effect of symbol timing offset simulation.
2  % Symbol timing offset within and outside the CP is evaluated.
3  % This script can be run for various timing offset values
4
5  % Symbol timing offset within the CP
6  N=256;
7  params.fft_size=256;
8  used_subc_indices=1:1:256;
9  CP_size=1/4;
10 symbols=2*(randn(1,N)>0)-1;
11 to_ifft=zeros(1,N);
12 to_ifft(used_subc_indices)=symbols(used_subc_indices);
13 time=ifft(to_ifft);
14 time_cp=[time(end-N*CP_size+1:end) time];
15 rx=time_cp(N*CP_size+1:end);
16 subplot(2,2,1)
17 rx_f=fft(rx);
18 plot(real(rx_f),imag(rx_f),'o')
19 axis([-1.1 1.1 -1.1 1.1])
20 title('(a) no timing offset')
21 subplot(2,2,2)
22 t_offset=-3;
23 rx=time_cp(N*CP_size+1+t_offset:end+t_offset);
24 rx_f=fft(rx);
25 plot(real(rx_f),imag(rx_f),'o')
26 title('(b) with timing offset within CP')
27 axis([-1.1 1.1 -1.1 1.1])
28 subplot(2,2,3)
29 plot(real(rx_f.*conj(symbols)))
30 title('(c) Rotation of the effective CFR due to timing offset')
31
32 % Symbol timing offset after the CP
33 num_OFDM_symbols=2;
34 frame=[];
35 N=256;
36 CP_size=1/4;
37
38 for ii=1:num_OFDM_symbols
39  symbols=2*(randn(1,N)>0)-1;
40  tx_symbols(ii,:)=symbols;
41  used_subc_indices=1:1:256;
42  to_ifft=zeros(1,params.fft_size);
43  to_ifft(used_subc_indices)=symbols(used_subc_indices);
44  time=ifft(to_ifft);
45
46  time_cp=[time(end-N*CP_size+1:end) time];
47  frame=[frame time_cp];
48 end
49
50 t_offset=3;
51 rx=frame(N*CP_size+1+t_offset:N*CP_size+t_offset+N);
52 rx_f=fft(rx);
53 subplot(2,2,4)
54 plot(real(rx_f),imag(rx_f),'o')
55 title('(d) with timing offset outside CP')
```

As mentioned earlier, the use of CP in OFDM avoids the ISI. However, there is another advantage of the use of CP. As discussed above, when there is a timing offset in the synchronization, and if the timing offset falls within the ISI-free part of the CP interval, the system still maintains orthogonality. If ZP is used instead of CP, the system has immunity against ISI, but it cannot maintain the orthogonality if there is any timing offset in the synchronization. Therefore, the use of CP is preferred in OFDM systems compared to the use of ZP in general. The following Matlab script compares the CP and ZP when there is a symbol timing offset in an OFDM system.

```
1  % Comparing the difference between CP and ZP
2  % when there is symbol timing offset with various offset values
3  params.fft_size=256;
4  num_OFDM_symbols=100;
5  frame=[];
6  frame1=[];
7  N=256;
8  CP_size=1/4;
9
10 for ii=1:num_OFDM_symbols
11  symbols= qammod( floor(rand(1,N)*4) ,4);
12  used_subc_indices=1:1:256;
13  to_ifft=zeros(1,params.fft_size);
14  to_ifft(used_subc_indices)=symbols(used_subc_indices);
15  time=ifft(to_ifft);
16
17  time_cp=[time(end-N*CP_size+1:end) time]; % using CP
18  time_zp=[zeros(1,N*CP_size) time];        % using guard time
19  frame=[frame time_cp];
20  frame1=[frame1 time_zp];
21 end
22
23 t_offset=-9;
24 rx=frame(N*CP_size+1+t_offset:N*CP_size+t_offset+N);
25 rx_f=fft(rx);
26 scatterplot(rx_f) % with timing offset
27
28 rx1=frame1(N*CP_size+1+t_offset:N*CP_size+t_offset+N);
29 rx_f1=fft(rx1);
30 scatterplot(rx_f1) % with timing offset
```

7.2.5.3 Sampling Clock Offset

The clock timing difference between the transmitter and receiver causes a sampling clock (or sometimes referred to as sample timing) error. The sampling clock error can be ignored for a small number of subcarriers or a low number of symbols within a given subframe. The sampling error causes subcarrier and symbol-dependent rotation in the received symbols. When there is a sampling error, the recovered symbols can be related to the transmitted symbols as follows:

$$Y_m(k) = X_m(k)e^{\frac{j2\pi mk\xi N_S}{N}}, \tag{7.25}$$

where ξ is the relative clock deviation of the reference oscillator, $N_s = N + N_g$ with N_g being the number of samples used for the CP. It should be noted that the rotation increases as the subcarrier and symbol index grows. In other words, the effect of sampling clock error is more in the higher indexed subcarriers and the later symbols of the subframe.

Figure 7.17 shows an example of the effect of sampling clock offset in OFDM systems. The sampling clock offset effect increases while moving away from the center subcarrier (i.e. toward the edge carriers). Also, the average error vector magnitude (EVM) over all subcarriers increases as the symbol index increases.

The sampling clock offset can be simulated by resampling the time signal and creating an intentional clock offset. For example, the following Matlab script simulates an OFDM system with a sampling clock offset of 900 ppm.

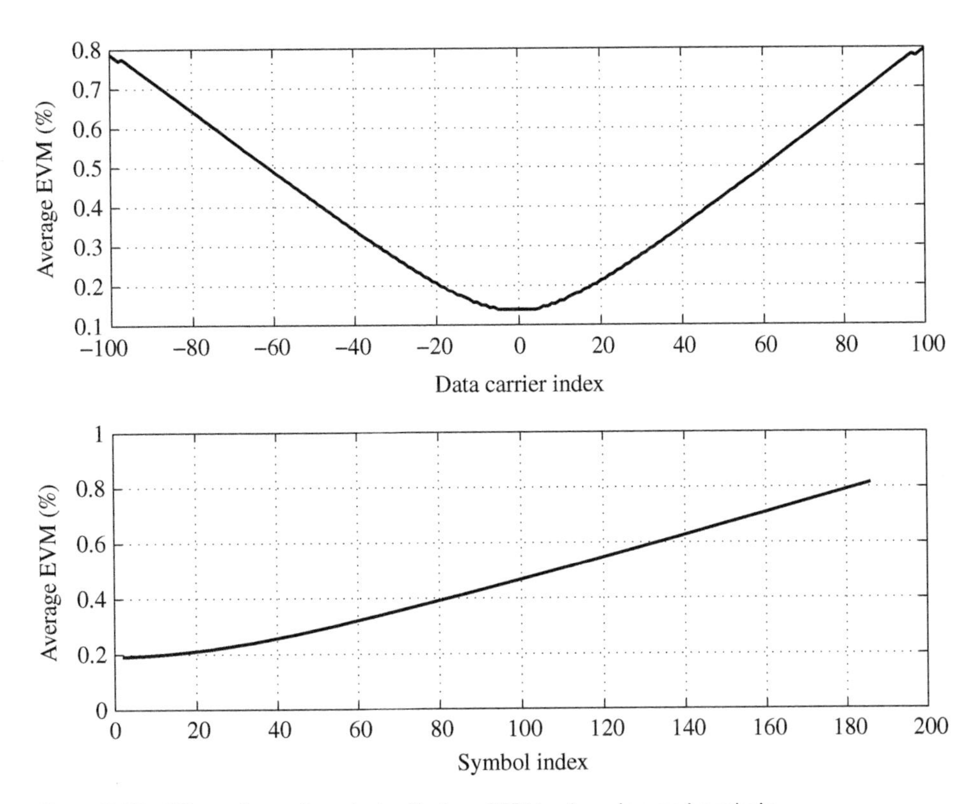

Figure 7.17 Effect of sampling clock offset on OFDM subcarriers and symbols.

```
1   % Generation of sampling clock offset.
2   % Its effect is evaluated on the constellation diagram.
3
4   num_OFDM_symbols=100;
5   frame=[];
6   N=1024;
7   CP_size=1/4;
8
9   for ii=1:num_OFDM_symbols
10      symbols= qammod( floor(rand(1,N)*4) ,4);
11      used_subc_indices=1:1:N;
12      to_ifft=zeros(1,N);
13      to_ifft(used_subc_indices)=symbols(used_subc_indices);
14      time=ifft(to_ifft);
15      time_cp=[time(end-N*CP_size+1:end) time];
16      frame=[frame time_cp];
17  end
18
19  %Introduce sample clock offset
20  frame1=resample(frame,10000,10009);
21  rx=frame1(N*CP_size+1:N*CP_size+N);
22  rx_f=fft(rx);
23  %Plot the constellation diagram
24  scatterplot(rx_f)
```

7.2.5.4 Phase Noise

The phase noise is a random process as similar to Doppler spread, and it causes three major problems in OFDM systems: common phase offset, power degradation, and ICI. Assuming an OFDM signal, which is only affected by phase noise ($\phi(n)$) at the receiver, the received time-domain signal can be expressed as follows:

$$y_m(n) = x_m(n)e^{j\phi_m(n)}. \tag{7.26}$$

Assuming that the phase offset is small (i.e. $e^{j\phi(n)} \approx 1 + j\phi(n)$), the recovered symbols is expressed in the following form:

$$Y_m(k) \approx \underbrace{X_m(k)\left\{1 + j\frac{1}{N}\sum_{n=0}^{N-1}\phi_m(n)\right\}}_{Common\ phase\ term} + \underbrace{\frac{j}{N}\sum_{r=0,r\neq k}^{N-1} X_m(r)\sum_{n=0}^{N-1}\phi_m(n)\cdot e^{j(2\pi/N)(r-k)n}}_{ICI\ term}. \tag{7.27}$$

The common phase term in Eq. (7.27) introduces a rotation to the constellations. This rotation is the same for all subcarriers, and it is representative of the average phase noise. When the phase noise is small, the common phase term is the dominant phase noise effect, and it accumulates over time (i.e. the variance increases over time). The problems associated with this term can be avoided easily using a carefully implemented pilot-based tracking.

The last term in Eq. (7.27) represents the leakage from neighboring subcarriers (i.e. ICI) and illustrated in Figure 7.18. This term cannot be corrected since both phase offset ($\phi_m(n)$) and input data sequence ($X_m(n)$) are random. Therefore, it causes signal-to-noise ratio (SNR) degradation of the overall system. The only way to reduce the interference due to this term is to improve the performance of the oscillator with the associated cost increase. The ICI power due to the phase noise can be approximately modeled as follows [23]:

$$P_{ICI} \cong \frac{\pi B_{3\,\text{dB}} T_s E_s E_H}{3}, \tag{7.28}$$

where $B_{3\,\text{dB}}$ represents the two-sided 3 dB bandwidth (i.e. the frequency spacing between 3-dB points of its Lorentzian power spectral density [23, 24]), T_s is the OFDM symbol duration, and E_s and E_H represent the symbol energy and channel power, respectively, which are often normalized to unity. Accordingly, the signal-to-phase-noise-interference-ratio (SPNIR) can be expressed as follows [23]:

$$SPNIR = \frac{1 - \frac{\pi B_{3\,\text{dB}} T_s}{3}}{\frac{\pi B_{3\,\text{dB}} T_s}{3}}. \tag{7.29}$$

The numerator in the above equation includes signal power degradation due to phase noise. It should also be noted that Eq. (7.29) assumes no other noise and interference in the system.

7.2.5.5 PA Nonlinearities

PAs can operate as linear devices only for limited input power, and they distort the transmitted signal beyond a certain input power. The linear and nonlinear operating regions of a PA can be defined with respect to its 1 dB compression point, where the output power of a PA is reduced by 1 dB as shown in Figure 7.19a. Also, the saturation effect can be seen clearly in Figure 7.19b by checking the signal power variance between the PA input and PA output. To minimize its power consumption and operate at its highest efficiency, a PA is ideally operated close to its saturation point. However, beyond the saturation point, the PA nonlinearity causes several problems, such

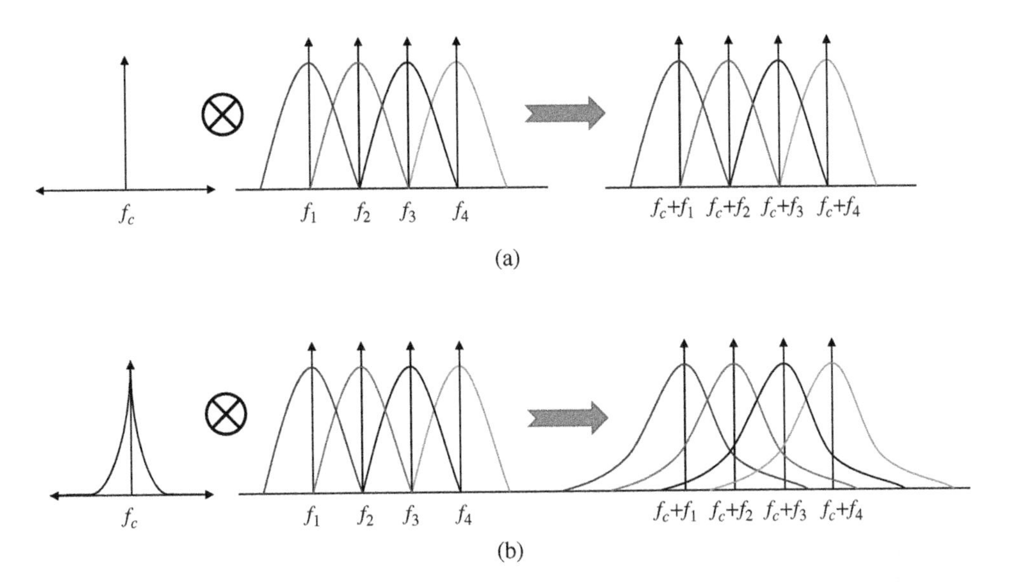

Figure 7.18 Effect of phase noise on OFDM subcarriers: (a) ideal local oscillator; (b) practical local oscillator.

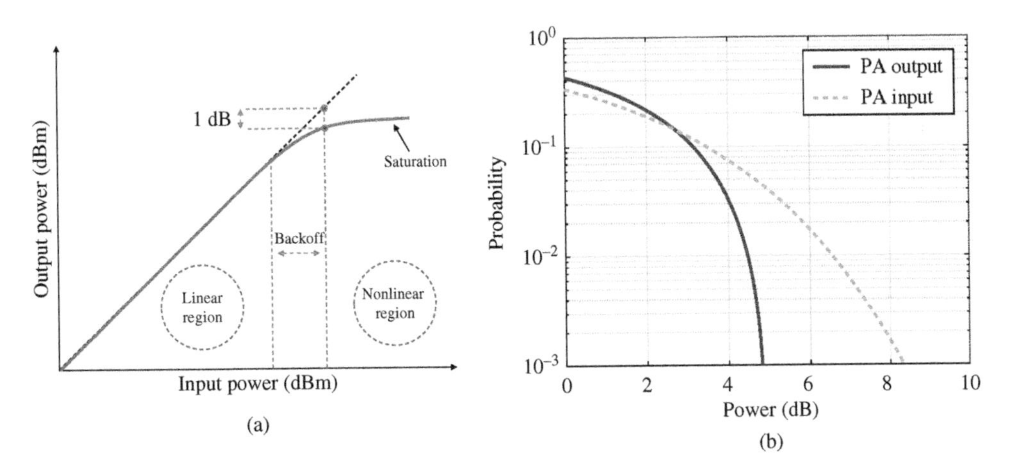

Figure 7.19 PA characteristics: (a) PA input power versus output power characteristics curve. (b) Power CCDF curve of PA input and PA output.

as amplitude-to-amplitude (AM-AM) distortion, amplitude-to-phase (AM-PM) distortion, spectral regrowth, harmonic distortion, intermodulation distortion, SNR degradation, and modulation inaccuracy. For example, the power back-off technique is widely used in current wireless technologies to remedy the problems due to wide signal dynamic ranges. However, this technique sacrifices efficiency and increases power consumption. On the other hand, baseband linearization techniques are utilized to pre-distort the signal and compensate for the nonlinear effects of PAs.

One of the classical and commonly used nonlinear PA model is Saleh's PA model [25]. It is a simple nonlinear model without memory, and it is defined by only two parameters α and β. The model uses two functions to model the AM-AM and AM-PM characteristics of nonlinear amplifiers. It should be noted that α and β are different for each function. For a given PA, these two parameters (i.e. α and β) can be extracted using a least-squares approximation to minimize the relative error between the target PA measurements and the predictions by the model. The amplitude and phase equations determining the distortions can be modeled as follows:

$$A(r) = \frac{\alpha_A r}{1+\beta_A r^2}, \tag{7.30}$$

and

$$P(r) = \frac{\alpha_P r^2}{1+\beta_P r^2}, \tag{7.31}$$

where $r(t)$ represents the envelope of the PA input signal. The input signal is expressed as follows:

$$s_{in}(t) = r(t)\cos(2\pi f_c t + \psi(t)), \tag{7.32}$$

where f_c is the carrier frequency and $\psi(t)$ is the phase of the input signal. The output of PA can be derived as follows:

$$s_{out}(t) = A(r(t))\ \cos(2\pi f_c t + \psi(t) + P(r(t))). \tag{7.33}$$

The effect of PA nonlinearities in OFDM is more significant compared to single-carrier systems, since OFDM systems have a higher PAPR that requires the PAs to be operated in the linear region. The nonlinear PAs cause distortions to the OFDM signal both in-band and out-of-band, which have a severe impact on the communication performance. To see these distortions, an OFDM signal is generated and passed through PA using the Saleh's PA model above in the following Matlab script.

```
1  %The effect of PA nonlinearities on the OFDM spectrum
2  i_ind = [1, 3.2, 5.0, 10];
3  for i=1:4
4       num_OFDM_symbols=1000;
5       frame_1=[];
6       N=1024;
7       CP_size=1/4;
8       mod_order=16;
9
10      for ii=1:num_OFDM_symbols
11          symbols= qammod( floor(rand(1,N)*mod_order) ,mod_order);
12          used_subc_indices=256:1:N-256;
13          to_ifft=zeros(1,N);
14          to_ifft(used_subc_indices)=symbols(used_subc_indices);
15          tx_symbols(ii,:)=to_ifft;
16          time=ifft(to_ifft);
17          time_cp=[time(end-N*CP_size+1:end) time];
18          frame_1=[frame_1 time_cp];
19      end
20
21      frame=frame_1*i_ind(i); % change the input power
22
23      % PA parameters - Saleh's non-linear PA model
24      alpha_A=11.534; beta_A= 1.6242;
25      alpha_P=11.431; beta_P=39.071;
26
27      amp_f=(alpha_A*abs(frame))./(1+beta_A*abs(frame).^2);
28      angle_f=(alpha_P*abs(frame).^2)./(1+beta_P*abs(frame).^2);
29      f_dist=(amp_f./abs(frame)).*frame.*exp(j*angle_f);
30
31      [Pxx1,F]=pwelch(f_dist,[],[],2048,1,'twosided');
32      plot(F,10*log10(Pxx1),'LineWidth',1.1)
33      xlabel('Normalized Frequency')
34      ylabel('Power (dB/Hz)')
35      grid on; hold on
36  end
```

The effect of PA nonlinearity is shown on the constellation in Figure 7.20a. The constellation is noisy due to the in-band interference created by the PA. Furthermore, Figure 7.20b illustrates

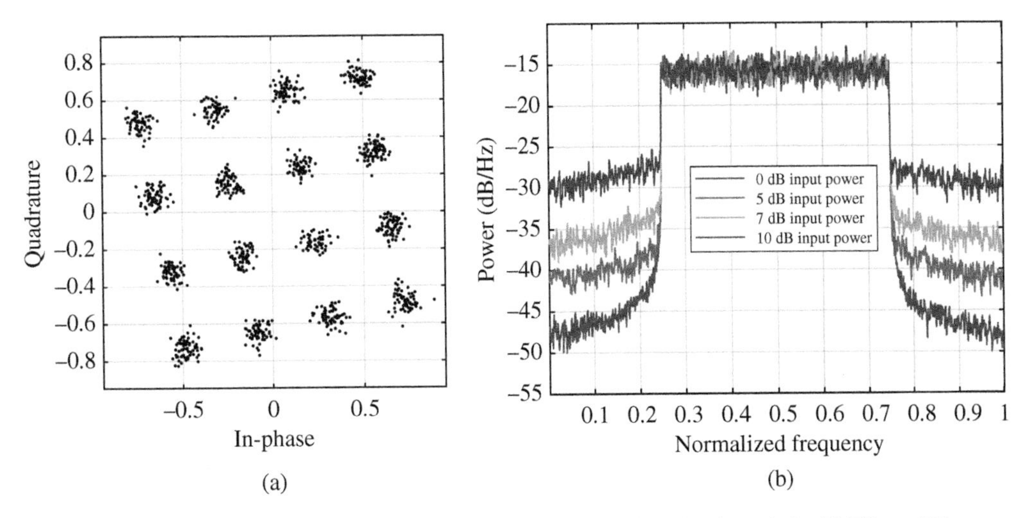

Figure 7.20 (a) Effect of PA nonlinearities on the constellation of received symbols. (b) Effect of PA nonlinearities on the spectrum.

the spectrum of the PA output. The PA nonlinearities lead to spectral regrowth and out-of-band distortions accordingly for the OFDM signal under test. To overcome PA nonlinearities, the PAPR of OFDM should be reduced using various techniques such as amplifier back-off and various pre-coding strategies.

7.2.5.6 I/Q Impairments

One of the major impacts of the in-phase and quadrature-phase (I/Q) modulators in OFDM systems is that they introduce interference between the modulated symbols on the image carriers. The image carriers are the carriers, which are equally spaced with respect to the direct current (DC) carrier. For example, for an FFT size of N, if the DC carrier is indexed as the carrier number $N/2$, the carriers $N/2+k$ and $N/2-k$ are called as image carriers where k is the carrier away from the DC carrier. Therefore, the received signal at every carrier of interest (let's say kth carrier) is dependent on the symbol transmitted on that carrier as well as the symbol transmitted on the opposite image carrier (i.e. $-k$th carrier). As a result, the image carrier causes interference on the carrier of interest. Depending on the transmission on the image carrier, it causes self- or multi-access interference. If the transmission on the image carrier belongs to the same user, it causes self-interference. For example, in a time division multiple access (TDMA)-based OFDM system, the type of interference is the self-interference. On the other hand, if the image carrier transmission belongs to another user, it causes multi-access interference. For example, in OFDMA systems, the interference can be both self- or multi-access interference.

I/Q modulator modulates the in-phase "I" and quadrature-phase "Q" components with high-frequency carriers. The transmitted signal that goes through the I/Q vector modulator experiences several levels of signal distortion due to imperfections in the I/Q modulator. The major I/Q impairments can be classified as I/Q offset, I/Q gain imbalance, and I/Q quadrature-error. I/Q offset, which is also called as I/Q origin offset or carrier leakage, indicates the magnitude of the carrier feedthrough. I/Q offset can be observed as an offset in the constellation. Gain mismatch or gain imbalance results in the amplitude of one channel being smaller than the other. The baseband model of I/Q origin offset and I/Q gain imbalance that are introduced into time-domain transmitted signal can be represented as follows:

$$y_m(n) = I_s\Re\{x_m(n)\} + jQ_s\Im\{x_m(n)\} + I_o + jQ_o, \tag{7.34}$$

where $\Re\{x\}$ represents the real part of x, $\Im\{x\}$ represents the imaginary part of x, and I_s and Q_s represent the gains in I and Q branches, respectively. Similarly, I_o and Q_o represent the offsets in I and Q branches, respectively. When $I_s = Q_s$, the I/Q imbalance becomes zero, and when both I_o and Q_o are zero, there is no I/Q offset.

Quadrature skew error indicates the orthogonal error between the I and Q signals. Ideally, I and Q channels should be exactly orthogonal (i.e. 90° apart). When the orthogonality is not ideal, then a quadrature error is observed. For example, the following model shows the quadrature offset that is introduced into the time-domain signal:

$$y_m(n) = \Re\{x_m(n)\} + \Im\{x_m(n)\}\sin(\zeta) + j\Im(x_m(n))\cos(\zeta), \tag{7.35}$$

where ζ is the quadrature error. When ζ is equal to 0°, $y_m(n)$ becomes equal to $x_m(n)$ (i.e. no quadrature error).

The effect of I/Q gain imbalance and I/Q quadrature error on OFDM is illustrated in Figure 7.21.

If the DC carrier, which is not used in various standards, is ignored, the transmitter I/Q impairments, excluding the origin offset, can be represented in a unified form in the frequency-domain transmitted samples as follows:

$$\hat{X}_m(k) = \frac{X_m(k)}{2}\{I_s + Q_s e^{-j\zeta}\} + \frac{X_m(-k)}{2}\{I_s - Q_s e^{-j\zeta}\}. \tag{7.36}$$

As can be seen from the above equation, the transmitted signal after the I/Q modulator at the kth carrier depends not only on the symbol that is allocated to the kth carrier but also to the symbol that is allocated to the $-k$th carrier. If these interferences are not handled properly at the receiver, they cause performance degradation. The I/Q modulator impairments also cause distortion in the signal amplitude and phase. However, this can be handled by a proper channel estimator, as the distortion effect can be folded into the effective channel frequency response. It should be noted that the I/Q impairments are assumed to be due to the transmitter I/Q modulator. Since the receiver I/Q modulator effects are known at the receiver, they can always be precomputed and precompensated at the receiver.

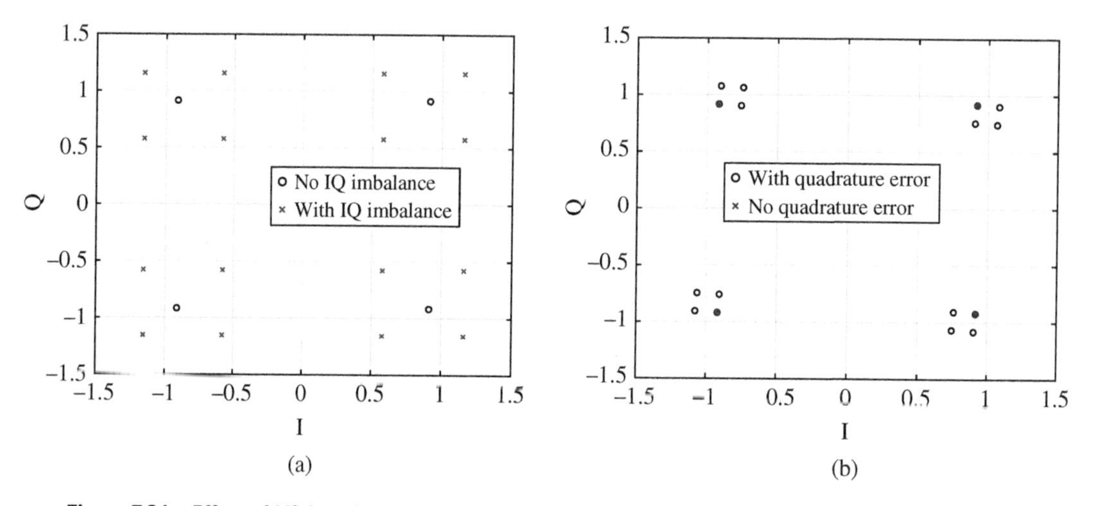

Figure 7.21 Effect of I/Q impairments on OFDM: (a) IQ gain imbalance, (b) IQ quadrature error.

The following Matlab script simulates the I/Q impairments on OFDM. The effect of the I/Q impairments can be visualized through this script, either individually or collectively.

```
1  % Understanding the effect of I/Q impairments on OFDM
2  % Various I/Q impairements can be tested individually or together.
3  num_OFDM_symbols=100;
4  frame=[];
5  N=1024;
6  CP_size=1/4;
7
8  for ii=1:num_OFDM_symbols
9    symbols= qammod( floor(rand(1,N)*4) ,4);
10   used_subc_indices=1:1:N;
11   to_ifft=zeros(1,N);
12   to_ifft(used_subc_indices)=symbols(used_subc_indices);
13   time=ifft(to_ifft);
14   time_cp=[time(end-N*CP_size+1:end) time];
15   frame=[frame time_cp];
16 end
17
18 % I/Q impairments
19 I_offset=0; Q_offset=0;
20 I_scale=2; Q_scale=1;
21
22 % DC offset and I/Q gain imbalance
23 zz=mean(abs(real(frame)));
24 tx_signal = (I_offset + sqrt(-1)*Q_offset)*zz + ...
25 I_scale*real(frame) + sqrt(-1)*Q_scale*imag(frame);
26
27 % Quadrature error
28 quad_angle=10; % degree
29 angle_radian=quad_angle*pi/180;
30
31 tx_signal=real(tx_signal)+imag(tx_signal)*sin(angle_radian) + ...
32  sqrt(-1)*imag(tx_signal)*cos(angle_radian);
33 rx=tx_signal(N*CP_size+1:N*CP_size+N);
34 rx_f=fft(rx);
35 scatterplot(rx_f)
```

7.2.6 Summary of the OFDM Design Considerations

Engineers/researchers need to consider several design parameters into account when designing an OFDM system that can be summarized as follows:

- CP should be larger than maximum excess delay ($T_g \geq \tau_{max}$) of the channel to prevent ISI.
- To increase the spectrum efficiency, CP duration should not be a significant portion of original OFDM signal duration, i.e. $T_g << T_s$.
- To reduce the impact of ICI, maximum Doppler spread should be within a manageable portion of subcarrier spacing, i.e. $f_{d\,max} << \Delta f$, suggesting that the OFDM symbol duration should not be larger than the coherence time of the channel.
- To minimize the impact of ICI against frequency offset, phase noise, etc., the subcarrier spacing should not be chosen small, which is conflicting with the goal of making the subcarrier spacing less than the coherence bandwidth of the channel to ensure that each subchannel experiences frequency flat fading channel.
- FFT size and number of used subcarriers should be chosen wisely by considering PAPR, total allocated bandwidth, and desired data rate into account. Increasing the number of used carriers increases PAPR, bandwidth, and data rate.
- Bit rate per subcarrier and the modulation order to be used depend on the signal-to-interference-plus-noise ratio (SINR) allowed, which is related to all the parameters above. Higher-order modulations are more sensitive to ISI and ICI.

7.2.7 Coherent versus Differential OFDM

The coherent OFDM signaling and detection require its receiver to estimate the channel accurately. The absolute phase information in each subcarrier needs to be known precisely. Also, the timing synchronization accuracy must be within a level so that the phase of the channel frequency response does not change too fast, as it makes the channel tracking more difficult. On the other hand, differential signaling and detection does not require channel estimation as long as the channel (both amplitude and phase) is approximately constant within differential interval. The differential interval depends on the type of differential modulation. For OFDM, differential modulation can be applied either in time or in frequency. In the time-domain approach, differential modulation is applied on two consecutive OFDM symbols on the same subcarrier. On the other hand, in the frequency-domain approach, differential modulation is applied in two subsequent subcarriers of the same OFDM symbol.

Although differential modulation does not require the channel or phase information and provide significant saving in spectral efficiency and power by not transmitting training symbols and pilot symbols, the performance of differential modulation is approximately 3 dB worse compared to coherent modulation. Moreover, if the channel between the two consecutive differential interval changes due to various reasons, such as Doppler spread, oscillator frequency change, or timing jitter, the performance of the differential demodulator becomes even worse. It should be noted that as long as the change is the same for both symbols, the differential detector is not affected by the change. Therefore, the differential detector is very robust to common phase and amplitude changes between consecutive data symbols either in time or frequency. However, any change that affects one symbol more than the other has an impact on the performance of the differential detector. For example, in the time-domain approach, the time variation of the channel between two OFDM symbols due to Doppler spread affects the phase and amplitude of the consecutive symbols differently. Similarly, random phase noise affects the two symbols differently, especially when the normalized linewidth (two-sided 3 dB bandwidth) of the oscillator is not negligibly small. In the frequency-domain approach, if the coherence bandwidth of the channel is less than or in the neighborhood of twice the subcarrier spacing, it indicates that the channels between two consecutive subcarriers are different and affecting the performance of the differential detector. It should also be noted that the coherence bandwidth of the channel should not be less than a single subcarrier spacing, as discussed earlier. Similarly, the frequency-domain approach is not robust to the timing offset as the timing offset introduces phase rotation in the frequency domain and leads to the subsequent subcarrier phases different from each other. More timing offset creates more difference and degradation in the performance of the frequency-domain differential detector.

In the case of the time-domain approach, the first OFDM symbol of a frame should be known to allow the differential encoder to start with an initial value. Similarly, in the frequency-domain approach, the first subcarrier of each OFDM symbol should be a known pilot carrier. The following Matlab script simulates an OFDM system with differential encoding using both the time-domain and frequency-domain approaches. The maximum excess delay of the channel, Doppler spread of the channel, and other system parameters can be changed to observe the performances under various scenarios. As can be seen from Figure 7.22 clearly, the time-domain approach is very sensitive to the Doppler spread. Doppler spread is a result of the channel variation in the time domain, and hence the channel between two consecutive OFDM symbols is different in this channel condition. Even though Doppler spread creates ICI in both the time-domain and frequency-domain approaches, the demodulator performance is lower in the time-domain approach due to

```
1  % Comparison of differential OFDM techniques
2  SNR=[0 2 4 6 8 10 15 20 30];% dB
3  N=64; % FFT size
4  CP_size=1/4; % CP length
5  G=N*CP_size;
6  NN=5000; % Frame length
7  frame_t=[];frame_f=[];
8  symbols_init=ones(1,N); % reference symbol
9  for kk=1:NN
10     symbols=randn(1,N)>0;
11     tx_syms(kk,:)=symbols;
12     % Time-domain differential modulation
13     sym_t=mod(symbols_init+symbols,2);
14     % Differential encoding in time domain
15     symbols_init=sym_t;
16     time_t=ifft(sym_t*2-1)*sqrt(N); %IFFT
17     time_cp_t=[time_t(end-N*CP_size+1:end) time_t];
18     frame_t=[frame_t time_cp_t];
19     % Frequency-domain differential modulation
20     sym_init=1; % Reference carrier
21     for ll=1:N
22         sym_f(ll)=mod(sym_init+symbols(ll),2);
23         sym_init=sym_f(ll);
24     end
25     % Differential encoding in frequency domain
26     time_f=ifft(sym_f*2-1)*sqrt(N); %IFFT
27     time_cp_f=[time_f(end-N*CP_size+1:end) time_f];
28     frame_f=[frame_f time_cp_f];
29 end
30 f_d_normalized=0.2; L=5;% number of channel taps
31 CHAN=rayleighchan(1/N,f_d_normalized,[1:1:L]/N,ones(1,L)/sqrt(L));
32 %Tap gains are normalized to 1 and delays are sample spaced (Ts/N)
33 y1=filter(CHAN,frame_t);
34 y2=filter(CHAN,frame_f);
35 noise = 1/sqrt(2)*(randn(1,(N+G)*NN) + j*randn(1,(N+G)*NN));
36 for kk=1:length(SNR)
37     r1 = y1 + noise*10^(-SNR(kk)/20);
38     r2 = y2 + noise*10^(-SNR(kk)/20);
39     r11=vec2mat(r1,N+G);
40     r22=vec2mat(r2,N+G);
41     r111=r11(:,G+1:G+N);
42     r222=r22(:,G+1:G+N);
43     b1=fft(r111.').';
44     b2=fft(r222.').';
45     b11=-(sign(real(conj(b1(1:end-1,:)).*b1(2:end,:)))-1)/2;
46     b22=-(sign(real(conj(b2(:,1:end-1)).*b2(:,2:end)))-1)/2;
47     [aa,bb]=find(b11 ≠ tx_syms(2:end,:));
48     BER_time(kk)=length(aa)/((NN-1)*N);
49     [aa,bb]=find(b22 ≠ tx_syms(:,2:end));
50     BER_freq(kk)=length(aa)/(NN*(N-1));
51 end
52 figure(1);clf
53 semilogy(SNR,BER_time,'k-o'); hold;
54 semilogy(SNR,BER_freq,'r-*');
55 xlabel('SNR (dB)'); ylabel('BER');
56 legend('Time-domain approach','Frequecy-domain approach')
```

the significant variation of the channel between two consecutive OFDM symbol intervals. On the other hand, the frequency-domain approach is more sensitive to delay spread. The frequency-domain differential detector observes different channel phase and amplitude over two consecutive subcarriers as the maximum excess delay of the channel increases, due to the decrease in the channel frequency correlation. As a result, the performance of the differential detector decreases significantly, even if the maximum excess delay of the channel is less than the cyclic prefix duration.

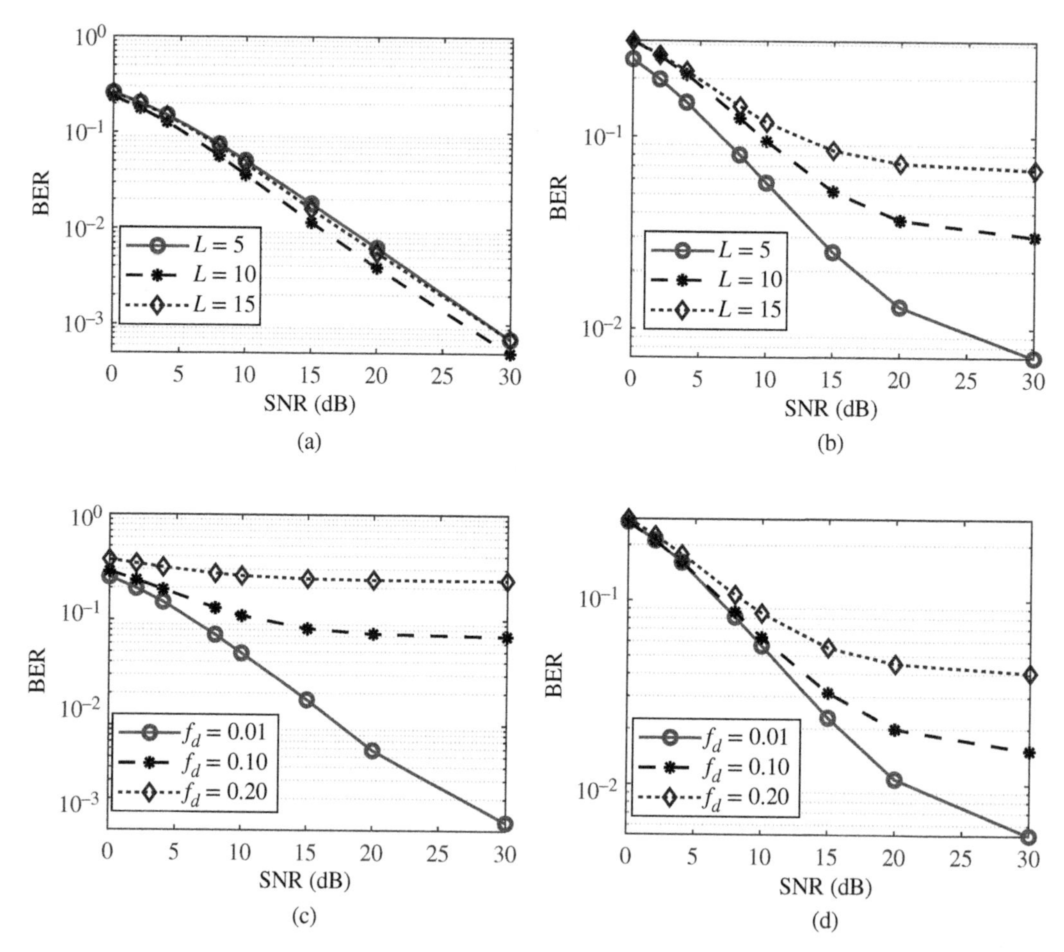

Figure 7.22 Performance of differential OFDM demodulators in various channel conditions. It should be noted that L represents the number of channel taps, and f_d represents the normalized Doppler spread of the channel. (a) Time-domain approach ($f_d = 0.001$). (b) Frequency-domain approach ($f_d = 0.001$). (c) Time-domain approach ($L = 5$). (d) Frequency-domain approach ($L = 5$).

7.3 OFDM-like Signaling

7.3.1 OFDM Versus SC-FDE

single carrier frequency domain equalization (SC-FDE) is an alternative to OFDM for broadband communication systems, and it also employs the equalization process in the frequency domain [26]. Unlike OFDM, where each data symbol occupies a narrow channel, the data symbols are not mapped from the frequency domain to the time domain (i.e. no FFT operation) at the transmitter in SC-FDE. As a result, each symbol occupies the whole bandwidth. However, similar to OFDM, the data symbols are grouped in blocks, and each block of data symbols are cyclically extended with CP or ZP before the transmission. At the receiver, the block of data symbols is transferred to the frequency domain to employ equalization after the CP removal operation. Once the data block is equalized in the frequency domain, it is then converted back to the time domain for demodulation

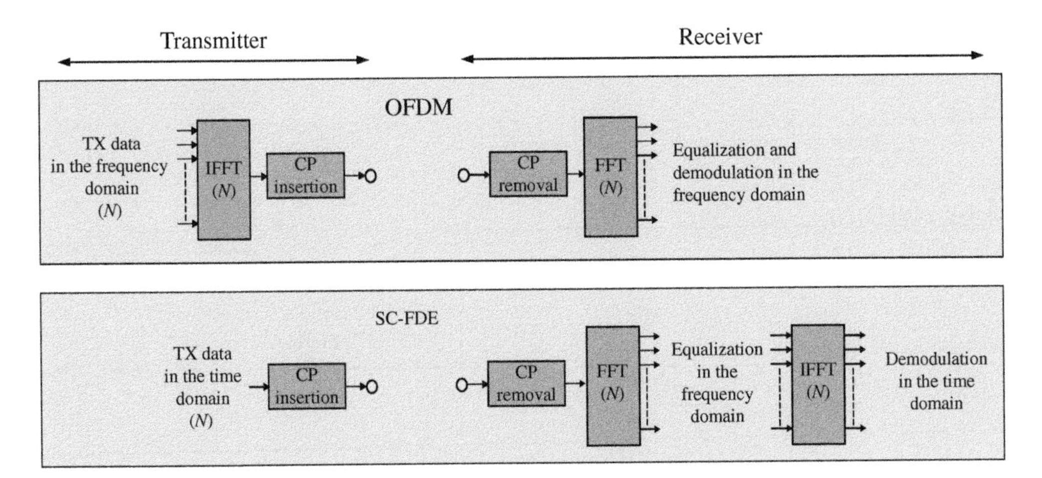

Figure 7.23 Comparison of basic OFDM and SC-FDE transceiver block diagrams.

of the data symbols. In other words, both FFT and IFFT operations are employed at the receiver of SC-FDE. The comparison of OFDM and SC-FDE block diagrams is shown in Figure 7.23.

One of the most significant advantages of SC-FDE over OFDM is its low PAPR because of the single-carrier transmission. The low PAPR property of SC-FDE is especially fascinating for uplink transmission in cellular radio. Another advantage of SC-FDE for uplink transmission is the complexity shift from the transmitter of a mobile handset (by not performing IFFT at the transmitter) to the base station receiver. On the other hand, SC-FDE does not allow subcarrier-based adaptation capabilities, which are very powerful for optimizing the link capacity.

7.3.2 Multi-user OFDM and OFDMA

In the previous sections, a single-user OFDM system is considered, where the available channel is utilized by a single user. It should be noted that OFDM by itself is not a multi-access technique. However, it can be combined with existing multiple accessing methods to allow multiple users to access the available channel. For example, TDMA, frequency division multiple accessing (FDMA), code division multiple access (CDMA), and CDMA-based schemes are common multiple accessing techniques for OFDM systems. A mixture of TDMA and FDMA, known as OFDMA, is also possible, and it has been successfully applied to mobile WiMAX, 4G LTE, and 5G NR.

In contrast to OFDM, where each symbol is used by only one user, OFDMA divides the the available resource into smaller units so that multiple users can send data simultaneously using non-overlapping resource blocks in the time-frequency domain. Figure 7.24 shows the allocation of the subcarriers to different users. OFDMA allows subchannel-based scheduling of the users to the subcarriers based on their channel quality. Especially, it increases the capacity of the system significantly for slowly varying channels. Various subcarrier allocation strategies are investigated for OFDMA systems. Also, in the WiMAX standard, several modes of subcarrier allocation are included. Users can be assigned to the subcarriers in a distributed or block-wise manner. In the distributed allocation schemes, strong frequency diversity can be obtained when the user's data are channel-coded. It is due to distributing the user's information over the whole bandwidth, which is much larger than the coherence bandwidth of the channel in these schemes. On the other hand,

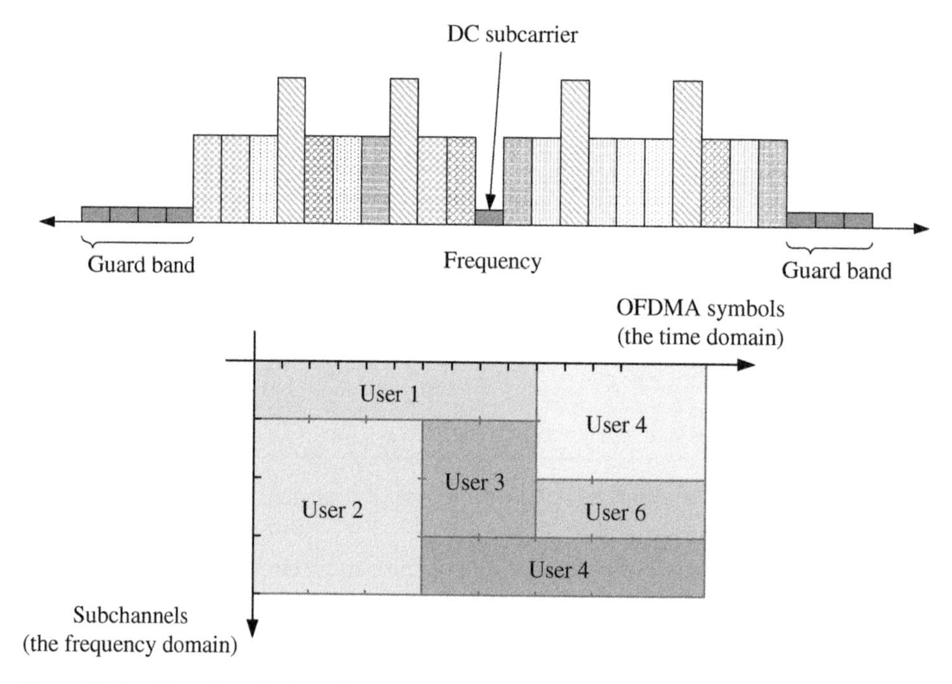

Figure 7.24 Allocation of subcarriers to multiple users in OFDMA.

a band-adaptive modulation and multiuser diversity, where a subchannel group is assigned to the user with the best channel quality in that subchannel group, can be employed easily in the block-wise assignment schemes.

7.3.3 SC-FDMA and DFT-S-OFDM

Single carrier frequency division multiple access (SC-FDMA) and DFT-Spread-OFDM (DFT-S-OFDM) have been selected for uplink multiple accessing schemes of LTE and LTE-advanced, whereas OFDMA is still considered for downlink. One of the major concerns in the uplink is the PAPR of the transmitted signal, so that the PAs can be used in a power-efficient manner. Maintaining the great capabilities of OFDM in terms of scheduling multiple users to the subchannels flexibly and efficiently is desired while achieving the PAPR goal.

SC-FDMA splits the available channel into multiple orthogonal carriers and assigns users into these available subcarriers in a block-wise (a.k.a., localized mapping) or distributed (a.k.a., interleaved mapping) manner similar to OFDMA. Unlike OFDMA, where the data symbols are directly mapped to the subcarriers, a block of data symbols, which are originally interpreted as time symbols, are spread through a discrete Fourier transform (DFT) processing before mapping them to the corresponding subcarriers in SC-FDMA. Hence, all transmitted carriers corresponding to a user contain a component of each modulated data symbol. In other words, a signal sample that is mapped to each subcarrier is a linear combination of all data symbols of a block that corresponds to a user. A comparison of OFDM and SC-FDMA block diagrams is shown in Figure 7.25. As seen in the figure, the first step of SC-FDMA transmission is to spread the block of M data symbols with an M-point DFT operation. Afterward, these spread data symbols are mapped to M out of N subcarriers ($N \geq M$) like in OFDMA. The rest of the transmitter design is similar to OFDMA, such as the N point inverse discrete Fourier transform (IDFT) operation followed by the CP addition.

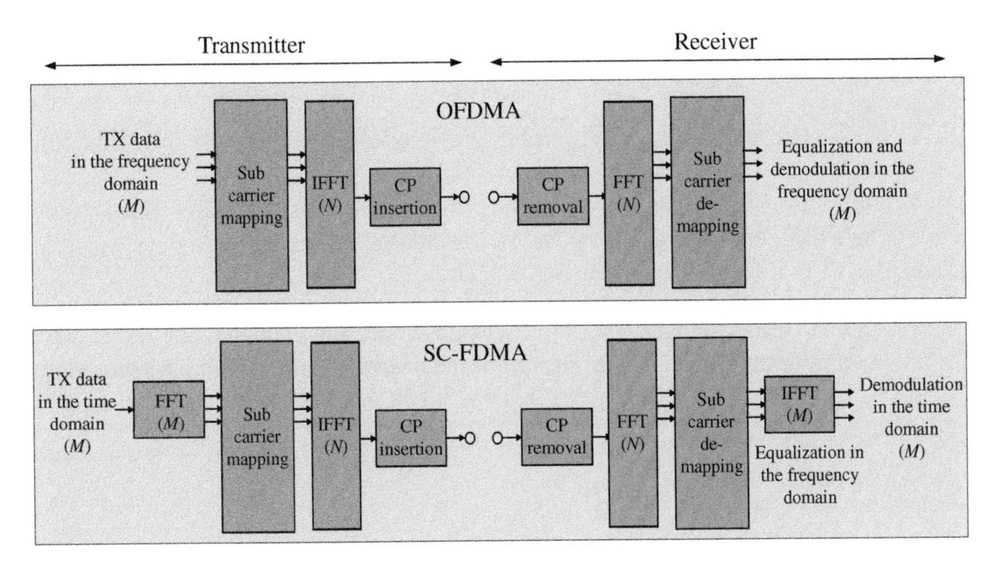

Figure 7.25 Comparison of basic OFDMA and SC-FDMA transceiver block diagrams.

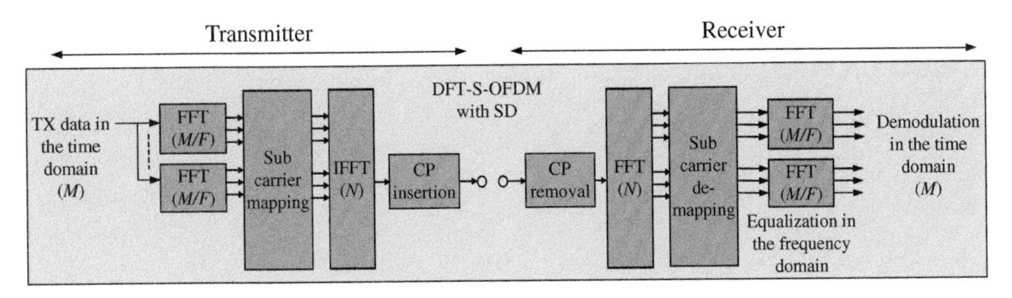

Figure 7.26 Block diagram of a basic DFT-S-OFDM with SD transceiver.

In LTE uplink, the mapping of the spread user data to the corresponding subcarriers is done with block-wise resource allocation (a.k.a., localized mapping), as it enables better PAPR properties. The interleaved resource allocation (a.k.a., distributed mapping) has larger PAPR but provides better frequency diversity. A more flexible version of DFT-S-OFDM for the uplink transmission is also possible. In this more flexible version, the user's data block is divided into a number of subblocks, and each of these subblocks are spread individually with a smaller size of DFT operations. This scheme is referred as DFT-S-OFDM with spectrum division (SD) [27], which is shown in Figure 7.26. As can be seen in the figure, the user data block size of M is divided into F separate blocks wherein each block there are M/F data symbols. Each of these F data subblocks is passed through spreading operation with DFT processing. Then, each spread data block is mapped to subcarriers in a local manner within a single DFT subblock but in a distributed manner between subblocks. In this way, a balance between PAPR reduction and also frequency diversity is achieved.

OFDM-like waveform schemes can be unified under an umbrella as follows:

- $N = M$ and $F = 1$: This scheme is called as SC-FDE. At the transmitter, FFT and IFFT cancels each other (i.e. no IFFT or FFT but the CP still exits). At the receiver, FFT is used for transforming the data to the frequency domain and equalizing it in the frequency domain. Afterward, the equalized data are demodulated in the time domain.

- $N > M$, N is an integer multiple of M, and $F = 1$: This scheme is called as SC-FDMA, and a user uses only a part of the spectrum. Two resource allocation methods are available as localized and interleaved mappings.
- $N > M > F$, N is an integer multiple of M, and $F > 1$: This scheme is called DFT-S-OFDM with SD.
- $N = M = F$: This scheme is the regular OFDM. In essence, at the transmitter only IFFT is used, FFT is non-functional (i.e. FFT size is 1).
- $N > M$ and $M = F$: This scheme is the regular OFDMA.

The following Matlab script simulates the above schemes under one umbrella. The PAPR characteristics and other performance-related results can be obtained using this simplified simulation.

```
1  %% Basic simulation of OFDM, OFDMA, SCFDE, SCFDMA, and DFT-S-OFDM
2  mapping_scheme=1;
3  num_OFDM_symbols=1000;
4  SNR=20; % dB
5  N=256;  % total BW
6  %M=N;   % it corresponds to SC-FDM
7  M=32;   % how many carriers that the user is assigned
8  spreading=N/M; % The spreading factor in SC-FDMA
9  F=4;    % number of DFTs that is used for the user
10 CP=1/4;
11 G=N*CP;
12 mod_order=4;
13 frame=[];
14
15 for kk=1:num_OFDM_symbols
16     symbols=qammod(floor(rand(1,M)*mod_order), mod_order); %QPSK
17     %symbols=(randn(1,M)>0)*2-1; %BPSK
18
19     if (M>F)
20         V_symbols=vec2mat(symbols,M/F,F);
21         Xk=fft(V_symbols.',M/F);
22         Xkk=reshape(Xk,1,M);
23         else
24         Xkk=symbols;
25     end
26
27     if(mapping_scheme==1)
28         % Interleaved
29         Xi=upsample(Xkk,spreading);
30         xi=ifft(Xi,N);
31         time=xi*sqrt(sum(abs(symbols).^2)/sum(abs(xi).^2));
32     elseif(mapping_scheme==2)
33         % Localized
34         Xl=[Xk zeros(1,length(Xkk)*spreading)];
35         xl=ifft(Xl,N);
36         time=xl/sqrt(mean(abs(xl).^2));
37     end
38
39     time_cp=[time(end-G+1:end) time];
40     frame=[frame time_cp];
41 end
42
43 L=5;% number of channel taps
44 h_cir=(randn(1,L)+j*randn(1,L))/sqrt(2); % uniformly dist. PDP
45 % the channel is asumed to be static
46 noise=(randn(1,length(frame))+j*randn(1,length(frame)))/sqrt(2);
47 received = filter(h_cir,1,frame) + 10^(-SNR/20)*noise;
48
49 % CP removal
50 rec_mat=vec2mat(received,N+G);
51 rec_time=rec_mat(:,G+1:end);
52 rec_freq=fft(rec_time.');
53
54 %Freq. domain equalization
```

```
55  H_cfr= fft(h_cir,N); % perfect channel knowledge is assumed
56  H_mtrx=repmat(H_cfr.',1,num_OFDM_symbols);
57  rec_equalized=conj(H_mtrx).*rec_freq./abs(H_mtrx).^2;
58
59  % For SC-FDE, SC-FDMA, and DFT-S-OFDM,
60  % the data is converted back to the time-domain
61
62  if(mapping_scheme==1)
63      % Interleaved
64      Xd=downsample(rec_equalized,spreading);
65  elseif(mapping_scheme==2)
66      % Localized
67      Xd=rec_equalized(1:M,:);
68  end
69
70  if (M>F)
71      Xdd=zeros(size(Xd));
72      for ll=1:F
73          dummy=Xd( (ll-1)*M/F+1:ll*M/F,:) ;
74          dummy2=ifft(dummy,M/F);
75          Xdd((ll-1)*M/F+1:ll*M/F,:)=dummy2;
76      end
77  else
78      Xdd=Xd;
79  end
```

7.4 Case Study: Measurement-Based OFDM Receiver

In this section, the digital baseband transceiver of an OFDM system is studied in detail, and the emphasis is on the measurement-based receiver design. Even though the system parameters are selected based on the WMAN standard (i.e. IEEE 802.16-2004), the model is similar to many other standards. Hence, rather than focusing on specific standard features, the goal is to provide a fairly universal transmitter and receiver structure. Simple block diagram of a transmitter and receiver design is illustrated for measurement-based performance evaluation. Within the receiver blocks, each measurement point and its impact on the performance measurement are described in detail. The performances under various noise and impairment scenarios are discussed as well. Also, the usage of different measurement techniques for identifying different sources of noise and impairments is addressed in this section.

In the case study, an OFDM system with 256 carriers is considered. However, different subcarrier sizes can be implemented in a similar way. It should be noted that in terms of operating carrier frequency and transmission bandwidth, wireless standards provide a wide variety of options. However, the baseline PHY receiver algorithm design, which is based on the I/Q samples, is not affected by the variation in carrier frequency and bandwidth. On the other hand, advanced receiver algorithms might take advantage of these variations and employ better solutions depending on the operating frequency and bandwidth. In this study, a generic baseline receiver algorithm design, which works reasonably well over different operational frequencies, is provided. Improvements in receiver algorithms, such as taking *a priori* information into account, are possible, and the readers are encouraged to enhance them after successful completion of this case study.

7.4.1 System Model

In this section, the transmitter signal structure including, the frame format (downlink and uplink subframes), OFDM symbol formats (both for preamble and data symbols), baseband transmitter blocks, and a basic transmitted signal model, are described.

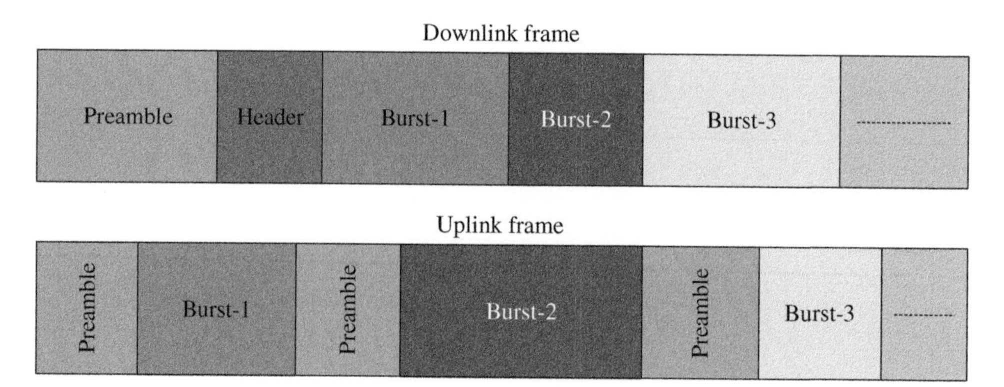

Figure 7.27 Typical downlink and uplink frame formats.

7.4.1.1 Frame Format

The frame format in this study is similar to a traditional packet-based structure that uses a preamble and header followed by data bursts. In a standard downlink frame, as shown in Figure 7.27, the base station transmits a preamble, header, and multiple downlink bursts that are assigned to different users. It should be noted that midambles, which are not shown in the figure, can also be inserted optionally before some bursts. In the uplink, a different preamble is used for each uplink burst transmitted by a user. It should also be noted that each uplink transmitter is assigned a time slot to transmit its burst in a time division multiplexing (TDM) manner. Similarly, in the downlink, a TDM-type multiplexing is employed, as can be seen from the figure. Even though more complex frame structures are possible, like in 5G NR, the format in Figure 7.27 is used for the sake of simplicity. The details are provided in the following sections.

7.4.1.2 OFDM Symbol Format

Each preamble, header, and burst is made up of one or more OFDM symbols. Modulation on the OFDM carriers is BPSK, QPSK, 16-QAM, or 64-QAM. Depending on the link quality between the transmitter and receiver, an appropriate modulation is selected for individual data bursts. There is an ability to use different modulation formats on each data burst. The information bits are mapped into data symbols, depending on the selected modulation type. Afterward, the serial data symbols are demultiplexed into parallel blocks, and IFFT is applied to these parallel blocks to obtain the time-domain OFDM symbols.

There are three types of subcarrier assignments, namely, data, pilot, and null. The number of subcarriers is 256 for OFDM-256. In a regular OFDM-256 symbol, 200 subcarriers are used for data and pilot, and the remaining 56 carriers are nulled for providing guard bands and dealing with carrier leakage, which renders the DC carrier unusable as well. Of the 200 subcarriers, 8 of these are used as pilot subcarriers, and these pilots are inserted at regular intervals among the other data subcarriers, which make up the remaining 192 active data carriers. This exemplary subcarrier assignment for an OFDM-256 symbol is shown in Figure 7.28.

7.4.1.3 Baseband Transmitter Blocks and Transmitted Signal Model

Figure 7.29 shows a basic digital baseband OFDM transmitter block diagram. The information bits from the upper layers of the protocol stacks are first passed through a channel coding, which consists of three blocks: randomizer, FEC block, and interleaver. Data randomization is performed on each burst of data on the downlink and uplink. The FEC can employ various schemes, such as

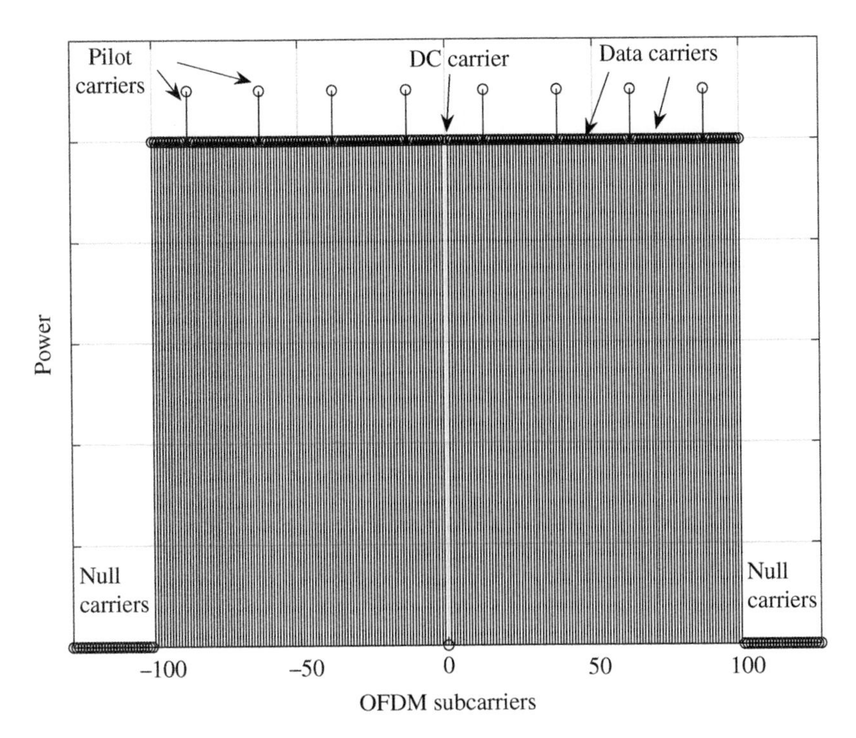

Figure 7.28 Exemplary subcarrier assignment for an OFDM-256 symbol.

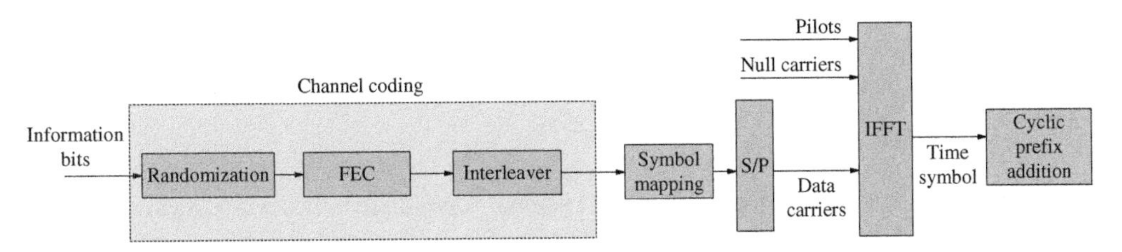

Figure 7.29 A basic digital baseband OFDM transmitter block diagram.

Reed–Solomon (RS) code, rate-compatible convolutional code, block-turbo coding (BTC), and convolutional turbo coding (CTC). All encoded data bits are interleaved by a block interleaver with a block size corresponding to the number of coded bits per allocated subchannels per OFDM symbol. The interleaver is defined by a two-step permutation. The first step ensures that adjacent coded bits are mapped onto nonadjacent subcarriers. The second permutation ensures that adjacent coded bits are mapped alternately onto less or more significant bits of the constellation. Thus, it avoids long runs of less reliable bits. The implementation of a specific channel coding scheme is out of scope in this case study.

The bits after the channel coding operation are mapped to symbols, depending on the selected modulation type. As mentioned before, four different modulation options are available as BPSK, QPSK, 16-QAM, and 64-QAM. Afterward, the data symbols are mapped onto allocated data subcarriers (i.e. the 192 data subcarriers shown in Figure 7.28) in the order of increasing frequency index. Also, the pilot subcarriers (i.e. the eight subcarriers shown in Figure 7.28) are multiplexed into these data carriers in order to constitute the OFDM symbol. It should be noted that BPSK modulation is used for the modulation of pilot carriers.

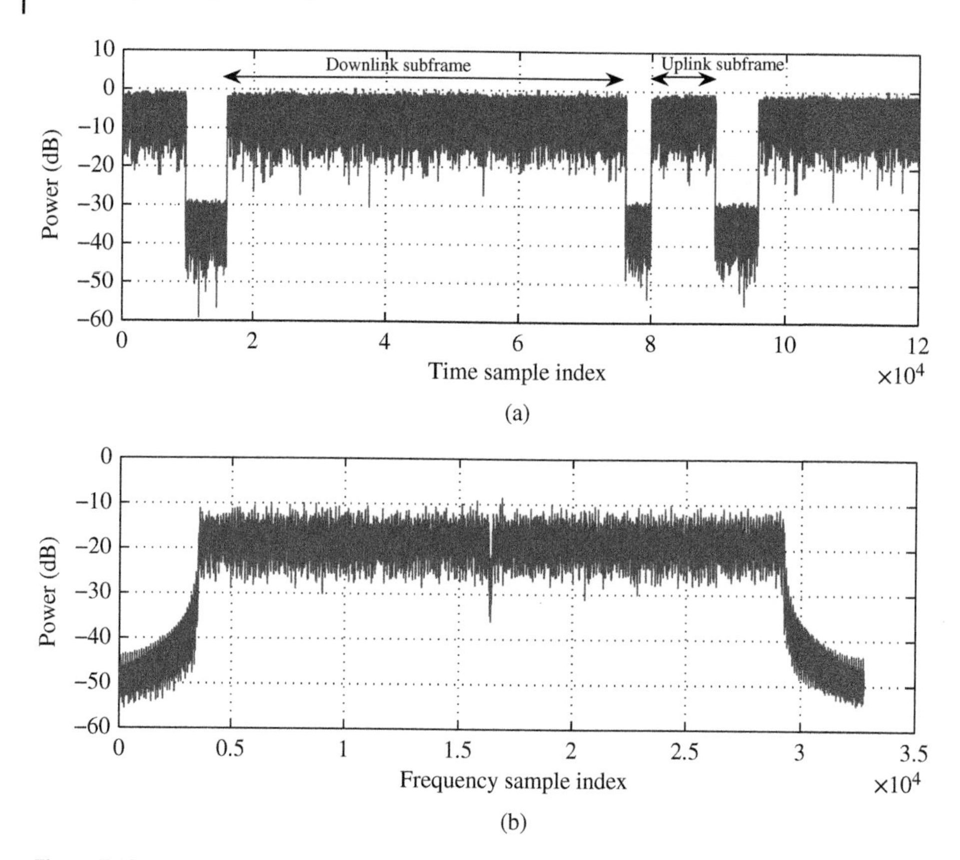

Figure 7.30 OFDM signal within a TDD frame. (a) Tx signal in the time domain. (b) Tx signal in the frequency domain.

Once all the subcarriers that form a full OFDM symbol is populated, the serial data symbols are converted to parallel blocks. Afterward, an IFFT operation is performed to these parallel blocks to obtain the time-domain OFDM symbols. The rest of the transmitter blocks are standard OFDM modulation blocks, as described in the previous sections. The transmitted OFDM signal within a time division duplexing (TDD) frame is illustrated in time and frequency domains in Figure 7.30. Both downlink and uplink RF bursts consist of multiple user bursts in a TDM manner.

7.4.1.4 Received Signal Model

The received signal without the I/Q impairments can be modeled as follows:

$$Y_m(k) = \gamma_m H(k) X_m(k) e^{\Phi_m^{cm} + \Phi_m^{cd}(k)} + z_m(k), \tag{7.37}$$

where Φ_m^{cm} represents the common phase error, which could be caused by uncompensated frequency offset (i.e. the remaining portion after the frequency offset compensation step) and other possible impairment sources as described earlier, $\Phi_m^{cd}(k)$ represents the subcarrier-dependent phase rotation, which is mainly caused by sample timing offset, and γ_m is the common gain offset, which can occur due to the gain variation of the amplifiers. All the other noise sources are either folded into the AWGN term, $z_m(k)$, if they are additive, or into the channel frequency response $H(k)$ if they are multiplicative.

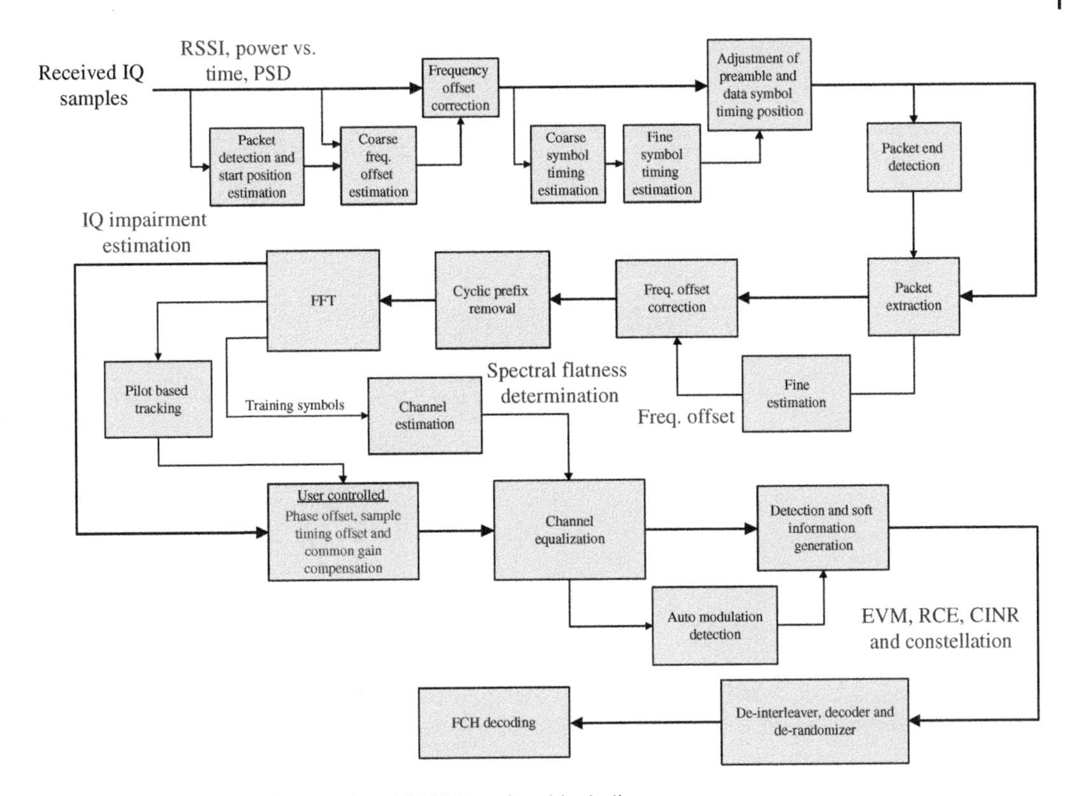

Figure 7.31 A basic digital baseband OFDM receiver block diagram.

7.4.2 Receiver Structure and Algorithms

Figure 7.31 shows the receiver block diagram. It should be noted that the receiver is designed for test and measurement purposes, and hence there are some differences from the actual receiver algorithms used in a real product. It should also be noted that the demonstrated receiver is a digital baseband receiver that processes the I/Q data provided by the analog front-end of any receiver or vector signal analyzer. The received signal is passed through the channel emulator to add the desired impairments for test purposes that are mentioned earlier. Therefore, the received signal represents I/Q samples that include all possible impairments due to an actual hardware (both at transmitter and receiver front-ends) as well as the intentional impairments that are included in the channel emulator.

The transmitter can be any signal source that is transmitting the standard signal as described above. Also, it can be hardware such as a vector signal generator (VSG) or a simulator. The simulated transmitter can include functions like various modulation and channel coding options, different CP alternatives, uplink and downlink burst generation, multiple sampling rate and bandwidth options, flexible number of symbols selection within a burst, and ability to generate random and standard test data. However, as mentioned above, it is not necessary to use the transmitter simulator. The main goal is to test third-party transmitter products with the proposed receiver. However, it is desirable to have the option of using synthetic data from the simulated transmitter as well for debugging and cross-checking purposes. Independent of the transmitter, as long as it is

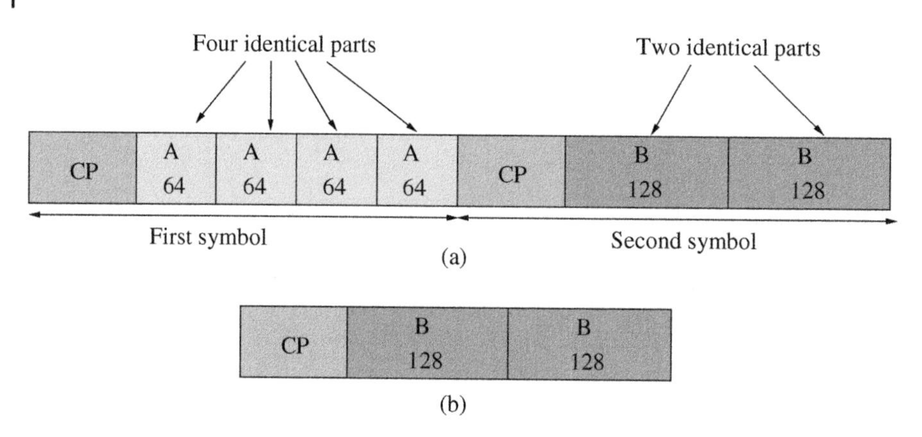

Figure 7.32 The long and short preamble structures. (a) Long preamble. (b) Short preamble.

a standard-based one with the parameters described above, the proposed receiver functions properly. The blocks that are necessary to test and measure the transmitter quality are illustrated in Figure 7.31.

The baseband receiver, which is simulated in Matlab, performs the following operations: find the RF burst and packet edge (i.e. starting point) of the burst; estimate and correct the coarse frequency offset; estimate and perform coarse and fine symbol time synchronization; estimate packet (i.e. RF burst) end position, and extract useful packet information; perform fine frequency offset estimation and correction of the remaining frequency offset; remove CP and convert the time-domain signal to the frequency-domain symbols; estimate the channel frequency response (CFR) and correct symbol rotations due to common and subcarrier-dependent phase offset; employ frequency-domain equalization using the CFR; detect the symbols and obtain soft symbol values to be used by the channel decoder; apply de-interleaver/decoder/de-randomizer; decode the frame control header (FCH) field and check the cyclic redundancy check (CRC). In the following section, a brief explanation of each block is provided. It should be noted that some of the blocks are standard signal processing blocks such as de-interleaving/decoder/de-randomization, FFT, and CP removal. Therefore, these blocks are not discussed in detail since these discussions are widely available in the literature.

In many wireless communication systems, training sequences are inserted within data symbols to help synchronization and channel estimation. The downlink subframe begins with two OFDM symbols used for synchronization and channel estimation at the subscriber station. These two symbols together represent the preamble of the downlink subframe and are referred to as the "long preamble." The uplink subframe begins with one OFDM symbol that is used at the base station for synchronization to the individual subscriber station. This single uplink symbol is referred to as the "short preamble." Figure 7.32 shows the long and short preamble structures. The first symbol in the long preamble is composed of every fourth OFDM subcarrier (i.e. 50 out of 200 in total). Therefore, the time-domain signal has four repeated parts. While the first symbol in the long preamble is useful for coarse signal acquisition, it is not sufficient for detailed channel measurement and correction. Therefore in the downlink subframes, the first symbol is followed by another of the same length, containing alternate active carriers. The second symbol in the time domain has two repeated parts.

7.4.2.1 Packet Detection

Packet detection is employed to check whether there is a useful packet or not and also to find the starting point of that packet. The repeating structure of the training sequences is used in this search. Two sliding windows are used. The first window is used to calculate the autocorrelation between the received signal and a delayed version of it. The amount of the delay, D as shown in the equations below, is equal to the length of the repeating sequence, depending on whether it is downlink or uplink. The second window is used to obtain the received signal power, which is used to normalize the decision statistics so that the decision variable does not depend on the instantaneous power. The size of both windows is the same, M. The values of the first and second windows can be expresses as follows:

$$w_1(n) = \sum_{k=0}^{M} y(n+k)y^*(n+k+D), \tag{7.38}$$

$$w_2(n) = \sum_{k=0}^{M} y(n+k+D)y^*(n+k+D), \tag{7.39}$$

and the decision variable is

$$d(n) = \frac{w_1(n)}{w_2(n)}, \tag{7.40}$$

where $y(n)$ is the received signal and M is the window size. If the magnitude of $d(n)$ exceeds a threshold, it is assumed that there is an incoming packet starting at the point where $d(n)$ exceeds that threshold. The selection of the threshold is a design criterion, and there is a trade-off between the false alarm and miss rates.

An additional modification can be introduced to the above method, including the use of different D in the downlink ($D = 128$) and uplink ($D = 64$), and an adaptive threshold, which starts with a high threshold and than gradually decreases. In addition, the following $w_2(n)$ calculation improves the performance:

$$w_2(n) = 0.5\sum_{k=0}^{M} y(n+k+D)y^*(n+k+D) + y(n+k)y^*(n+k). \tag{7.41}$$

Figure 7.33 shows a sample output of the packet detection algorithm after autocorrelation and moving average filtering operations. The output samples are tested against a threshold. Whenever these samples exceed a predetermined threshold, the existence of a packet is declared, and that point is assumed as the starting point of the packet. It should be noted that Figure 7.33 is obtained for a high SNR value ($SNR = 80$ dB), where the correlation of the noise is close to zero, and the correlation of the repeated parts are close to one. For low SNR values, the peak decreases. It should also be noted that the correlator output around the peak is not like an impulse or a narrow pulse, and it is pretty broad. Therefore, the output of this packet detection provides only a rough idea about where the packet starts.

7.4.2.2 Frequency Offset Estimation and Compensation

Frequency offset is estimated using the training sequences. The Moose method [28] is utilized in this case study. The average phase difference between two identical parts of the training sequences is calculated and then normalized to obtain the frequency offset. The average phase difference can be calculated as follows:

$$\Delta\phi_{av} = -\frac{1}{M}\sum_{n=0}^{M-1} \angle\{y(n+k)y^*(n+k+D)\}. \tag{7.42}$$

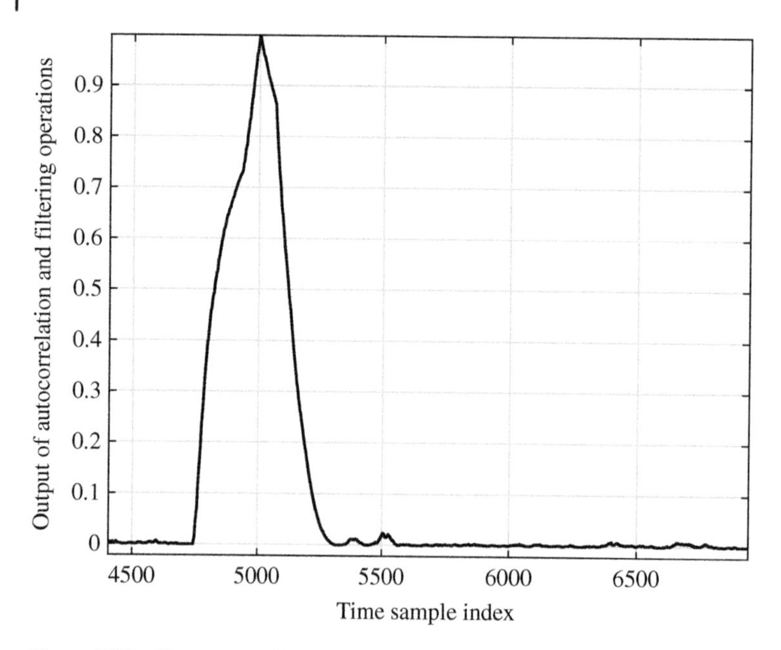

Figure 7.33 The output of the autocorrelation and moving average filtering operations in the packet detection algorithm.

Afterward, the mean phase value is used to calculate the frequency offset by the following equation:

$$f_o = \frac{\Delta\phi_{av}}{2\pi \frac{D}{N}}. \tag{7.43}$$

For coarse frequency offset estimation in the downlink, two blocks with a length-64 in the middle of the first symbol ($M = D = 64$) are used. There are two advantages of performing this operation: (i) using only shorter blocks of 64, larger frequency offsets can be compensated and (ii) using middle blocks, the effect of inaccuracy for coarse timing on frequency offset estimation can be reduced. In the uplink, there is no other choice than employing a short preamble (where $M = D = 128$).

For fine frequency offset estimation both in the uplink and downlink, the second symbol is used, where better noise averaging can be obtained using a block of 128 (i.e. $M = D = 128$). It should be noted again that in the uplink, there is no other choice than using this symbol.

Once the frequency offset is calculated, the received time samples can be rotated in the opposite direction of the estimated frequency offset to compensate for the effect of the frequency offset as follows:

$$y_{corrected}(n) = y(n)e^{j2\pi n\frac{f_o}{N}}. \tag{7.44}$$

7.4.2.3 Symbol Timing Estimation

Timing synchronization refers to finding the exact timing instant of the beginning of each OFDM symbol. Unless the correct timing is known, the receiver cannot remove the cyclic prefixes at the right timing instant and separate individual symbols correctly before the FFT operation. The timing synchronization algorithm basically fine-tunes the rough symbol timing obtained by the packet detection algorithm. Fine symbol timing is calculated using the short training sequence. Cross-correlation between the received signal and known reference is calculated. In this case, the optimum timing position is obtained when the cross-correlation value is maximum. However, it

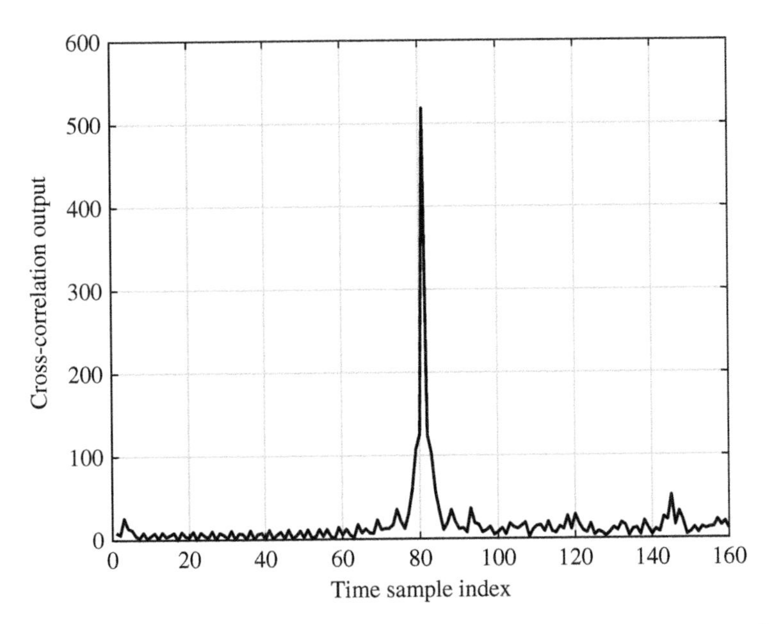

Figure 7.34 The output of the cross-correlation during the symbol timing estimation considering a flat fading channel.

is possible to observe multiple peaks due to multipath components. Also, the largest peak might not correspond to the first multipath component. Therefore, additional fine-tuning of the symbol timing is needed for multipath channels. Joint channel estimation and symbol timing estimation is used in this case study for fine-tuning of the symbol timing estimate.

Figure 7.34 shows the cross-correlation between the received signal and known reference. The peak, where maximum correlation occurs, clearly shows the correct timing point for a flat fading channel (i.e. a single multipath component). However, when the channel is frequency-selective, multiple peaks are observed, as shown in Figure 7.35.

7.4.2.4 Packet-end Detection and Packet Extraction

This block detects the end of a packet, and it is especially crucial for measurement purposes. In other words, it is not an integral part of a regular receiver. If the FCH information content is not available, this is a critical block that detects when the useful information in the RF burst ends. In this way, the noise part of the received signal is not decoded or measured. If FCH content is available, or some other side information about the RF burst is given as an input, the packet end detection calculation can be turned off.

The algorithm takes advantage of the gaps before and after the RF burst to determine the end of the RF burst. Before the RF burst is turned on, a short gap, which constitutes noise purely, is used to calculate the noise power. Since the beginning of the RF burst is detected, the signal-plus-noise-power after the starting point of the RF burst can also be calculated. It should be noted that the data part after the preamble is used for this purpose. A short period, such as one OFDM symbol length of data, is enough for the calculation of signal-plus-noise-power. Using these two measurements, a decision metric to find the end of the packet can be developed.

First, the SNR using the noise power measured before the packet start and signal power after the packet start are determined as follows (it should be noted that the noise power during the packet transmission is ignored assuming that signal power is much larger than the noise power):

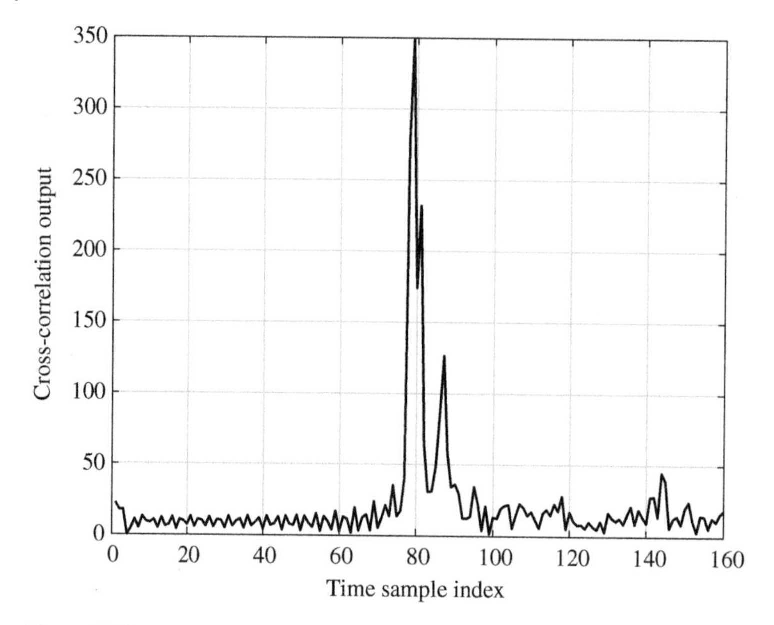

Figure 7.35 The output of the cross-correlation during the symbol timing estimation considering a frequency-selective channel.

$$SNR \approx \frac{\sum_{i=data\ start\ position}^{W+data\ start\ position} |y(i)|^2}{\sum_{i=packet\ start\ position-W}^{packet\ start\ position} |y(i)|^2}, \tag{7.45}$$

where W represents a window of samples. In this case, it is the number of samples for one OFDM symbol length, which is 256. The SNR over a non-overlapping block of samples is monitored for all the received samples. The non-overlapping blocks are used for a lower computational complexity; however, sliding windows are also possible. Afterward, the SNR values for each block are compared with the SNR value given above. When the SNR drops below a certain threshold value, the end of the package is declared. The choice of the threshold value is a design criterion.

7.4.2.5 Channel Estimation and Equalization

Channel estimation is an integral part of many coherent wireless communication receivers. Although there are many advanced channel estimation techniques available in the literature [29], a very simple channel estimator is employed in this case study. Also, the channel is assumed to be almost constant during a data packet here. A simple least squares (LS) channel estimator with a sliding window noise averaging is used for channel frequency estimation. The LS estimates can be given as follows:

$$\hat{H}(k) = \frac{Y(k)}{X_{pilot}(k)}, \tag{7.46}$$

where X_{pilot} are known carrier symbols in the preamble. The preamble symbol with two identical repetitive blocks is used for channel estimation. As mentioned earlier, one subcarrier is skipped (i.e. nulled) after a pilot carrier in this training symbol. Therefore, the total number of known pilot carriers is 100. Interpolation is required to find the channel values in the nulled carriers, and a simple linear interpolation is used in the receiver. After the interpolation step, a sliding window filtering is applied to reduce the noise effect in the estimates. It should be noted that the sliding window size needs to be chosen carefully. If the window size is large, the channel estimates might not be good for highly dispersive channels. On the other hand, if the window size is small, optimal noise suppression can not be obtained.

During the data transmission, channel equalization is performed to remove the channel effect from the received samples. The channel equalization can be performed as follows:

$$\hat{Y}(k) = \frac{Y(k)\hat{H}^*(k)}{|\hat{H}(k)|^2}. \tag{7.47}$$

7.4.2.6 Pilot Tracking

In pilot tracking, common phase estimation, sample timing estimation, and common gain tracking are performed. The phase rotation in the received signal is estimated and corrected for each OFDM frame using eight pilot symbols. The average phase difference between transmitted and received symbols is found by multiplying the received pilots with the conjugate of reference pilots and finding the phase of the result. Once the phase rotation is estimated, it can be corrected simply. Common gain error is also estimated similarly by comparing the gain differences between the transmitted and received symbols. The sample timing error is a little bit tricky, which needs to take into account the variation of the phase both in time and frequency domains. In the receiver, which is design for this case study, the phase differences in time and frequency domains are used to calculate the sample timing error.

The pilot tracking is one of the most critical elements of the receiver design. There are many possible improvements that can be made to the current pilot tracking design. It is also possible to combine pilot tracking with data-aided channel estimation for further improving the receiver design.

7.4.2.7 Auto-modulation Detection

The receiver must know the modulation type at the transmitter to detect the transmitted symbols correctly. This information could be included in the FCH content. However, if the FCH is not available, or the receiver cannot decode it, blind modulation detection is necessary. Blind modulation detection has been traditionally approached in two ways, namely, pattern recognition and decision-theoretic approaches. The objective of the blind modulation detection is to determine the modulation type with the information conveyed by the least possible number of received samples. The only empirical data provided by the received noisy samples is the distance to the closest legitimate constellation point of all possible modulation schemes. Therefore, the distribution of these empirical data or errors should be utilized to make a statistical inference of the used modulation type.

One of the techniques for blind modulation detection is based on the decision-theoretic approach [30]. The mean Euclidean distances between the received samples and all the closest legitimate constellation points of all possible modulation schemes are calculated. The average Euclidean distance for different hypotheses can be calculated as follows:

$$e_m = \frac{\sum_k |Y_m(k) - \hat{Y}_m(k)|^2}{K} \quad m = BPSK,\ QPSK,\ etc., \tag{7.48}$$

where Y is the received sample, $\hat{Y}$ is the hypothesis of the received sample, m is the index of the modulation, and K is the number of samples used for averaging. This scheme, which minimizes the average Euclidean distance e, is chosen for demodulation Eq. (7.48). However, there is always a bias toward the higher-order modulation schemes irrespective of the actual modulation used.

The reason for this is that there are closer legitimate points for higher modulation schemes, which would yield lower errors.

In order to compensate for the bias, information-theoretic approaches are proposed in the literature, such as Akaike information criteria and Bayesian information criteria. A similar technique is used in the receiver of this study based on exhaustive measurements. Similar to the above approach, first, the error terms corresponding to each modulation hypothesis are calculated. Afterward, a correction term is introduced as follows:

$$\hat{e}_m = \log_2(e_m) + \Psi \log_2(M_m), \tag{7.49}$$

where Ψ is a constant, which is 0.9 in this case, and M_m is the constellation size of the mth modulation hypotheses. The resulting detected modulation is the one that minimizes $\hat{e}_m$.

7.4.3 FCH Decoding

The FCH decoding is very useful since the transmission parameters, which define the burst profiles, such as modulation and coding used, and burst length, are contained in FCH. More accurate test and measurement capabilities can be achieved with the ability of FCH decoding. FCH decoding requires demodulation, de-randomization, decoding, and de-interleaving of the received samples. In addition, the CRC decoder is also needed to check whether the content of the decoded bits is correct or not. In other words, almost a full standard-based receiver is needed to be able to decode the FCH content.

There could be several options for measurement purposes. In one option, FCH can be decoded like a regular receiver. In another option, the FCH content can be assumed to be known by the receiver via offline sharing. Also, the FCH content can be blindly estimated, which requires very advanced receiver algorithm designs.

7.4.4 Test and Measurements

Most of the wireless standards provide various transmitter and receiver test requirements. These requirements need to be tested, and conformance should be demonstrated. Therefore, for the case study, the following measurement capabilities are incorporated in the receiver in addition to the ones discussed in more detail later:

- EVM in % or dB
- EVM vs. symbol number
- EVM vs. subcarrier number
- Spectral flatness
- Crest factor
- Peak, average, and minimum EVM
- RSSI
- CINR, SNR
- BER, FER

These measurements are standard measurements that are widely discussed in the literature and heavily utilized in practice. The reader is referred to the Chapter 2 for further details. However, the following measurements are discussed briefly in this chapter:

- **Frequency error:** Provides the frequency error measurement, which is the measurement of the difference of the carrier frequencies generated by the local oscillators at the transmitter and receiver. Rather than the absolute frequency error, the normalized frequency error is a more appropriate value. The frequency error is often measured from the time-domain signal. However, it is also possible to measure the frequency error using frequency-domain samples. In this case study, time-domain samples are used for the frequency error measurement.
- **Sample clock error:** Provides the sampling clock difference at the transmitter and receiver. This measurement in an OFDM system is often performed during the pilot tracking period. Since the sampling clock introduces a phase rotation that depends on the subcarrier and OFDM symbol index, the variation of the rotation can be used to estimate the sample clock error. Sample clock error is popularly estimated using frequency-domain samples after the channel equalization step.

The accuracy of the receiver algorithms affects measurement performance. For example, if the channel estimation algorithm is not designed properly, one might observe worse spectral flatness measure, and then, might conclude that the filters that are used at the transmitter do not have good spectral properties. However, the problem might not be the filter that is used at the transmitter or receiver, and it could be the channel estimation algorithm that is used. Similarly, one might observe a large EVM at the constellation due to the improper receiver algorithm design. Ideally, EVM should reflect errors due to the device under test (DUT), not due to the low-performance receiver algorithms. Therefore, in the testing and measurement world, it is desired to implement "THE" optimal performance receiver algorithms in order to reduce the errors contributed by these algorithms. On the other hand, engineers and researchers have to be careful not to increase the computational complexity and measurement delay. Both fast and accurate measurements are preferable.

For the same reason, when the DUT is being measured, the other impairments caused by other parts of the transceiver chain need to be compensated or calibrated. The calibration or compensation should be done in such a way that the impairments caused by the DUT are not compensated for unintentionally.

Another important point is regarding the location in the receiver chain where a particular measurement should be performed. Ultimately, the goal is to identify the impairment caused by the DUT performing the measurements. If the receiver algorithms correct or change the structure of the impairment, then reliable measurements are not possible. For example, if there is a sample clock compensation algorithm at the receiver, and if an engineer/researcher tries to measure the sample clock error after the compensation, he/she cannot be able to identify the error. It was an obvious and easy example, but, for some tricky measurements, such as I/Q impairments measurements, one must know where to measure in order to obtain the most accurate results. It is possible to make a measurement in two different locations and get similar results. However, in most cases, there is a preferable point where a specific measurement makes the most sense for obtaining better performance, less computational complexity, and other possible reasons.

7.5 Conclusions

In this chapter, OFDM is discussed in detail, along with a comprehensive performance evaluation under various channel conditions and impairments. There is no perfect waveform yet that fits all requirements of diverse services. OFDM and its variants have proven to be effective waveforms that satisfy the requirements of many applications. The flexibility of OFDM has enabled it to be the waveform choice for 5G standards. It is not clear at this point whether more alternatives will coexist with OFDM in 6G and beyond. However, the coexistence of different flexible waveforms should be considered for next-generation communication systems. Unlike the previous standards, future standards will support high flexibility to fully exploit the potentials of future communications systems.

References

1 T. Hwang, C. Yang, G. Wu, S. Li, and G. Y. Li, "OFDM and its wireless applications: a survey," *IEEE Transactions on Vehicular Technology*, vol. 58, no. 4, pp. 1673–1694, May 2009.

2 A. F. Demir, M. Elkourdi, M. Ibrahim, and H. Arslan, "Waveform design for 5G and beyond," in *5G Networks: Fundamental Requirements, Enabling Technologies, and Operations Management.*Wiley, 2018, ch. 2, pp. 51–76.

3 X. Lin, J. Li, R. Baldemair, J. T. Cheng, S. Parkvall, D. C. Larsson, H. Koorapaty, M. Frenne, S. Falahati, A. Grovlen, and K. Werner, "5G new radio: unveiling the essentials of the next generation wireless access technology," *IEEE Communications Standards Magazine*, vol. 3, no. 3, pp. 30–37, Sept. 2019.

4 Y. Kim, Y. Kim, J. Oh, H. Ji, J. Yeo, S. Choi, H. Ryu, H. Noh, T. Kim, F. Sun, Y. Wang, Y. Qi, and J. Lee, "New radio (NR) and its evolution toward 5G-advanced,"*IEEE Wireless Communications*, vol. 26, no. 3, pp. 2–7, June 2019.

5 S. Parkvall, E. Dahlman, A. Furuskar, and M. Frenne, "NR: the new 5G radio access technology," *IEEE Communications Standards Magazine*, vol. 1, no. 4, pp. 24–30, Dec. 2017.

6 S. Lien, S. Shieh, Y. Huang, B. Su, Y. Hsu, and H. Wei, "5G new radio: waveform, frame structure, multiple access, and initial access," *IEEE Communications Magazine*, vol. 55, no. 6, pp. 64–71, June 2017.

7 Z. E. Ankarali, B. Peköz, and H. Arslan, "Flexible radio access beyond 5G: a future projection on waveform, numerology, and frame design principles," *IEEE Access*, vol. 5, pp. 18 295–18 309, 2017.

8 A. A. Zaidi, R. Baldemair, H. Tullberg, H. Bjorkegren, L. Sundstrom, J. Medbo, C. Kilinc, and I. D. Silva, "Waveform and numerology to support 5G services and requirements," *IEEE Communications Magazine*, vol. 54, no. 11, pp. 90–98, Nov. 2016.

9 T. van Waterschoot, V. L. Nir, J. Duplicy, and M. Moonen, "Analytical expressions for the power spectral density of CP-OFDM and ZP-OFDM signals,"*IEEE Signal Processing Letters*, vol. 17, no. 4, pp. 371–374, Apr. 2010.

10 A. F. Demir, Q. H. Abbasi, Z. E. Ankarali, A. Alomainy, K. Qaraqe, E. Serpedin, and H. Arslan, "Anatomical region-specific in vivo wireless communication channel characterization," vol. 21, no. 5, pp. 1254–1262, Sept. 2017.

11 X. Huang, J. A. Zhang, and Y. J. Guo, "Out-of-band emission reduction and a unified framework for precoded OFDM," *IEEE Communications Magazine*, vol. 53, no. 6, pp. 151–159, June 2015.

12 H. Lin, "Flexible configured OFDM for 5G air interface," *IEEE Access*, vol. 3, pp. 1861–1870, 2015.

13 A. F. Demir and H. Arslan, "Inter-numerology interference management with adaptive guards: a cross-layer approach," *IEEE Access*, vol. 8, pp. 30378–30386, 2020.

14 B. Farhang-Boroujeny, "OFDM versus filter bank multicarrier," *IEEE Signal Processing Magazine*, vol. 28, no. 3, pp. 92–112, May 2011.

15 T. Weiss, J. Hillenbrand, A. Krohn, and F. K. Jondral, "Mutual interference in OFDM-based spectrum pooling systems," in *2004 IEEE 59th Veh. Technol. Conf.*, vol. 4, Milan, May 2004, pp. 1873–1877.

16 E. Bala, J. Li, and R. Yang, "Shaping spectral leakage: A novel low-complexity transceiver architecture for cognitive radio," *IEEE Vehicular Technology Magazine*, vol. 8, no. 3, pp. 38–46, 2013.

17 A. Sahin, I. Guvenc, and H. Arslan, "A survey on multicarrier communications: Prototype filters, lattice structures, and implementation aspects,"*IEEE Communications Surveys Tutorials*, vol. 16, no. 3, pp. 1312–1338, Third 2014.

18 A. Şahin and H. Arslan, "Edge windowing for OFDM based systems," *IEEE Communications Letters*, vol. 15, no. 11, pp. 1208–1211, Nov. 2011.

19 Y. Rahmatallah and S. Mohan, "Peak-to-average power ratio reduction in OFDM systems: a survey and taxonomy," *IEEE Communications Surveys Tutorials*, vol. 15, no. 4, pp. 1567–1592, Apr. 2013.

20 H. Ochiai and H. Imai, "On the distribution of the peak-to-average power ratio in OFDM signals,"*IEEE Transactions on Communications*, vol. 49, no. 2, pp. 282–289, Feb. 2001.

21 T. Wang, J. G. Proakis, E. Masry, and J. R. Zeidler, "Performance degradation of OFDM systems due to Doppler spreading," *IEEE Transactions on Wireless Communications*, vol. 5, no. 6, pp. 1422–1432, June 2006.

22 Y. Li and L. J. Cimini, "Bounds on the interchannel interference of OFDM in time-varying impairments,"*IEEE Transactions on Communications*, vol. 49, no. 3, pp. 401–404, Mar. 2001.

23 S. Wu and Y. Bar-Ness, "OFDM systems in the presence of phase noise: consequences and solutions," *IEEE Transactions on Communications*, vol. 52, no. 11, pp. 1988–1996, Nov. 2004.

24 A. Demir, A. Mehrotra, and J. Roychowdhury, "Phase noise in oscillators: a unifying theory and numerical methods for characterization," *IEEE Transactions on Circuits and Systems I: Fundamental Theory and Applications*, vol. 47, no. 5, pp. 655–674, May 2000.

25 A. A. M. Saleh, "Frequency-independent and frequency-dependent nonlinear models of TWT amplifiers," *IEEE Transactions on Communications*, vol. 29, no. 11, pp. 1715–1720, 1981.

26 D. Falconer, S. L. Ariyavisitakul, A. Benyamin-Seeyar, and B. Eidson, "Frequency domain equalization for single-carrier broadband wireless systems," *IEEE Communications Magazine*, vol. 40, no. 4, pp. 58–66, 2002.

27 L. Liu, T. Inoue, K. Koyanagi, and Y. Kakura, "Uplink access schemes for LTE-advanced," *IEICE Transactions on Communications*, vol. 92, no. 5, pp. 1760–1768, 2009.

28 P. H. Moose, "A technique for orthogonal frequency division multiplexing frequency offset correction," *IEEE Transactions on Communications*, vol. 42, no. 10, pp. 2908–2914, 1994.

29 M. K. Ozdemir and H. Arslan, "Channel estimation for wireless ofdm systems," *IEEE Communications Surveys Tutorials*, vol. 9, no. 2, pp. 18–48, 2007.

30 T. Keller and L. Hanzo, "Blind-detection assisted sub-band adaptive turbo-coded OFDM schemes," in *1999 IEEE 49th Vehicular Technology Conference (Cat. No.99CH36363)*, vol. 1, Houston, TX, May 1999, pp. 489–493.

8

Analysis of Single-Carrier Communication Systems

Hüseyin Arslan

Department of Electrical Engineering, University of South Florida, Tampa, FL, USA
Department of Electrical and Electronics Engineering, Istanbul Medipol University, Istanbul, Turkey

Single-carrier communication systems have a rich and long history in the design of radio-access technologies. Even though multi-carrier communication systems gained significant interest lately for broadband radio access, there are still a lot of reasons to use single-carrier waveforms for some applications, channel, and radio frequency conditions. In this chapter, some of the fundamental concepts related to single-carrier communication will be discussed. The focus will be on the practical aspects of the implementation, analysis, and performance measurement of single-carrier systems.

8.1 A Simple System in AWGN Channel

A simple single-carrier communication system is shown in Figure 8.1. The system consists of a digital baseband transmitter, wireless radio channel, and digital baseband receiver. In this section, a single-carrier communication system in additive white Gaussian noise (AWGN) channel will be described. In the upcoming sections, this simple model will be extended to more complex multipath channel conditions.

A basic transmitter system block diagram for single-carrier systems is shown in Figure 8.2. A block of desired number of information bits is generated, representing the digital data source. These bits are mapped to data symbols using the preferred modulation option and order. These data symbols are upsampled (with the desired over sampling rate) and passed through a baseband pulse-shaping filter to contain the transmitted signal in a finite and reasonable bandwidth. These samples along with training symbols are used to form the burst/packet. The packet is then passed through the radio channel. Note that this block diagram is specific for single-carrier time-domain equalization-based waveforms. Depending on the waveform and signaling type, there will be some variation in the transmitter blocks. For example, for single-carrier frequency-domain equalization, cyclic extension for the data block is needed. For code division multiple access (CDMA) type of waveforms, spreading with orthogonalization and long codes is needed.

Figure 8.3 shows a simplified baseband receiver block diagram for single-carrier systems. The transmitted signal reaches the receiver after passing through the radio channel. Here we focus only on the baseband digital processing assuming that the received signal passed through the receiver radio frequency (RF) front-end and analog-to-digital converter (ADC) to obtain the complex in-phase and quadrature-phase (I/Q) samples. The received oversampled I/Q data is first passed through a matched filter. These filtered samples are then used to find the packet start position. This is needed to find a rough reference point so that we can identify the training

Wireless Communication Signals: A Laboratory-based Approach, First Edition. Hüseyin Arslan.

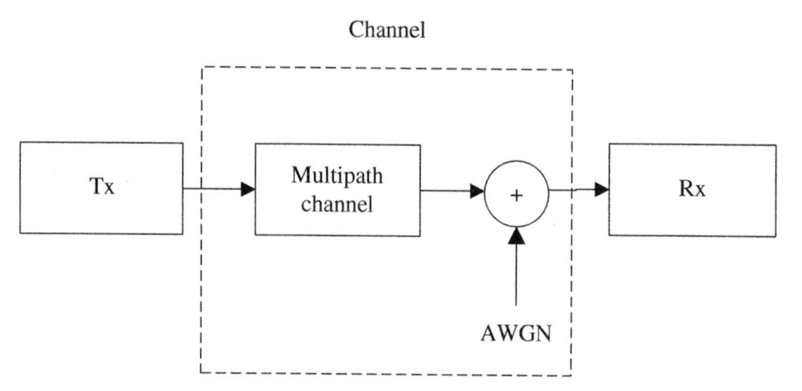

Figure 8.1 A simple communication system model.

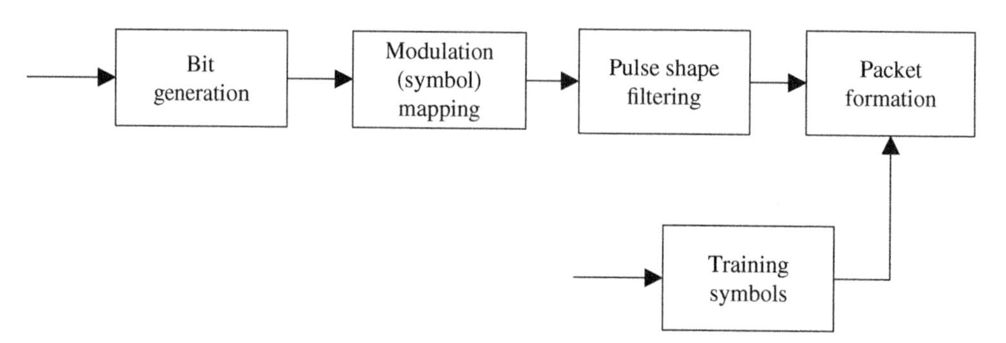

Figure 8.2 A basic transmitter system block diagram for single-carrier systems.

and data symbols' starting points. Using this reference point, we can find frequency offset and compensate it. The coarse synchronization is not sufficient for processing the data accurately. There is a need for fine-tuning the synchronization point, which is achieved through fine symbol and sample timing estimation. Using this reference point, we can extract the data signal and process the received samples. In most cases, we employ fine timing synchronization jointly with channel estimation. While joint synchronization and estimation is optimal in performance, it increases the computation complexity. Hence, this approach is not always used. For coherent reception, we will need to estimate the channel tap delays and coefficients which are performed through channel estimation block. Once the channel coefficients are estimated, we can equalize the received signal to compensate for the channel effect. Equalization is the most critical part of the receiver processing. The details of the equalization process are discussed later in this chapter. After the equalization, the samples are used for detecting the transmitted symbols and bits.

The receiver shown in Figure 8.3 is specific for single-carrier time-domain equalization type signaling. For other waveforms and signaling types, there will be some variations in the receiver chain and blocks. For example, Rake receivers in CDMA systems are quite different than what is illustrated in the figure. Every waveform has its own features and receivers, though there are several common blocks. The details and differences are described later in this chapter.

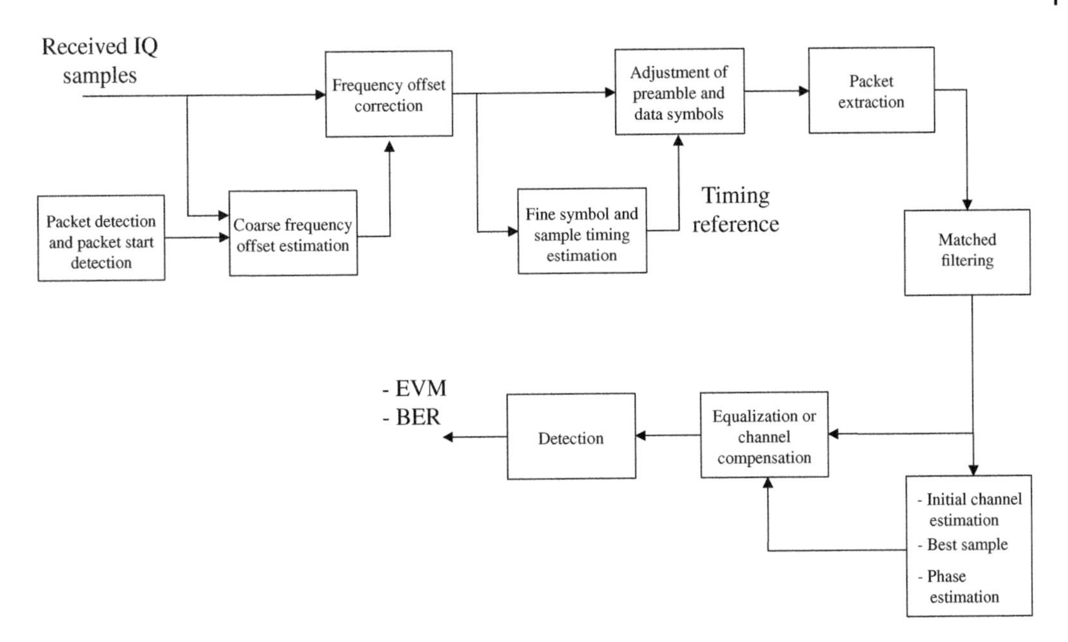

Figure 8.3 A basic measurement base receiver system block diagram for single-carrier systems.

A complex baseband-equivalent system model is considered. The complex values correspond to in-phase (cosine) and quadrature (sine) components of the radio signal. At the transmitter, a sequence of digital symbols is transmitted using pulse shaping, giving

$$x(t) = \sum_{k} b_k f(t - kT), \tag{8.1}$$

where b_k corresponds to the sequence of symbols, $f(\tau)$ is the pulse shape as a function of delay τ, and T is the symbol period. The complex symbols are obtained from the transmitted bits after source and channel encoding blocks. The encoded and possibly interleaved bits are mapped to symbols. The mapping depends on the modulation used at the transmitter. In current digital communication systems, various modulation options are used. Examples would be binary phase shift keying (BPSK), Gaussian minimum shift keying (GMSK), quadrature phase shift keying (QPSK), quadrature amplitude modulation (QAM), eight-state phase shift keying (8-PSK), 16-state quadrature amplitude modulation (16-QAM), 64-state quadrature amplitude modulation (64-QAM), and so on (see Chapter 6 for the details of these modulation options). The mapped bits produce modulated symbols in the form of complex numbers that will later be used to generate I (in-phase) and Q (quadrature phase) waveforms. Before applying the baseband (pulse shaping) filter to these symbols, the data need to be upsampled. Upsampling is necessary as the filtered signal will often have more bandwidth than the symbol rate. Usually, the transmitted signal has an excess bandwidth, i.e. the bandwidth of the signal is more than the symbol rate. The amount of excess bandwidth depends on the filter shape as described in previous chapters. For example, if a raised cosine filter is used, filter roll-off (α) has a significant impact on bandwidth, $Bandwidth = (1 + \alpha) * (symbol\ rate)$. Due to this excess bandwidth, the sampling rate should be more than the symbol rate. Therefore, in order to satisfy the Nyquist criteria, data symbols need to be upsampled (above Nyquist rate) and applied to a digital pulse-shaping filter with the same

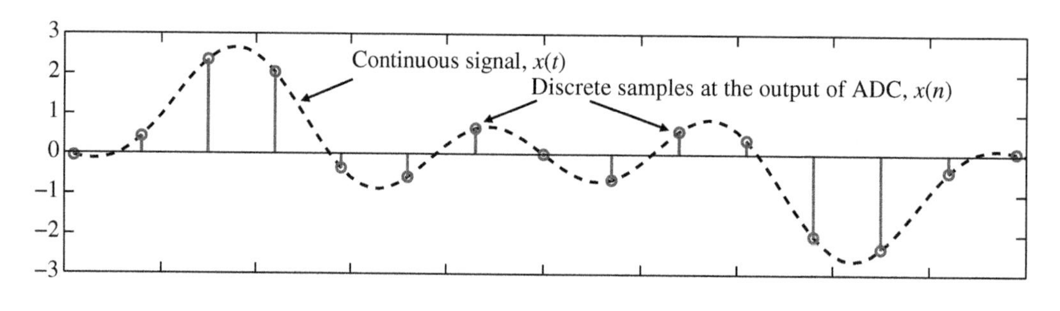

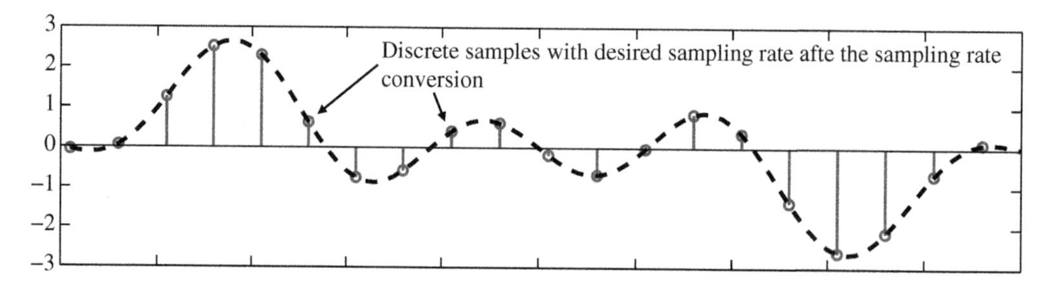

Figure 8.4 Sample rate conversion of the ADC output.

sampling rate. Oversampling to above Nyquist rate also relaxes the requirements for a reconstruction filter in the actual digital-to-analog converter (DAC). It is common to use upsampling rate of four (i.e. four samples per symbol). After the DAC and passing the signal through transmitter and receiver front-end, assuming a basic AWGN channel, the received matched-filtered signal can be written as:

$$r(t) = \sum_k b_k h(t - kT) + z(t), \tag{8.2}$$

where $h(t)$ is the effective filter, which is the combined transmit and receiver pulse-shaping filter. Here, we assume that matched filtering is applied in analog domain. Often, the matched filtering is applied to discrete samples after ADC. ADC produces samples more than the symbol rate for the same reason given above. Especially for SDR-type receivers, where multi-rate processing is needed due to the wide variety of signaling techniques that the receiver needs to process, it might not always be possible to get samples as integer multiples of symbol rate from ADC. If this is the case, the received samples can be passed through a sampling rate conversion unit where the data received at the output of ADC is resampled to generate the desired sampling rate, which is an integer multiple of the symbol rate (see Figure 8.4). As long as the output of ADC satisfies the Nyquist criteria, resampling to an integer multiple of symbol rate maintains the integrity of the signal.

After receiver filtering and coarse synchronization, where the processing might be preferred with the oversampled data, we often downsample the received signal to the symbol rate for the rest of the processing. Choosing optimal sample phase that provides best signal-to-noise ratio (SNR) is a critical factor. If the combined filter is Nyquist filter, then by choosing the optimal sampling phase and employing symbol-spaced sampling with this optimal sample phase, the received signal can be reduced to the following form:

$$\boxed{r_k = b_k + z_k}, \tag{8.3}$$

where z_k represents the filtered noise samples. Note that z_k corresponds to a sequence of complex Gaussian noise samples, which may be correlated depending on the pulse shape autocorrelation function. As can be seen, the signal model in AWGN channel is a very simple model and the receivers to demodulate the signal in this channel are straightforward.

The best sample phase can be obtained using training sequences and by employing maximum likelihood estimation. For all possible sample hypotheses, the average error between what is actually received and what is ideally expected can be calculated. The sample phase that minimizes the mean squared error (MSE) will lead to the best sample phase. If a training sequence is not available, the best sample phase can be found using cyclostationarity analysis. For each sample phase hypothesis, we can calculate the average power of the receiver signal $E\{|r_k|^2\}$. The sampling phase that provides the maximum average power would be the best sample phase. Note that in both of these techniques, we chose the best offset among the possible sample offset values. This does not mean that the chosen offset value is optimal. In computer simulations, the best might mean the optimal, but this is not necessarily the case when samples are received from the output of actual ADC hardware. The possible sampling phases might not contain the optimal sample phase as the data is oversampled with only a few samples per symbol. There is only one position that will provide the optimal inter-symbol interference (ISI) free samples, this is the point where the eye is wide open in the eye diagram. However, our samples might not contain this optimal sample position (see Figure 8.5). In this case, there will be some ISI in the received signal depending on the oversampling ratio that we used at the receiver. Increasing the oversampling rate reduces the ISI value, as the receiver will more likely pick the best sample phase that is closer to the optimal sample phase (i.e. as the oversampling rate increases, the difference between optimal and best sampling phase decreases). Alternatively, we can take the output of ADC and apply sample phase correction or timing recovery algorithm. Timing recovery is desired not only for maximizing SNR but also minimizing ISI by sampling each symbol at a specific point (time) where it has zero contribution from other symbol which is an inevitable case for Nyquist pulse shaping.

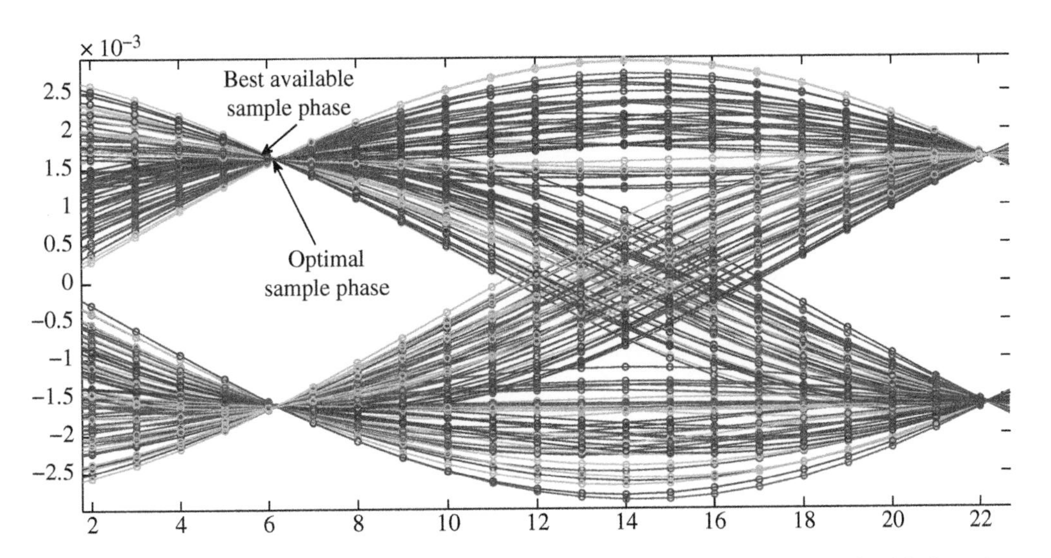

Figure 8.5 Illustration of the best and optimal sample phase in eye diagram. The data for this figure is obtained from a real hardware with oversampling ratio of 16 (16 samples per symbol) with raised cosine filtering. Even with the large oversampling ratio, the best sample phase is not exactly on the top of the optimal sample phase. However, thanks to large oversampling ratio, the difference between optimal and the best available sampling phase is minor. Therefore, the ISI observed will be negligible.

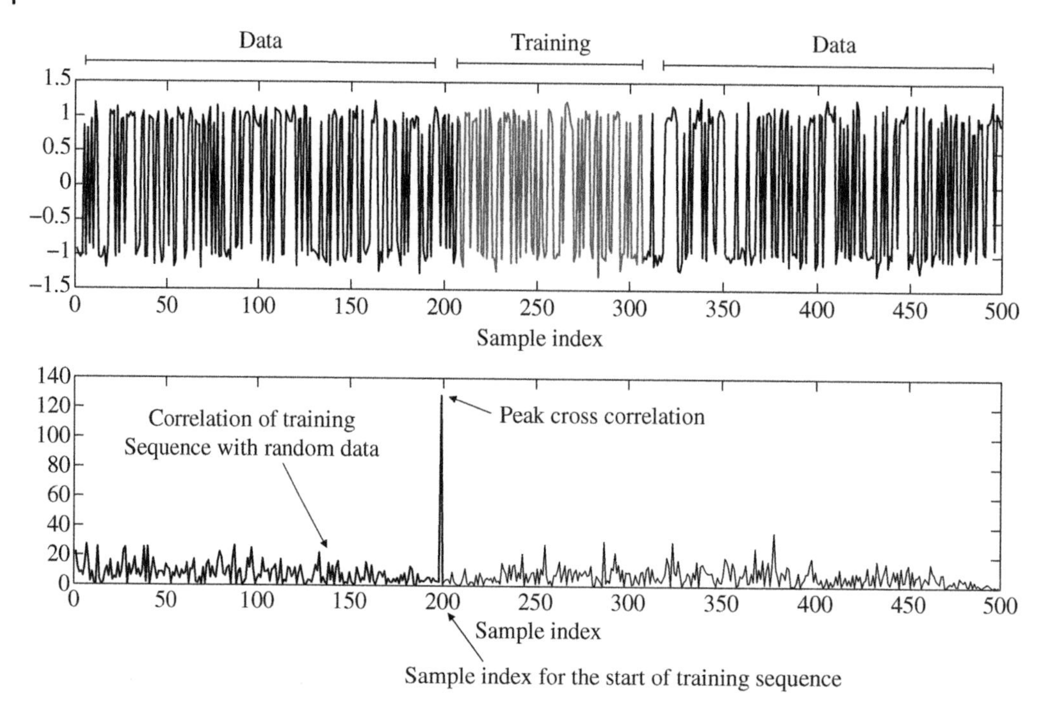

Figure 8.6 A simple example that shows the cross-correlation of a received signal, which includes the training sequence (along with data sequence), with the local copy of the training sequence. If the training sequence is chosen to have good correlation properties, the cross-correlation provides a nice peak that can be used to obtain symbol timing reference point estimation. In this specific figure, an m-sequence of length 127 is used. Here, the symbol timing reference is the beginning of the training sequence in the received burst.

In addition to finding optimal sample phase, obtaining the reference point in the received signal is critical. This is referred to as symbol timing synchronization in the literature. Finding the reference point allows the receiver to know which symbol within the frame is being processed. Symbol timing estimation is a critical and integral part of all communication systems. Usually, transmitter sends a known training sequence (with very good autocorrelation properties) along with the unknown data symbols to be able to identify this reference point in the received frame. Since the receiver also knows the training sequence, by correlating the received samples with the known sequence, the reference point can be obtained easily (see Figure 8.6). For a sequence with perfect autocorrelation properties, the correlation output will provide a single peak at the reference position in AWGN and flat fading channels. When the channel is dispersive, then there will be a peak corresponding to each multipath component.

If there is a frequency offset between transmitter and receiver carrier frequencies, then this will cause rotation of the samples from the original constellations as shown below:

$$\hat{r}_k = r_k e^{j2\pi k \frac{f_o}{f_s}}, \tag{8.4}$$

where the rotation in each symbol will be dependent on the frequency offset value, f_o, and f_s represents the sampling frequency. With phase offset in the system, the received signal will be rotated depending on the phase offset value θ as given below:

$$\hat{r}_k = r_k e^{j\theta}. \tag{8.5}$$

There is a linear relation between sample time index and frequency offset; meaning that with increasing sample time index, the amount of rotation (i.e. the phase difference between the received and ideal) in each symbol increases as well. Therefore, this offset should be estimated and compensated correctly to be able to demodulate the signal. There are various algorithms available in the literature exploiting this property to find the frequency offset value. Especially for the systems transmitting repeating training sequences along with the data, the use of the phase change among these repeating sequences can be exploited to find frequency offset. When the repeating training sequences are far from each other in such a way that the phase might make a whole 2π rotation between two consecutive known sequences, then a single sequence can be used to estimate the linear phase change and hence estimate the frequency offset that causes this change. In addition to using training sequences, the phase difference (and phase growth) can also be tracked in data mode using data-directed tracking. After each symbol decision, assuming the decisions are correct, the phase differences between what is received and ideal symbols (corresponding to the detected symbol) are compared. Then, these phase differences are used to calculate the frequency offset. Again, the linear phase change can be exploited to relate the phase growth in time to frequency offset. There are some issues that one needs to be cautious about in data-directed mode. Since the data symbols are unknown, we need to make sure that the phase growth does not reach to a value where it crosses the decision boundary. This problem can be resolved by unwrapping the phase values when they exceed the decision boundary. When the phase growth reaches around the decision boundary, the possibility of incorrect symbol decision increases, hence affecting the phase estimates. This situation creates a ping-pong effect on the estimated phases (phase difference comes and goes between positive and negative peak values). For QAM type of modulations, the use of low

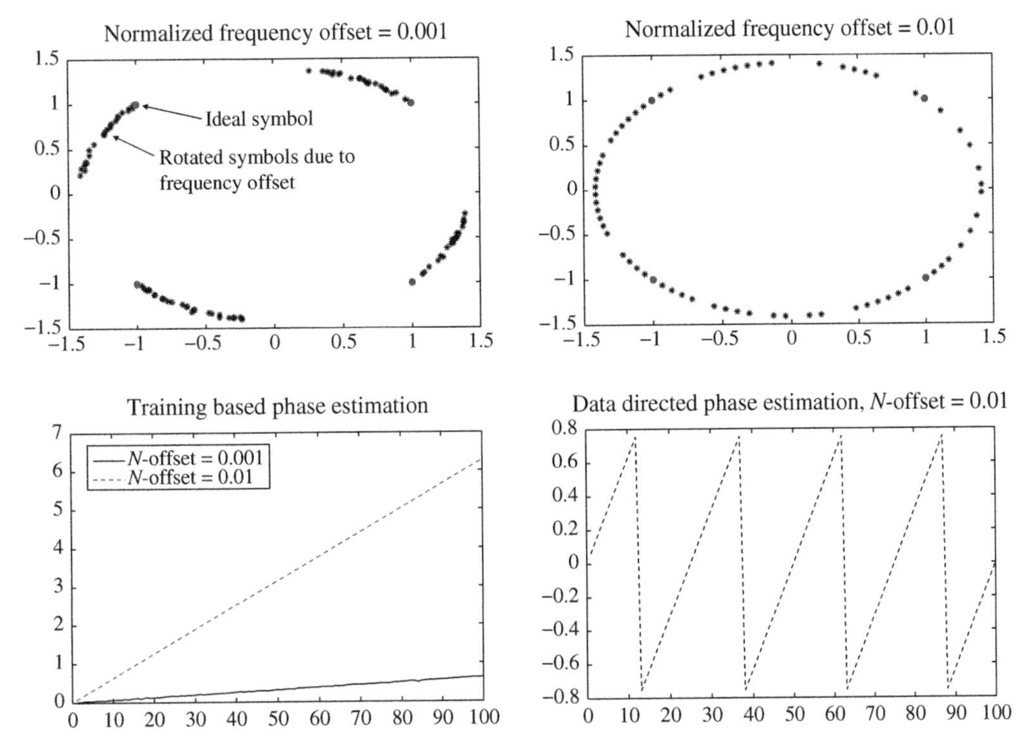

Figure 8.7 Impact of frequency offset in the phase of the received samples. Estimation of the phase when the data is known (training mode) and when the data is estimated (data directed mode) are a1lso given.

SNR constellations might also create incorrect phase estimates. This problem can be avoided by only using the corner constellation points. These samples that correspond to the corner states have the highest amplitude value among all other constellation points. But, this will reduce the statistics on estimating the phase value as the probability of receiving only the corner states is less (e.g. in rectangular 64-QAM, four points are in the corner, the probability of having a corner state is 1/16). The following Matlab code provides a simple test for the impact of frequency offset with different normalized (with symbol rate) frequency offset values over 100 symbols interval. The phase estimates in training mode and data-directed mode are also calculated. The corresponding plots are shown in Figure 8.7.

```
% Impact of frequency offset over a data block.
% Observe the effect of: block size, normalized frequency offset,
% modulation order, and noise variance
N=1e2; % block size in terms of number of symbols
M=4; % modulation order
symb=qammod(floor(rand(1,N)*M),M); %  modulated symbols
fo=0.001; % normalized frequency offset wrt symbol rate
symb_rx=symb.*exp(j*2*pi*[1:N]*fo); % symbols with frequency offset
noise=0.01*(randn(1,N)+j*randn(1,N)); % AWGN noise
symb_rx=symb_rx+noise; % symbols with noise and frequency offset

subplot(2,2,1)
plot(real(symb_rx),imag(symb_rx),'*')
hold on
plot(real(symb),imag(symb),'o')
title('normalized freq. offset=0.001')
subplot(2,2,3)
ph_e=symb_rx.*conj(symb); % phase error in time (wrt to the exact symbols)
plot(unwrap(angle(ph_e)))
hold on

fo=0.01;
symb_rx=symb.*exp(j*2*pi*[1:N]*fo);
subplot(2,2,2)
plot(real(symb_rx),imag(symb_rx),'*')
hold on
plot(real(symb),imag(symb),'o')
title('normalized freq. offset=0.01')
phase_diff=symb_rx.*conj(symb);
subplot(2,2,3)
plot(unwrap(angle(phase_diff)),'--')
legend('N-offset=0.001','N-offset=0.01')
title('training based phase estimation')

symbs_dtct=qamdemod(symb_rx,M); % detected noisy and impaired symbols
symbs_rcntrct=qammod(symbs_dtct,M); % reconstructed symbols
ph_e_dd=symb_rx.*conj(symbs_rcntrct); % phase error in data directed mode
subplot(2,2,4)
plot(unwrap(angle(ph_e_dd)),'--')
title('data directed phase estimation')
end
```

Estimating the phase offset is much simpler than the frequency offset. Since the phase offset will rotate the constellation, we only need to estimate the amount of rotation and correct the received samples back to their original locations using the estimated rotation (or phase value). If there are some known symbols in the transmitted signal, by simply comparing the phase of the received signal with that of actual known symbols, we can estimate the phase difference:

$$\hat{\theta} = \angle\{r_k b_k^*\} = \angle\{|b_k|^2 e^{j\theta} + v_k\}, \tag{8.6}$$

where $\angle\{\}$ and v_k represent the angle operator and noise, respectively. If there are more known symbols and if the phase is constant across them, the estimated phase can be averaged to increase the estimation performance.

Sample clock error is one of the most critical impairments in this model and the one that will be described in the next section (flat fading multipath channel). Clock error shifts the optimal sample phase position. Even if the optimal sample phase is found during the training period, due to clock error, the optimal sample phase will be shifted. Similar to frequency offset that creates a linear phase difference between what is received and what is ideal, the sample clock error creates a linearly increasing sampling phase offset between the optimal (ideal) phase (where the eye is wide open) and the estimated phase (during the training). Unless the receiver tracks changes in the optimal sampling phase, in spite of the Nyquist pulse shaping, ISI will be observed in the received symbol spaced samples. The amount of ISI will increase as the clock error accumulates in time. The increase in error vector magnitude (EVM) due to clock error can be observed by looking at the EVM-versus-time plot. Timing recovery methods are popularly used to compensate the impact of sample clock error. There are several methods available in the literature [1, 2], such as band-edge timing recovery method and Gardner's timing recovery method.

The following code describes the impact of sample clock error in a very simple setup. A clock error of 500 ppm is generated. Symbol and sample offset estimation are done using a training sequence that is inserted at the beginning of the frame (preamble). But, due to the sample clock error, the best sample index position will be shifted. EVM versus time measurement is an excellent tool to see this effect. The corresponding plots are given in Figure 8.8

```
1  % Impact of sample clock error over a block of data symbols.
2  % Observe the effect of: block size, clock error
3  N=1e4; % block size in terms of number of symbols
4  sps=8; % oversampling rate (i.e. number of samples per symbol)
5  fltr=rcosine(1,sps,'normal',0.3); % raised cosine filter
6  M=4; % modulation order
7  symb_tr=50; %length of training symbols
8  symb_tr=qammod(floor(rand(1,symb_tr)*M),M); % training symbols
9  symb=qammod(floor(rand(1,N)*M),M); % data symbols
10 frame=[symb_tr symb]; % whole frame
11 frame_up=upsample(frame,sps); % upsampled frame
12 sgnl_tx=conv(fltr,frame_up); % filtered signal
13
14 subplot(2,2,1)
15 % re-sampling the signal to introduce clock error
16 sgnl_tx=resample(sgnl_tx,10000,10000-2); % 200ppm (can change it +/-)
17 dummy=vec2mat(sgnl_tx(100:end-100),sps*2);
18 plot(real(dummy(1:end-1,:).'))
19 title('Eye diagram of the whole block of data')
20 subplot(2,2,2)
21 plot(real(dummy(1:50,:).'))
22 title('Eye diagram of only some part of the data block')
23 % find the optimal sample phase at the beginning
24 % first find symbol reference with any arbitrary sample phase
25 sgnl_dwn_tr=downsample(sgnl_tx,sps);
26 for kk=1:50
27     corr(kk)=abs(mean(sgnl_dwn_tr(kk:kk+symb_tr-1).*conj(symb_tr)));
28 end
29 [a,b]=max(corr);
30 idx=b-2; % step back two symbols
31 % now find the optimal sample phase
32 for N_tr=1:sps*3 % searching over a neighborhood of 3 symbol period
33     sgnl_dwn=downsample(sgnl_tx(idx*sps+N_tr:(symb_tr+idx)*sps-1+N_tr),sps);
34     error(N_tr)=mean(abs(sgnl_dwn-symb_tr).^2);
35 end
36 [c,d]=min(error);
37 idx_best=d;
38 symb_rx=downsample(sgnl_tx(idx*sps+idx_best:end),sps);
39 symb_dtct=qammod(qamdemod(symb_rx,M),M);
40 EVM_time=abs(symb_rx(1:N+symb_tr)-symb_dtct(1:N+symb_tr)).^2;
41 subplot(2,2,[3,4])
42 plot(EVM_time)
43 xlabel('Symbol Index'); ylabel('EVM')
44 title('EVM versus time')
```

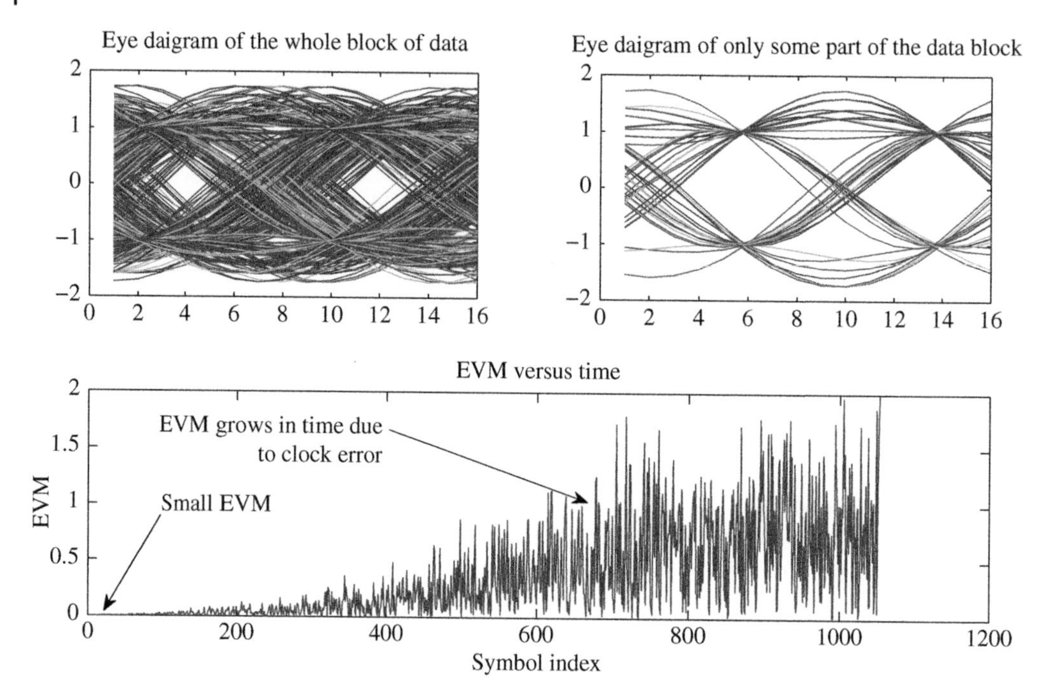

Figure 8.8 Illustration of the impact of the sample clock error.

8.2 Flat Fading (Non-Dispersive) Multipath Channel

When the transmitted signal passes through a multipath radio channel, where the relative delays among these multipath components are small compared to the transmitted symbol period, different "images" of the *same* symbol arrive at the same time, adding either constructively or destructively. The overall effect is a random, fading channel response. This type of fading is often referred to as flat fading (non-dispersive) channel and modeled with a single tap (complex channel coefficient). The received samples in this case can be represented as:

$$\boxed{r_k = h_k b_k + z_k}, \tag{8.7}$$

where h_k represents the time-varying complex channel coefficient. In case of time-invariant channel, the time index in the channel coefficient can be dropped (simply h).

Here, a simple coherent receiver for the case of a one-tap channel is given. Assume the information symbol b_k is either +1 or −1 (BPSK), the information can be recovered using

$$\hat{b}_k = \text{sgn}(\text{Re}\{\hat{h}_k^* r_k\}), \tag{8.8}$$

where $\hat{h}_k$ is an estimate of the channel coefficient and $\text{Re}\{\cdot\}$ denotes the real part of a complex number. When QPSK is used, each symbol represents two bit values. One of the bit values is recovered using Eq. (8.8), and the other bit value is recovered using a similar expression in which the imaginary part is taken instead of the real part.

Multiplying the received value by the conjugate of the channel coefficient estimate removes the phase rotation introduced by the channel and weights the value proportional to how strong the channel coefficient is, which is important when soft information is used in subsequent forward error correction (FEC) decoding. The following Matlab code represents a simple simulation to evaluate the performance of BPSK modulation in AWGN and flat fading channels and the corresponding plot is given in Figure. 8.9. Note that the channel is generated independently at each symbol. Examples for time correlated channel amplitudes are given in Figure 8.10 with different normalized Doppler values.

```
1  % Simulation of BPSK transmission in AWGN and flat fading channel.
2  N=1e6; % block size in terms of number of symbols
3  SNR=[0:20]; % desired SNR values
4  symb=2*(randn(1,N) > 0)-1; % BPSK symbols
5  chnl = (randn(1,N)+j*randn(1,N))/sqrt(2);
6  % independent realization in each symbol.
7  % i.e. channel is time-invariant within a symbol and
8  % independent (uncorrelated) across different symbols
9  noise=(randn(1,N)+j*randn(1,N))/sqrt(2);
10 for kk=1:length(SNR)
11     noise_var=10^(-SNR(kk)/10) ;
12     sgnl_rx_AWGN=symb+noise*sqrt(noise_var);
13     sgnl_rx_fading=chnl.*symb+noise*sqrt(noise_var);
14     det_AWGN=sign(real(sgnl_rx_AWGN));
15     BER_AWGN(kk)=(length(find(symb ≠ det_AWGN)))/N;
16     det_fading=sign(real(sgnl_rx_fading.*conj(chnl)));
17     BER_fading(kk)=(length(find(symb ≠ det_fading)))/N;
18 end
19 figure(1)
20 semilogy(SNR,BER_AWGN,'r-')
21 hold
22 semilogy(SNR,BER_fading,'k(*-\hspace{-.15pc}-*)')
23 xlabel('SNR (dB)')
24 ylabel('BER')
25 legend('AWGN channel','fading channel')
```

Channel estimation in flat fading channels is pretty straightforward, especially if there are some known symbols in the transmitted signal. By simply removing the known modulation information from the received signal, an estimate of the channel coefficient can be obtained as:

$$\hat{h}_k = \frac{r_k}{b_k} = h_k + u_k, \tag{8.9}$$

where u_k represents the noise samples. If there are more of the known symbols and if the channel is static across these known symbols, the estimated channel can be averaged to increase the estimation performance by averaging the noise out. When the channel is time-varying, we need to be careful in averaging these estimates. If averaging is done over a large window, this will cause error between actual channel and estimated channel due to the variation of the channel over the window. The window size needs to be chosen carefully so that maximum noise averaging is achieved while also being able to track the variation of the channel. There are various techniques to track the channel, like least-mean-square (LMS) algorithm, recursive least squares (RLS) algorithm, Kalman filtering, Kalman-LMS, etc. [3]. The details of various tracking approaches can be found in [4]. In general, the tracking performance can be improved by knowing the variation of the channel (Doppler spread information) and the SNR value. These two values can be used for selecting the window size for sliding windowing-based channel tracking. Similarly, the step size in LMS algorithm and forgetting factor in RLS algorithm can be adapted based on the SNR and Doppler information.

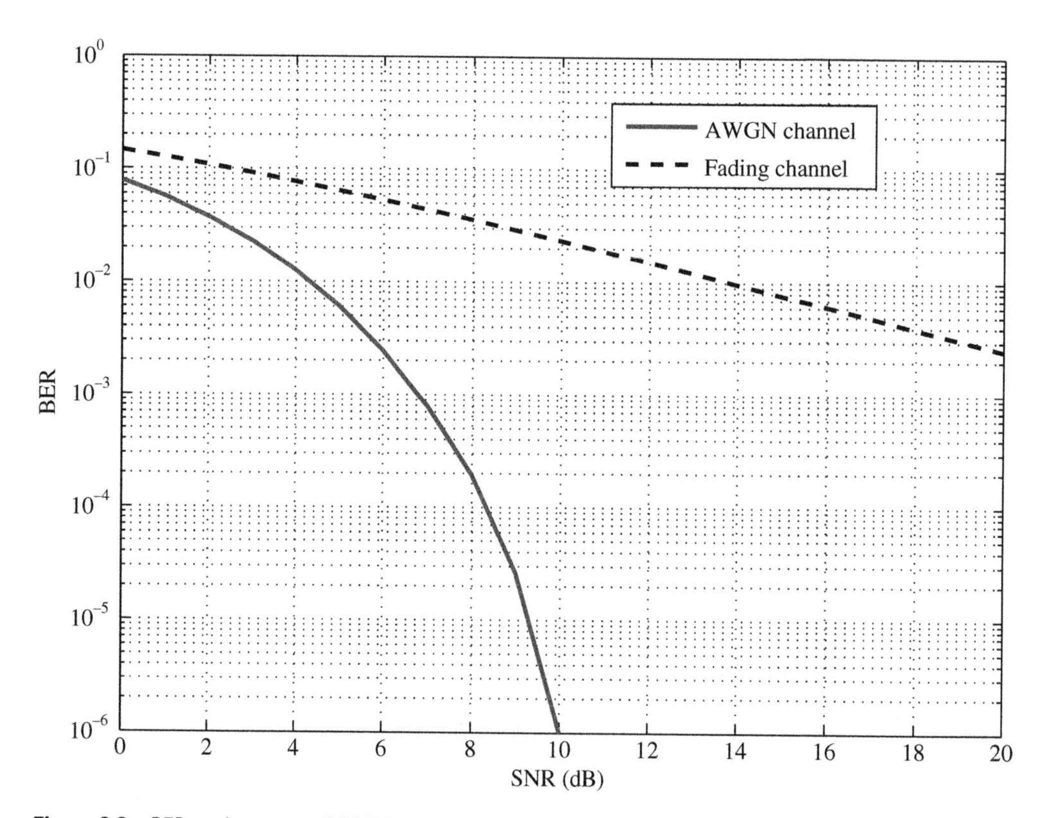

Figure 8.9 BER performance of BPSK modulation in AWGN channel and in flat fading channel.

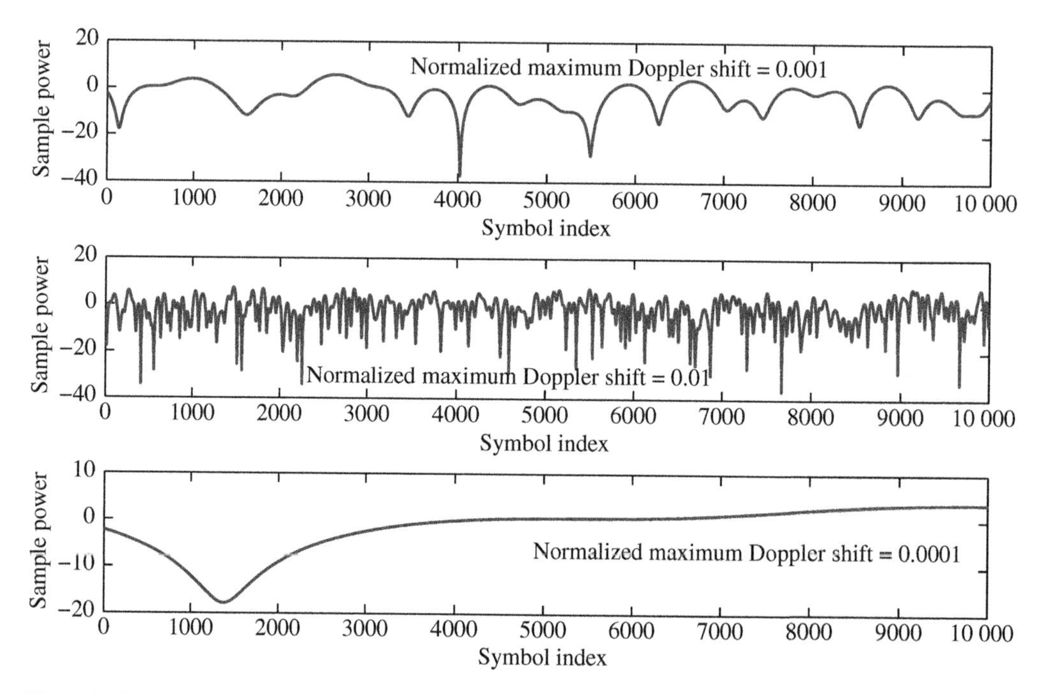

Figure 8.10 Time-correlated fading waveforms for three different normalized maximum Doppler shifts.

Often in time division multiple access (TDMA)-based wireless communication systems, where the received signal is bursty and there are training sequences in the burst (e.g. global system for mobile communications (GSM) and enhanced data rates for GSM evolution (EDGE) [5–7]), channel tracking is not needed if the burst size is short and the vehicle speed is low since the variation of the channel within the burst is not significant. For example, in GSM, the channel can be estimated during the training period that is inserted into the middle of the burst (called midamble), then the estimated channel parameters are used to demodulate the rest of the burst without tracking. GSM case does not really fit to flat fading channel situation as GSM bandwidth is more than the coherence bandwidth of the channel in wide area network (and GSM does not use Nyquist pulse shape), making the effective channel response frequency selective. However, GSM is given here just as an example of the situation where the channel tracker might not be necessary. The following code and the corresponding plots in Figure 8.11 illustrate the discussion above.

An interesting thing to note here while discussing the variation of the channel is that both frequency offset that we discussed earlier and multipath channel cause the variation of the phase (frequency offset) and amplitude (multipath channel) of the signal. If we can track the channel, we do not only compensate the channel effect but also the frequency offset. Since large frequency offset values might make the channel vary more than it actually is, it would be better if we can estimate the frequency offset value separately and compensate the received samples with the estimated offset, so that the channel tracker can be implemented optimally with more noise averaging

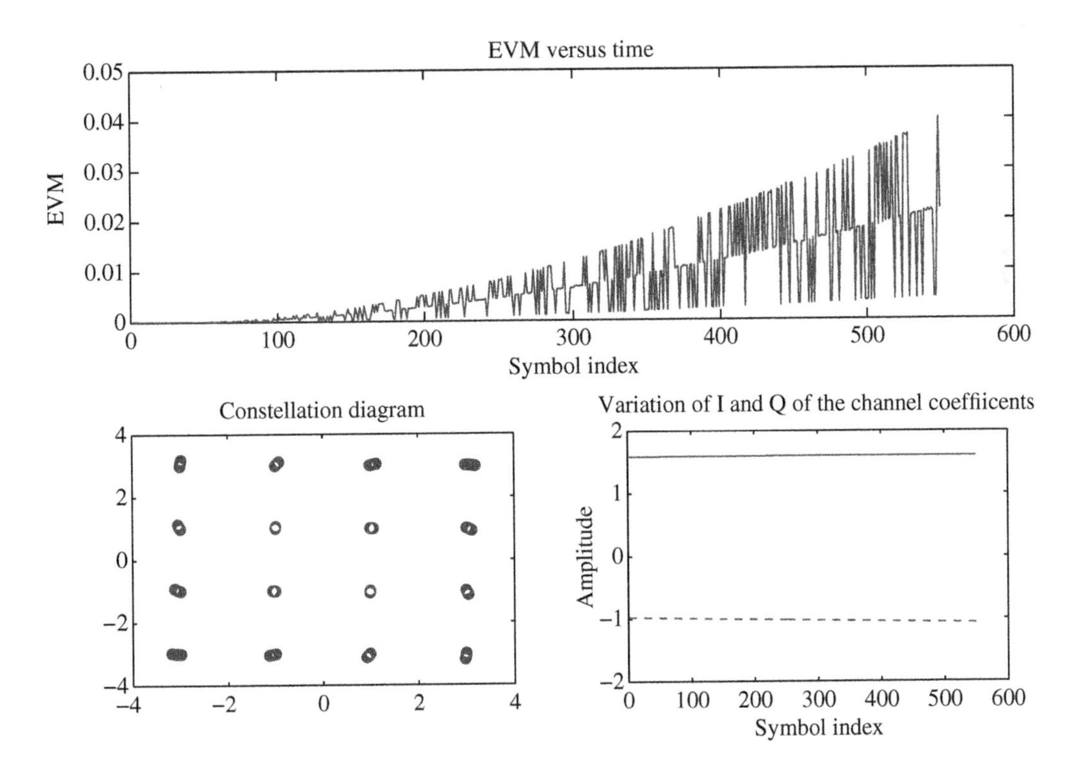

Figure 8.11 Impact of the time variation of channel without using channel tracker. Channel estimated at the beginning of the frame using training, and the estimated value is used for the rest of the data in the frame. Relative Doppler spread is not significant here. As can be seen in the figure, the constellation is not terribly bad. For significant Doppler spread values, channel trackers are needed.

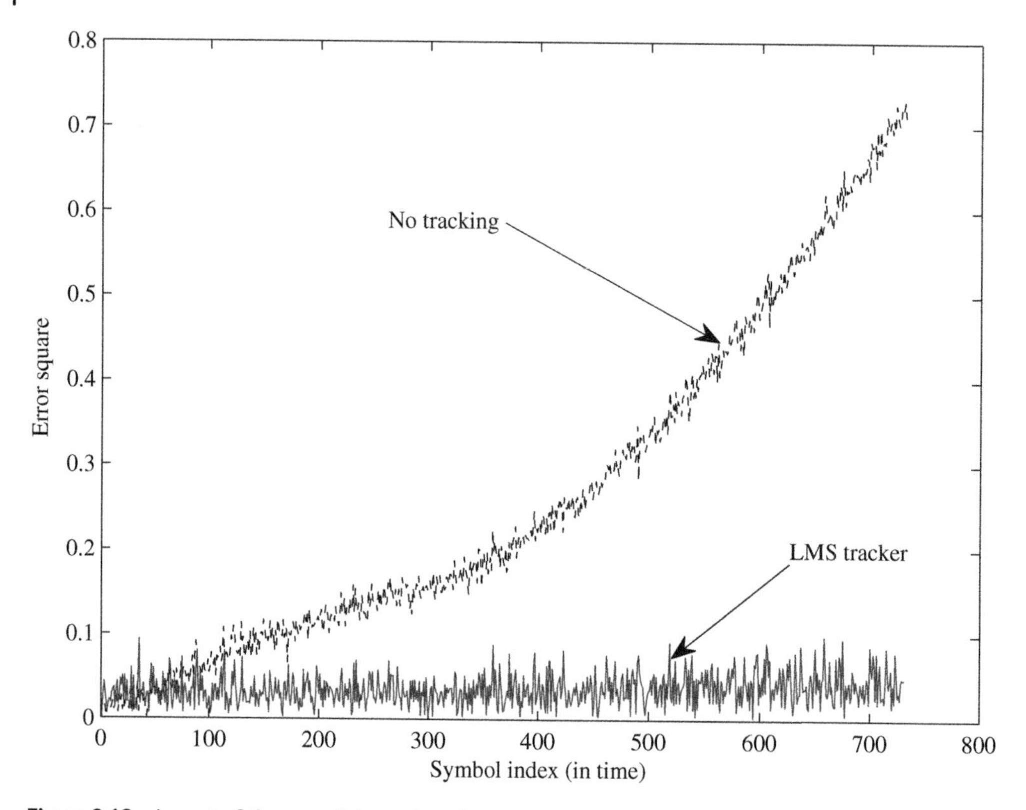

Figure 8.12 Impact of the use of channel tracker.

capability. Therefore, folding large frequency offset values into the channel and estimating the effective channel (with the frequency offset folded into it) is not always a good idea, as it puts too much stress on the channel tracker algorithm and the channel estimates will be more noisy (worse MSE performance). Then, the question is how can we estimate the frequency offset separately in time-varying multipath channels. The problem is not as difficult as it looks, thanks to the linear variation of the phase due to frequency offset as opposed to the random variation due to the multipath channel. If we have enough samples, in the long run, we can separate the variation of the phase due to frequency offset from the multipath channel.

Similar to the above discussion, channel estimation can also be used to compensate the phase offset. By estimating the effective (cumulative) channel, the effect of the phase offset can be folded into the channel response, and hence one can automatically compensate phase offset while removing the effect of the channel from the received samples. Therefore, from now on, in this chapter of the book, the phase offset and frequency offset will be folded into the channel estimation and tracking and considered together as estimation and compensation of the effective channel response.

The following Matlab code compares the performance of the receiver with and without channel tracker when the channel is varying in time. In this specific code, the variation is due to the Doppler spread of the channel. Figure 8.12 shows the EVM versus time plot for an LMS channel tracker and without using a channel tracker. As can be seen, when a channel tracker is used, the time variation of the channel can be tracked and hence the received signal can be corrected adaptively.

```
1  % Impact of the time variation of the channel over
2  % a burst when a channel tracker is not used.
3  % Observe effect of: with different Doppler shift, block size, noise power
4  M=16;% modulation order
5  N=500;% block size in terms of number of symbols
6  symb_tr=50; %length of training symbols
7  co=rayleighchan(1, 5e-5); %  fading with a given normalized Doppler shift
8  reset(co)
9  chnl=filter(co,ones(1,N+symb_tr)); % time correlated fading
10 symb_tr=qammod(floor(rand(1,symb_tr)*M),M);  % training symbols
11 symb=qammod(floor(rand(1,N)*M),M); % data symbols
12 frame=[symb_tr symb]; % frame formation
13 noise=(randn(1,N+50)+j*randn(1,N+symb_tr))*0.001; % noise with desired power
14 sgnl_rx= frame.*chnl + noise; % received signal with noise
15
16 % estimate the single tap channel over the training period
17 chnl_est = mean(sgnl_rx (1:symb_tr)./symb_tr) ;
18 % use the estimated channel to equalize the signal
19 sgnl_eq = sgnl_rx*conj(chnl_est) /abs(chnl_est).^2;
20
21 symb_dtct=qammod(qamdemod(sgnl_eq,M),M); % detected symbols
22 EVM_time=abs(sgnl_eq-symb_dtct).^2;
23 subplot(2,2,[1,2])
24 plot(EVM_time)
25 xlabel('symbol index')
26 ylabel('EVM')
27 title('EVM versus time')
28 subplot(2,2,3)
29 plot(real(sgnl_eq),imag(sgnl_eq),'o')
30 title('constellation diagram')
31 subplot(2,2,4)
32 plot(real(chnl))
33 hold
34 plot(imag(chnl),'—')
35 title('variation of I and Q of the channel coefficients')
36 xlabel('symbol index')
37 ylabel('amplitude')
```

8.3 Frequency-Selective (Dispersive) Multipath Channel

When the transmitted signal passes through a multipath radio channel where the relative path delays are on the order of a symbol period or more, then images of *different symbols* arrive at the same time. The dispersive channel, which is often referred as frequency selective channel, can be modeled as a linear, FIR (finite impulse response) filter. The resulting signal is received in the presence of noise, giving,

$$y(t) = \sum_{\ell=0}^{L-1} c(\ell, t)x(t - \tau(\ell)) + w(t), \tag{8.10}$$

where $c(\ell, t)$ and $\tau(\ell)$ are the ℓth complex, time-varying channel coefficient and ℓth delay, respectively, and L is the number of channel taps. The noise term $w(t)$ is assumed to be white, complex (circular) Gaussian noise.

The delays are often assumed to be equally spaced, i.e. $\tau(\ell) = \ell T/M$ where M is an integer, and the spacing (T/M) for accurate modeling depends on the bandwidth of the system [8]. Typically, M is 1 (symbol-spaced channel modeling) or 2 (fractionally spaced channel modeling). For simplicity, symbol-spaced channel modeling is assumed, though extension to fractionally spaced channel modeling is possible.

As in flat fading, the coefficients represent the result of different multipath signal images added together, constructively or destructively. They are well modeled as random variables.

Specifically, they are modeled as uncorrelated, zero-mean complex Gaussian random variables [9]. This corresponds to "Rayleigh" fading, where channel tap magnitudes (amplitudes) are Rayleigh-distributed. Also, the phase of the channel taps are uniformly distributed.

```
1  % Performance of receiver in time varying channel (w/ & w/o tracker)
2  % Observe effect of: SNR, Doppler shift, block size
3  SNR=30; % desired SNR value in dB
4  N=500;% block size in terms of number of symbols
5  symb_tr=[-1 1 1 -1 1 -1 -1 1 -1 -1 -1 -1 ...
6      1 -1 1 -1 1 1 1 -1 1 1 -1 -1 -1 1 1 1 1 1 -1]; % training symbols
7  symb_tr=length(symb_tr);
8  co=rayleighchan(1, 1e-4); %  fading with a given normalized Doppler shift
9  reset(co)
10 chnl=filter(co,ones(1,N+symb_tr)); % time correlated fading
11 symb=2*(randn(1,N)>0)-1 ; % generate data symbols
12 frame=[symb_tr symb];
13 noise=(randn(1,N+symb_tr)+j*randn(1,N+symb_tr))/sqrt(2);
14 sgnl_rx=frame.*chnl + noise*10^(-SNR/20);
15 % estimate the single tap channel over the training period
16 chnl_est = mean(sgnl_rx (1:symb_tr)./symb_tr) ;
17 % use initially estimated channel to equalize the signal without tracker
18 rx_corrected=sgnl_rx*conj(chnl_est) /abs(chnl_est).^2;
19 EVM_time=abs(rx_corrected-sign(real(rx_corrected)));
20
21 % Channel tracker using LMS
22 mu_fixed=0.1 ;% step size
23 chnl_trk_lms(1)=chnl_est; % initial channel estimate
24 symb_dtct(1:symb_tr)=symb_tr;
25 for ii=2:length(sgnl_rx)
26     error(ii)=sgnl_rx(ii-1)-symb_dtct(ii-1)*chnl_trk_lms(ii-1);
27     chnl_trk_lms(ii)=chnl_trk_lms(ii-1) + ...
28         mu_fixed*conj(symb_dtct(ii-1)) * error(ii);
29     if(ii>symb_tr)
30         symb_dtct(ii)=sign(real(sgnl_rx(ii)*conj(chnl_trk_lms(ii))));
31     end
32 end
33 plot(EVM_time,'k—')
34 hold
35 plot(abs(error),'r')
```

As the mobile transmitter or receiver moves, the phases of each multipath signal image change. This changes how the multipath images add together, so that the channel coefficient varies with time. This time variation is characterized by an autocorrelation function. The Jakes' model [10], which is commonly used, assumes the Gaussian channel coefficients have the following autocorrelation function:

$$R_\ell(\tau) = E\{c^*(\ell,t)c(\ell,t+\tau)\} = \sigma_\ell^2 J_0(2\pi f_D \tau), \tag{8.11}$$

where superscript "*" denotes complex conjugation, subscript ℓ denotes the ℓth channel coefficient, τ is the autocorrelation delay, σ_ℓ^2 is the mean-square value of the channel coefficient, f_D is the Doppler spread, and $J_0(\cdot)$ is the zeroth-order Bessel function of the first kind.

The Doppler spread is proportional to the radio carrier frequency and the speed of transmitter or receiver. At a high vehicle speed of 100 km/h, the carrier frequencies of 900 MHz and 2 GHz correspond to Doppler spread values of 83 and 185 Hz, respectively. The ability to track channel variation depends on how fast the channel changes from symbol to symbol. In general, the symbol rate is much higher than the Doppler spread, so that the channel coefficient value is highly correlated from symbol to symbol, making channel tracking possible. Also, the Doppler spread is considered constant, because the speed of the transmitter or receiver changes slowly relative to the transmission rate. However, it is possible to model the time variation of the Doppler spread [11].

At the receiver, $y(t)$ is passed through a filter matched to the pulse shape and sampled, giving received samples

$$r_k = \int f^*(\tau)y(\tau + kT_s)d\tau, \quad k = 0, 1, \ldots \tag{8.12}$$

where $T_s = T/M = T$ is the sampling period. Because the fading varies slowly from symbol to symbol, it can be approximated as constant over the pulse shape. With this approximation, substituting Eq. (8.10) into Eq. (8.12) gives

$$\boxed{r_k = \sum_{j=0}^{J-1} h_k(j) b_{k-j} + z_k}, \tag{8.13}$$

where $h_k(j)$ is the jth composite channel coefficient, reflecting the influence of the transmit filter, the radio channel, and the receive filter, i.e.

$$h_k(j) = \sum_{\ell=0}^{L-1} c(\ell, kT) R_{ff}(jT - \ell T), \tag{8.14}$$

where $R_{ff}(\tau)$ is the pulse shape autocorrelation function. Observe that at the receiver, there is ISI, as the received samples contain the current symbol b_k as well as interference from previous symbols. The number of composite coefficients (J) needed to accurately model r_k depends on L and the shape of $R_{ff}(\tau)$. For the special case of root-Nyquist pulse shaping, $J = L$, $h_k(j) = c(j, kT)$, and the noise

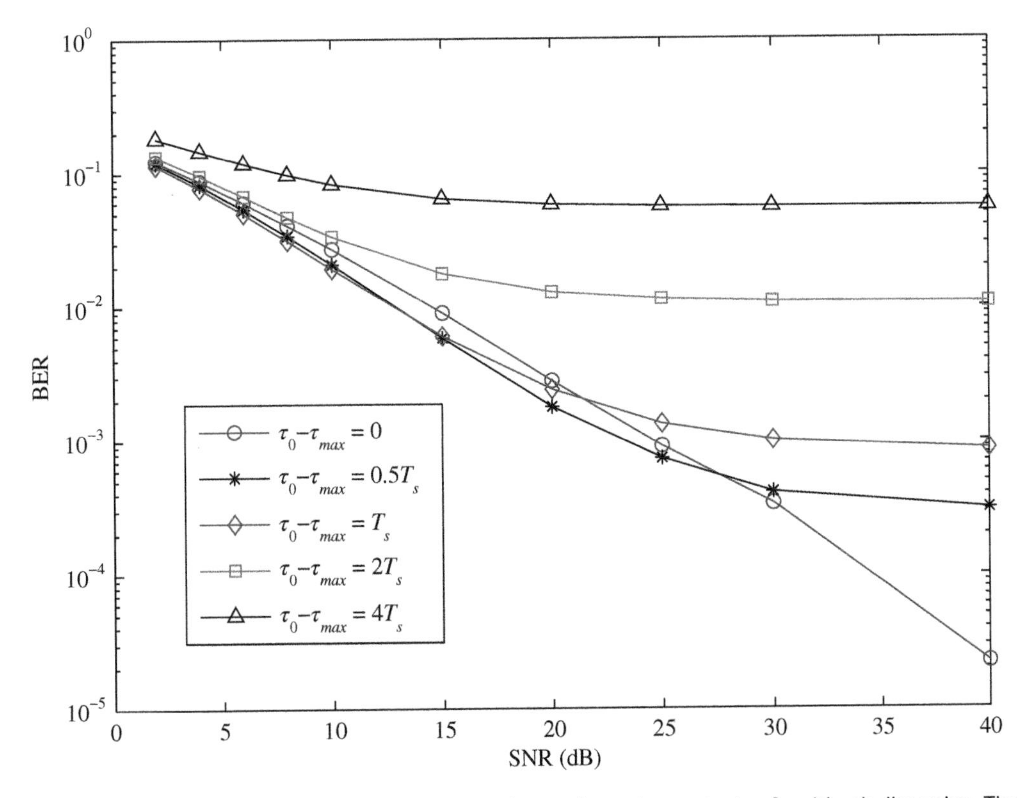

Figure 8.13 BER performance of a single-tap receiver under various extents of multipath dispersion. The receiver locks into the strongest multipath coefficient using the training sequence and maximum likelihood-based fine synchronization algorithm. The channel coefficient of the strongest channel is estimated using the training sequence and received symbols are phase-corrected using the estimated channel to be able to demodulate the transmitted bits.

samples z_k are uncorrelated. Note that the above argument is only true if the medium response coefficients are symbol-spaced. For fractionally spaced multipath coefficients, the effective channel response will have components due to pulse shape as well, even if the combined (Tx and Rx) pulse shape is root-raised cosine. In practice, multipath coefficients are not necessarily spaced at each symbol interval; the multipath delays are random and can be at any arbitrary spacing. Hence, when the channel is dispersive, the use of raised cosine pulse shaping cannot guarantee ISI free samples at the receiver, on the contrary, it can make the channel look more dispersive than it actually is due to slowly decaying side lobes of the raised cosine filtering.

The following Matlab code simulates the impact of multipath dispersion when an equalizer is not employed at the receiver. The receiver employs simple symbol and sample timing estimation along with single-tap channel estimation followed by channel phase correction and symbol detection. The multipath dispersion level (i.e. maximum excess delay of the channel) can be changed and its impact can be observed at the receiver. Constant (time-invariant) channel coefficients are used at each burst, and independent realizations are generated across different bursts (this is also referred as block-fading model in the literature). Multipath power delay profile is assumed to be uniform with independent sample-spaced multipath coefficients, and each coefficient is randomly fading with the Rayleigh amplitude distribution. The impact of the ISI due to multipath dispersion on bit-error-rate (BER) performance is given in Figure 8.13. As can be seen, as the maximum excess delay of the channel increases, ISI also increases, and hence decreases the performance of the system. Notice that when SNR increases, ISI component becomes dominant, and it causes irreducible BER performance (creating error floors at large SNR values). Figures 8.14 and 8.15

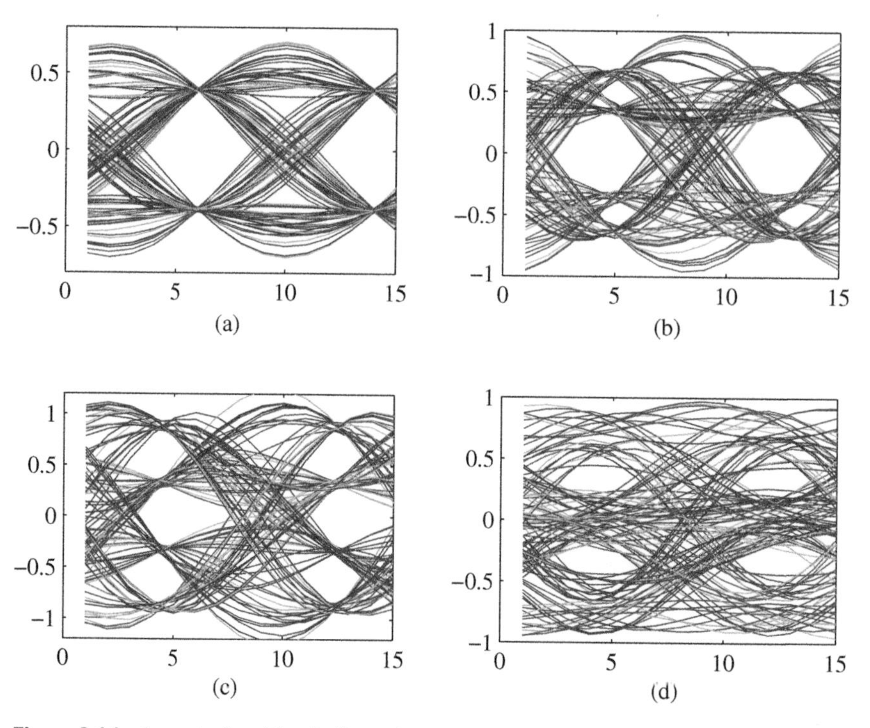

Figure 8.14 Impact of multipath dispersion and ISI on the eye diagram. The eye diagram of in-phase component is shown. Only a single-channel realization is given here for illustration purpose. Each subplot is drawn from an independent realization. (a) $\tau_0 - \tau_{max} = 0$. (b) $\tau_0 - \tau_{max} = 0.5T_s$. (c) $\tau_0 - \tau_{max} = T_s$. (d) $\tau_0 - \tau_{max} = 2T_s$.

show the impact of dispersion on eye and constellation diagrams, respectively. In these figures, the SNR is very large (300 dB), so that ISI effect can be observed without the impact of the noise. Only a single realization is shown in both figures. In the eye diagram, the effect of dispersion can be seen as eye closing. Similarly, in the constellation diagram, the impact of the dispersion can be seen clearly with the samples scattered away from the perfect constellation point.

```
1   % Impact of multipath dispersion without using equalizer
2   % Observe effect of: SNR, delay spread
3   SNR=[2 4 6 8 10 15 20 25 30 40];% desired SNR values in dB
4   sps=8;
5   symb_tr=[-1 1 1 -1 1 -1 -1 1 -1 -1 -1 -1 ...
6       1 -1 1 -1 1 1 1 -1 1 1 -1 -1 -1 1 1 1 1 1 -1];
7   % 31 symbol length training sequence generated using m-sequence
8   N_tr=length(symb_tr);
9   fltr = rcosine(1,sps,'sqrt',0.3);
10  N=340; %block size in terms of number of symbols
11  NN=100; % number of slots to run
12  LL=[1 4 8 16 32];% number of channel taps, dispersion, (symbol spaced)
13  for vv=1:length(LL)
14      L=LL(vv);
15      for zz=1:NN
16          symb=2*(randn(1,N)>0)-1; % random BPSK symbols
17          frame=[symb_tr symb zeros(1,10)];
18          u_frame=upsample(frame,sps);
19          sgnl_tx=filter(fltr,1,u_frame);
20
21          %********* Channel ***********
22          cir=(randn(1,L)+j*randn(1,L))/sqrt(2)/sqrt(L);
23          % channel is static over the burst. Block fading model is assumed
24          sgnl_rx_fading=filter(cir,1,sgnl_tx);
25          N_tr=length(sgnl_rx_fading);
26          noise = 1/sqrt(2)*(randn(1,N_tr) + j*randn(1,N_tr));
27          for kl=1:length(SNR)
28              sgnl_rx_noisy = sgnl_rx_fading + noise*10^(-SNR(kl)/20);
29              % ********* Receiver **********
30              % Matched filtering
31              rec_filtered=conv(sgnl_rx_noisy,fltr);
32              % fine synchronization
33              for N_tr=1:sps*10 %serching over a neighborhood
34                  sgnl_dwn=downsample(rec_filtered(N_tr:N_tr*sps-1+N_tr),sps);
35                  corx(N_tr)=  abs(sum(sgnl_dwn.*conj(symb_tr)));
36              end
37              [c,d]=max(corx);
38              best_ind=d;
39              symb_rx=downsample(rec_filtered(best_ind:end),sps);
40              chan_est=mean(symb_rx(1:N_tr)/symb_tr);
41              symb_rx=symb_rx(N_tr+1:N+N_tr);
42              symb_dtct=sign(real(symb_rx*conj(chan_est)));
43              BER(zz,kl)=length(find(symb_dtct≠symb))/N;
44          end
45      end
46      ber_av=mean(BER); ber_all(vv,:)=ber_av
47  end
```

8.3.1 Time-Domain Equalization

As illustrated in previous section, channel dispersion and ISI impact the receiver performance significantly. Due to the ISI, in frequency-selective channels, channel equalization is necessary to be able to demodulate transmitted symbols. Time-domain adaptive equalization has an extensive history [12, 13]. In narrowband wireless communication systems, decision feedback equalizer (DFE) and maximum likelihood sequence estimation (MLSE) are two commonly used forms of time-domain equalization. These approaches are preferred over linear equalization, which is sensitive to spectral nulls in the channel response [9]. To understand the basic principles of these

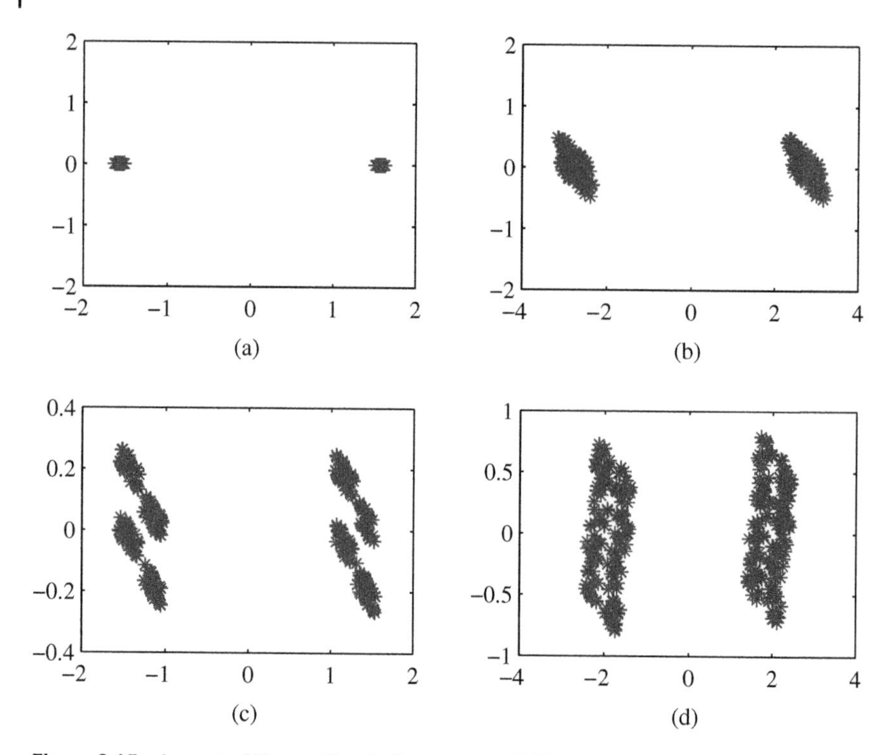

Figure 8.15 Impact of the multipath dispersion and ISI on the constellation. Only a single-channel realization is given here for illustration purpose. Each subplot is drawn from an independent realization. Constellations are shown after the channel phase compensation. (a) $\tau_0 - \tau_{max} = 0$. (b) $\tau_0 - \tau_{max} = 0.5T_s$. (c) $\tau_0 - \tau_{max} = T_s$. (d) $\tau_0 - \tau_{max} = 2T_s$.

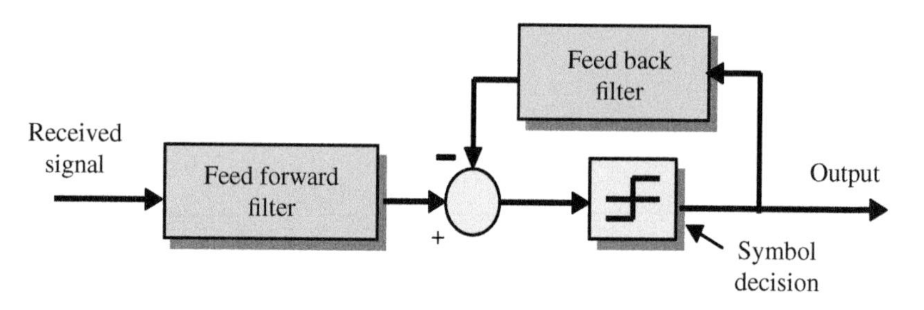

Figure 8.16 Decision feedback equalization.

approaches, consider the simple case of BPSK symbols ($b_k = \pm 1$) and a two-tap channel model, such that

$$r_k = h_k(0)b_k + h_k(1)b_{k-1} + z_k. \tag{8.15}$$

The DFE, shown in Figure 8.16, consists of feedforward filter, feedback filter, and decision device [9]. Conceptually, the feedforward filter tries to collect all signal energy for b_k (which appears in both r_k and r_{k+1}) while suppressing ISI from subsequent symbols (e.g. b_{k+1}). The feedback filter removes ISI from previous symbols (e.g. b_{k-1}). Notice that the feedforward filter introduces delay between when the first image of a symbol arrives and when that symbol is decided.

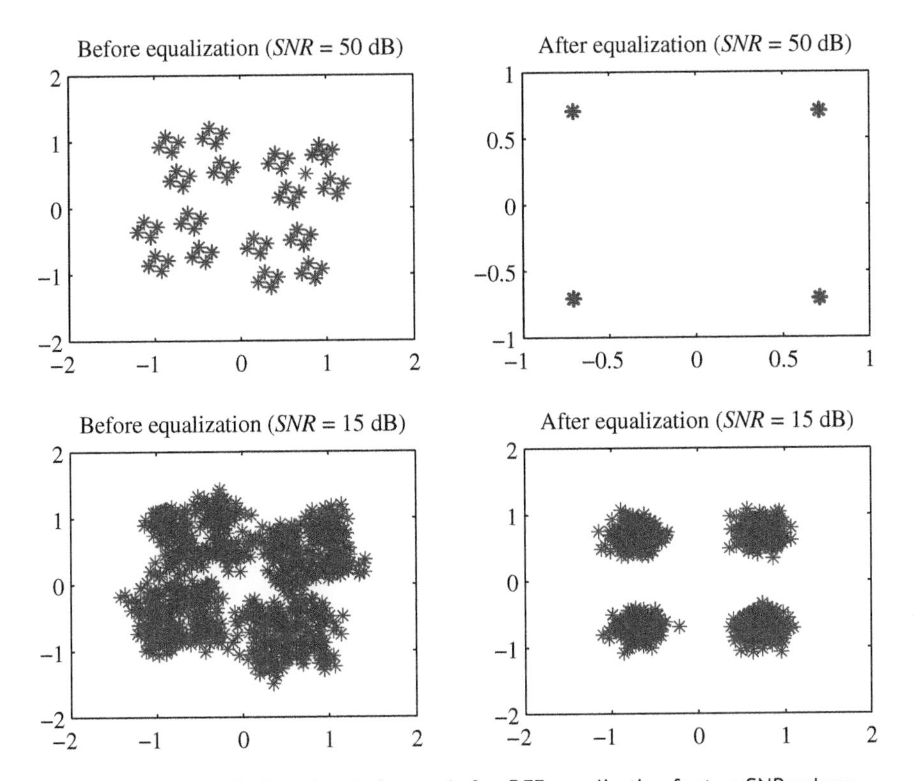

Figure 8.17 Constellation plots before and after DFE equalization for two SNR values.

A simple Matlab code on explaining DFE equalization is given below for a specific three-tap time-invariant channel realization. A single-tap feedforward and two-tap feedback filters are employed for this specific channel. Figure 8.17 shows the constellation plots before and after the equalization for two different SNR values.

With MLSE, the likelihood of the received data samples, conditioned on the symbol values, is maximized [9]. The conditional log-likelihood of the kth data sample, assuming z_k is Gaussian and uncorrelated in time, is related to the following metric or cost function:

$$M(\hat{b}_k, \hat{b}_{k-1}) = |r_k - (h_k(0)\hat{b}_k + h_k(1)\hat{b}_{k-1})|^2. \tag{8.16}$$

For different symbol sequence hypotheses ("paths"), this metric is accumulated, generating a path metric. Once all the data samples have been processed, the sequence corresponding to the smallest path metric determines the detected sequence. Intuitively, the metric indicates how well the model of the received data fits the actual received data.

A brute force search of all possible sequences would require high computational complexity. However, it is possible to determine the smallest metric sequence through a process of path pruning known as the Viterbi algorithm [14]. This involves defining a set of "states" corresponding to a set of paths. Tentative decisions can be made after some delay D. The path "history" (sequence of symbol values) corresponding to the state with the best path metric after processing the kth sample can be used to determine the $(k - D)$th symbol value.

The Viterbi algorithm is best explained using the trellis in Figure 8.18. Let $b_0 = -1$ be a known training symbol and $k > 0$ correspond to unknown data. For $k = 1$, there are two possible values for $M(b_1, b_0)$ corresponding to the two possible values for b_1. These two metrics correspond to the two branches shown in Figure 8.18 between $k = 0$ and $k = 1$. These two branches end in two "states" corresponding to the possible values for b_1. At $k = 2$, there are four possible branches, two from each state, corresponding to the possible values for the pair (b_1,b_2). Observe that two branches end at the state corresponding to $b_2 = +1$ and two end at the state corresponding to $b_2 = -1$. At each state, only the "best" sequence of the two is kept, the other will be "pruned." The best is determined by candidate "path" metrics.

```
1  % A simple DFE equalization
2  % Observe effect of: SNR, power delay profile (CIR)
3  SNR=20; % desired SNR values in dB
4  N=1000; %block size in terms of number of symbols
5  symb=(2*(randn(1,N)>0)-1+j*(2*(randn(1,N)>0)-1))/sqrt(2);
6  % random QPSK symbols
7  symb(1:2)=[1+j 1+j]/sqrt(2); % 2 initial symbols are known to start eq.
8  cir=[5-j 2+0.5*j  0.6-0.3*j]; % 3 tap CIR - symbol spaced
9  cir=cir/sqrt(sum(abs(cir).^2)); % normalize the channel
10 rx=filter(cir,1,symb); % signal after the channel
11 noise=(randn(1,length(rx))+j*randn(1,length(rx)))/sqrt(2);
12 sgnl_rx_noisy = rx + noise*10^(-SNR/20); % noisy received signal
13 
14 ff=conj(cir(1))/abs(cir(1)).^2; % feedforward part of the channel
15 fb=cir(2:3)* ff; % feedback part of the channel
16 
17 % DFE equalization
18 prev_symbols = fliplr(symb(1:2));
19 for kk=3:N
20     symb_eq(kk)= sgnl_rx_noisy(kk)*ff - prev_symbols*fb.';
21     symb_dt(kk)=(sign(real(symb_eq(kk)))+j*sign(imag(symb_eq(kk))))/sqrt(2);
22     prev_symbols=[symb_dt(kk) prev_symbols(1:end-1)];
23 end
24 b1_errors=length(find(real(symb(3:end))≠ real(symb_dt(3:end))));
25 b2_errors=length(find(imag(symb(3:end))≠ imag(symb_dt(3:end))));
26 ber= (b1_errors+b2_errors)/2/(N-2);
27 
28 figure(1);clf
29 plot(real(sgnl_rx_noisy),imag(sgnl_rx_noisy),'*')
30 title('Scatterplot before equalization')
31 figure(2);clf
32 plot(real(symb_eq(3:end)),imag(symb_eq(3:end)),'*')
33 title('Scatterplot after equalization')
```

For example, at the state corresponding to $b_2 = +1$, the two sequences considered are $\{b_1 = +1, b_2 = +1\}$ and $\{b_1 = -1, b_2 = +1\}$. The corresponding candidate metrics are $M(b_2 = +1, b_1 = +1) + M(b_1 = +1, b_0 = -1)$ and $M(b_2 = +1, b_1 = -1) + M(b_1 = -1, b_0 = -1)$. Whichever candidate metric is the smallest determines which "path" is kept and which "path metric" is accumulated. Thus, each state has a "path history" that determines previous symbol values.

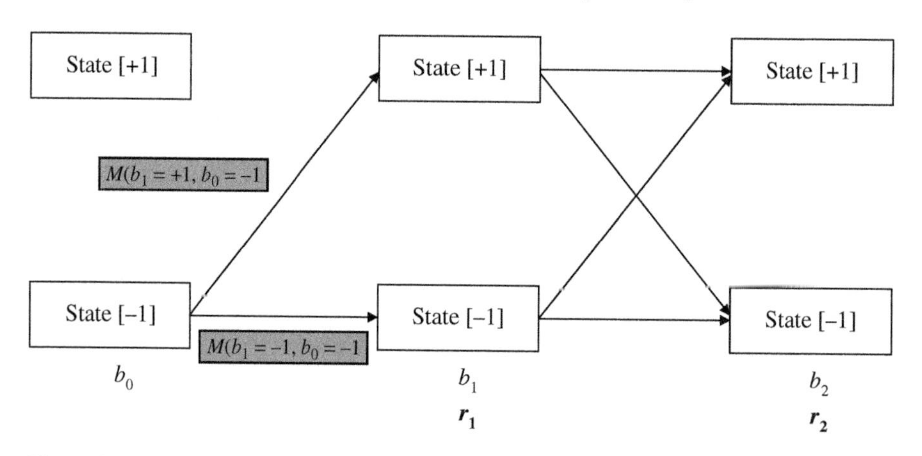

Figure 8.18 Trellis for Viterbi algorithm.

Ideally, the process continues until all the data samples have been processed. Then, the state with the smallest metric determines the detected sequence. In practice, a decision can be made after some delay D. The path history corresponding to the state with the best path metric after processing the kth sample can be used to determine the $(k - D)$th symbol value.

The following Matlab code simulates MLSE equalizer with Viterbi algorithm using BPSK modulation and two-tap channel model as described above.

```
% A simple MLSE equalization with two-tap chanel using BPSK modulation
% Observe effect of: SNR, power delay profile (CIR)
SNR=20;% desired SNR values in dB
N=5000; % number of data symbols
L=2; % number of channel taps
symb_tr=[-1 1 1 -1 1 -1 -1 1 -1 -1 -1 -1 ...
    1 -1 1 -1 1 1 1 -1 1 1 -1 -1 -1 1 1 1 1 1 -1];
tail=[1 1 1 1 1]; % add 5 tail symbols
N_tr=length(symb_tr); tt=length(tail);
symb=2*(randn(1,N)>0)-1 ; % generate data symbols
frame=[symb_tr symb tail]; % frame formation
cir=(randn(1,L)+j*randn(1,L))/sqrt(2)/sqrt(L);
% Channel is randomly generated over each burst
% channel is static over the burst. Block fading model
rx=filter(cir,1,frame); % pass the signal through channel
noise=(randn(1,length(rx))+j*randn(1,length(rx)))/sqrt(2);
sgnl_rx = rx + noise*10^(-SNR/20); % add noise with the desired SNR

%***** MLSE equalization with perfect channel knowledge
%Initializing
Eq_o1_old=[];Eq_o2_old=[];
Surv_M2=0; Surv_M1=1000; % start at known state
%last symbol of training is -1 (which is second state).
for i=32:length(sgnl_rx)
    % Branch metrics
    M11=abs(sgnl_rx(i)-(cir(1)+cir(2)))^2;
    M12=abs(sgnl_rx(i)-(cir(1)-cir(2)))^2;
    M21=abs(sgnl_rx(i)-(-cir(1)+cir(2)))^2;
    M22=abs(sgnl_rx(i)-(-cir(1)-cir(2)))^2;
    % Accumulated metrics
    Ac_M11=Surv_M1+M11;
    Ac_M12=Surv_M2+M12;
    Ac_M21=Surv_M1+M21;
    Ac_M22=Surv_M2+M22;
    % Finding surviving metrics for each state
    [a,b]=min([Ac_M11 Ac_M12]); Surv_M1=a;
    if b==1; Eq_o1_new=[Eq_o1_old 1];
    else; Eq_o1_new=[Eq_o2_old -1];
    end
    [a,b]=min([Ac_M21 Ac_M22]);Surv_M2=a;
    if b==1; Eq_o2_new=[Eq_o1_old 1];
    else; Eq_o2_new=[Eq_o2_old -1];
    end
    Eq_o1_old=Eq_o1_new;Eq_o2_old=Eq_o2_new;
end
ber=length(find(Eq_o1_new(2:N+1)≠symb))/length(symb)
```

8.3.2 Channel Estimation

As noticed from the previous section, the equalization process (both in DFE and MLSE) requires estimation of equalizer tap coefficients. These equalizer coefficients are related to channel tap coefficients. Therefore, estimation of channel coefficients is a critical part of the receiver design. There are various techniques for channel estimation. If the channel is time-invariant or can be assumed to be static over the duration of a burst in TDMA-type of communication (like in GSM and EDGE, where the slot duration is short, 577 microseconds, so that the channel response over the time slot is approximately constant for reasonable vehicle speeds [15, 16]), then the channel coefficients can be estimated over the known training symbols, and then these estimated channel coefficients can be used to equalize the rest of data symbols. There are various techniques for estimating the channel coefficients using training symbols as described in detail in [4].

Least squares (LS) channel estimation is a very commonly used technique when a training sequence is available [17–20].

We can rewrite Eq. (8.13) in vector and matrix forms. Assuming a total of K received samples, the following column vectors are defined as: $\mathbf{r} = [r_0\ r_1\ \cdots\ r_{K-1}]^T$, $\mathbf{h} = [h(0)\ h(1)\ \cdots\ h(J-1)]^T$, and $\mathbf{z} = [z_0\ z_1\ \cdots\ z_{K-1}]^T$. Then, Eq. (8.13) can be expressed as:

$$\boxed{\mathbf{r} = \mathbf{B}\mathbf{h} + \mathbf{z}}, \tag{8.17}$$

where $\mathbf{B}$ is a $K \times J$ matrix whose rows correspond to different shifts of the transmitted sequence of symbols.

The LS channel coefficient estimate is defined by

$$\hat{\mathbf{h}}_{LS} = \arg\min_{\mathbf{h}} (\mathbf{r} - \mathbf{B}\mathbf{h})^H(\mathbf{r} - \mathbf{B}\mathbf{h}), \tag{8.18}$$

where $K = N - (J - 1)$ samples are used to form $\mathbf{r}$. By differentiating with respect to each channel coefficient and setting the result to zero, the following closed-form expression for the LS channel estimate can be obtained as:

$$\hat{\mathbf{h}}_{LS} = \left(\mathbf{B}^H\mathbf{B}\right)^{-1}\mathbf{B}^H\mathbf{r}\,. \tag{8.19}$$

Substituting Eq. (8.17) into Eq. (8.19), the LS channel estimate can be expressed as:

$$\hat{\mathbf{h}}_{LS} = \mathbf{h} + \left(\mathbf{B}^H\mathbf{B}\right)^{-1}\mathbf{B}^H\mathbf{z}. \tag{8.20}$$

From Eq. (8.20), the mean-squared-error (MSE) matrix, assuming white noise, is given by $(\mathbf{B}^H\mathbf{B})^{-1}\sigma_z^2$. Thus, the estimation error variance depends on noise power, length of the training sequence, number of channel taps, and correlation properties of training sequences [17].

The estimation error variance is minimized when $\mathbf{B}^H\mathbf{B} = (N - (J - 1))\mathbf{I}$ [18]. This property defines "ideal" autocorrelation properties for a training sequence used with LS channel estimation, as the off-diagonal elements of $\mathbf{B}^H\mathbf{B}$ are correlations between different portions of the training sequence. With ideal autocorrelation properties, the estimation error variance simplifies to $\sigma_z^2/(N - (J - 1))$ and the MSE matrix becomes

$$MSE_{LS} = \left(\frac{\sigma_z^2}{N - (J - 1)}\right)\mathbf{I} \tag{8.21}$$

Observe that the error in different coefficient estimates is uncorrelated. Also, the estimation variance is inversely proportional to $N - (J - 1)$.

In the following Matlab code, LS channel estimation is implemented in time-invariant multipath channels. A block-fading model is used where the channel is constant over a burst but independent across different bursts. A symbol-spaced multipath channel model is considered. The transmitted symbols are passed through the channel filter and the signal is received along with AWGN noise. The channel estimates are used in the MLSE equalizer. The performance of the MLSE equalizer with a three-tap multipath channel is shown in Figure 8.19. The performance of the receiver (with LS channel estimation and MLSE equalization) in dispersive channel is compared with the performance in flat fading channel (with a single-tap receiver and practical single-tap channel estimator). As can be seen, when an MLSE equalizer is employed, the receiver performs better in dispersive channel compared to flat fading channel due to the diversity obtained in these channels.

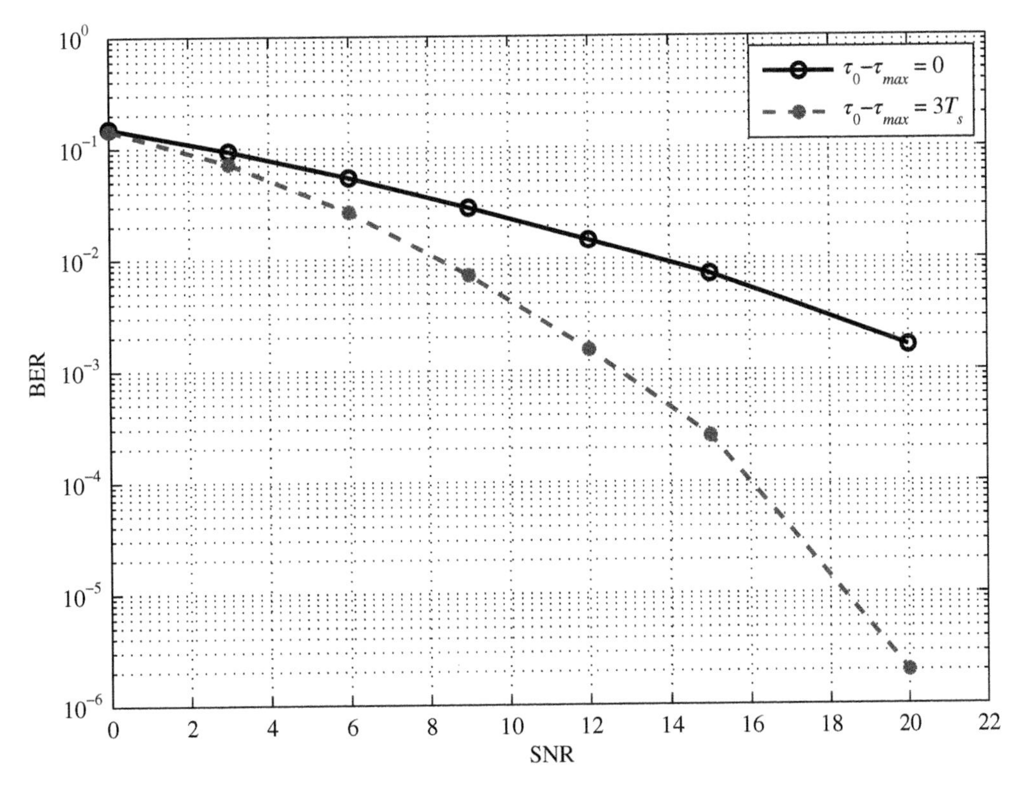

Figure 8.19 Performance in flat fading and dispersive channels with LS channel estimation and MLSE equalization.

```
1   % LS channel estimation and MLSE equalization
2   % in dispersive time invariant multipath channels
3   SNR=[0 3 6 9 12 15 20 30];% SNR values in dB
4   N=500; % number of data symbols
5   NN=1000; % number os slots to run
6   L=3; % number of channel taps
7   symb_tr=[-1 1 1 -1 1 -1 -1 1 -1 -1 -1 -1 ...
8       1 -1 1 -1 1 1 1 -1 1 1 -1 -1 -1 1 1 1 1 1 1 -1];
9   tail=[1 1 1 1 1]; % add 5 tail symbols
10  N_tr=length(symb_tr);
11  for pp=1:NN
12      symb=2*(randn(1,N)>0)-1 ; % generate data symbols
13      frame=[symb_tr symb tail]; % frame formation
14      cir=(randn(1,L)+j*randn(1,L))/sqrt(2)/sqrt(L);
15      % Channel is randomly generated over each burst
16      % channel is static over the burst. Block fading model
17      rx=filter(cir,1,frame); % pass the signal through channel
18      noise=(randn(1,length(rx))+j*randn(1,length(rx)))/sqrt(2);
19
20      for kk=1:length(SNR)
21          sgnl_rx = rx + noise*10^(-SNR(kk)/20); % add noise
22          % Channel estimation (LS estimation)
23          aa=convmtx(symb_tr(:),L);
24          aa=aa(L:N_tr,:);
25          cir_est = inv(aa' * aa) * aa' * sgnl_rx(L:N_tr).';
26          %***** MLSE equalization (using Matlab function) with LS chan. est.
27          symb_eq=mlseeq(sgnl_rx,cir_est,[1 -1],3,'rst');
28          sym_indx=N_tr+1:N+N_tr;
29          ber(pp,kk)=length(find(symb_eq(sym_indx)≠symb))/length(symb);
30      end
31  end
32  semilogy(SNR, mean(ber)); xlabel('SNR'); ylabel('BER')
```

8.3.3 Frequency-Domain Equalization

As wireless systems are becoming broadband, multipath dispersion of several microseconds makes it difficult and very complex to employ time-domain equalization techniques that are discussed in previous subsection, since the ISI could span over the duration of several dozens of data symbols. In fact, this was one of the main driving forces for the integration of multi-carrier-based technologies (like orthogonal frequency division multiplexing [OFDM]) in broadband communication systems as discussed in detail in Chapter 7. Multicarrier systems, such as OFDM, shift the complex equalization process from time-domain to frequency-domain where a simple complex multiplication operation is employed at each carrier. A single-carrier alternative of employing equalization in frequency-domain works similar to the one in OFDM, where the data symbols are grouped in blocks and each block of data symbols are cyclically extended with cyclic prefix (CP) or guard interval before the transmission. At the receiver, after the removal of CP, the block of data symbols are transferred to frequency domain to employ equalization [21–23]. Once the data block is equalized in frequency domain, then it is converted back to time for the demodulation of data symbols. The block size and cyclic prefix length requirements are similar to the OFDM modulation. Unlike OFDM, in single carrier frequency domain equalization (SC-FDE), data symbols are not mapped from frequency to time at the transmitter (no fast Fourier transform [FFT]), hence each data symbol occupies the whole bandwidth. Therefore, SC-FDE has a lower peak-to-average power ratio (PAPR) compared to OFDM. Another difference between SC-FDE from OFDM is that the frequency diversity is exploited by the averaging over the whole frequency band that is inherent in the equalization process. Therefore, unlike OFDM, channel coding might not be necessary when there is spectral fading.

At the receiver, after removing the CP, if we take discrete Fourier transform (DFT) of the received signal block of $r_m(k)$, in frequency domain, we get the following for each block:

$$R_m(\kappa) = H_m(\kappa)B_m(\kappa) + Z_m(\kappa), \tag{8.22}$$

where $H_m(\kappa)$ is the DFT of the channel impulse response of the mth block, $B_m(\kappa)$ are the DFT of the mth data block, and $Z_m(\kappa)$ are the noise terms. In order to be able to equalize the received signal in frequency domain, the channel frequency response needs to be estimated. If we assume $\hat{H}_m(\kappa)$ is the estimate of the channel frequency response, and employ zero-forcing equalization at the receiver, the equalizer coefficients will be $\frac{1}{\hat{H}_m(\kappa)}$. Hence, after the equalization, the frequency samples can be represented as:

$$R_m^{ZF} = R_m(\kappa)\frac{1}{\hat{H}_m(\kappa)} = \frac{H_m(\kappa)}{\hat{H}_m(\kappa)}B_m(\kappa) + \frac{Z_m(\kappa)}{\hat{H}_m(\kappa)}, \tag{8.23}$$

with perfect channel estimation, the above equation can be re-written as

$$R_m^{ZF} = B_m(\kappa) + \frac{Z_m(\kappa)}{\hat{H}_m(\kappa)}. \tag{8.24}$$

After the equalization is achieved, the data samples in frequency domain are converted back to time for the demodulation and decoding process. Note that the equalized symbols might suffer from the noise enhancements when there are deep fades in the channel frequency response. As can be seen from the above equation, noise samples are multiplied with the inverse of the channel frequency response. When there is a deep fade in the channel frequency response, the noise will be multiplied with a large number, and after inverse discrete Fourier transform (IDFT) operation, this enhanced noise sample will be spread to all data symbols in the block.

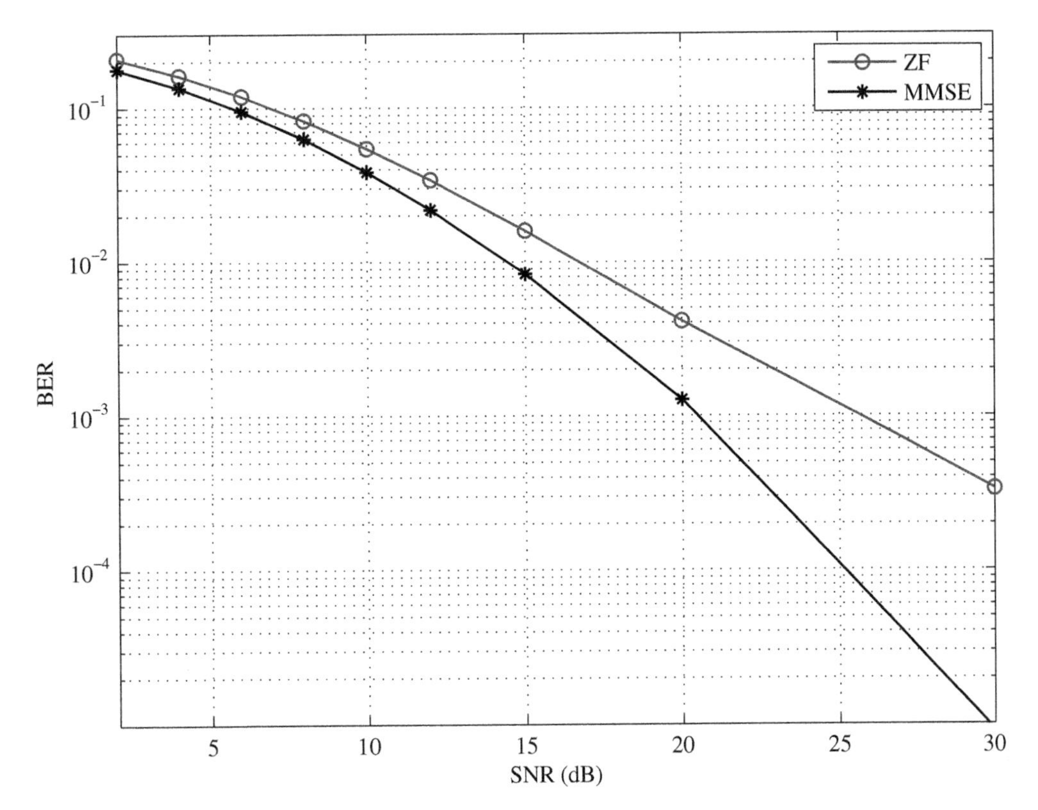

Figure 8.20 Performance of SC-FDE with known channel and with BPSK-modulated data symbols. Both ZF and MMSE equalization are shown. As can be seen, MMSE equalizer works better assuming the knowledge of noise variance. Note that the power loss due to the CP is not considered here for SNR adjustment.

The goal of zero-forcing equalization is to flatten the channel frequency response (or mitigate ISI). However, as described above, this goal is achieved with the penalty of noise enhancement. Therefore, this goal needs to be balanced so that the noise power is not enhanced. The noise enhancement problem can be partly alleviated by employing a minimum mean square error (MMSE) equalizer, which minimizes the expected mean squared error between transmitted and detected symbols at the equalizer output. Hence, MMSE provides a better balance between spectrum flatness (ISI mitigation) and noise enhancement. The equalizer coefficients in MMSE equalizer can be given as: $\frac{\hat{H}_m^*(\kappa)}{|\hat{H}_m(\kappa)|^2+\sigma_o^2}$, where σ_o^2 represents the noise variance. The equalized samples can be given as:

$$R_m^{MMSE} = \frac{B_m(\kappa)\hat{H}_m^*(\kappa)}{|\hat{H}_m(\kappa)|^2 + \sigma_o^2} + \frac{Z_m(\kappa)\hat{H}_m^*(\kappa)}{|\hat{H}_m(\kappa)|^2 + \sigma_o^2}. \tag{8.25}$$

The following Matlab code simulates SC-FDE with both zero-forcing and MMSE equalization. The corresponding plots are given in Figure 8.20.

Like in OFDM, SC-FDE suffers from RF front-end impairments. Even though SC-FDE is more robust to power amplifier nonlinearities due to lower PAPR, time-varying impairments like Doppler spread, frequency offset, and random phase noise can severely impact the performance of SC-FDE like in OFDM. These impairments and their impact have been well studied in OFDM, but, until the writing on this chapter, there is very limited amount of research available on the impact of time-varying impairments on SC-FDE. Even though one might think that the time-varying impairments could impact the performance of SC-FDE similar to OFDM as they both employ equalization in frequency domain, in practice, there are several differences. One obvious difference is that in SC-FDE, the modulated data symbols are in time (not in frequency). Therefore, after the equalization, the equalized samples need to be transferred back to time. As a result, before the demodulation of the data symbols, the impact of the time-varying impairments can be partly compensated which was not the case in OFDM. However, the impact of these time-varying impairments, after frequency-domain equalization and transferring back to time, might not be modeled the same as before. The impact of frequency-domain equalization on these impairments needs to be modeled properly before the suppression of their effect. At this point, this is an open research area; therefore, in this chapter, we will not discuss it further. Similar to time-varying impairments, other RF front-end impairments like I/Q modulation impairments (I/Q gain imbalance, DC offset, and quadrature error), sampling clock error and common phase noise need to be investigated further for SC-FDE.

```
1  % Simulation of a simple SC-FDE. Both ZF and MMSE equalization implemented
2  N=256; % block size in terms of number of symbol
3  L=5; % Channel impulse response length
4  NN=15000; % number of data blocks to run
5  SNR=[2 4 6 8 10 12 15 20 30];
6  CP_size=1/4; % CP ratio
7
8  for ll=1:NN
9      symb=2*(randn(1,N)>0)-1;
10     frame=[symb(end-N*CP_size+1:end) symb]; % Add CP to the data block
11
12     %*** Channel
13     cir=(randn(1,L)+j*randn(1,L))/sqrt(2)/sqrt(L);
14     % Channel is randomly generated over each block
15     % channel is static over the block. Block fading model
16     rx=filter(cir,1,frame); % pass the signal through channel
17     noise=(randn(1,length(rx))+j*randn(1,length(rx)))/sqrt(2);
18
19     for kk=1:length(SNR)
20         sgnl_rx = rx + noise*10^(-SNR(kk)/20); % add noise
21         %*** receiver
22         sgnl_rxo=sgnl_rx(N*CP_size+1:end); % remove CP
23         sgnl_rxF=fft(sgnl_rxo)*sqrt(N); % transfering to frequency domain
24         CFR=fft(cir,N)*sqrt(N); % Obtaining CFR from CIR
25         % frequency domain equalization (Zero forcing & MMSE)
26         sgnl_eq_ZF=sgnl_rxF./CFR;
27         sgnl_eq_MMSE=sgnl_rxF.*conj(CFR)./(abs(CFR).^2+10^(-SNR(kk)/10));
28         % going back to time domain
29         sgnl_eq_ZF_t=ifft(sgnl_eq_ZF)/sqrt(N);
30         sgnl_eq_MMSE_t=ifft(sgnl_eq_MMSE)/sqrt(N);
31         % Symbol detection
32         symb_dtct_ZF=sign(real(sgnl_eq_ZF_t));
33         symb_dtct_MMSE=sign(real(sgnl_eq_MMSE_t));
34         ber_ZF(ll,kk)=length(find(symb_dtct_ZF≠symb))/length(symb);
35         ber_MMSE(ll,kk)=length(find(symb_dtct_MMSE≠symb))/length(symb);
36     end
37 end
38 figure(1);clf
39 semilogy(SNR,mean(ber_ZF),'r-o')
40 hold
41 semilogy(SNR,mean(ber_MMSE),'k-*')
42 xlabel('SNR (dB)')
43 ylabel('BER')
```

Another issue that is worth mentioning here is that in OFDM, when there is spectral fading (in one or some of the carriers), the impact is local in those frequencies. The frequency-domain equalization (or channel correction) does not impact SNR. In other words, the SNR before and after the equalization is the same for these carriers where data symbols are embedded into them. As a result, noise enhancement is not an issue during the equalization process. On the other hand, in SC-FDE, the impact of equalization is not local. It impacts all the data symbols (which are in time) within the block. As a result, noise enhancement is a very critical issue when linear equalization is employed in SC-FDE. Due to the impact of noise enhancements of linear equalizers, recently, nonlinear equalization techniques are being considered along with frequency-domain linear equalization [22]. This is also a critical area in SC-FDE that requires more research.

8.4 Extension of Dispersive Multipath Channel to DS-CDMA-based Wideband Systems

A similar system model that is used for narrowband systems in dispersive channels can be used for direct spread code division multiple access (DS-CDMA) systems. For these systems, the pulse shape $f(\tau)$ is replaced with the convolution of the chip sequence and a chip pulse shape. Basically, each symbol is represented by a sequence of N_c chips, so that $T = N_c T_c$, where T_c is the chip period. We refer to N_c as the spreading factor, which is typically a large integer (e.g. 64, 128) for speech applications. In the DS-CDMA type of systems, the chip sequence might change each symbol period, so that the overall symbol pulse shape is time-dependent ($f(\tau)$ is replaced by $f_k(\tau)$). Also, channel tap delays are typically modeled in the order of the chip period T_c, not the symbol period T as the multipath resolvability is increased by spreading the signal over wider bandwidth. Thus, $\tau(\ell) = \ell T_c/M$. As in the narrowband case, we assume $M = 1$.

Like the narrowband case, the receiver correlates to the symbol pulse shape. For each symbol period k, it produces a "despread" value for each of the L channel taps, i.e.

$$r_{k,\ell} = \int f_k^*(\tau) y(\tau + \tau(\ell)) d\tau \quad \ell = 1, \dots, L. \tag{8.26}$$

In practice, this involves filtering matched to the chip pulse shape followed by correlation (despreading) using the chip sequence for symbol k (see Figure 8.21). Assuming the spreading factor is large enough, the contribution from adjacent symbols (ISI) can be ignored. Thus, for symbol period k, the despread value for the ℓth channel tap can be modeled as:

$$r_{k,\ell} \approx b_k c(\ell, kT) R_{f_k f_k}(kT_c - \ell T_c) + z_{k,\ell}. \tag{8.27}$$

Typically $R_{f_k f_k}(iT_c) \approx \delta(i)$ (e.g. when N_c is large), so that

$$r_{k,\ell} \approx c(\ell, kT) b_k + z_{k,\ell} = h_k(\ell) b_k + z_{k,\ell}. \tag{8.28}$$

Comparing Eqs. (8.28)–(8.13), we see that the DS-CDMA case can be treated as L separate flat channels.

The RAKE receiver [9] combines signal energy from each signal image to form a decision variable. For BPSK modulation, this gives the detected bit value

$$\hat{b}_k = \text{sgn}\left(\text{Re}\left\{ \sum_{\ell=1}^{L} \hat{h}^*(\ell, kT) r_{k,\ell} \right\} \right). \tag{8.29}$$

Observe that the despread values are weighted by the conjugates of the channel coefficient estimates, then added together. Thus, channel estimates are needed to combine the signal images properly.

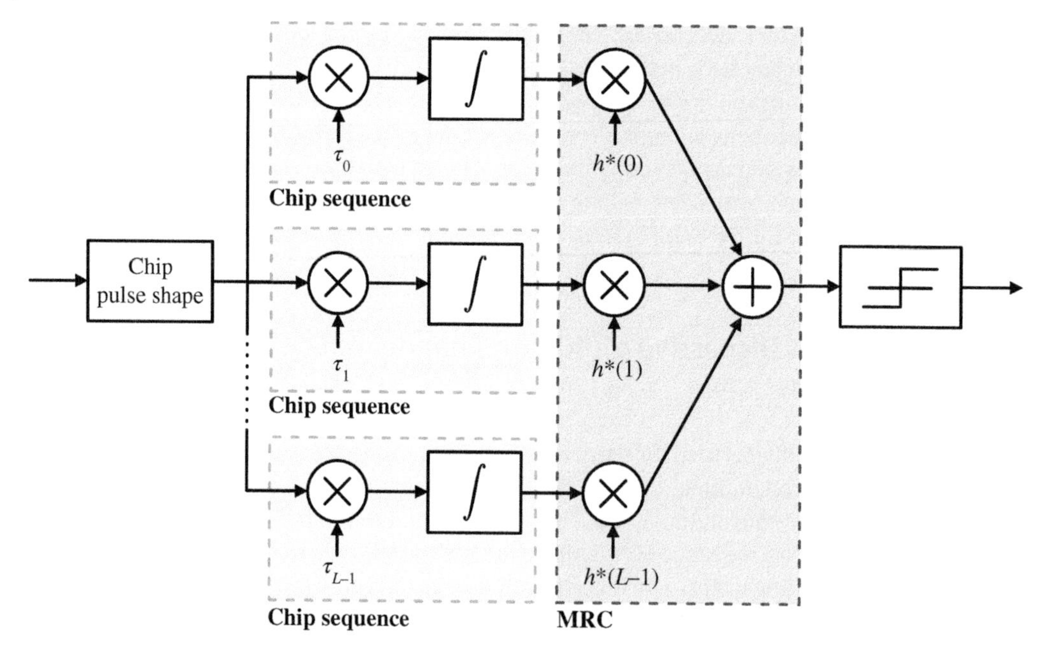

Figure 8.21 Rake receiver.

In a coherent Rake receiver, the placement of fingers on the path delays of multipath dispersive channel and estimating the corresponding multipath coefficients are critical. One way to get around the finger delays estimation is to place fingers in each chip interval (chip-spaced finger). This is possible due to pulse shaping which makes the multipaths within a chip duration to be correlated, and hence relaxing the requirement of the exact estimation of the multipath delays. However, in this case, it is required to assign more fingers than necessary (compared to the case where exact finger locations are estimated) to be able to collect maximum amount of energy that is distributed to multiple paths. As we discussed above, the channel coefficients estimation problem is relatively easy as for each finger it is similar to estimating flat fading channel that is described earlier.

In the following Matlab code, simulation of Rake receiver on a very basic CDMA system is illustrated. The simulation parameters are derived from the Time Division Synchronous Code Division Multiple Access (TD-SCDMA) standard[1] [24], where time division duplexing (TDD) is combined with synchronous CDMA. Only one slot is simulated in an oversimplified manner to avoid complexity and confusion and keep the focus on the simulation of CDMA concept and Rake reception. The multipath delays are chosen to be chip-spaced in the simulation. At the receiver, multipath delays and corresponding complex coefficients are assumed to be known. In practice, both the delays and coefficients need to be estimated using the known midamble sequence (or using pilots in other CDMA standards) which is 144 chips long. At the receiver, a Rake reception is employed with maximal ratio combining.

1 TD-SCDMA was recommended as a radio transmission interface to the International Telecommunication Union (ITU) and consequently declared one of the 3G standards.

```
% Simulation of Rake receiver on a very basic CDMA system.
% Parameters are derived from TD-SCDMA standard
% ********* SYSTEM PARAMETERS ************************
scrambling_code=[-1 1 -1 -1 -1 1 -1 -1 1 -1 1 1 -1 1 -1 -1];
% only the first code out of 128 scrambling codes is used here
channel_code=[   [1 -1 -1 1 -1 1 1 -1 1 -1 -1 1 -1 1 1 -1]*j
    [1 -1 -1 1 -1 1 1 -1 -1 1 1 -1 1 -1 -1 1]*-1];
% The last two orthogonalization codes out of 16 codes are used
Ncs=352*2; % number of chips in one slot
Tc=1/(1.28 * 10^6);  % chip duration
Qs=16;  % Using only the spreading factor of 16 (standard supports OVSF)
SNR=10; % desired SNR values in dB
Ts= Qs*Tc;  % Symbol duration
Max_num_users=16;
% In this simulation, since only two codes are given, we can have 2 users
User_codes_all=[1];% 2]; % the spreading codes that the users use.
usr_sel=1; % This is for selected analysis
% ************** TRANSMITTER ************************
slot=zeros(1,864); % only one slot is simulated for the sake of simplicity
for jk=1:length(User_codes_all) % up to 16 users can be multiplexed
    User_code=User_codes_all(jk);
    % one slot of user data generation
    N = (Ncs / Qs )*2;% number of data symbols to generate
    bts(jk,:) = randn(1,N) >0 ;

    %******** Modulation with QPSK mapping
    bts_o= reshape(-(2*bts(jk,:)-1),2,length(bts(jk,:))/2);
    symb=(bts_o(1,:) + j * bts_o(2,:))*exp(j*pi/4)/sqrt(2);

    %******** spreading
    spread_code(jk,:)=channel_code(User_code,:);
    symb_spread = ...
        reshape((symb(:)*spread_code(jk,:)).',1,Ncs);

    %******** Scrambling
    scr_code=repmat(scrambling_code(1,:).*(j).^[1:Qs],1,Ncs/Qs);
    symb_scrambled = symb_spread .* scr_code ;

    chips_midamble(jk,:)=(randn(1,144)>0)*2-1;
    % using random Midamble. In the standard Midamble sequences are defined
    %****** Slot formation
    % [DATA(352) MIDAMBLE(144) DATA(352) GUARD(16)] total size is 864 chips
    slot=slot + [symb_scrambled(1:Ncs/2)    chips_midamble(jk,:) ...
        symb_scrambled(Ncs/2+1:end) zeros(1,16)];
end

%******* Channel
L=3; % Channel impulse response length in chip spaced
cir=(randn(1,L)+j*randn(1,L))/sqrt(2)/sqrt(L); % uniform delay profile
% Channel is randomly generated over each block
% channel is static over the block. Block fading model
rx=filter(cir,1,slot); % pass the signal through channel
noise=(randn(1,length(rx))+j*randn(1,length(rx)))/sqrt(2);
sgnl_rx = rx + noise*10^(-SNR/20); % add noise

for LL=1:L % RAKE RECEIVER (L correlators)
    %******* extract the data from slot (demultiplexing)
    data_r=[sgnl_rx(LL:352+LL-1) sgnl_rx(352+LL+144:352+LL+144+352-1)];
    %******* De-scrambling
    sgnl_Dscramble = data_r.* conj(scr_code);
    %******* remove the effect of channel
    sgnl_eq=sgnl_Dscramble.*conj(cir(LL));
    % channel tap delays and coefficients are assumed to be known
    % for maximum ration combining, we need to weight the correlator output
    %******* De-spreading
    sp_code=spread_code(usr_sel,:);
    x1x=reshape(sgnl_eq,Qs,length(sgnl_eq)/Qs);
    rec_Dspread(LL,:) = (x1x.' * conj(sp_code(:)))/Qs ;
end

sgnl_combined = mean(rec_Dspread); % combining the correlator outputs
%******* Demodulator (QPSK)
```

```
73  y1=sgnl_combined*exp(-j*pi/4);
74  b0=abs((sign(real(y1))-1)/2); % the first detected bit - b0
75  b1=abs((sign(imag(y1))-1)/2); % the second detected bit - b1
76  bits_rec=reshape(([b0(:) b1(:)]).',1,2*length(b0));
77
78  BER=length(find(bts(usr_sel,:)≠ bits_rec))/length(bits_rec)
```

8.5 Conclusions

In this chapter, a basic single-carrier communication system is studied. The transmitter and receiver blocks for various channel and impairment conditions are discussed. Single-carrier systems offer several advantages compared to multi-carrier systems even if the latter has gained more popularity recently. Especially, when the time variation of the channel and/or impairments are critical, single-carrier waveform offers the ability to track and compensate these variations as it offers better processing resolution in time. At the end of the day, there is no waveform that suits perfectly to all channel and impairment conditions. Ultimately, different waveforms offer various trade-offs and they all have their preferred scenarios. It is possible that the future radio access technologies would offer a variety of these waveform options for more flexible system design. Hence, the combination of multiple waveforms can be used in a hybrid manner to satisfy multiple application simultaneously.

References

1 N. Jablon, C. Farrow, and S.-N. Chou, "Timing recovery for blind equalization," in *Proc. Twenty-Second Asilomar Conference on Signals, Systems and Computers*, vol. 1, Pacific Grove, CA, 31 Oct.–2 Nov. 1988, pp. 112–118.

2 F. Gardner, "A BPSK/QPSK timing-error detector for sampled receivers," *IEEE Transactions on Communications*, vol. 34, no. 5, pp. 423–429, May 1986.

3 A. H. Sayed, *Adaptive Filters*. Hoboken, NJ: Wiley Inter-science, 2008.

4 H. Arslan and G. E. Bottomley, "Channel estimation in narrowband wireless communication systems," *Wireless Communications and Mobile Computing*, vol. 1, pp. 201–219, Apr./June 2001.

5 ETSI, "Overall requirements on the radio interface (s) of the UMTS," Tech. Rep. ETR/SMG-21.02, Tech. Rep., 1997.

6 A. Furuskar, S. Mazur, F. Muller, and H. Olofsson, "EDGE: enhanced data rates for GSM and TDMA/136 evolution," *IEEE Personal Communications*, vol. 6, no. 3, pp. 56–66, Jun. 1999.

7 ETSI. GSM 05.02:, *Digital Cellular Telecommunications Systems (Phase 2+): Multiplexing and Multiple Access on the Radio Path*, 1996/1997.

8 H. L. Van Trees, *Detection, Estimation, and Modulation Theory: Radar-Sonar Signal Processing and Gaussian Signals in Noise*. Malabar, FL: Krieger Publishing Co., 1992.

9 J. G. Proakis, *Digital Communications*. McGraw Hill, 1995.

10 W. Jakes, *Microwave Mobile Communications*, 1st ed. 445 Hoes Lane, Piscataway, NJ: IEEE Press, 1993.

11 A. Aghamohammadi, H. Meyr, and G. Ascheid, "Adaptive synchronization and channel parameter estimation using an extended Kalman filter," *IEEE Transactions on Communications*, vol. 37, pp. 1212–1218, Nov. 1989.

12 R. W. Lucky, "A survey of the communication theory literature: 1968-1973," *IEEE Transactions on Information Theory*, vol. 19, pp. 725–739, Nov. 1973.

13 J. G. Proakis, "Adaptive equalization for TDMA digital mobile radio," *IEEE Transactions on Vehicular Technology*, vol. 40, pp. 333–341, May 1991.

14 G. D. Forney, "The Viterbi algorithm," *Proceedings of the IEEE*, vol. 61, pp. 268–277, Mar. 1973.

15 Y.-J. Liu, M. Wallace, and J. W. Ketchum, "A soft-output bidirectional decision feedback equalization technique for TDMA cellular radio," *IEEE Journal on Selected Areas in Communications*, vol. 11, no. 7, pp. 1034–1045, Sept. 1993.

16 K. Jamal, G. Brismark, and B. Gudmundson,"Adaptive MLSE performance on the D-AMPS 1900 channel," *IEEE Transactions on Vehicular Technology*, vol. 46, no. 3, pp. 634–641, Aug. 1997.

17 S. N. Crozier, D. D. Falconer, and S. A. Mahmoud, "Least sum of squared errors (LSSE) channel estimation," *IEEE Proceedings-F*, vol. 138, pp. 371–378, Aug. 1991.

18 R. A. Ziegler and J. M. Cioffi, "Estimation of time-varying digital radio channels," *IEEE Transactions on Vehicular Technology*, vol. 41, pp. 134–151, May 1992.

19 A. Gorokhov, "On the performance of the Viterbi equalizer in the presence of channel estimation errors," *IEEE Signal Processing Letters*, vol. 5, pp. 321–324, Dec. 1998.

20 D. Dzung,"Error probability of MLSE equalization using imperfect channel estimation," pp. 19.4.1–19.4.5, Denver, CO, 23–26 June 1991.

21 H. Sari, G. Karam, and I. Jeanclaude, "Transmission techniques for digital terresial TV broadcast channels," *IEEE Communication Magazine*, vol. 33, pp. 100–109, Feb. 1995.

22 N. Benvenuto, R. Dinis, D. Falconer, and S. Tomasin, "Single carrier modulation with nonlinear frequency domain equalization: an idea whose time has come again," *IEEE Proceedings*, vol. 98, no. 1, pp. 69–96, Jan. 2010.

23 D. Falconer, S. Ariyavisitakul, A. Benyamin-Seeyar, and B. Eidson, "Frequency domain equalization for single-carrier broadband wireless systems," *Communications Magazine, IEEE*, vol. 40, no. 4, pp. 58–66, Apr. 2002.

24 B. Li, D. Xie, S. Cheng, J. Chen, P. Zhang, W. Zhu, and B. Li, "Recent advances on TD-SCDMA in China," *IEEE Communications Magazine*, vol. 43, no. 1, pp. 30–37, Jan. 2005. doi:10.1109/MCOM.2005.1381872.

9

Multiple Accessing, Multi-Numerology, Hybrid Waveforms

Mehmet Mert Şahin[1] *and Hüseyin Arslan*[1,2]

[1] *Department of Electrical Engineering, University of South Florida, Tampa, FL, USA*
[2] *Department of Electrical and Electronics Engineering, Istanbul Medipol University, Istanbul, Turkey*

In wireless communication systems, system resources are distributed among multiple users to enable what is called multiple accessing [1, 2]. This chapter sheds light on the issue of how to intelligently share the resources in a hyperspace whose dimensions are time, frequency, space, beam, and code. Popular multiple access schemes are revisited and visualized with different tools, and basic implementation procedures are provided to understand the concept thoroughly.

So far, different multiple access techniques have been deployed for wireless communication technologies, and they become a prominent element in developing wireless communication systems [3]. In the first generation, cellular networks were employed with frequency division multiple accessing (FDMA), where the frequency band is divided into frequency channels and assigned to users. In the second generation, time division multiple access (TDMA) was used where the new dimension, time, is introduced. The transmission duration is divided into time slots and assigned to each user. In the third generation, code division multiple access (CDMA) was deployed, which utilizes larger bandwidth by spreading the modulated symbols. The fourth and fifth generations of cellular systems are deployed with the orthogonal frequency division multiple access (OFDMA) whose physical resources are time slots and frequency subcarriers. OFDMA technology gives flexibility in resource allocation and keeps the available information bandwidth at the desired band. Additionally, directional antennas often obtained by employing antenna array processing, add the angular dimension that can also be used to channelize the signal space. This multiple access technique is called as space-division multiple access (SDMA). Table 9.1 categorizes several multiple accessing schemes that are used in several standards and wireless technology generations. More details will be given inside the sections about these deployments.

This chapter also provides a comparison between random access schemes. It also gives insight into which specific scheduled or random access technique to apply. It emphasizes the real wireless system development problem of how to design a multiuser system depending on the system requirements, traffic characteristics of users in the network, performance measurements, channel characteristics, and other interfering systems operating in the same bandwidth. Today's wireless networks are also discussed to present the evolution of wireless communication and multiple access.

9.1 Preliminaries

This section explains the terminology used throughout the chapter, including duplexing schemes, downlink, and uplink transmission.

Wireless Communication Signals: A Laboratory-based Approach, First Edition. Hüseyin Arslan.

Table 9.1 Multiple access techniques in different wireless technologies.

MA technique	Physical resources	Standardization
FDMA	Frequency	1G, AMPS, NMT, Wake-up radio
TDMA	Time slots	2G, GSM, IS-54
CDMA	Time slots/PN codes	3G, WCDMA, IS-95, CDMA2000
OFDMA	Time/frequency	4G-LTE, 5G-NR

9.1.1 Duplexing

Radio frequency (RF) transmission is grouped in three different categories regarding its transmission and reception functionality, such as simplex, half-duplex, and full-duplex. A simplex RF system is a radio technology that allows only one-way communication from a transmitter to a receiver. A half-duplex RF system provides both transmission and reception functionality, but not simultaneously. The communication is bidirectional over the same frequency, but unidirectional during the communication period. A full-duplex system providing high spectral efficiency in a cost-efficient manner allows communication at the same time and frequency resources. However, the self-interference issue, which is depicted in Figure 9.1, is a restrictive challenge that needs to be solved. Encouraging techniques for self-interference mitigation and more insights about full-duplex communication can be found in [4] and references therein.

Frequency division duplexing (FDD) provides two distinct bands of frequencies for every user to perform two-way traffic between the base station (BS) and user equipment (UE). On the other hand, time division duplexing (TDD) exploits the time slots to provide two-way traffic. An advantage of TDD is the channel reciprocity property that is widely used in current cellular standards. The main idea is that bi-orthogonal channels are typically symmetrical in their channel gains, so channel measurement made in one direction can be used to estimate the channel in the other direction. Therefore, uplink pilots can be used to estimate the channel in the downlink transmission.

Moreover, TDD utilizes the spectrum more efficiently than FDD. FDD cannot be used in an environment where the service provider does not have enough bandwidth to provide the required guard band between transmit and receive channels. On the other hand, FDD introduces no additional time delay and latency as channels are always open. In contrast, additional latency is added in the

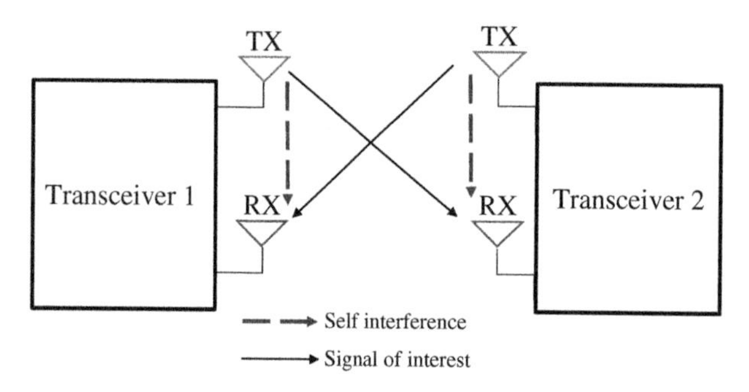

Figure 9.1 Full duplex communication.

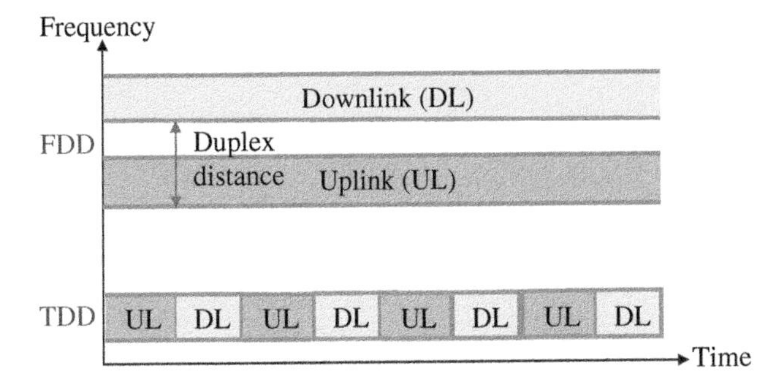

Figure 9.2 Time and frequency division duplexing.

TDD to ensure that transmit and receive links do not overlap. Resource sharing in the uplink and downlink transmission for FDD and TDD transmission can be seen in Figure 9.2.

9.1.2 Downlink Communication

In the downlink transmission, as shown in Figure 9.3, BS transmits a common signal which consists of the aggregate sum of all UEs signal in that cell. It is assumed that there is no multipath effect during the transmission. Each UE is physically located in a different position in the cell and thus experiences different channel effects, including path loss, shadowing, and one-tap fading, denoted by h^k. The additive white Gaussian noise (AWGN) in each UE receiver is represented by w^k.

The received signal from the downlink channel by the kth user transmitting its signal with power P^k is represented as:

$$y^k = h^k \left(\sum_{k=1}^{K} P^k x^k \right) + w^k, \tag{9.1a}$$

and

$$\sum_{k=1}^{K} P^k = P_{total}, \quad 1 \leq k \leq K. \tag{9.1b}$$

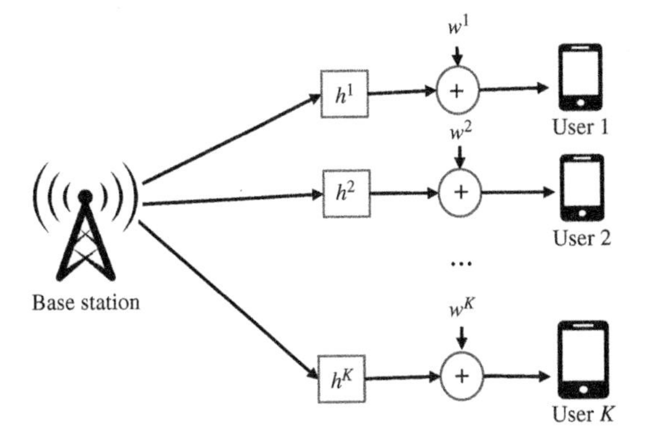

Figure 9.3 Downlink communication.

where the total number of users is denoted by K and P_{total} stands for total power that BS can transmit. As it can be seen for kth user in Eq. (9.1), both signal $x^k(t)$ and interference $x^j(t)$, $j \neq k$, are distorted by the same channel $h^k(t)$.

Some of downlink wireless communication types can be exemplified as:

- radio and television broadcasting,
- transmission link from satellite to multiple ground stations,
- transmission link from a base station to the mobile terminals in a cellular system.

9.1.3 Uplink Communication

In the uplink transmission, as shown in Figure 9.4, each UE emits its signal that comes across with different fading channel. These individual signals are superimposed at the BS with its AWGN denoted by w. The received signal y at the BS can be represented as:

$$y = \sum_{k=1}^{K} h^k P^k x^k + w, \quad 1 \leq k \leq K, \tag{9.2}$$

where each user has an individual power constraint P^k. Since the signals are sent from different transmitters, these transmitters must coordinate if signal synchronization is required. Even if the transmitted powers P^k are the same, received powers associated with other users will be different since their channel gains (h^k) are random depending on large- and small-scale fading effects.

Examples of uplink wireless communication include the followings:

- data transmission from any device to a wireless LAN access point,
- transmissions from ground stations to a satellite,
- transmissions from mobile terminals to a base station in cellular systems.

9.1.4 Traffic Theory and Trunking Gain

In this section, how many subscribers can be covered with a wireless communication system by one BS will be investigated. It is described in a very simplified way how to compute the required number of communication channels so that a given number of users can be served with a sufficient quality. Trunking is the process of the user assignment to channels on demand where each cellular cell has a pool of channels. When the user requires service, a channel is allocated to that user. When the user no longer requires service, the channel returns to the pool in order to be allocated to next users. The concept of trunking can be applicable to any wireless network that channelizes the hyperspace resources.

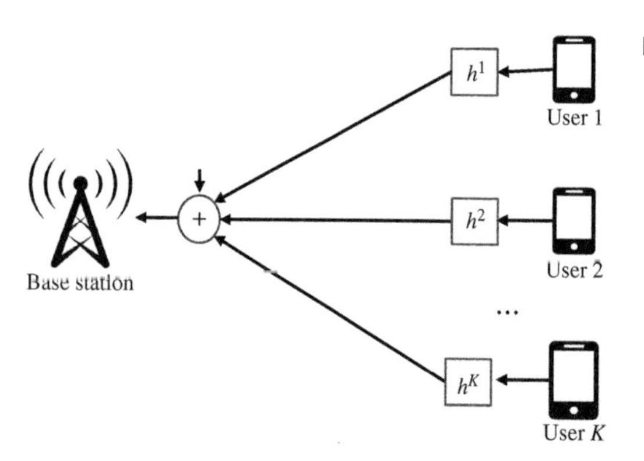

Figure 9.4 Uplink communication.

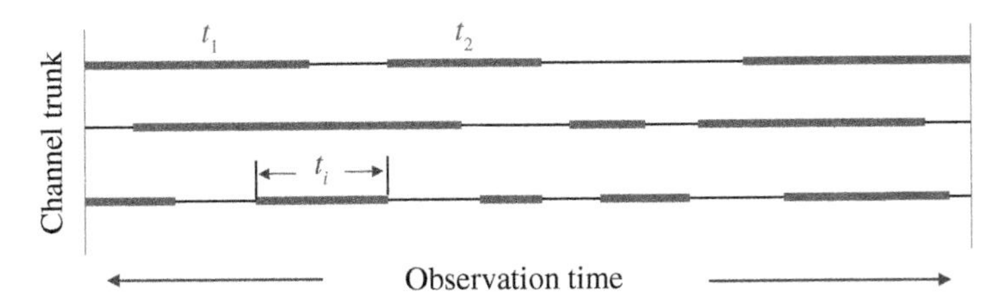

Figure 9.5 Channel trunk with several random occupations.

Here, it is assumed that a system is designed purely for speech communication. For the following, the offered traffic is defined as the product of the call arrival rate ($1/s$) and the average call holding time (s); the offered traffic is usually said to have the unit **Erlang**, a traffic flow unit, though this is a dimensionless quantity [5]. The channel trunk with random occupations by different services can be depicted as in Figure 9.5. The essential equations for traffic theory can be defined as follows:

$$T_{hold} = \sum_{i=1}^{N_{call}} t_i/N_{call}, \tag{9.3}$$

$$A_u = \lambda T_{hold} \quad \text{Erlang}, \tag{9.4}$$

$$A_{total} = UA_u \quad \text{Erlang}, \tag{9.5}$$

where T_{hold}, t_i, and N_{call} are average holding time, occupation time of ith call, and total number of calls, respectively. Let A_u and A_{total} denote traffic intensity per user and total system traffic for U users, respectively, where λ denotes the average call request rate per user. Here one unit Erlang means that one channel is occupied continuously.

Traffic theory aims to accommodate a large number of users with a small number of channels. It exploits the statistical behavior of users. In a trunked system, when a user requests service and all channels are already in use, the user is blocked. Grade of service (GOS) is the measure of the ability of a user to access the trunked system during the busiest hour. To meet the GOS requirement of the system, it is crucial to estimate the maximum required capacity and to allocate a proper number of channels. Traffic tables depict how many channels are required for a minimum GOS. One example of a traffic table can be found in the Appendix.

The number and duration of calls depend on the time of day. Therefore, the busy hour, which is usually around 10 am and 5 pm, can be defined as the hour when most calls are made. The traffic during that busy hour determines the required network capacity. Moreover, the spatial distribution of users is time-variant. While business districts (city centers) usually see a lot of activity during the daytime, suburbs and entertainment districts experience more traffic during the nighttime. So, telephoning habits change over the years. While in the late 1980s, calls from cellular phones were limited to a few minutes, now hour-long calls have become quite common.

For the computation of the blocking probability of a simplified system. We make the following assumptions:

- The time slots when calls are placed are statistically independent.
- The duration of calls is an exponentially distributed random variable.
- If a user is rejected, his/her next call attempt is made statistically independent of the previous attempt (i.e. behaves like a new user).

Such a system with these assumptions is called an Erlang-B system [6]. The probability of call blocking in the Erlang-B system can be shown to be

$$p_{block} = \frac{T_{tr}^{N_{sc}-1}/N_{sc}!}{\sum_{k=0}^{N_{sc}} T_{tr}^{k}/k!}, \tag{9.6}$$

where N_{sc} is the number of speech channels per cell, and T_{tr} is the average offered traffic.

```
1  Nc = [5 10, 20, 50,100]; %number of speech channels
2  T_tr = logspace(-1,2,50); %average offered traffic
3  figure()
4  for i = 1:length(Nc)
5      for j = 1:length(T_tr)
6          Pr_block(i,j) = T_tr(j)^Nc(i)/factorial(Nc(i))...
7          /sum(T_tr(j).^[0:Nc(i)]./factorial(0:Nc(i)))
8      end
9      loglog(T_tr,Pr_block(i,:))
10     hold on
11 end
12 legend('Nc=5','Nc=10','Nc=20','Nc=50','Nc=100')
13 grid on; ylim([1e-3 1])
```

It can be seen in Figure 9.6 that the ratio of required channels to offered traffic is very high, if N_{sc} is small, especially for low blocking probabilities. Assuming the required blocking probability of 1%, the ratio of possible offered traffic to available channels is about 0.9 for $N_{sc} = 50$.

In the research of next-generation wireless networks, including 5G, promising solutions come up to meet the requirements of applications such as streaming, computing, gaming, communicating, and storage. Definitely, these diverse applications and immense usage of wireless bands lead researchers to investigate the traffic models of 5G and beyond 5G. The details about the use case and related traffic models can be found in [7].

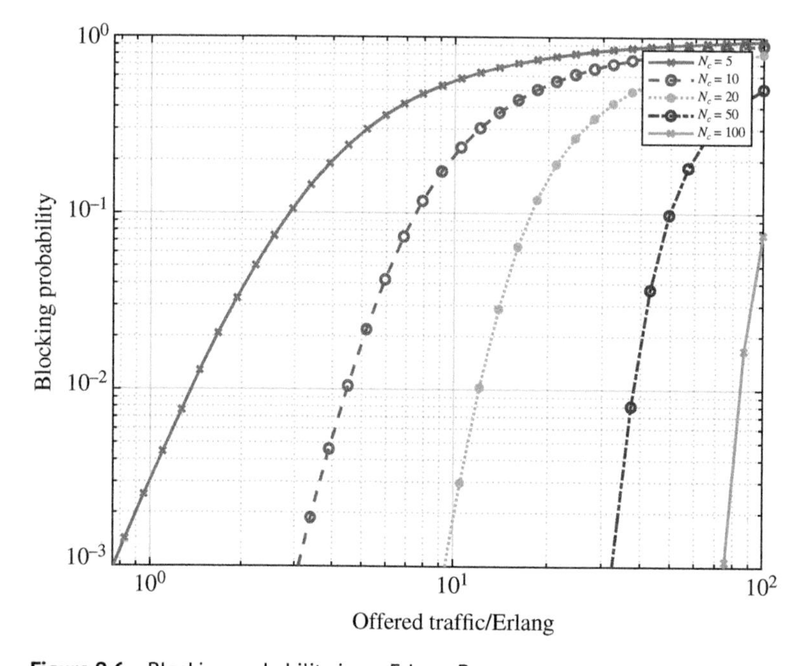

Figure 9.6 Blocking probability in an Erlang-B system.

9.2 Orthogonal Design

Currently, most of the wireless communication standards prefer to design orthogonal frames in order to manage the interference between the users. Figure 9.7 depicts several orthogonal schemes separated in different hyperspaces that are commonly used in wireless communication standards.

9.2.1 TDMA

In TDMA structure, users allocate different time slots separated orthogonally, which means that users' transmission does not partially or fully overlap with each other in the time domain. Time slots often have guard bands in time to separate different user's slots because synchronization errors and multipath causes inter-symbol interference (ISI) [8]. In a TDMA frame, the preamble contains the address and synchronization information that both the base station and subscribers use to identify each other. Guard times are utilized to allow synchronization of the receivers between different slots and frames.

TDMA properties can be listed as follows:

- Data transmission for users of a TDMA system is not continuous, but it occurs in bursts. This kind of transmission results in low battery consumption because the subscriber transmitting unit can be turned off when it is not in use.
- High synchronization overhead is needed.
- Since each TDMA transmissions are slotted, receivers need to be synchronized for each data burst.

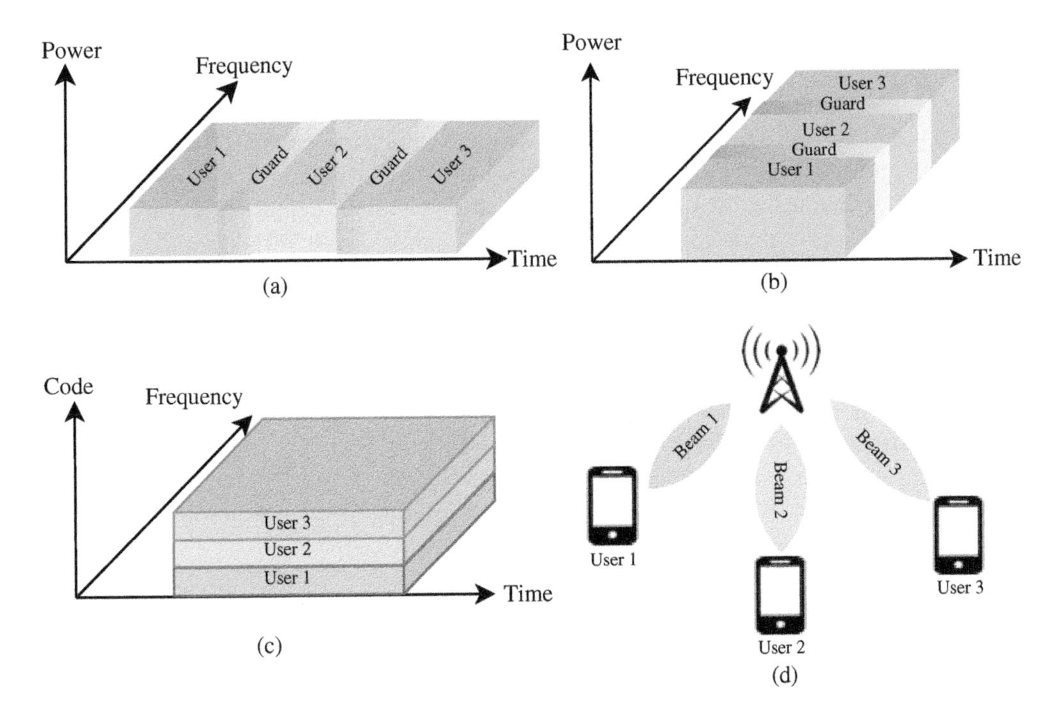

Figure 9.7 (a) Time division multiple access (TDMA). (b) Frequency division multiple access (FDMA). (c) Code division multiple access (CDMA). (d) Space division multiple access (SDMA).

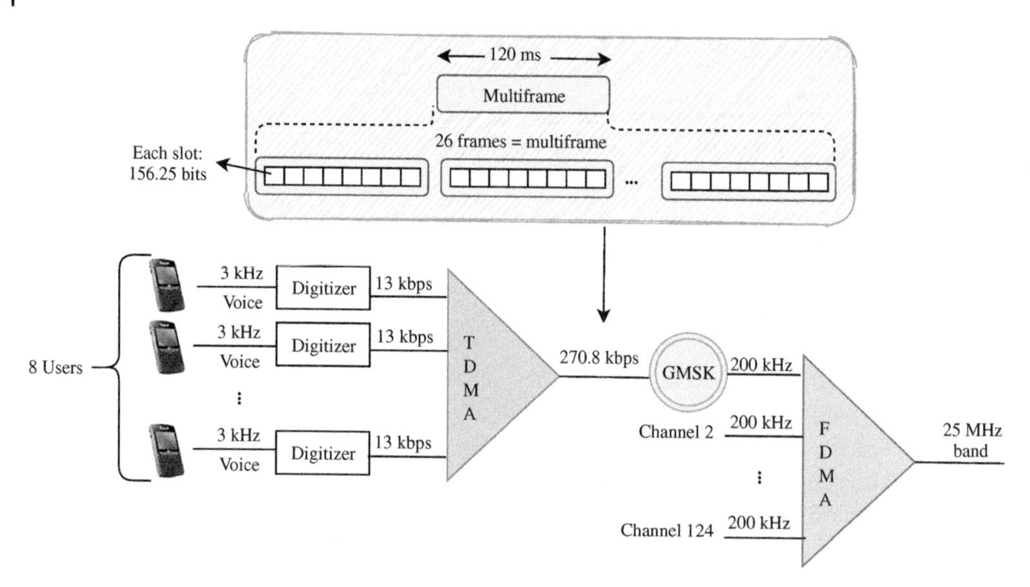

Figure 9.8 GSM frame structure including TDMA and FDMA.

- Guard slots are necessary to separate users in TDMA. Also, guard bands in the time domain allow time delay variations of uplink transmissions.
- Adaptive and complex equalization is usually necessary for TDMA systems because high transmission rates require long data duration as compared to FDMA systems.

Different TDMA wireless standards have other TDMA frame structures. In Figure 9.8, global system for mobile communications (GSM) structure developed for 2G wireless communication systems is depicted using a combination of TDMA and FDMA. 25 MHz is allocated with 124 channels with a 200-kHz band. Then, each channel is split into eight parts using TDMA for each user. A frame consisting of 8 slots carries 1250 bits. Twenty-six frames form a multiframe. For each user, each 3-kHz voice channel is digitized and compressed to a 13-kbps digital signal. After adding overheads and forward error coding bits, the channel data rate becomes 270.8 kbps.

9.2.2 FDMA

FDMA is usually combined with the FDD, where the total frequency band are split into subchannel each user. A user send its data through a subchannel by modulating via a carrier wave at the frequency of the subchannel. The frequency band generally includes guard bands between them to compensate for imperfect filters, adjacent channel interference, and spectral spreading due to Doppler and phase noise [5]. Advantages of the FDMA system can be expressed as follows:

- The synchronization of each user in FDMA transmission is simple. Once synchronization has been established during the call setup, it is easy to maintain it utilizing a simple tracking algorithm, as transmission occurs continuously.
- The symbol time is large as compared to the average delay spread. It implies that the amount of intersymbol interference is low and, thus, little or no equalization is required in narrowband FDMA systems.

Some difficulties in FDMA implementation are listed as follows:

- FDMA requires tight RF filtering to minimize adjacent channel interference.
- Jitters in the carrier frequency result in adjacent channel interference. High spectral efficiency also requires the use of very steep filters to extract the desired signal. Therefore, guard bands are generally used to separate bands.
- The BS needs to transmit multiple speech channels, each of which is active during the whole time. Typically, a BS uses 20–100 frequency channels. If these signals are amplified by the same power amplifier, third-order modulation products lying at undesirable frequencies can be created due to the nonlinear region of the amplifier. Therefore, either a separate amplifier is utilized for each speech channel or a highly linear amplifier is used for the composite signal. Each of these solutions makes a BS more expensive. Also, nonlinearities in the power amplifier cause signal spreading in the frequency domain.
- If an FDMA channel is not in use, then it sits idle and cannot be used by other users to increase or share capacity.

9.2.3 Code Division Multiple Access (CDMA)

In the CDMA, the signals of different users are modulated using bi-orthogonal or non-orthogonal spreading codes [2]. The resulting spread signals simultaneously occupy the same time and bandwidth, as shown in Figure 9.7c.

In these multiuser systems, the interference between users is determined by the cross-correlation of their spreading codes. The design of spreading codes typically has either good autocorrelation properties to mitigate ISI or good cross-correlation properties to mitigate multiuser interference. However, there is usually a trade-off between optimizing the two features. Thus, the best choice of code design depends on the number of users in the system and the severity of multipath and interference. Downlink transmission typically uses orthogonal spreading codes such as Walsh–Hadamard codes, although ISI can result in the case of having of multipath channel. Besides, the uplink transmission uses non-orthogonal codes to increase the number of connectivity. More details about non-orthogonal transmission in uplink will be given in Section 9.3.

The receiver performs a time correlation operation to detect only the specific desired codeword [8]. All other codewords appear as noise due to decorrelation. In CDMA, the power of multiple users at a receiver determines the noise floor after decorrelation. If the power of each user within a cell is not controlled such that they do not appear equal at the base station, then the *near-far problem* occurs.

In existing CDMA systems, the received spreading codes are not entirely orthogonal, and the correlation process cannot be so efficient. The desired signal does not significantly distinguish itself from interfering users whose effect can be modeled as increased background noise. There are some suboptimal detection schemes to cope with inter-user interference (IUI). *Rake receiver* mentioned in Chapter 8 is one of them, which do not extract all CDMA codes in parallel. A parallel extraction method called joint detection unit is the optimal multiuser detection tool mentioned in [9] and references therein.

In CDMA, stronger received signal levels raise the noise floor at the base station, thereby decreasing the probability that weaker signals will be demodulated. Power control is used in most CDMA implementations to combat the near-far problem. Power control is provided by each base station in a cellular system. It assures that each mobile within the base station coverage area provides the same signal level to the base station receiver. Power control is implemented at the base station by rapidly sampling the received signal strength indicator (RSSI) levels of each mobile and then sending a power change command over the forward radio link. CDMA properties can be summarized as follows:

- Multipath fading may be substantially reduced because the signal is spread over a large spectrum. If the spread spectrum bandwidth is greater than the coherence bandwidth of the channel, the inherent frequency diversity will mitigate the effects of small-scale fading [8].
- Channel data rates are supposed to be very high in CDMA systems. Consequently, the symbol (chip) duration is concise and usually much less than the channel delay spread. Since pseudo-noise (PN) sequences have low autocorrelation, signals coming along with multipath which are delayed by more than a chip will appear as noise. A *Rake receiver* can be used to improve reception by collecting time-delayed versions of the required signal.
- In CDMA systems, since all the cells share the same spectrum, soft handoffs are possible. Multiple base stations can simultaneously decode the mobile's data, with the switching center choosing the best reception among them. Soft handoff provides another level of diversity to the users [1].
- Since signals for the different users in the cell are all transmitted at the BS, it is possible to make the users orthogonal to each other, something that is more difficult to do in the uplink, as it requires chip-level synchronization between distributed users. It reduces but does not remove intra-cell interference, since the transmitted signal goes through multipath channels and signals with different delays from different users still interfere with each other.

Figure 9.9 shows the code-domain power distribution under the time-invariant frequency-selective channel with AWGN. The number of taps and strength of noise determines the distribution of code-domain power. More taps cause more uncontrollable leakage on the symbols; therefore, code-domain power decreases. Moreover, code length also affects the code-domain

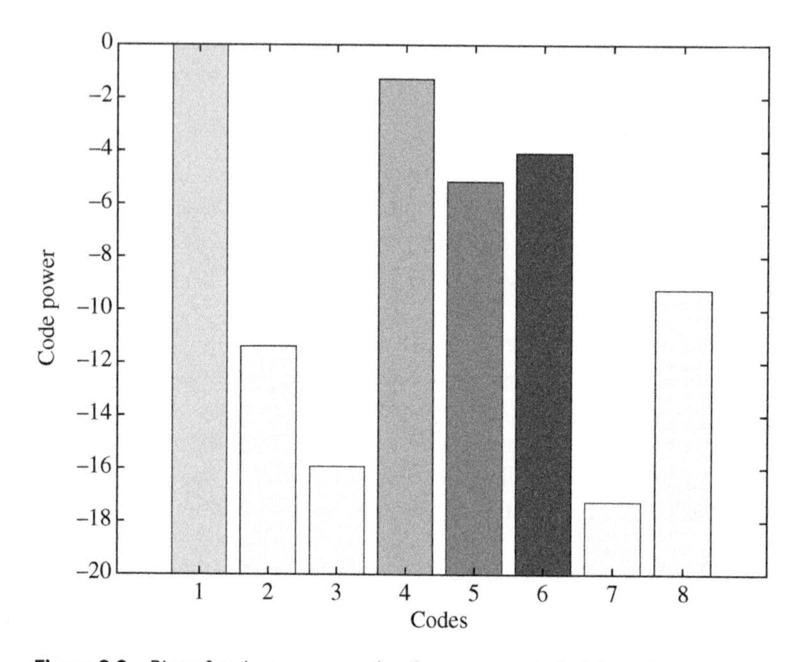

Figure 9.9 Plot of code powers serving four users out of eight codes.

power, which gives resistance to multipath. Details for the generation of CDMA signals are given in Chapter 8.

```
spread_code=[1 1 1 1 1 1 1 1 ; 1 1 1 1 -1 -1 -1 -1;
   1 1 -1 -1 1 1 -1 -1; 1 1 -1 -1 -1 -1 1 1; 1 -1 1 -1 1 -1 1 -1;
   1 -1 1 -1 -1 1 -1 1; 1 -1 -1 1 1 -1 -1 1; 1 -1 -1 1 -1 1 1 -1];
% There are 8 orthogonalization codes
Ncs=352*2; % number of chips in one slot
Qs=8; % Using only the spreading factor of 8 (orthogonal)
SNR=20; % desired SNR values in dB
M=4; %modulation order (QPSK)
mid=144; %length of midamble
Max_num_users=8; User_all=4;
ord=sort(randperm(8,4));
code_order = spread_code(ord,:);
% the spreading codes that the users use.
% ************** TRANSMITTER ***********************
slot=zeros(1,864);
for u=1:User_all % up to 16 users can be multiplexed
    % one slot of user data generation
    bits(u,:) = randi([0 1],1,(Ncs / Qs)*log2(M));
    %******** Modulation with QPSK mapping
    symb=qammod(bits(u,:).',M,'InputType','bit',...
        'UnitAveragePower',true);
    %******** spreading
    symb_spread=reshape((symb(:)*code_order(u,:)).',1,Ncs);

    chips_midamble(u,:)=(randn(1,144)>0)*2-1;
    %****** Slot formation
    %[DATA(352) MIDAMBLE(144) DATA(352) GUARD(16)]
    slot=slot + [symb_spread(1:Ncs/2) chips_midamble(u,:)...
        symb_spread(Ncs/2+1:end) zeros(1,16)];
end
%******* Channel
L=2; % Channel impulse response length in chip spaced
cir=(randn(1,L)+1i*randn(1,L))/sqrt(2)/sqrt(L); % uniform
% Channel is static and randomly generated over the block.
rx=filter(cir,1,slot); % pass the signal through channel
noise=(randn(1,length(rx))+1i*randn(1,length(rx)))/sqrt(2);
sgnl_rx = rx + noise*10^(-SNR/20); % add noise

for usr_sel = 1:Max_num_users
    for LL=1:L % RAKE RECEIVER (L correlators)
        %******* extract the data from slot (demultiplexing)
        data_r=[sgnl_rx(LL:Ncs/2+LL-1) sgnl_rx(Ncs/2+LL+...
            mid:Ncs/2+LL+mid+Ncs/2-1)];
        %******* remove the effect of channel
        sgnl_eq=data_r.*conj(cir(LL));
        % channel tap delays and coefficients are known
        %******* De_spreading
        %sp_code=code_order(usr_sel,:);
        sp_code=spread_code(usr_sel,:);
        x1x=reshape(sgnl_eq,Qs,length(sgnl_eq)/Qs);
        rec_Ds(LL,:) = (x1x.' * conj(sp_code(:)))/Qs;
    end
    s_comb(:,usr_sel) = mean(rec_Ds,1); %combine
end
E=10*log10(mean(abs(s_comb).^2)/max(mean(abs(s_comb).^2)));
b=bar(E);b(1).BaseValue = -20;
```

9.2.4 Frequency Hopped Multiple Access (FHMA)

Frequency hopped multiple access (FHMA) is a digital multiple access system in which the carrier frequencies of the individual users are varied in a pseudo-random fashion within a wideband channel. The digital data is broken into uniformly sized bursts which are transmitted on different carrier frequencies. The instantaneous bandwidth of any transmission burst is much smaller than the total spread bandwidth. The pseudo-random change of the carrier frequencies of the user randomizes the occupancy of a specific channel at any given time, thereby allowing for multiple access over a wide range of frequencies. In the FHMA receiver, a locally generated PN code is used to synchronize the receivers' instantaneous frequency with that of the transmitter. At any given point in time, a frequency-hopping signal only occupies a single, relatively narrow channel [8]. FHMA properties are listed as follows:

- FHMA systems often employ energy-efficient constant envelope modulation. Inexpensive receivers may be built to provide noncoherent detection of FHMA.
- A frequency-hopped system provides a level of security when a large number of channels are used, since an illegitimate receiver that does not know the pseudo-random sequence of frequency slots must dynamically search for the signal that it wishes to intercept.
- Frequency hopping is also used in cellular systems to average out interference from other cells.

The spectrogram of the bluetooth signal, which utilizes FHMA, is depicted in Figure 9.10 generated via MATLAB code provided below.

```
1  numPackets = 10;          % Number of packets to generate
2  sps = 16;                 % Samples per symbol
3  messageLen = 2000;        % Length of message in bits
4  phyMode = 'LE1M';         % Select mode
5  channelBW = 2e6;          % Channel spacing (Hz) as per standard
6  symbolRate = 1e6;
7  x = [];
8  winL = 1000; noverlap = 500; nfft = 8192;
9  sampleRate = symbolRate*sps*10;
10 % Loop over the number of packets, generate a BLE waveform
11 rng default;
12 for packetIdx = 1:numPackets
13     message = randi([0 1],messageLen,1); % Message bits
14     chanIndex = randi([0 39],1,1);   % Channel index
15     if(chanIndex ≥37)
16         % Access address for periodic advertising channels
17         accessAdd = [0 1 1 0 1 0 1 1 0 1 1 1 1 1 0 1 1 0 0 ...
18                           1 0 0 0 1 0 1 1 1 0 0 0 1]';
19     else
20         % Random access address for data channels
21         accessAdd = [0 0 0 0 0 0 0 1 0 0 1 0 0 ...
22             0 1 1 0 1 0 0 0 1 0 1 0 1 1 0 0 1 1 1]';
23     end
24     waveform = bleWaveformGenerator(message,'Mode',phyMode,...
25                     'SamplesPerSymbol',sps,...
26                     'ChannelIndex',chanIndex,...
27                     'AccessAddress',accessAdd);
28     frequencyOffset = channelBW*chanIndex;
29     waveform = waveform.* exp(1i*2*pi*frequencyOffset*...
30     (1:length(waveform)).'/sampleRate);
31     x = [x; waveform];
32 end
33 [s,f,t]=spectrogram(x,winL,noverlap,nfft,sampleRate,'yaxis');
34 imagesc(t/1e-6,f/1e6,10*log10(abs(s))); ylim([0, 100])
35 set(gca,'YDir','normal'); cc = colorbar;
36 ylabel(cc,'Power/frequency (dB/Hz)');
37 hColourbar.Label.Position(1) = 3;
```

9.2.5 Space Division Multiple Access (SDMA)

SDMA uses direction (angle) as another dimension in signal space, which can be channelized and assigned to different users. It is generally done with directional or phased-array type antennas.

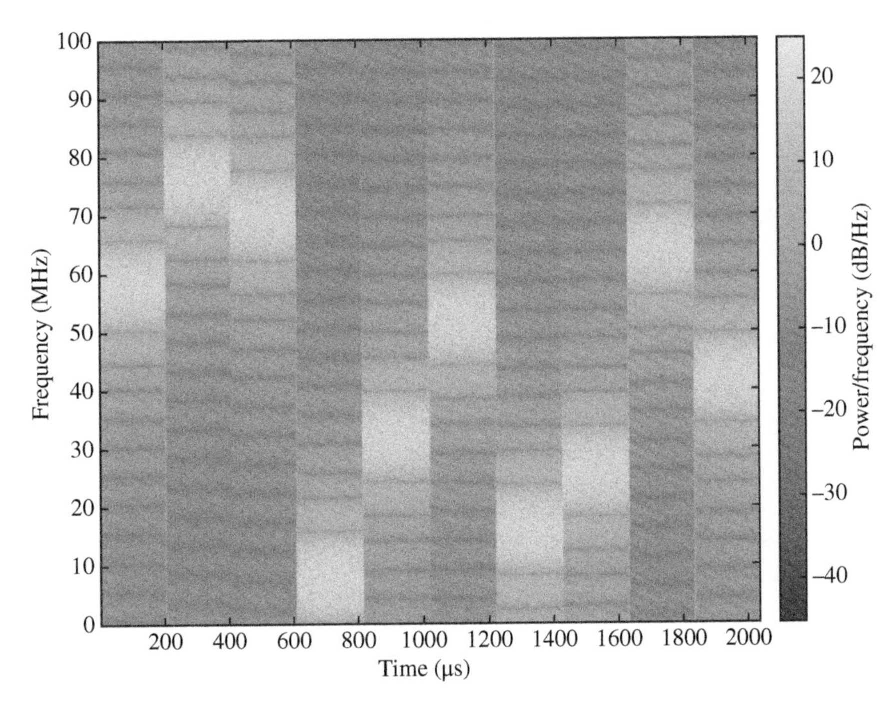

Figure 9.10 Spectrogram of Bluetooth signal having frequency hopping pattern.

Orthogonal channels can be assigned to each user if the angular separation between users exceeds the angular resolution of the directional antenna. If directionality is obtained using an antenna array, a precise angular resolution requires a large array size [2]. Directionality can also be obtained with the sectorization of antenna arrays. Moreover, physical separation can be regarded as another way of SDMA where two transmitters can use the same portion of radio resource if both are not within the radio range of the same receiver. For example, transmitters in different cells can use the same frequency, which is called the reuse of frequency.

The multiple antennas can be used to increase data rates through multiplexing or to improve bit-error-rate (BER) performance through diversity. It can also be used to create pencil-like beams to send a signal to a specific direction or user. In multiple-input multiple-output (MIMO) systems, the transmit and receive antennas can both be used for diversity gain. Multiplexing exploits the structure of the channel gain matrix to obtain independent signaling paths that can be used to send independent data. The spectral efficiency gains coming from multiple antennas often require accurate knowledge of the channel at the receiver and sometimes at the transmitter as well. In addition to spectral efficiency gains, ISI and IUI can be reduced using smart antenna techniques [2]. The multiplexing gain of a MIMO system results from the fact that a MIMO channel can be decomposed into a number r of parallel independent channels. By multiplexing independent data onto these independent channels, we get an r-fold increase in data rate in comparison to a system with just one antenna at the transmitter and receiver. The increased data rate is called multiplexing gain.

9.2.5.1 Multiuser Multiple-input Multiple-output (MIMO)

MIMO systems can be utilized in a cellular scenario where the BS communicates with multiple users at the same allocated resource block. The system model is depicted in Figure 9.11 where a single BS with N_{BS} antenna elements communicates each with K UEs with N_{UE}^{k} antenna elements.

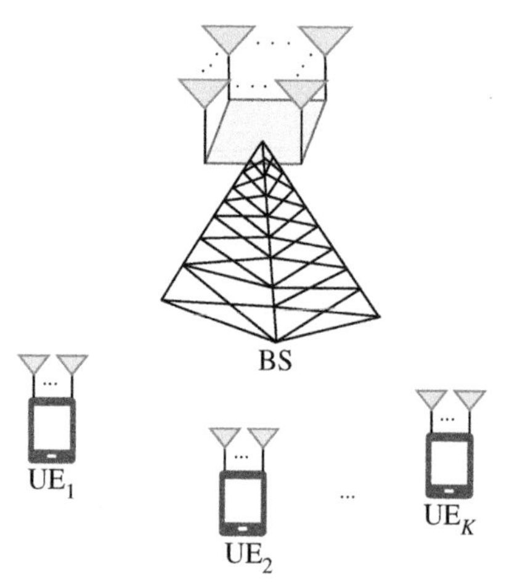

Figure 9.11 Multi user MIMO system configuration.

It is generally assumed that $N_{BS} > N_{UE}^k$, with the constraint that $N_{BS} < \sum_{k=1}^{K} N_{UE}^k = N_{UE}$ which is the general case occurring in cellular systems. For the multiuser MIMO systems, the key points can be emphasized as follows:

- Scheduling considering the behavior of the wireless channel can significantly reduce the probability of a breakdown of the overall capacity. If there are more UEs than BS antennas, then an intelligent scheduling algorithm can greatly reduce the possibility that the channel of users are linearly dependent.
- In real wireless systems, it is not expected that many UEs collaborate to decode many data streams. With channel state information at the transmitter (CSIT), however, the BS can null out undesired streams in the direction of different UEs and thus transmit more data streams while still enabling a good signal-to-interference-plus-noise ratio (SINR) at each UE.

Considering downlink transmission, the received signal vector $\mathbf{y}$ is

$$\mathbf{y} = \mathbf{H}\mathbf{x} + \mathbf{w}, \tag{9.7}$$

where the n_T dimensional vector $\mathbf{x}$ denotes the collection of signals transmitted from jth transmitter x_j. The matrix $\mathbf{H}$ is the MIMO channel with N_{BS} transmit antennas at the BS and n_{UE} total receive antennas. The UEs cannot cooperate for the decoding of the data streams. Therefore, each user has access to several elements of $\mathbf{y}$ not the whole vector. There are various ways of precoding at the T_X to eliminate interference at the UE, where the most common one is dirty paper code [10] requiring CSIT. The desired signal can be precoded in such a way that R_X does not see any interference in the presence of interference knowledge. The dirty paper analogy relates the interference in a communication channel to dirt that is present on a piece of paper. The signal is the ink, which is chosen based on interference, dirt.

The model Eq. (9.7) assumes that the transmission is over flat-fading or narrowband channel. As a complementary waveform to MIMO design, orthogonal frequency division multiplexing (OFDM), which is discussed in Chapter 7, splits the wideband channel into many narrowband channels where MIMO processing can be implemented appropriately.

A simple way of dealing with IUI is by precoding the symbol vector $\mathbf{d}$ with the pseudo-inverse of the channel matrix as follows:

$$\mathbf{x} = \mathbf{H}^H(\mathbf{H}\mathbf{H}^H)^{-1}\mathbf{d}. \tag{9.8}$$

At the receiver, this approach results in $\mathbf{y} = \mathbf{d} + \mathbf{w}$. This technique is called channel inversion [11]. This method has a drawback that in case of power constraint, an ill-conditioned channel matrix when inverted will require a large normalization factor that will dramatically reduce the SNR at the receivers.

```
1  Nrx = 10; %rx antenna
2  Nuser = 10; %user number
3  Ntx = 10; %tx antenna
4  SNR = -2:10; %dB
5  N_data = Nrx; %symbol length
6  M = 4; %modulation order
7  for ii = 1: 1000 %MONTECARLO
8      for jj = 1: length(SNR)
9          snr=SNR(jj); noiseP=1; sPow=10^(snr/10);
10         bits = randi([0 1],N_data*log2(M),1);
11         d = qammod(bits,M,'InputType','bit',...
12         'UnitAveragePower',true);
13         H = sqrt(1/2)*(randn(Nrx,Ntx)+1i*randn(Nrx,Ntx));
14         x = pinv(H)*d; %Channel Inversion
15         x = sqrt(sPow/mean(abs(x).^2))*x; %normalize power
16         noise = sqrt(noiseP/2).*(randn(Nrx,1)+1i*randn(Nrx,1));
17         x2 = H'*inv(H*H'+(Nuser/sPow)*...
18         eye(size(H,1)))*d; %Regularized
19         x2 = sqrt(sPow/mean(abs(x2).^2))*x2; %normalize
20         y = H*x + noise;
21         y2 = H*x2 + noise;
22         dS = qamdemod(y,M,'OutputType','bit',...
23         'UnitAveragePower',true);
24         dS2 = qamdemod(y2,M,'OutputType','bit',...
25         'UnitAveragePower',true);
26         ber(jj,ii) = mean(abs(dS-bits));
27         ber2(jj,ii) = mean(abs(dS2-bits));
28     end
29 end
30 BER = mean(ber,2);
31 BER2 = mean(ber2,2);
32 figure(1);semilogy(SNR,BER);
33 hold on;semilogy(SNR,BER2);
```

If it is assumed that white noise exists with power constraint P, then the regularized inversion for downlink will be as follows:

$$\mathbf{x} = \mathbf{H}^H(\mathbf{H}\mathbf{H}^H + K/P\mathbf{I})^{-1}\mathbf{d}, \tag{9.9}$$

where the loading factor K/P maximizes the SINR at the receiver. The BER performance of both two techniques is demonstrated in Figure 9.12. Both methods are designed to achieve the SINR value that is identical for each user. However, in the case of dense cells with many users, each user may have different design parameters leading to different interference ratios. Therefore, a more sophisticated technique needs to be used to design multiuser MIMO systems. Readers interested in the advanced implementations of multiuser MIMO regarding user's requirements may look at the references [5, 11–14].

9.3 Non-orthogonal Design

In view of emerging applications such as the Internet of Things (IoT), and in order to fulfill the need for massive numbers of connections with diverse requirements in terms of latency and throughput, 5G and beyond cellular networks are experiencing a paradigm shift in design philosophy: shifting from orthogonal to non-orthogonal design in waveform and multiple access. It is well known that

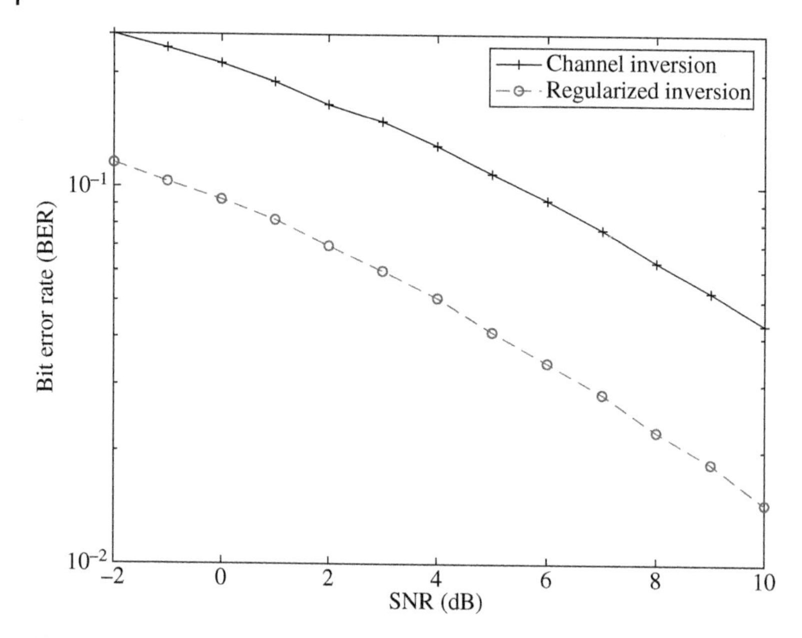

Figure 9.12 Uncoded probability of error for channel inversion and regularized inversion.

the number of served users in all orthogonal multiple access (OMA) techniques is inherently limited by the number of resources, i.e. the number of spreading orthogonal codes in CDMA and the number of resource blocks in OFDMA [3]. However, the concept of non-orthogonal multiple access (NOMA) has been proposed to overcome the limitation of orthogonal resources.

9.3.1 Power-domain Non-orthogonal Multiple Access (PD-NOMA)

Power-domain NOMA exploits the channel gain differences between the users for multiplexing via power allocation [3]. To illustrate the capacity region of NOMA systems, consider a single-cell network with two users, as depicted in Figure 9.13. The capacity regions of OMA, power-controlled OMA, and NOMA are compared in Figure 9.14. For the OMA model, TDMA is considered. The time

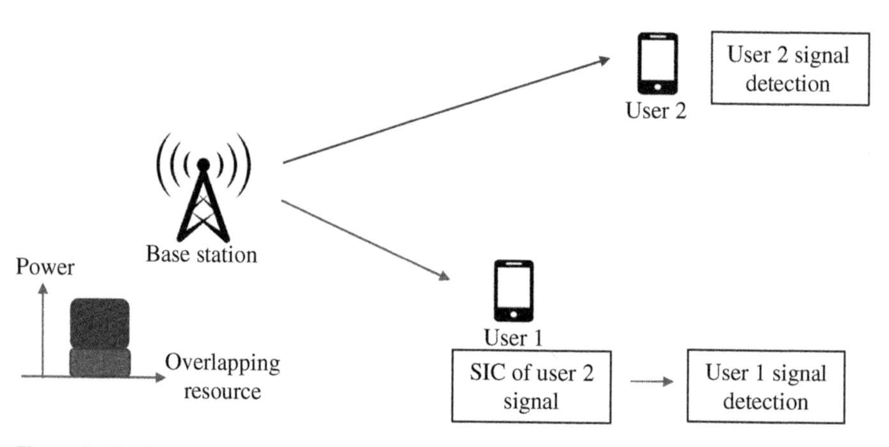

Figure 9.13 Power-domain NOMA.

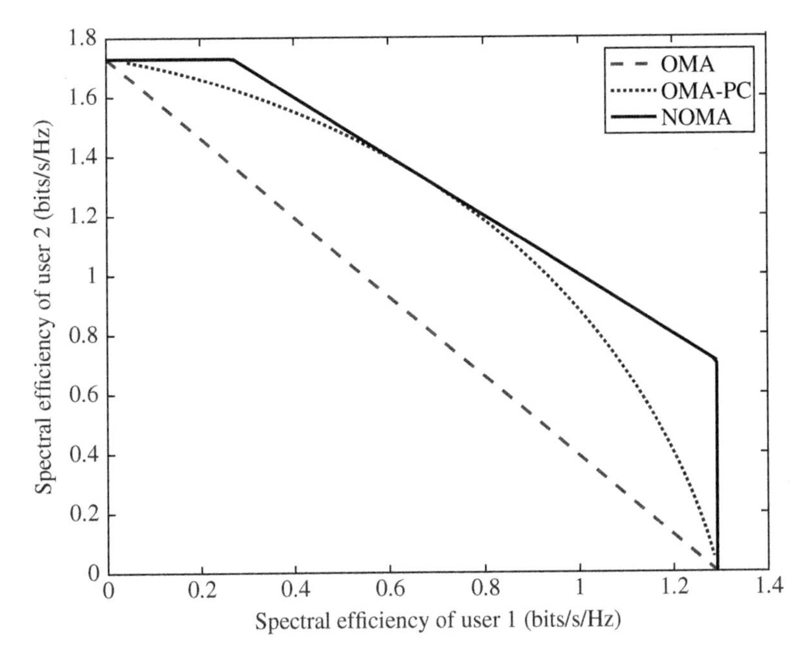

Figure 9.14 Best achievable capacity regions in two user uplink for different multiple access schemes with $SNR_1 = 6.9897$ dB and $SNR_2 = 10$ dB.

slot is divided by two fractions α and β where $\alpha + \beta = 1$. In OMA scenario, power control of user 1 and 2 become as follows:

$$R_1 = \alpha \log_2(1 + |h_1|P_1), \tag{9.10a}$$

$$R_2 = \beta \log_2(1 + |h_2|P_2). \tag{9.10b}$$

With the power control mechanism, R_1 and R_2 become

$$R_1 = \alpha \log_2(1 + \frac{|h_1|P_1}{\alpha}), \tag{9.11a}$$

$$R_2 = \beta \log_2(1 + \frac{|h_2|P_2}{\beta}). \tag{9.11b}$$

In NOMA, capacity regions are bounded as

$$R_1 \leq \log_2(1 + |h_1|P_1), \tag{9.12a}$$

$$R_2 \leq \log_2(1 + |h_2|P_2), \tag{9.12b}$$

$$R_1 + R_2 \leq \log_2(1 + |h_1|P_1 + |h_2|P_2), \tag{9.12c}$$

where these bounds are obtained with the assumption of perfect successive interference cancellation (SIC), where the strong received signal is totally removed from the superimposed signal to decode the other user's signal.

9.3.2 Code-domain Non-orthogonal Multiple Access

Code-domain NOMA uses user-specific sequences for sharing the entire radio resource [3]. In the literature, there are several different techniques regarding code domain NOMA, such as sparse code multiple access (SCMA) [15], interleave division multiple access (IDMA) [16], pattern division multiple access (PDMA) [17], low density spreading multiple access (LDS-MA) [18]. In this section, the most common one, which is SCMA, will be investigated. In this scheme, the encoding operations, for both the uplink and downlink, involve replacing the quadrature amplitude modulation (QAM) modulator with an SCMA modulator, which maps the coded bits directly to the multidimensional codeword from the standardized codebooks. The decoding operations replace the single-user channel equalization and QAM de-mapper of existing LTE receiver with an SCMA demodulator that jointly detects the superposed data layers and output separate log-likelihood ratio (LLR) results to the turbo decoders of each layer [15].

```
1  timeFrac = linspace(0,1,100);
2  fSNR = 5; %linear SNR
3  sSNR = 10; %linear SNR
4  %OMA
5  R1 = timeFrac*(1/2*log2(1+fSNR));
6  R2 = (1-timeFrac)*(1/2*log2(1+sSNR));
7
8  figure(); plot(R1,R2,'r')
9  hold on
10 %OMA with power control
11 R1_pc = timeFrac.*(1/2*log2(1+fSNR./timeFrac));
12 R2_pc = (1-timeFrac).*(1/2*log2(1+sSNR./(1-timeFrac)));
13 plot(R1_pc,R2_pc,'b')
14 %NOMA
15 R1_R2 = ones(1,numel(timeFrac))*(1/2*log2(1+sSNR+fSNR));
16 R1_noma = linspace(0,(1/2*log2(1+fSNR)));
17 R2_noma = min(R1_R2-R1_noma,(1/2*log2(1+sSNR)));
18 R2_nomaRev = linspace(0,(1/2*log2(1+sSNR)));
19 R1_nomaRev = min(R1_R2-R2_nomaRev,(1/2*log2(1+fSNR)));
20 plot(R1_noma,R2_noma,'k'); plot(R1_nomaRev,R2_nomaRev,'k')
21 legend('OMA','OMA-PC','NOMA')
```

SCMA is capable of supporting overloaded access over the coding domain, hence increasing the overall rate and connectivity. By carefully designing the codebook and multidimensional modulation constellations, the coding and shaping gain can be obtained simultaneously. In an SCMA system, users occupy the same resource blocks in a low-density way, which allows affordable low multiuser joint detection complexity at the receiver. The sparsity of signal guarantees a small collision even for a large number of concurrent users, and the spread-coding like code design brings good coverage and anti-interference capability due to spreading gain as well [3].

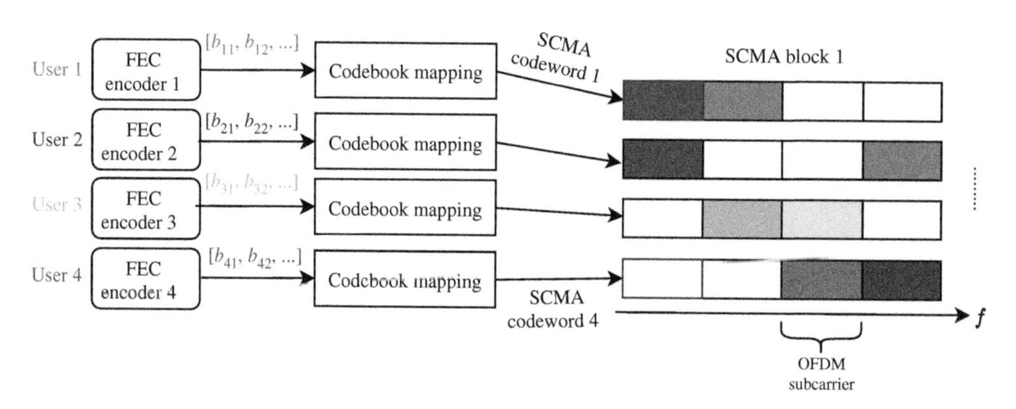

Figure 9.15 SCMA system model.

An SCMA transmitter system model can be illustrated in Figure 9.15 where K synchronous users multiplexing over N share orthogonal resources. Each SCMA codebook maps the coded bits $b_{k,n}$ to a N-dimensional complex codeword. Generated codewords are superimposed over the air transmission in uplink or at the transmitter in downlink, constituting an SCMA block. This multiple access process is similar to that of CDMA, where the spread signals are replaced with the SCMA codewords. Multiuser detection is carried out at the receiver to recover the colliding codewords. Beside, SCMA uses sparse spreading to reduce interlayer interference, so that more codewords collisions can be tolerated with low receiver complexity [3].

In order to strike a good balance between link performance (close to maximum-likelihood detection and robust to channel imperfection) and implementation complexity (the complexity of the receiver), the features of SIC with message-passing algorithm (MPA) can be combined to have a SIC-MPA receiver which developed to reduce the complexity of conventional MPA receiver. Specifically, MPA is first applied to a limited number of layers, so that the number of colliding layers over each resource element does not exceed a given threshold value. Then, the successfully decoded MPA layers are removed by SIC, and the procedure continues until all layers are successfully decoded, or no new data layer gets successfully decoded by MPA. Due to the fact that MPA is used for a limited number of layers instead of all the layers, the decoding complexity is greatly reduced [19].

9.4 Random Access

Most data applications do not require continuous transmission. Actually, the data of users are generated at random time instances so that dedicated channel assignment can be extremely inefficient. Moreover, most systems have many more total users consisting of active or idle users than can be accommodated simultaneously, so at any given time, channels can only be allocated to users that need them. Random access strategies are used in such systems to use available channels efficiently.

If packets from different users overlap in time, the collision occurs, in which case both packets may be decoded unsuccessfully. Packets may also be decoded in error as a result of noise or other channel impairments. The probability of a packet decoding error is called the packet error rate. The performance of random access techniques is typically characterized by throughput T of the system. The throughput, which is unitless, is defined as the ratio of the average rate of packets successfully transmitted divided by the channel packet rate $R_p = R/N$, where N is the number of bits in one packet and R is the channel data rate in bits per second [2]. Here, some random access techniques are explained.

9.4.1 ALOHA

There are two types of ALOHA-based random access. Pure (unslotted) ALOHA is asynchronous, where users transmit data packets as soon as they are formed. It has low throughput under heavy loads where the maximum throughput is 18% of incoming packets. Slotted ALOHA is synchronous where time is assumed to be slotted in time slots of duration τ. Users can only start their packet transmissions at the beginning of the next time slot after the packet has formed, leading to no partial overlap of transmitted packets. Its maximum throughput increases to 36%. It is used in GSM to reserve a time slot for a voice connection. ALOHA is preferred for wide-area applications. The transmission pattern of the ALOHA can be seen in Figure 9.16.

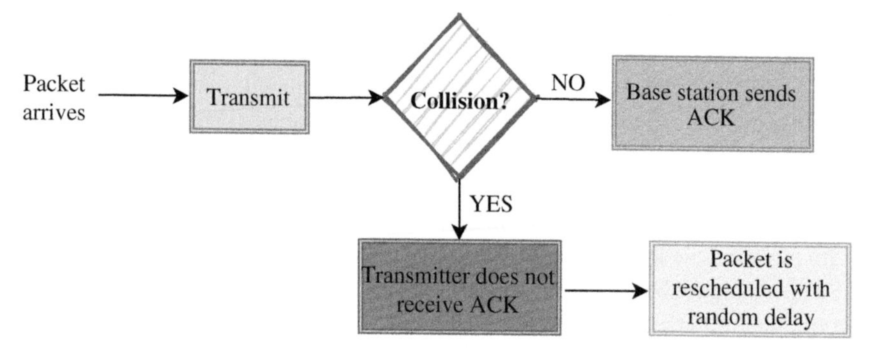

Figure 9.16 ALOHA based random access.

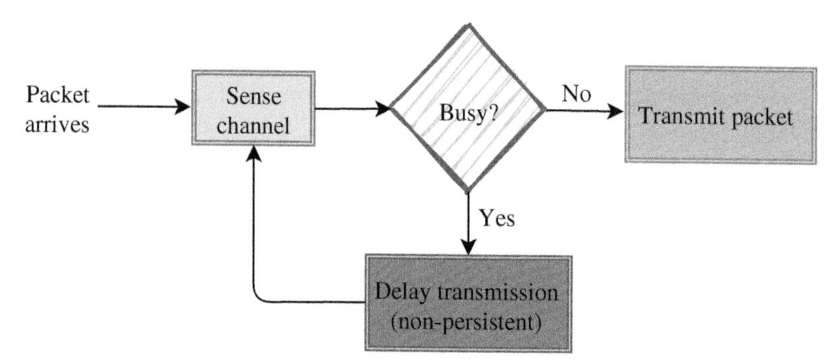

Figure 9.17 CSMA based random access.

ALOHA is extremely inefficient owing to collisions between users, which leads to very low throughput. That's why different random access techniques are investigated and used in modern wireless systems.

9.4.2 Carrier Sense Multiple Accessing (CSMA)

Carrier sense multiple accessing (CSMA) is a protocol called listen before the talk. Collisions can be reduced by CSMA, where users sense the channel and delay transmission if they detect that another user is currently transmitting. Still, collisions may occur if transmitters cannot sense the other transmission due to large propagation delay. There is a lower probability of collision and higher throughput than ALOHA. The transmission pattern of the CSMA can be seen in Figure 9.17.

There are two types of carrier sensing, which are nonpersistent and persistent. In nonpersistent carrier sensing, the terminal senses the channel after a random waiting period once it senses a busy channel. In the persistent case, the terminal senses the channel until the channel becomes free. Once it is detected that the channel is free, the terminal can transmit immediately or randomly with probability p_t. Still, CSMA has some issues such as the hidden terminal problem and exposed terminal problem due to the nature of wireless channels. More explanations about these problems can be found in [2].

9.4.3 Multiple Access Collision Avoidance (MACA)

In multiple access collision avoidance (MACA), a sender transmits an request to send (RTS) message to its intended receiver before the data transmission. The data is transmitted only after the

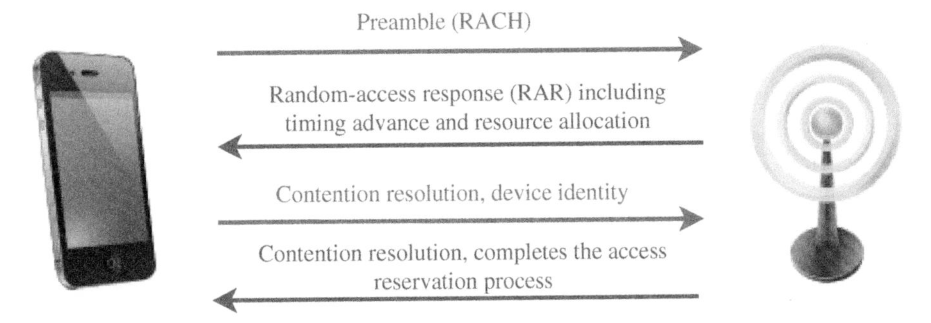

Figure 9.18 Four step RACH procedure.

reception of a clear to send (CTS) message from the receiver, which is sent after the reception of successful RTS. MACA also requires that any user that overhears a RTS or CTS packet directed elsewhere inhibits its transmitter for a specified time to avoid the collision. Its use is particularly useful when large message sizes are to be transmitted, but less interesting when small packets (comparable to the RTS/CTS packets) are to be transmitted [20]. MACA introduces overhead on the channel with RTS and CTS packet transmissions.

9.4.4 Random Access Channel (RACH)

The random access channel (RACH) is adopted in the standardization of both 3G, 4G-LTE, and 5G-NR that is used for dynamic reservation of resources on the uplink [21]. It is the first message transmitted by UE when the power of UE is turned on. Before UE decides to send a RACH signal, there are many preconditions to be met, as described in follows. Users first randomly transmit short packet preambles and then wait for a positive acquisition indicator from the base station prior to the transmission of the complete message [3]. The random access procedure consists of a four-message handshake between the UE and the BS, which can be seen in Figure 9.18.

The aim of the third and fourth messages to resolve the potential collisions due to simultaneous transmissions of the same preamble from multiple devices within the cell.

9.4.5 Grant-free Random Access

In order to meet the strict demands of next-generation wireless systems such as low latency and massive connectivity, the contention-based grant-free (GF) random access can be considered as a promising solution for uplink transmission. In this scheme, a UE is able to send a packet immediately upon traffic arrival without needing a scheduling request. It is different from the grant-based (GB) uplink transmission in the existing wireless standards, where a UE has to send a scheduling request upon traffic arrival and get a dynamically scheduled uplink grant from the BS before the UE starts the uplink data transmission. Figure 9.19 illustrates the GF contention-based uplink transmission scheme with two users.

Hence, the GF multiple access scheme addresses high connectivity, low latency, signaling overhead reduction, and UE energy-saving issues. Therefore, it is applicable to the small packet transmission and ultra-reliable, low-latency communication (URLLC) in 5G and IoT networks. Detailed explanations can be found in [3, 22].

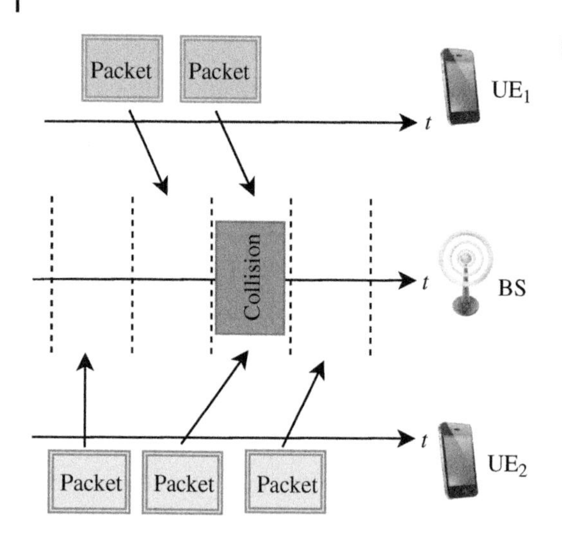

Figure 9.19 GF contention based transmission.

9.5 Multiple Accessing with Application-Based Hybrid Waveform Design

Here various waveform structures to serve multiple applications are explained. In Section 9.5.1, different lattice designs leading to the multi-numerology OFDM concept are investigated. The main aim of having various lattice structures in time and frequency domain is to serve a different kind of applications with various requirements. On the other hand, a novel concept called radar-sensing and communication convergence is examined by the perspective of the hybrid waveforms concept in Section 9.5.2. Moreover, Section 9.5.2 introduces the coexistence of different waveforms assigned to each user in the same coverage area for 6G and beyond wireless systems.

9.5.1 Multi-numerology Orthogonal Frequency Division Multiple Access (OFDMA)

5G NR is designed to operate from 450 MHz to 100 GHz with a wide range of deployment scenarios, while supporting a variety of services. Even single-numerology OFDMA was chosen as the multiple-access scheme for the downlink and uplink for previous 3GPP standards; it is not possible for single numerology to satisfy the requirements of various use cases. Therefore, NR defines a family of OFDM numerologies for various frequency bands and deployment scenarios. OFDM numerology is characterized by a subcarrier spacing and a cyclic prefix (CP). The requirements for the OFDM subcarrier spacing is determined based on the carrier frequency, phase noise, delay spread, and Doppler spread. The use of smaller subcarrier spacing would result in either large error vector magnitude (EVM) due to phase noise or more stringent requirements on the local oscillator. The small subcarrier spacing further leads to performance degradation in high Doppler scenarios. The required CP overhead and thus anticipated delay spread sets an upper limit for the subcarrier spacing. A large subcarrier spacing would result in unwanted overhead due to CP. The maximum fast Fourier transform (FFT) size of the OFDM modulation, along with subcarrier spacing, determines the channel bandwidth. Based on these observations, the subcarrier spacing should be as small as possible, while the system is still robust against phase noise, and Doppler spread and supports the desired channel bandwidth. From the network perspective, multiplexing of different numerologies over the same NR carrier bandwidth is possible in TDMA and/or FDMA manner in

Table 9.2 Supported OFDM parameters in 5G-NR.

μ	Subcarrier spacing (kHz) $\Delta f = 2^{\mu} \times 15$	Cyclic prefix length μs	OFDM symbol length μs
0	15	4.69	71.35
1	30	2.34	35.68
2	60	1.17	17.84
3	120	0.57	8.92
4	240	0.29	4.46

Source: Ahmadi [23].

the downlink and uplink. From the UE perspective, multiplexing of different numerologies is performed in TDMA and/or FDMA manner within or across a subframe. Regardless of the numerology used, the lengths of the radio frame and subframe are always 10 and 1 ms, respectively [23]. Having a flexible OFDMA system is crucial to deploy a wide variety of 5G services efficiently [24]. Therefore, the novel 5G standard introduces flexible OFDM to support various spectral deployments. The numerologies in the 5G standard can be seen in Table 9.2. Specifically, the sub-carrier spacing (SCS) options are 15, 30, 60, 120, and 240 kHz. Also, the length of CP varies according to the application requirements [21].

```
1  %multi-numerology OFDMA
2  Q                     = 2; % scs difference
3  N                     = [2048 2048/Q]; %Number of FFT points
4  CP                    = [144 144/Q]; %CP length
5  num_OFDM_symbols      = 2; %OFDM symbol count
6  scs                   = [30e3 Q*30e3];   % Subcarrier Spacing
7  sampleRate            = N(1)*scs(1); %sample rate
8  totalBW               = 40e6; % total allocated bandwidth
9  BWusers               = [18e6 18e6]; %allocated bandwidth for each user
10 gBW                   = 4e6; %guard band
11 gSc                   = floor(gBW/scs(2)); %guard subcarriers
12 N_d                   = floor(BWusers./scs); %Data subcarriers
13 M                     = 4; %QPSK
14 used_subc_indices     = {100:N_d(1)+99,...
15   50+N_d(2)+gSc:50+N_d(2)+gSc+N_d(2)-1}; %data subcarriers
16 wL                    = 100; %window length
17
18 for jj = 1:100 % Montecarlo for smooth plot
19     for ii = 1:2
20         cp_add=sparse(1:CP(ii)+N(ii),[N(ii)-CP(ii)+1:N(ii)...
21           ,1:N(ii)],ones(1,CP(ii)+N(ii)),CP(ii)+N(ii),N(ii)); %cp addition
22         ssc  = zeros(N(ii),1); %subcarriers
23         bits = randi([0 1],N_d(ii)*log2(M),1); % bit generation
24         dataNorm = qammod(bits,M,'InputType','bit',...
25              'UnitAveragePower',true); %symbol generation
26         ssc(used_subc_indices{ii})  = dataNorm; % map symbols to subcarriers
27         time  = (N(ii))/sqrt(N_d(ii))*ifft(ssc); % time-domain signal
28         time_cp{ii} = cp_add*time; %cp addition
29     end
30     frameOFDM(jj,:)=time_cp{1}+[time_cp{2};...
31        zeros(CP(1)-CP(2)+N(2),1)]; %frame design
32     [ss(:,:,jj),f,t]= spectrogram(frameOFDM(jj,:),...
33         ones(wL,1)/sqrt(wL),0,N(1),sampleRate,'yaxis'); %spectogram
34 end
35 summedSpec =  mean(abs(ss),3);
36 imagesc(t/1e-6,f/1e6,10*log10(abs(summedSpec).^2)); %plotting
```

Examples for the aim of using multi-numerology OFDMA can be given by depending on propagation characteristics. For example, it is expected that the lower frequency bands will be used for large-area deployments with smaller SCS and associated larger subframe time duration. In comparison, higher frequency bands are expected to be used for the dense deployments with larger SCS and their associated smaller subframe time duration. Moreover, lower SCS is more suitable for massive machine type communication (mMTC), since they can support a higher number of simultaneously connected devices within the same bandwidth and require lower power, intermediate SCS are appropriate for enhance mobile broadband (eMBB), which requires both high data rate and significant bandwidth, and higher numerologies are more suitable for delay-sensitive applications pertaining to the URLLC service due to their shorter symbol duration [25]. Three different lattice structures consisting of multi-numerology OFDMA depending on the requirements of applications are shown in Figure 9.20. Though multi-numerology OFDMA is efficient in providing the required flexibility, this approach introduces a new kind of interference into the system known as inter-numerology interference (INI) [25]. Figure 9.21 depicts two different OFDMA structures with different SCSs. Since their SCS are different, the time duration that the OFDM signal allocates is also different.

9.5.2 Radar-Sensing and Communication (RSC) Coexistence

In contrast to 5G and earlier generations, the next generation of wireless systems requires environmental awareness to leverage adjustable radio parameters into communication. Therefore, radar-sensing systems will be integrated into the wireless networks to cope with specific applications and use cases such as autonomous vehicles and environment-aware access points. Also, it is expected that this new paradigm supports flexible and seamless connectivity [26]. A key-enabler of high-mobility networks is the ability of a node to track its dynamically changing environment continuously and react accordingly. Although state sensing and communication have been designed separately in the past, power, spectral efficiency, hardware costs encourage the integration of these two functions, such that they are operated by sharing the same frequency

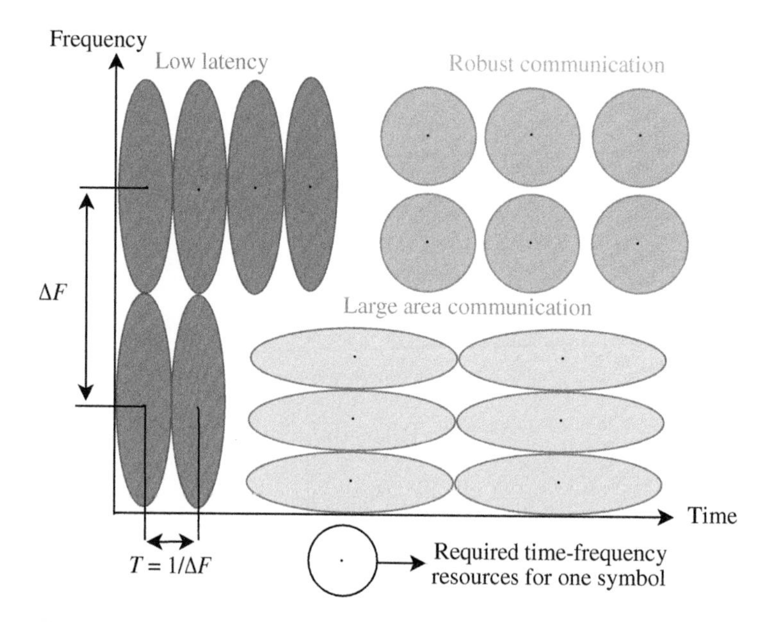

Figure 9.20 Multi-numerology OFDM depending on application requirements.

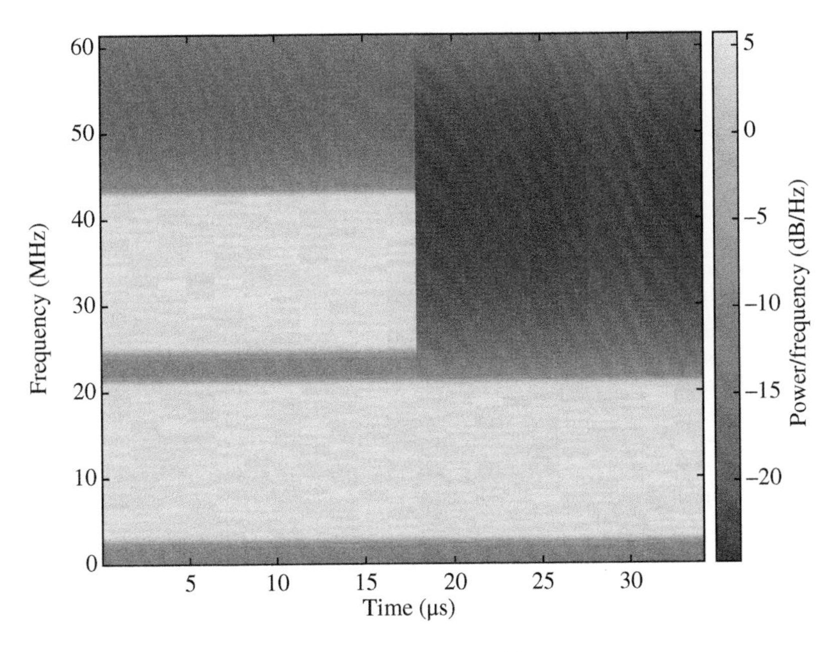

Figure 9.21 Spectogram of Multi-numerology OFDMA.

band and hardware [27]. Figure 9.22 depicts the possible use cases of this appealing technology in the environment.

Table 9.3 categorizes the existing techniques pointing joint radar-sensing and communication use of wireless systems and technologies. It explains how to utilize sources such as time, frequency, code, and polarization. More details can be found in the recent survey [28]. Existing works on joint radar and communication can be roughly classified into three classes. The first class considers some

Figure 9.22 Realization of radar-sensing and communication (RSC).

Table 9.3 Existing techniques used in joint radar-sensing and communication.

System	Feature	Advantage	Disadvantage
RCS with separated waveforms	RCS signals are separated in the radio resources where hardware and software architectures are partially shared.	• Small inter application interference. • Almost independent design for the waveforms.	• Low spectrum efficiency. • Low order of integration. • Complex hardware.
Coexisting RCS	This system uses separated signals but share the same spectrum.	• Higher spectrum efficiency.	• Inter waveform interference. • Cooperation and complicated signal processing.
Joint RCS	A joint transmitted signal design is utilized in RCS.	• Fully shared transmitter. • Exchange information to support of each function. • Coherent sensing.	• Requirement for full-duplex capable transceiver. • Joint optimization of waveform.
Passive sensing	Received RF signal is used for sensing at a particularly designed receiver using extensive radar signal processing techniques.	• Without requiring any change to existing structure. • Higher spectrum efficiency.	• Tough synchronization. • Limited sensing capability when the waveform is optimized for communication.

resource-sharing approach, such that time, frequency, or space resources are split into either radar or data communication. The second class uses a common waveform for both radar and communication. This approach includes information-embedded radar waveforms as well as the direct usage of standard communication waveforms applied to radar detection [29]. Also, as a third technique, the authors propose the superimposition of two different waveforms, which are frequency modulated continuous wave (FMCW) and OFDM, to serve both purposes [30]. Details about how to separate waveforms and how to exchange information between radar and communication to improve the performance can be found in the paper.

To sum up, the use of higher frequency ranges, wider bandwidths, and massive antenna arrays leads to enhanced sensing solutions with very fine range, Doppler and angular resolutions, as well as localization to a cm-level degree of accuracy. Moreover, new materials, device types, and reconfigurable intelligent surface (RIS) will allow network operators to reshape and control the electromagnetic response of the environment. At the same time, machine learning and artificial intelligence will leverage the unprecedented availability of data and computing resources to tackle the biggest and hardest problems in both wireless communication and radar-sensing systems. Therefore, the convergence of these two applications will continue to create new research areas and innovations.

9.5.3 Coexistence of Different Waveforms in Multidimensional Hyperspace for 6G and Beyond Networks

Waveform design is one of the core components of the physical layer in wireless communication systems. Basically, waveform can be defined as a physical signal that contains information. These

Table 9.4 Fundamental points on the waveform design.

Parameter	Feature
Data symbols	They are a set of complex numbers representing information bits.
Redundant symbols	They can be utilized for precoding, guard utilization and artificial noise generation.
Lattice structure	It represents locations of samples in hyperspace. It is a multidimensional resource mapping and each mapped sample shows a location of one resource element.
Pulse shape	The form of symbols in the signal plane is defined by the pulse-shaping filters. The shape of filters determines how the energy is spread over the multidimensional hyperspace.
Frame structure	It can be defined as a packaging of multiple user information because the waveform design is the process of generating the collective physical signal, which occupies the hyperspace, corresponding to multiple users.

signals occupy physical resources in multidimensional hyperspace consisting of time, frequency, space, code, power, and beam. The main components of the waveform design procedure are shown in Table 9.4, including data and redundant symbols, lattice structure, pulse shape, and frame structure. Waveform designs employ various parameters under these main components. The detailed explanation on waveform design and its parameters can be found in [31–33].

Possible spacings between lattice points are defined by the numerology structure of a waveform. Numerology includes a set of parameters for a specific lattice structure of a waveform and 5G NR is standardized based on multiple numerologies of CP-OFDM on the time-frequency plane. For 5G NR, parameters that define numerology type of the waveform are subcarrier spacing, CP duration, inter-numerology guard band, roll-off factor, filter coefficients, slot duration, number of symbols in one slot, number of slots per subframe, and frame length.

For 6G and beyond wireless system studies, coexistence of different waveforms and numerologies attracts researchers' attention in order to meet the potential future requirements of networks flexibly. One of the example architecture can be shown in Figure 9.23 where each user utilizes different parameters that define its waveform and related numerology. The assignment of waveform parameters for each user is done at transmission point (TP) considering the user feedback and the other information acquired in different network layers considering latency, reliability, security, and privacy metrics. However, multiplexing different waveforms may cause performance degradation phenomenon, including new forms of interference such as INI and inter-waveform interference (IWI), scheduling complexity, and signaling overhead. Therefore, the state-of-the-art optimization mechanisms need to be developed to compensate or exploit the adverse effects on multiplexing of different waveforms concept in 6G and beyond wireless networks.

9.6 Case Study

Here the real-time multiple access scenario will be described with relative requirements for different users. For the computer simulations or hands-on implementations, the procedures described below may be taken into consideration. Saying that there are three different users in a wireless cell, the design constraints and requirements can be listed as follows:

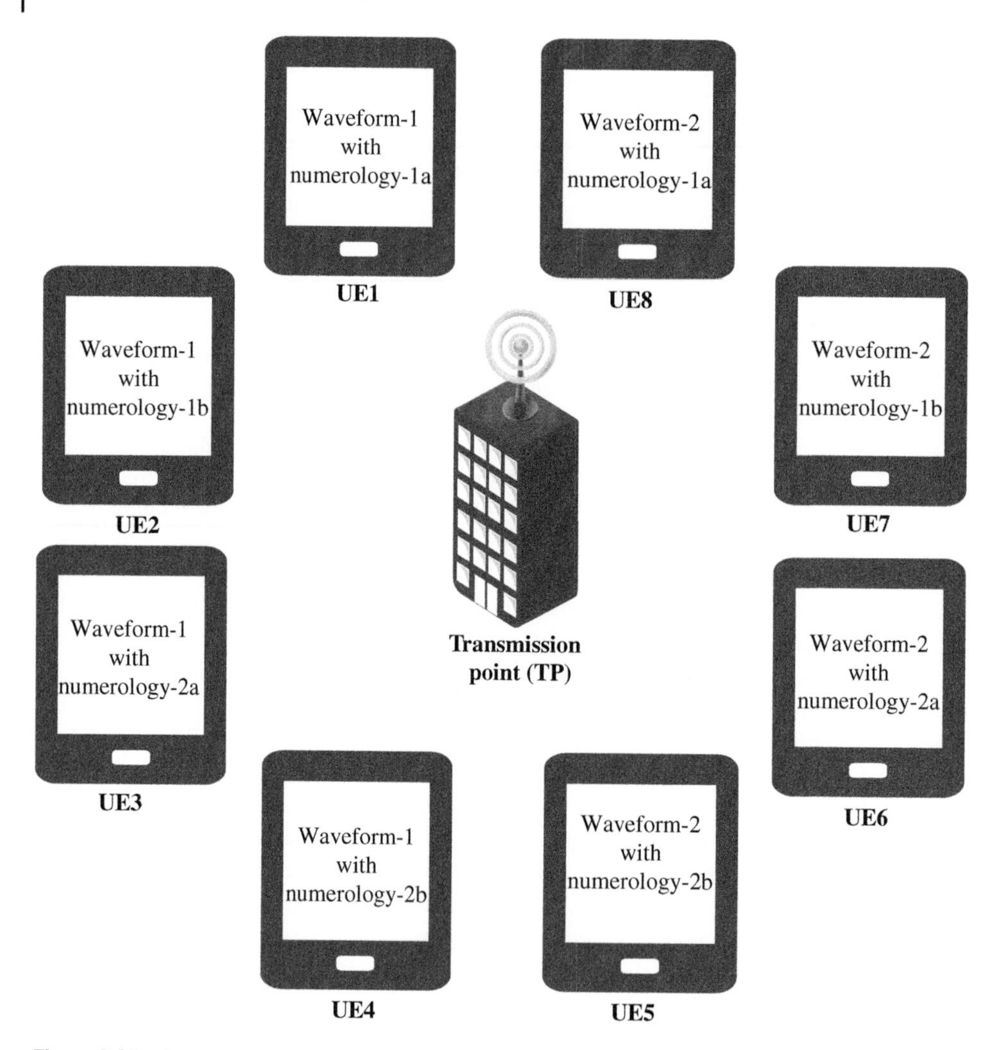

Figure 9.23 Different waveform and numerology assignments for each user in the same coverage area.

- The aim is to have a maximum achievable rate considering requirements of users.
- The total allocated bandwidth is 30 MHz.
- The transmission should be investigated burst by burst.
- The frame length should not be longer than 1 ms.
- Different waveforms can be used in the transmission.
- Uplink or downlink transmission can be considered.
- Transmission can be orthogonal or non-orthogonal.

One of the examples for the case study is given in Figure 9.24. It is based on non-orthogonal multiple access, including two different waveforms OFDM and OFDM-index modulation (IM). Details about the scheme and algorithms to extract the information embedded in waveforms can be found in [34].

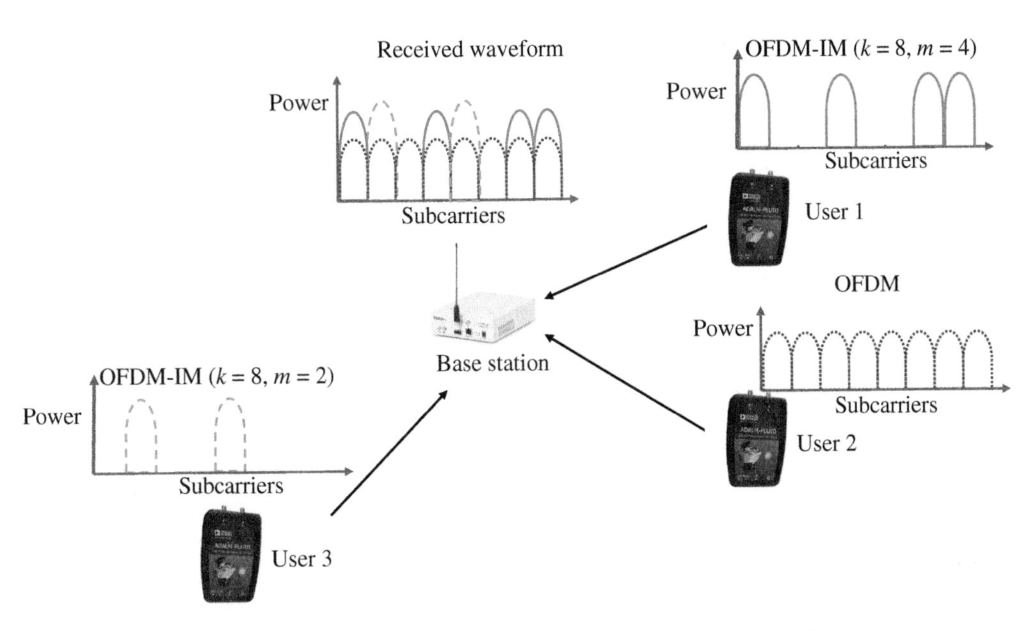

Figure 9.24 An example for case study.

Appendix: Erlang B table

		Probability of call blocking			
		0.01	0.015	0.02	0.03
Number of trunks	1	0.010	0.015	0.020	0.031
	2	0.153	0.190	0.223	0.282
	3	0.455	0.536	0.602	0.715
	4	0.870	0.992	1.092	1.259
	5	1.361	1.524	1.657	1.877
	6	1.913	2.114	2.277	2.544
	7	2.503	2.743	2.936	3.250
	8	3.129	3.405	3.627	3.987
	9	3.783	4.095	4.345	4.748
	10	4.462	4.808	5.084	5.529

References

1 D. Tse and P. Viswanath, *Fundamentals of Wireless Communication*. Cambridge: Cambridge University Press, 2013.

2 A. Goldsmith, *Wireless Communications*. USA: Cambridge University Press, 2005.

3 M. Vaezi, H. V. Poor, and Z. Ding, *Multiple Access Techniques for 5G Wireless Networks and Beyond*. Springer, 2019.

4 A. Sabharwal, P. Schniter, D. Guo, D. W. Bliss, S. Rangarajan, and R. Wichman, "In-band full-duplex wireless: challenges and opportunities," *IEEE Journal on Selected Areas in Communications*, vol. 32, no. 9, pp. 1637–1652, 2014.
5 A. F. Molisch, *Wireless Communications*. Wiley, 2014.
6 J. F. Shortle, J. M. Thompson, D. Gross, and C. M. Harris, *Fundamentals of Queueing Theory*. Wiley, 2018.
7 J. Navarro-Ortiz, P. Romero-Diaz, S. Sendra, P. Ameigeiras, J. J. Ramos-Munoz, and J. M. Lopez-Soler, "A Survey on 5G usage scenarios and traffic models," *IEEE Communications Surveys Tutorials*, vol. 22, no. 2, pp. 905–929, 2020.
8 T. S. Rappaport, *Wireless Communications: Principles and Practice*. Publishing House of Electronics Industry, 2012.
9 V. Sergio, *Multiuser Detection*. Cambridge University Press, 2011.
10 C. B. Peel, "On "dirty-paper coding"," *IEEE Signal Processing Magazine*, vol. 20, no. 3, pp. 112–113, 2003.
11 Q. H. Spencer, C. B. Peel, A. L. Swindlehurst, and M. Haardt, "An introduction to the multi-user MIMO downlink," *IEEE Communications Magazine*, vol. 42, no. 10, pp. 60–67, 2004.
12 M. Schubert and H. Boche, "Solution of the multiuser downlink beamforming problem with individual SINR constraints," *IEEE Transactions on Vehicular Technology*, vol. 53, no. 1, pp. 18–28, 2004.
13 Q. H. Spencer, A. L. Swindlehurst, and M. Haardt, "Zero-forcing methods for downlink spatial multiplexing in multiuser MIMO channels," *IEEE Transactions on Signal Processing*, vol. 52, no. 2, pp. 461–471, 2004.
14 G. Caire and S. Shamai, "On the achievable throughput of a multiantenna Gaussian broadcast channel," *IEEE Transactions on Information Theory*, vol. 49, no. 7, pp. 1691–1706, 2003.
15 R1-162155: Sparse Code Multiple Access (SCMA) for 5G Radio Transmission, Huawei, HiSilicon, 3GPP TSG RAN WG1 Meeting #84 (2016).
16 K. Kusume, G. Bauch, and W. Utschick, "IDMA vs. CDMA: analysis and comparison of two multiple access schemes," *IEEE Transactions on Wireless Communications*, vol. 11, no. 1, pp. 78–87, 2012.
17 S. Chen, B. Ren, Q. Gao, S. Kang, S. Sun, and K. Niu, "Pattern division multiple access–a novel nonorthogonal multiple access for fifth-generation radio networks," *IEEE Transactions on Vehicular Technology*, vol. 66, no. 4, pp. 3185–3196, 2017.
18 M. AL-Imari, M. A. Imran, R. Tafazolli, and D. Chen, "Performance evaluation of low density spreading multiple access," in *2012 8th International Wireless Communications and Mobile Computing Conference (IWCMC)*, Limassol, 27–31 Aug. 2012, pp. 383–388.
19 R1-166098: Discussion on the Feasibility of Advanced MU-Detector, Huawei, HiSilicon, 3GPP TSG RAN WG1 Meeting #86 (2016).
20 P. Karn, "MACA: a new channel access method for packet radio," in *Proceedings of the ARRL/CRRL Ameteur Radio and 9th Computer Networking Conference*, 22 Sept. 1990.
21 S. Parkvall, E. Dahlman, and S. Johan, *5G NR: The Next Generation Wireless Access Technology*. Academic Press, 2018.
22 S. Doğan, A. Tusha, and H. Arslan, "NOMA with index modulation for uplink URLLC through grant-free access," *IEEE Journal of Selected Topics in Signal Processing*, vol. 13, no. 6, pp. 1249–1257, 2019.
23 S. Ahmadi, *5G NR: architecture, technology, implementation, and operation of 3GPP new radio standards*. Academic Press, an imprint of Elsevier, 2019.

24 Y. Liu, X. Chen, Z. Zhong, B. Ai, D. Miao, Z. Zhao, J. Sun, Y. Teng, and H. Guan, "Waveform design for 5G networks: analysis and comparison," *IEEE Access*, vol. 5, pp. 19 282–19 292, 2017.

25 A. B. Kihero, M. S. J. Solaija, and H. Arslan, "Inter-numerology interference for beyond 5G," *IEEE Access*, vol. 7, pp. 146 512–146 523, 2019.

26 A. Bourdoux, A. N. Barreto, B. van Liempd, *et al.*, "6G White Paper on Localization and Sensing," *ArXiv*, vol. abs/2006.01779, 2020.

27 L. Zheng, M. Lops, Y. C. Eldar, and X. Wang, "Radar and communication coexistence: an overview: a review of recent methods," *IEEE Signal Processing Magazine*, vol. 36, no. 5, pp. 85–99, 2019.

28 M. L. Rahman, J. A. Zhang, K. Wu, X. Huang, Y. Guo, S. Chen, and J. Yuan, "Enabling Joint Communication and Radio Sensing in Mobile Networks – A Survey," *ArXiv*, vol. abs/2006.07559, 2020.

29 D. Ma, N. Shlezinger, T. Huang, Y. Liu, and Y. C. Eldar, "Joint radar-communications strategies for autonomous vehicles," arXiv preprint arXiv:1909.01729, 2019.

30 M. M. Şahin and H. Arslan, "Multi-functional coexistence of radar-sensing and communication waveforms," *arXiv preprint arXiv:2007.05753*, 2020.

31 A. Yazar and H. Arslan, "A waveform parameter assignment framework for 6g with the role of machine learning," *IEEE Open Journal of Vehicular Technology*, pp. 1–1, 2020.

32 A. F. Demir, M. Elkourdi, M. Ibrahim, and H. Arslan, "Waveform design for 5G and beyond," *arXiv preprint arXiv:1902.05999*, 2019.

33 A. Sahin, I. Guvenc, and H. Arslan, "A survey on multicarrier communications: Prototype filters, lattice structures, and implementation aspects," *IEEE Communications Surveys Tutorials*, vol. 16, no. 3, pp. 1312–1338, 2014.

34 M. M. Şahin and H. Arslan, "Waveform-domain NOMA: the future of multiple access," in *2020 IEEE International Conference on Communications Workshops (ICC Workshops)*, 7–11 June 2020, pp. 1–6.

10

Wireless Channel and Interference

Abuu B. Kihero[1], *Armed Tusha*[1], *and Hüseyin Arslan*[1,2]

[1]*Department of Electrical and Electronics Engineering, Istanbul Medipol University, Istanbul, Turkey*
[2]*Department of Electrical Engineering, University of South Florida, Tampa, FL, USA*

The performance of any communication system is, to a large extent, determined by the medium utilized. This medium is referred to as communication channel and it may take a form of optical fiber, wireless link, or even a hard disk drive for a computer. In general, communication channels can be divided into two groups: wired and wireless channels. Wired channel is formed when there exists a solid connection that specifies a path of information flow from transmitter to the receiver. Wireless channel, on the other hand, lacks such tangible connection between communicating terminals. Their open nature makes wireless channels highly susceptible to noise, interference and other unpredictable time-varying impairments because of user mobility. Additionally, the behavior of wireless channels strongly depends on the propagation environment in which the communication is taking place. Different environments, such as urban, suburban, and indoor, respond differently to the signal propagating through them. Consequently, wireless channels pose severe challenge as a medium for a reliable communication. Therefore, successful design and optimization of a wireless communication system relies on the proper understanding of channel characteristics and their subsequent impact on the communication signal. To this end, this chapter sheds light on various critical aspects of wireless channel starting with explanation of the dominant propagation phenomena and their resultant effects on the system. The chapter then touches upon mechanisms employed for empirical measurement of these channel effects, as well as techniques commonly used to represent them in a way that is useful for system design (i.e. channel modeling). Finally, basic channel emulation techniques which are used to reproduce various propagation effects in the laboratory environment for system testing are also explained.

10.1 Fundamental Propagation Phenomena

Wireless propagation environments have varying degrees of complexity. In the simplest scenario, one may assume a free space between transmitting and receiving antennas such that only a line-of-sight (LOS) propagation path is present between them, as shown in Figure 10.1a. In such LOS propagation scenarios, the only impairment encountered by a communication signal is the path loss (i.e. attenuation of the signal power) due to atmospheric effects such as absorption. Path loss effect in the free space is a function of distance of separation between the transceiver nodes, and it is governed by the well-known Friis equation

$$P_r(d) = P_t \left[\frac{\sqrt{G_t G_r}\lambda}{4\pi d \sqrt{\xi}} \right]^2, \tag{10.1}$$

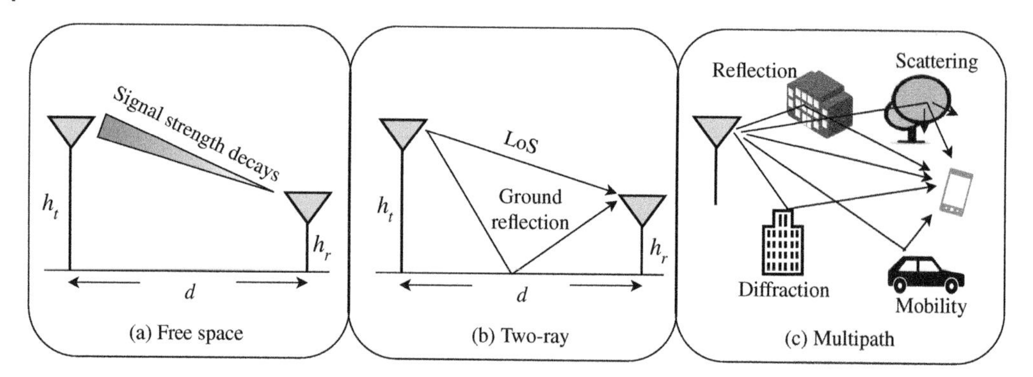

Figure 10.1 Propagation scenarios with different complexity levels considered in wireless communication.

where P_r is the signal power received by a receiver (Rx) at distance d from a transmitter (Tx). P_t and $G_{(\cdot)}$, are power of the transmitted signal and antenna gain, respectively. Parameter ξ models losses due to system's hardware imperfections and it is set to unity when a lossless system is assumed [1]. $\xi \geq 1$ implies that there are extra losses due to effects such as filters and antenna losses. The presence of the signal wavelength λ in Eq. (10.1) implies the dependence of the path loss effect on the frequency band of operation. For example, measurements have shown that, atmospheric absorption can cause up to 20 dB/km loss at millimeter wave (mmWave) frequency band compared to the 10^{-3} dB/km loss incurred at microwave band [2].

In the traditional communication systems that take place closer to the earth surface the assumption of single direct signal path between Tx and Rx as considered in the free space scenario rarely holds. Consider a typical propagation scenario given in Figure 10.1b. This scenario is better represented by a two-ray ground reflection model that extends the Friis equation to incorporate the effect of the ground reflected path. The modified Friis equation that represents the two-ray model is given as:

$$P_r(d) \approx \left[\frac{\sqrt{G_t G_r}\lambda}{4\pi d}\right]^2 \left[\frac{4\pi h_t h_r}{\lambda d}\right]^2 P_t = \left[\frac{\sqrt{G_t G_r} h_t h_r}{d^2}\right]^2 P_t, \tag{10.2}$$

where h_t and h_r are heights of the transmitting and receiving antennas. Compared to the conventional free space model, this model has been proved to be reasonably accurate for predicting received signal strength over large distances [1].

Now, if we consider a more realistic scenario in which the propagation environment is comprised of various objects, as shown in Figure 10.1c, even the two-ray model does not suffice to capture all important aspects of the signal propagation. In this case, LOS path propagation hardly exists. This is mainly due to severe signal obstructions by buildings, mountains, and foliage. In such circumstances, wireless coverage relies on non LOS (NLOS) propagation. Propagation mechanisms in NLOS scenarios are diverse but can be generally attributed to *reflection*, *scattering*, and *diffraction*, as shown Figure 10.1c:

- *Reflection* occurs when a signal hits an object whose surface irregularities have dimensions larger than signal's wavelength. Reflection is always accompanied by refraction (transmission of a signal through mediums of different densities). Strength of the reflected and refracted signal depends on the electromagnetic (EM) characteristics of the surface of incidence, and the whole phenomenon is governed by Snell's law.

- *Scattering* is not much different from reflection, but it happens when surface irregularities are of comparable dimension to the order of the wavelength of the transmitted signal. Scattering results in spreading of the signal over a wide area leading to high loss of energy of the signal.
- *Diffraction* refers to the bending of the signal when interacting with sharp edges of the surrounding objects.

Similar to the free space path loss, these physical propagation phenomena are also frequency-dependant. For example, smoothness of a surface is directly related to the signal wavelength. A surface considered to be smooth at microwave frequencies, leading to specular reflection, might no longer be smooth at millimeter wave frequencies, in which the signal will thus undergo scattering.

Essentially, these phenomena are quite useful as they allow the signal to propagate behind obstructions and thus facilitate coverage even in shadowed regions. However, due to these physical propagation phenomena, different replicas of the transmitted signal find their way to the intended receiver through different paths, a situation simply referred to as multipath propagation, as illustrated in Figure 10.1c.

10.2 Multipath Propagation

In the multipath propagation phenomenon, different replicas of the transmitted signal, also referred to as (MPC), arrive at the receiving terminal with different phases and strengths based on their own path charecteristics. Such randomly-phased MPCs combine constructively or destructively at the receiver to form a resultant signal with varying amplitude and phase, a phenomena known as *multipath fading*. In general, fluctuation of the received signal strength at the receiver can be characterized based on two different scales of observation:

- On a large-distance scale of about few hundred wavelength: In this case, a steady variation of the received signal strength is observed over a large area at a given distance from the transmitting terminal. This is known as *large-scale fading* and is influenced by path loss as well as terrain configuration and lofty man-made structures between communication terminals which may block or attenuate the signal power.
- On a very-short-distance scale of about few wavelength or short time duration: In this case, the received signal power fluctuates around a mean value on a scale comparable with one wavelength. This is termed as *small-scale fading* and it is due to the aforediscussed superposition of the randomly-phased MPCs at the receiver.

Further classification of fading phenomenon is summarized in Figure 10.2 and elaborated in the subsequent subsections.

10.2.1 Large-Scale Fading

Large-scale fading describes the gradual fluctuation of the mean field strength of the received signal over a local area [3]. Path loss and shadowing of the transmitted signal are two propagation effect attributed to the large-scale fading phenomenon. In essence, large-scale fading is a major concern for cellular system design as it determines spatial coverage of a given cell. Consequently, characterization of the path loss and shadowing effects as a function of distance and environment is of paramount importance to the cellular system designers. In some applications where terminal

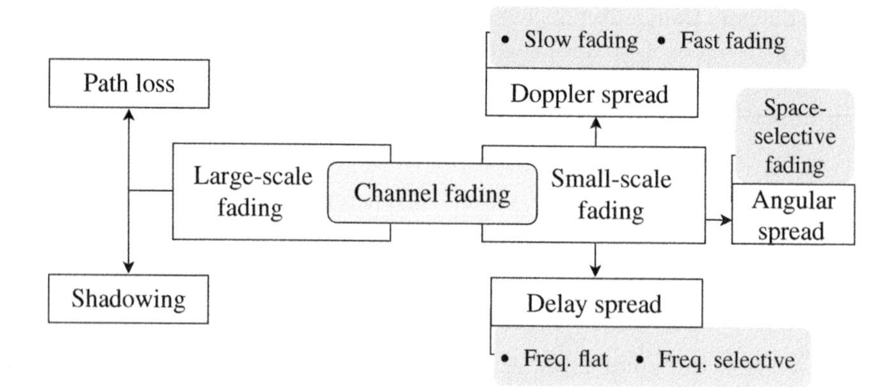

Figure 10.2 Classification of the channel fading characteristics.

placement is beyond the control of the system designer, like in the case of consumer wireless local area network (WLAN), wireless personal area network (WPAN), or in ad hoc networks, large-scale fading needs to be accounted for in the link budget and mitigated by high-level power control mechanisms.

10.2.1.1 Path Loss

In Section 10.1, a free space path loss phenomenon that considers a LOS transmission under atmospheric absorption was introduced. However, due to the propagation complexities in the terrestrial wireless systems, path loss depends on not only the carrier frequency and distance of separation between communication terminals, it is also a function of the nature of propagation environments. In this case, it is difficult to obtain a generalized model that characterizes path loss accurately across a range of different environment. In cases where system specification must be met precisely, or the best locations for base stations (BSs) deployment must be determined, complex ray-tracing (RT) techniques or exhaustive empirical measurement can be used to obtain accurate path loss models. Otherwise, simple models that capture the essence of signal propagation are enough for general trade-off analysis during system design. The following simplified path loss equation is commonly used

$$P_{loss}[\text{dB}](d) = P_{loss}[\text{dB}](d_0) + 10\eta \log\left(\frac{d}{d_0}\right), \quad d > d_0, \tag{10.3}$$

where $P_{loss}[\text{dB}](d_0)$ is the free space path loss at the reference distance d_0 from the transmitter, which is determined based on measurements close to the transmitter, and η is the path loss exponent which incorporates the effect of the propagation environment into the model. For the signal propagations that approximately follow free space, η is set to 2. For more complex environment, value of η is obtained via empirical measurement. Table 10.1 summarizes values of η for different propagation environment.

As we have mentioned, the path loss model given in Eq. (10.3) is an over-simplified model. Many wireless systems rely on more comprehensive models that have been developed via empirical measurement for performance analysis. Such empirical path loss models include Okumura and Hata models that predict signal strength in macrocell metropolitan environments. An extension to Hata model is the COST231 model, whose scope encompass medium cities and suburbs. Walfisch/Bertoni model which predicts average signal strength at street level [4].

Table 10.1 Path loss factor (η) for various scenarios mentioned.

Environment	Path loss exponent (η)
LOS in buildings	1.5 − 2
LOS free space	2
In factories	2 − 3
Urban area	2.7 − 3.5
Obstructed in buildings	4 − 6

Source: Modified from [1].

10.2.1.2 Shadowing

Findings from empirical studies suggested that the concept of path loss alone was not enough to account for the variation of the signal power at a given distance from the transmitter. The received signal strength was empirically observed to vary significantly about a mean value (determined by the path loss) between different locations equidistant from the transmitter. Such random variation is attributed to the attenuation and blockage effects by the randomly distributed objects and terrain features within the propagation environment. This effect was then referred to as *shadowing fading*. Due to the fact that location, size, and dielectric properties of the blocking objects, as well as the changes in the reflecting surfaces and scattering objects that cause such random attenuation are normally unknown, shadowing is generally statistically modeled to follow a log-normal distribution $Z \sim \mathcal{N}(\mu_z, \sigma_z^2)$ (in dB scale) given as

$$f_Z(z) = \frac{1}{\sqrt{2\pi\sigma_z^2}} e^{\frac{(z-\mu_z)^2}{2\sigma_z^2}}. \tag{10.4}$$

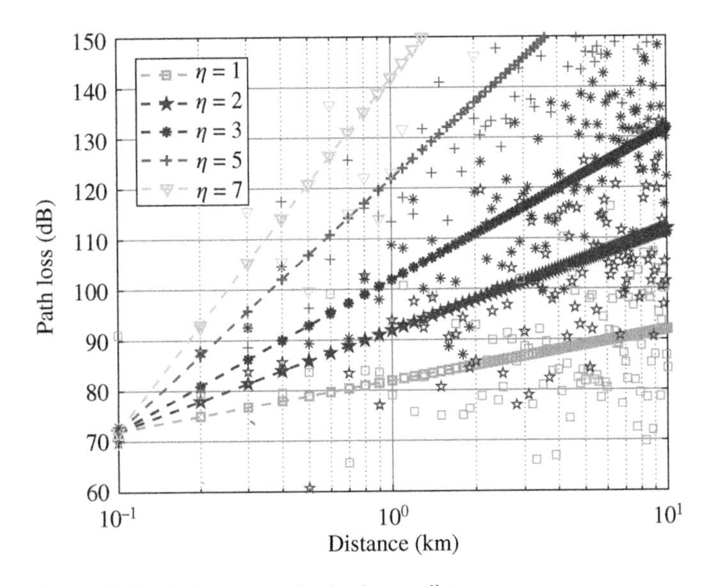

Figure 10.3 Path loss and shadowing vs distance.

Consequently, the superimposed path loss and shadowing effects model that fully captures the large-scale fading process is given by

$$P_{loss}[\text{dB}](d) = P_{loss}[\text{dB}](d_0) + 10\eta \log\left(\frac{d}{d_0}\right) + Z. \tag{10.5}$$

Such combined effect of path loss and shadowing as a function of distance and environment-based path loss exponent η is shown in Figure 10.3. Notice how the effect of large-scale fading increases with the increase in distance and complexity (increasing η) of the propagation environment. Matlab code associated with Figure 10.3 is given below:

```
1  % Path Loss Propagation (Friis Free space equation)
2  %% System parameters
3  Pt_dB = 0; % Transmit power (dB)
4  Gt_dBi = 8; % Transmit antenna gain (dBi)
5  Gr_dBi = 0; % Receiver antenna gain (dBi)
6  fc = 940*10^6; % Carrier frequency (Hz)
7  c  = 300*10^6; % Speed of light (m/s)
8  Hardware_Loss_dB = 8; % Hardware loss
9  d0 = 100; % Reference distance from Tx (m)
10 eta = [1,2,3,5,7]; % Path loss factor
11 %% System evaluation
12 Pt = 10^(Pt_dB/10);
13 Gt = 10^(Gt_dBi/10);
14 Gr = 10^(Gr_dBi/10);
15 d = [d0:100:10000]; % Distance between Tx and Rx (m)
16 sigma_dB = 11.8; % Variance of Shadowing (dB)
17 sigma = 10^(sigma_dB/10);
18 lamda = c/fc;
19 Hardware_Loss = 10^(Hardware_Loss_dB/10);
20 % Received Power (Free space)
21 pl_Pl={'g—s','r—p','b—*','m—+','c—v'};
22 pl_PlS={'gs','rp','b*','m+','cv'};
23 for ii = 1:length(eta)
24       Pr_d0 = Pt*(Gt*Gr*lamda^2)./((4*pi)^2.*(d0.^2)*Hardware_Loss);
25       Pr_d = Pr_d0*((d0./d).^eta(ii));
26       Pr_dB = 10*log10(Pr_d); % Received power (dB)
27       P_loss_dB = Pt_dB - Pr_dB;
28       % Received Power (Free space)
29       S=sigma*(randn(1,length(d))+j*randn(1,length(d)))/sqrt(2);
30       semilogx(d/1000,P_loss_dB,pl_Pl{ii},'LineWidth',1.5); hold on;
31       legendInfo{ii} = ['\eta = ' num2str(eta(ii))];
32 end
33 for ii = 1:length(eta)
34       Pr_d0         = Pt*(Gt*Gr*lamda^2)./((4*pi)^2.*(d0.^2)*Hardware_Loss);
35       Pr_d          = Pr_d0*((d0./d).^eta(ii));
36       Pr_dB         = 10*log10(Pr_d);
37       P_loss_dB        = Pt_dB - Pr_dB;
38       % Received Power (Free space)
39       S=sigma*(randn(1,length(d))+j*randn(1,length(d)))/sqrt(2);
40       P_loss_S_dB     = P_loss_dB + S;
41       semilogx(d/1000,P_loss_S_dB,pl_PlS{ii},'LineWidth',0.5);
42       legendInfo{ii} = ['\eta = ' num2str(eta(ii))];
43 end
44 xlabel(Distance (km));
45 legend(legendInfo);ylabel('Path Loss (dB)');
46 axis([d(1)/1000 d(end)/1000 60 150]);
```

10.2.2 Small-Scale Fading

As previously mentioned, small-scale fading, or simply multipath fading, refers to the rapid variation in the amplitude of the received signal due to the superposition of the MPCs at the receiver. As these randomly delayed and phased replicas of the transmitted signal add up either constructively or destructively, a composite received signal with fluctuating amplitude and phase is formed.

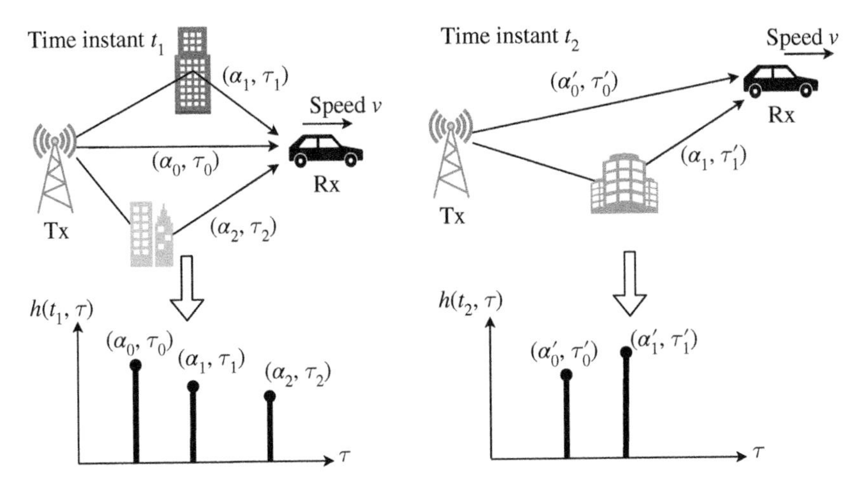

Figure 10.4 Illustration of how channel varies in the presence of mobility.

The fluctuation occurs very rapidly on the spatial order of signal wavelength which makes the small-scale fading problem difficult to deal with. For example, since channel varies very rapidly in the order of a typical duration of physical layer communication, it is hard to track channel gains precisely for system optimization unless significant amount of signaling overhead is employed. In additional to the multipath propagation phenomenon, characteristics of small-scale fading channel are also influenced by other factors such as properties of the transmitted signal (bandwidth, symbol period, etc.) and the presence of relative motion between communication terminals. The relative motion can be due to the mobility of one or both of the communication terminals. It can also spring from mobility of the surrounding objects within the environment. Figure 10.4 illustrates effect of mobility in the multipath environment. Due to the fact that number, strength and delays of the MPCs are tightly coupled to the distribution of scatterers in the propagation environment, the channel response $h(t, \tau)$ experienced by a mobile receiver varies as a function of receiver's location (which changes with time) as illustrated in Figure 10.4. The variable t represents time variation and τ represents multipath delays at a given instant t. In the next subsection, characterization of such time-varying multipath channel is discussed.

10.2.2.1 Characterization of Time-Varying Channels

Time-varying multipath channel can be intuitively visualized as a linear filter with a time-varying impulse response. That is, if an impulse is transmitted at the transmitter, it will be received as a train of impulses due to multipath propagation effect. In order to illustrate this analytically, lets assume a transmitted bandpass signal $x_p(t) = \Re\{x(t)e^{j2\pi f_c t}\}$, where f_c is the carrier frequency and $x(t)$ is the baseband signal transmitted at time instant t. The received bandpass signal, $y_p(t)$, that has travelled through a multipath channel can be represented as follows

$$y_p(t) = \Re\left\{\left(\sum_{i=0}^{L} \alpha_i(t)x(\tau - \tau_i(t))e^{-j2\pi f_c \tau_i(t)}\right)e^{j2\pi f_c t}\right\}, \tag{10.6}$$

where $\alpha_i(t)$ and $\tau_i(t)$ are amplitude and time delay of the i-th MPC at time instant t. L is the maximum number of the resolvable MPCs that can be detected by the receiver. From Eq. (10.6), the received baseband signal is equivalent to

$$y(t) = \sum_{i=0}^{L} \alpha_i(t)x(\tau - \tau_i(t))e^{-j2\pi f_c \tau_i(t)}. \tag{10.7}$$

Equation (10.7) can be rewritten as

$$y(t) = \sum_{i=0}^{L} \alpha_i(t) e^{-j\theta_i(t)} x(\tau - \tau_i(t)), \tag{10.8}$$

where $\theta_i(t) = j2\pi f_c \tau_i(t)$ is the phase of the i-th path. Further manipulation of Eq. (10.8) yields

$$y(t) = \underbrace{\sum_{i=0}^{L} \alpha_i(t) \delta(\tau - \tau_i(t) e^{-j\theta_i(t)}}_{h(t,\tau)} \times x(t), \tag{10.9}$$

where

$$h(t,\tau) = \sum_{i=0}^{L} \alpha_i(t) \delta(\tau - \tau_i(t) e^{-j\theta_i(t)} \tag{10.10}$$

is the impulse response of the channel and $\delta(\cdot)$ denotes Dirac delta function modeling delays of the MPCs. Note that $\theta_i(t)$ changes by 2π radians as soon as τ_i changes by $1/f_c$. Hence, the time variation of a channel is critical for high frequency signals such as mmWave and terahertz (THz) communications.

Due to the random nature of the received MPCs, $|h(t,\tau)|$ and $\theta_i(t)$ are usually modeled as a stochastic processes. The stochastic characterization describes how amplitude of the received signal changes with time. Based on the absence or presence of the dominant LOS component in the multipath scenario, Rayleigh or Rician distribution can be used to model amplitude variations of the multipath channel responses.

10.2.2.2 Rayleigh and Rician Fading Distributions

Rayleigh distribution is a well accepted model for representing multipath propagation effects in the absence of a dominant LOS component. When LOS component does not exist, the resulting small-scale fading process is known as Rayleigh fading. Consider a complex channel impulse response (CIR), $h(t)$ given as

$$h(t) = r(t) e^{j\theta(t)} = \varkappa_1(t) + j\varkappa_2(t), \tag{10.11}$$

where $r(t)$ and $\theta(t)$ are magnitude/envelope and phase of the instantaneous channel coefficients, with in-phase and quadrature-phase (I/Q) components $\varkappa_1(t)$ and $\varkappa_2(t)$, respectively. Assuming a rich multipath environment, by virtue of the central limit theorem (CLT), both $\varkappa_1(t)$ and $\varkappa_2(t)$ are random processes which, independently, follow a zero mean Gaussian distribution, such that $\varkappa_1 \sim \mathcal{N}(0, \sigma^2)$ and $\varkappa_2 \sim \mathcal{N}(0, \sigma^2)$. Their joint probability density function (PDF) is expressed as

$$f(\varkappa_1, \varkappa_2) = f(\varkappa_1) f(\varkappa_2) = \frac{1}{2\pi\sigma^2} e^{-\frac{\varkappa_1^2 + \varkappa_2^2}{2\sigma^2}}. \tag{10.12}$$

Rewriting Eq. (10.12) in terms of signal amplitude r and phase θ gives

$$f(r,\theta) = \frac{r}{2\pi\sigma^2} e^{-\frac{r^2}{2\sigma^2}}, \quad r \in [0, \infty], \quad \theta = [-\pi, \pi]. \tag{10.13}$$

Since r and θ are statistically independent random variables, then $f(r,\theta) = f_r(r) f_\theta(\theta)$ with PDFs of r and θ obtained as follows

$$f_r(r) = \int_{-\pi}^{\pi} \frac{r}{2\pi\sigma^2} e^{-\frac{r^2}{2\sigma^2}} d\theta = \frac{r}{\sigma^2} e^{-\frac{r^2}{2\sigma^2}}, \tag{10.14}$$

and

$$f_\theta(\theta) = \int_0^\infty \frac{r}{2\pi\sigma^2} e^{-\frac{r^2}{2\sigma^2}} dr = \frac{1}{2\pi}. \tag{10.15}$$

Therefore, as it can be concluded from Eqs. (10.14) and (10.15) above, the envelope and phase of the received signal at any time instant undergoes a Rayleigh probability distribution and uniform distribution, respectively.

If we consider a case with a dominant LOS component, the multipath phenomenon is better represented by Rician distribution. In this case, the means of the random processes $\varkappa_1(t)$ and $\varkappa_2(t)$ are no longer zero, they are rather determined by strength of the LOS component such that $\varkappa_1 \sim \mathcal{N}(\mu_1, \sigma^2)$ and $\varkappa_2 \sim \mathcal{N}(\mu_2, \sigma^2)$. The LOS component is given by $A_{LOS} = \sqrt{\mu_1^2 + \mu_2^2}$. Magnitude r of the received signal is now given by $r = \sqrt{(A_{LOS} + \varkappa_1)^2 + \varkappa_2^2}$ and its PDF is subsequently found as [5]

$$f_r(r) = \frac{r}{\sigma^2} e^{-\frac{r^2 + A_{LOS}^2}{2\sigma^2}} I_o\left(\frac{rA_{LOS}}{\sigma^2}\right), \tag{10.16}$$

where $I_0(\cdot)$ is the zero-th order modified Bessel function of the first kind. Characteristic of the Rician fading is usually quantified by the Rician K-factor which is given as the ratio of the strength of the LOS component to the power of the received random MPCs

$$K = \frac{A_{LOS}^2}{2\sigma^2}. \tag{10.17}$$

Essentially, Rayleigh fading is a special case of Rician fading with $K = 0$. As $K \implies \infty$, the propagation environment corresponds to a scenario with only a LOS component in which the signal is received from only one direction. PDFs of r and θ for various values of K are shown in Figure 10.5. Notice how the distribution of the phase change from being uniformly distributed at $K = 0$ (i.e. Rayleigh fading) and become dominated by the phase of the LOS component as K increases. Matlab code for obtaining these PDFs is given below:

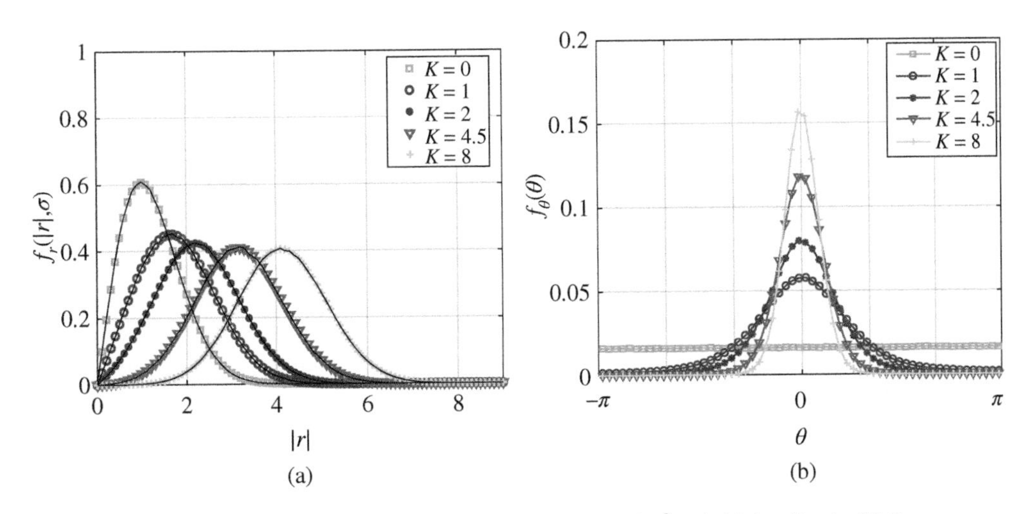

Figure 10.5 Analytical and approximated PDF for various K, and $\sigma^2 = 1$. (a) Amplitude. (b) Phase.

```
1  %Generation of Rician-Rayleigh RVs
2  %System parameters
3  nSamples=10^6;%Number of Samples
4  Var=1;%Variance
5  m1=[0 sqrt(2) sqrt(4) sqrt(9) sqrt(16)];%Mean
6  m2=[0 0 0 0 0];
7  quant=0.1;
8  rIndex=0:quant:9;%Some indexing values
9  A_LOS=sqrt(m1.^2 + m2.^2);
10 K=A_LOS.^2/(2*Var);% $K$-factor
11 %Theory PDF
12 pl_the={'gs','ro','b*','mv','c+'};
13 for ii=1:length(K)
14     f_r_theo=rIndex./(Var).*exp(-((rIndex.^2+A_LOS(ii)^2)...
15         ./(2*Var))).*besseli(0,rIndex.*A_LOS(ii)/(Var));
16     plot(rIndex,f_r_theo,pl_the{ii},'LineWidth',2);
17     legendInfo{ii} = ['K = ' num2str(K(ii))]; hold on;
18 end
19 %Simulation
20 pl_sim={'k-','k-','k-','k-','k-'};
21 for ii=1: length(K)
22     x1=sqrt(Var)*randn(1,nSamples)+ m1(ii);
23     x2=sqrt(Var)*randn(1,nSamples) + m2(ii);
24     r=x1+j*x2;%New RV
25     %PDF of signal envelope
26     [f_r_sim,rIndex_sim]=hist(abs(r),rIndex);
27     %Normalization of the PDF
28     plot(rIndex_sim,f_r_sim/(nSamples*quant),...
29         pl_sim{ii});hold on;
30 end
31 legend(legendInfo);grid on;xlabel('|r|');ylabel('f(|r|,\sigma)')
32 axis([0 rIndex(end) 0 1]);
```

```
1  %Effect of the multipaths' angle of arrival
2  pl_O={'g-s','r-o','b-*','m-v','c-+'};
3  figure;% Theory - Angle PDF
4  for ii=1:length(K)%Theory PDF
5      x1=sqrt(Var)*randn(1,nSamples)+ m1(ii);
6      x2=sqrt(Var)*randn(1,nSamples) + m2(ii);
7      r=x1+i*x2;%New RV
8      oIndex = [-pi:quant:pi];
9      [O_sim O_Index_sim] = hist(angle(r),oIndex);
10     plot(O_Index_sim,O_sim/sum(O_sim),...
11         pl_O{ii},'LineWidth',1);
12     hold on;
13 end
14 legend(legendInfo);grid on;
15 axis([-pi pi 0 max(O_sim/sum(O_sim))]); xlabel('\theta');
16 ylabel('f_{\Theta}(\theta)');
```

10.2.3 Time, Frequency and Angular Domains Characteristics of Multipath Channel

10.2.3.1 Delay Spread

As mentioned earlier, multipath propagation allows different replicas of the transmitted signal to arrival at the receiver via different routes with different delays. These random delays cause the received replicas to exhibit some kind of spreading when observed from time axis as illustrated in Figure 10.6. This is referred to as time dispersion or delay spread. Such spreading of the received signal in time domain is characterized by a power delay profile (PDP). PDP is a plot of the received power as a function of time delay, and it is found as an average of $|h(t,\tau)|^2$ measured around a local area. An example of PDP is given in Figure 10.7. From the PDP, some critical channel parameters that quantify the time dispersion of the channel can be extracted. These parameters include:

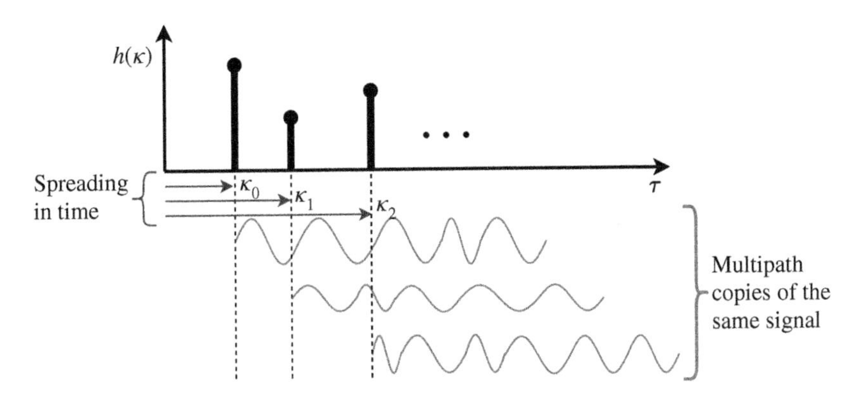

Figure 10.6 Illustration of the delay spread conception.

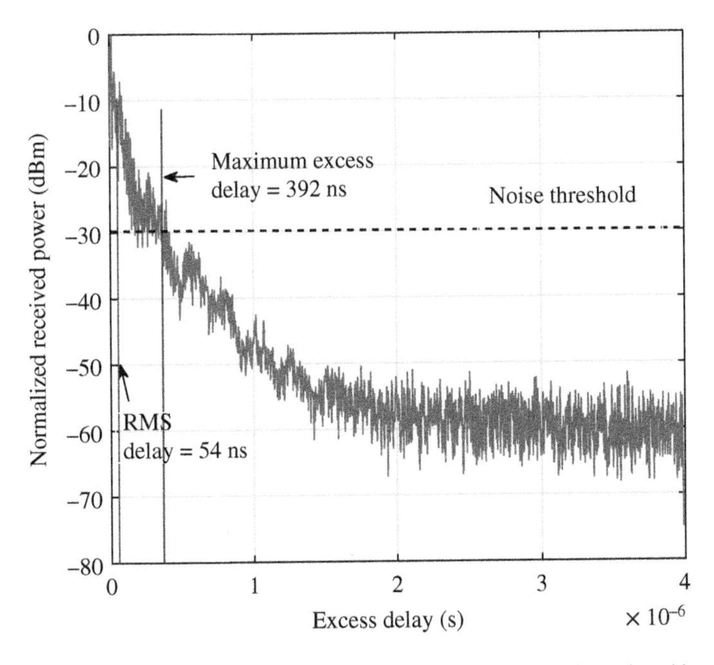

Figure 10.7 An example of a PDP for a multipath channel emulated by using a reverberation chamber.

- *First arrival delay* (τ_0): Refers to the delay of the MPC that has travelled the minimum propagation distance. It is taken as a reference and all other delay components coming after it are called *excess delays*.
- *Maximum excess delay* (τ_{max}): Time duration after which energy of the arriving MPCs falls below a certain threshold. The threshold depends on the sensitivity of the receiver as well as system's noise floor. τ_{max} is calculated as the difference between τ_0 and delay of the last arriving MPC (that is above the threshold). The maximum excess delay is not necessarily the best indicator of how a given system will perform on a channel. This is because of the fact that different channels with the same τ_{max} can exhibit quite different PDPs. Therefore, a more useful parameter is the root mean squared (RMS) delay spread described below.

- *RMS excess delay* (τ_{rms}): It is square root of the second moment of the PDP, given by

$$\tau_{rms} = \sqrt{\mathbb{E}\{\tau^2\} - \tau_{mean}^2}, \tag{10.18}$$

where

$$\mathbb{E}\{\tau^2\} = \frac{\int_0^\infty \tau^2 PDP(\tau)d\tau}{\int_0^\infty PDP(\tau)d\tau}$$

and τ_{mean} is the *mean excess delay* which is defined as the first moment of the PDP and it is given as

$$\tau_{mean} = \frac{\int_0^\infty \tau PDP(\tau)d\tau}{\int_0^\infty PDP(\tau)d\tau}. \tag{10.19}$$

τ_{rms} is commonly used to quantify the strength of inter-symbol interference (ISI) and thus determines complexity of the equalizer required at the receiver.

- *Coherence Bandwidth* (B_c): It is a statistical measure of a range of frequencies over which multipath channel has constant gain and linear phase response. In other words, coherence bandwidth is a range of frequencies over which two frequency components have strong potential for amplitude correlation. But if frequency separation between them is greater than B_c they are affected differently by the channel. It should be noted that, although coherence bandwidth is directly related to the delay spread of a channel, an exact analytical relationship between B_c and the above discussed delay spread parameters does not exist and one should resort to the signal analysis of the actual signal dispersion measurement in a particular channel in order to determine it. Nevertheless, some commonly used approximation of B_c from τ_{rms} are available in the literature. For example, if B_c is defined as the frequency interval over which channel's complex transfer function has a correlation of at least 0.9, B_c is approximated by $B_c \approx 1/50\tau_{rms}$. Similarly, for a correlation of at least 0.5, $B_c \approx 1/5\tau_{rms}$.

Based on the delay spread aspect of the multipath channels, two types of fading phenomena can be observed, namely *frequency-flat* and *frequency-selective* fading.

- *Frequency-flat fading*: A channel is said to exhibit frequency-flat property if bandwidth B_s of the transmitted signal is smaller than the coherence bandwidth of the channel, B_c, i.e. $B_s < B_c$. In this case, the received signal strength changes with time due to the fluctuation of the channel caused by multipath, but the spectrum of the transmitted signal is preserved. If T_s is the symbol duration of the transmitted signal, flat fading occurs when $T_s > \tau_{max}$. That is, all significant MPCs of the transmitted signal arrive within one symbol duration and they cannot be resolved. In this case, the channel behaves like a one tap channel. Flat fading channel are also referred to as narrowband channels.
- *Frequency-selective fading*: This is opposite of the frequency-flat case. It happens when the delays of the received MPCs of one symbols extend beyond the symbol duration, i.e. $T_s < \tau_{max}$. In this case, MPCs of the current symbol are received during symbol duration of the next symbol which leads to ISI. In other word, frequency-selective fading occurs when bandwidth of the transmitted signal exceeds channel's coherence bandwidth, i.e. $B_s > B_c$. Therefore, the spectrum of the received signal is distorted differently by the channel. Frequency-selective channels are also known as wideband channels. In this case, multipath components are said to be resolvable.

Figure 10.8 illustrates the relationship between signal's symbol duration, MPCs resolvability, and frequency selectivity phenomenon discussed above. The following Matlab code demonstrates how

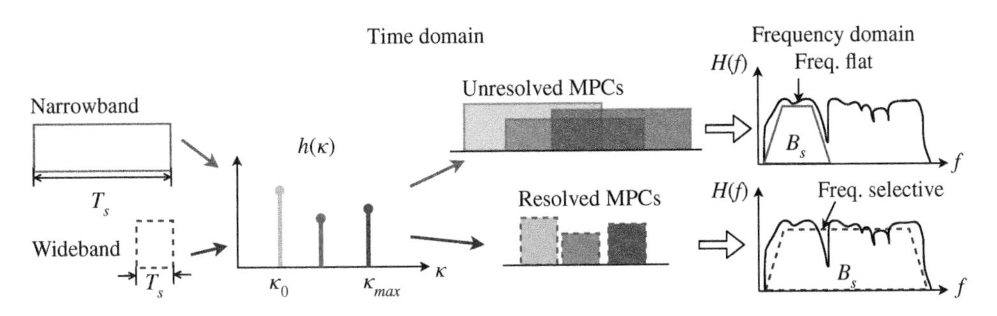

Figure 10.8 Relationship between signal bandwidth, resolvability and frequency selectivity.

to generate and visualize the effect of frequency-selective channel. Here we use a single carrier signal with root-raised cosine pulse shape with roll-off factor of 1.

```
1  % Demonstration of the Frequency selectivity with a single
2  % carrier Signal utilizing a root-raised Cosine filter
3  % Signal generation
4  close all; clear; clc
5  SymRate     = 50000;              % Hertz
6  OSR         = 8;                  % Oversampling rate
7  fs          = SymRate * OSR; % Ts = 1/fs = 2.5e-6s
8  ModSymbols = 2*((randn(2000,1))>0)-1; % Random BPSK symbols
9  Tx_Filter  = rcosdesign(1,10,OSR,'sqrt');
10 Tx_Signal  = conv(Tx_Filter,upsample(ModSymbols,OSR));
11 % Channel
12 TauMax = [1e-6, 20e-6]; % (In sec):1st case: FreqFlat
13                         %           2nd case: FreqSelective
14 for i = 1:length(TauMax)
15 N_taps           = floor(TauMax(i)*fs)+1; % Calculate
16                                            % Number of taps
17 TapDelays        = (0:1:N_taps)*(TauMax(i)/(N_taps));
18 TapGains         = (0:1:N_taps)*(-5/(N_taps));
19 RayLeighChann = comm.RayleighChannel(...
20                  'SampleRate',fs,'PathDelays',TapDelays, ...
21                  'AveragePathGains',TapGains, ...
22                  'NormalizePathGains',true);
23 Rx_Signal = RayLeighChann(Tx_Signal); % Pass the signal
24                                       % through the channel
25 % Plot Spectrum of the Rx signal
26 [Pxx,F] = pwelch(Rx_Signal,[],[],[],fs,'center');
27 figure
28 plot(F,10*log10(Pxx/max(Pxx)))
29 xlabel('Frequency (Hz)'); ylabel('Normalized Power (dB)')
30 grid on, box on;
31 end
```

The symbol duration, T_s is set as 2.5 μs. Two simulation cases are considered. In the first case, τ_{max} is set as 1 μs, which is smaller than the selected T_s. Therefore, this case demonstrates the flat fading. In the second case, τ_{max} is selected to be larger than T_s (about 20 μs) in order to achieve frequency-selective phenomenon. Spectrums of the received signal for both cases are shown in the Figure 10.9. Notice that, in the first case, shape of the root-raised cosine pulse is preserved. By using the same code, the effect of symbol rate can also be investigated by fixing the value of parameter `TauMax` and setting `SymRate` to different values.

10.2.3.2 Angular Spread

Angular spread refers to the spreading of the signal in space such that it arrives at the receiver from multiple direction regardless its original angle-of-departure (AoD). The directions from which

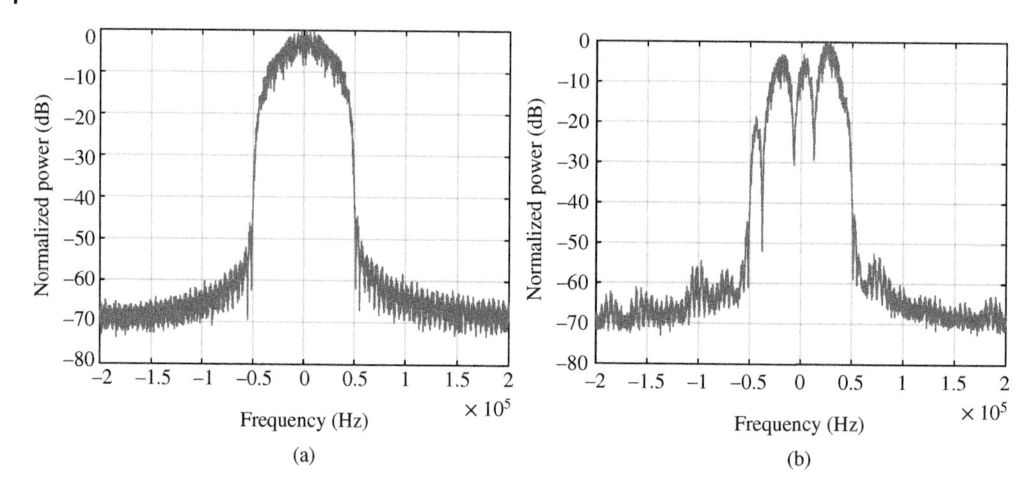

Figure 10.9 Spectrums of the received signals for the two considered cases. (a) Frequency flat. (b) Frequency selective.

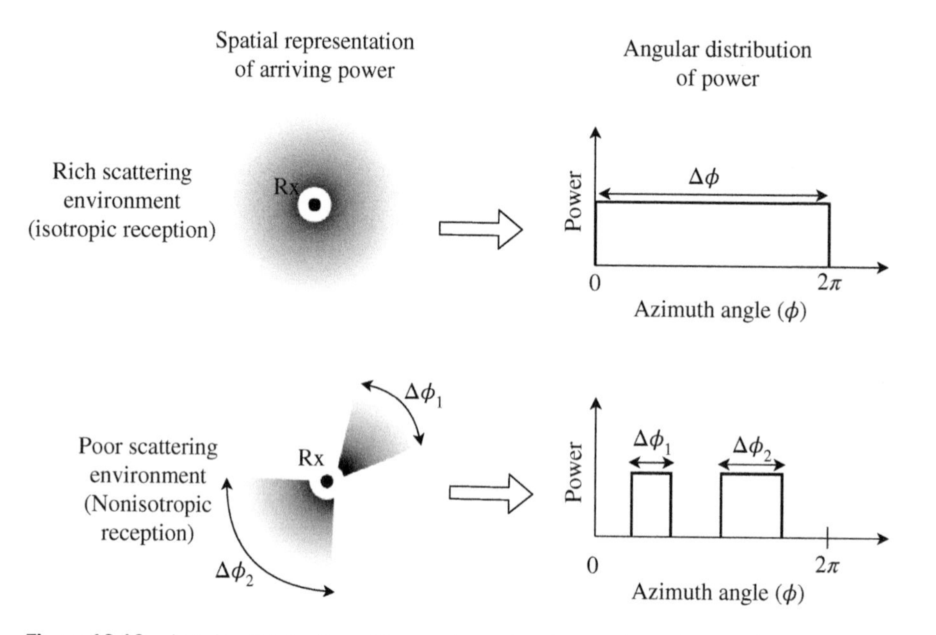

Figure 10.10 Angular distribution of the received signal power based on the AoA. $\Delta\phi$ denotes the Azimuth angular spread.

the signal is received, or simply the angle-of-arrival (AoA), is influenced by the distribution of scatterers in the environment. Rich scattering environments usually facilitate isotropic reception of the signal, whereas in poor scattering environments, the received signal may arrive from limited direction based on the location of the available scatters. A simple illustration of AoA is shown in Figure 10.10.

The dithered area in Figure 10.10 depicts the direction of arrival of the spatially scattered MPC. Therefore, position and orientation of the receiving antenna relative to the distribution of scatters can significantly impact strength of the received signal. Such variation of the received signal strength based on the receiver location or orientation is referred to as *space-selective fading*.

Space-selective fading is characterized by a *coherence distance*, D_c. Coherence distance defines the minimum distance by which any two antennas must be physically separated so that they receive uncorrelated signal. Characterization of angular spread is essential for multi antennas systems in which the highest capacity is achieved when the signals received by antenna elements are uncorrelated.

Angular spread is quantified by RMS azimuth angular spread (ϕ_{rms}). RMS angular spread of a channel refers to the statistical distribution of the angles from which energy of the received signal is arriving. Large ϕ_{rms} implies that the signal is arriving from many directions, which is mostly the case in rich scattering environment. Small ϕ_{rms} implies that the signal energy is more focused to the receiver. Large angular spread results in more spatial diversity which is generally exploited by multi antenna systems to enhance the channel capacity.

Relationship between D_c and ϕ_{rms} is approximated by

$$D_c = \frac{0.2\lambda}{\phi_{rms}}, \tag{10.20}$$

where λ is the wavelength of the signal. From the given relationship, it is clear that wireless systems operating at high frequency bands have shorter coherence distance, which allows many antenna elements to be packed together, even within a wireless devices with small form-factor.

10.2.3.3 Doppler Spread

Doppler shift (f_D) is a well-known phenomenon where the actual frequency, f_c, of a signal is modified to a new frequency $f = f_c + f_D$ due to the relative motion between transmitter and receiver.

For a signal arriving from direction ϕ at a mobile user with speed v, as illustrated in Figure 10.11, f_D is calculated as

$$f_D = \frac{v}{\lambda}\cos\phi = \frac{vf_c}{c}\cos\phi, \tag{10.21}$$

where λ and c are signal wavelength and speed of light, respectively.

In the case of multipath propagation, each MPC experiences different Doppler shift depending on its direction of arrival, which in turn leads to the spreading of the received MPCs along the frequency axis. This phenomenon is called Doppler spread. Essentially, the spreading in frequency causes broadening of the spectrum of the received signal. The resulting spectrum is called Doppler spectrum, $S(f)$. In simple terms, the Doppler spread is defined as a measure of the spectral broadening due to the relative motion between communication terminals or mobility of the surrounding

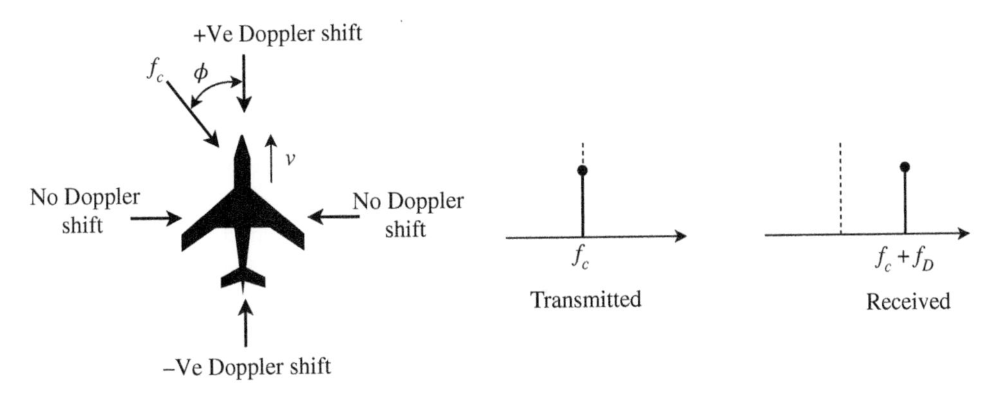

Figure 10.11 Doppler shift experienced by a moving vehicle.

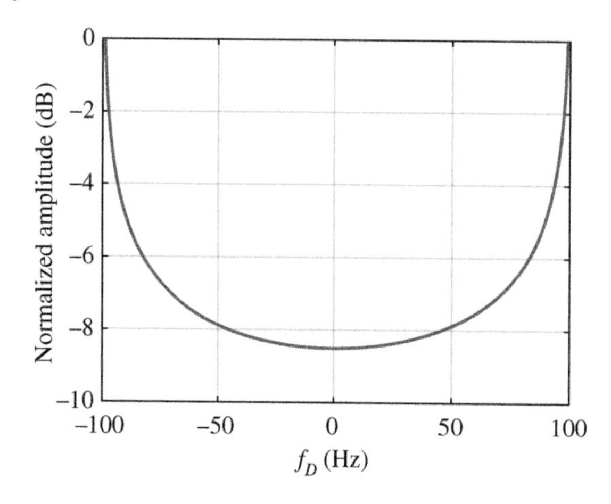

Figure 10.12 Classical Doppler spectrum by Jake's model for $f_{Dmax} = 100$ Hz.

objects. If we consider a rich scattering environment which facilitates isotropic signal reception, the Doppler spectrum of unmodulated carrier signal received at a mobile terminal is given by Jake's model

$$S(f) = \frac{1}{\pi f_{D\,max}\sqrt{1-\left(\frac{f-f_c}{f_{D\,max}}\right)^2}}, \tag{10.22}$$

where $f_{D\,max}$ is the maximum Doppler shift experienced by the signal, given as $f_{Dmax} = vf/c$. The spectrum is centered at f_c and it is zero outside the limit of $f_c \pm f_{Dmax}$. The Doppler spectrum approximated by Eq. (10.22) is classically bowl-shaped as shown in Figure 10.12. Matlab code that implements Eq. (10.22) to obtain Figure 10.12 is given below. Here, f_c is set to 0 Hz and $f_{Dmax} = 100$ Hz.

```
1  % Simulating a classical Doppler spectrum by using
2  % Jake's model
3  f_Dmax = 100;       % Hertz
4  f_D        = -f_Dmax+1:0.1:f_Dmax-1;
5  for i=1:1:length(f_D)
6  Sf(i)  =  (1/4*pi*f_Dmax)*(1/sqrt(1-(f_D(i)^2)/(f_Dmax^2)));
7  end
8  figure
9  plot(f_D,10*log10(Sf/max(Sf)),'linewidth',1.5);
10 xlabel('Frequency (Hz)'); ylabel('Normalized amplitude (dB)'); grid on;
```

Note that, due to the strong dependence of Doppler spread on the AoA (ϕ) as suggested by Eq. (10.21), non isotropic distribution of the arriving signal power results in the Doppler spectrum that is only part of the classical bowl-shaped spectrum shown in Figure 10.12. This fact is illustrated in Figure 10.13 for various cases of angular spread.

It should be noted that Doppler spectrum estimated by Eq. (10.22) does not hold for all environments. In some scenarios, where the condition of dense scattering is not satisfied, different spectral shape can be obtained. For instance, channel model for indoor environments assumes $S(f)$ to be a flat spectrum while in the aeronautical channels Gaussian shaped Doppler spectra can be observed.

The phenomenon of Doppler spread manifests itself in time domain of the received signal as time-selective fading. As previously explained in Section 10.2.2, the presence of mobility in the communication scenario causes the channel to vary with time. This variation in the channel response with time is referred to as *time selectivity*. The degree of selectivity of a time-varying channel is

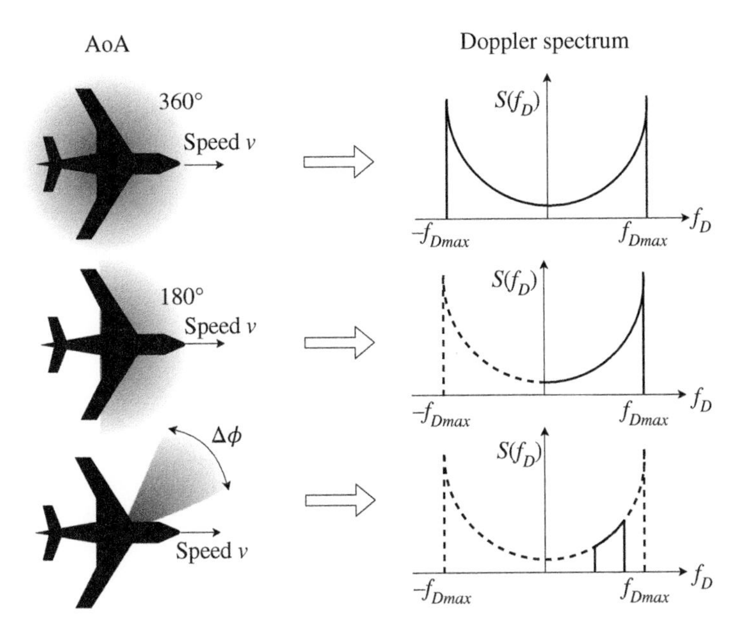

Figure 10.13 Illustration of the influence of AoA on the shape of the Doppler spectrum.

characterized by *coherence time*, τ_c. Coherence time is defined as a measure of the expected time duration over which the channel response remains fairly constant. Doppler spread, f_{Dmax}, and coherence time, τ_c, are reciprocally related. Their relationship is approximated by

$$\tau_c \approx \frac{1}{f_{Dmax}}. \tag{10.23}$$

Under time selectivity property of a channel, two types of fading processes can be observed depending on how fast the channel varies with respect to the rate of signal transmission:

- *Slow fading*: Occurs when coherence time of the channel is sufficiently long. That is, channel response varies at a rate much slower than that of the transmitted signal. Therefore, a signal experiences slow fading if

$$T_s << \tau_c \text{ or } B_s >> f_{Dmax}. \tag{10.24}$$

- *Fast fading*: In this case, the impulse response of the channel changes significantly within the symbol duration of the transmitted signal. In other words, the coherence time of the channel is much smaller than the symbol duration of the transmitted signal, i.e.

$$T_s >> \tau_c \text{ or } B_s << f_{Dmax}. \tag{10.25}$$

In the following Matlab code we show how to simulate time selectivity. The effect is visualized in time and frequency domain for three different speeds. The time domain envelope of the received signal is plotted to observe the variation of the channel due to mobility. As observed in Figure 10.14a, the rate of variation increases with speed. Similarly, for the frequency domain analysis, spectrums of the received signals are shown in Figure 10.14b. Again, as discussed above, bandwidth of the Doppler spectrum increases with the mobility speed.

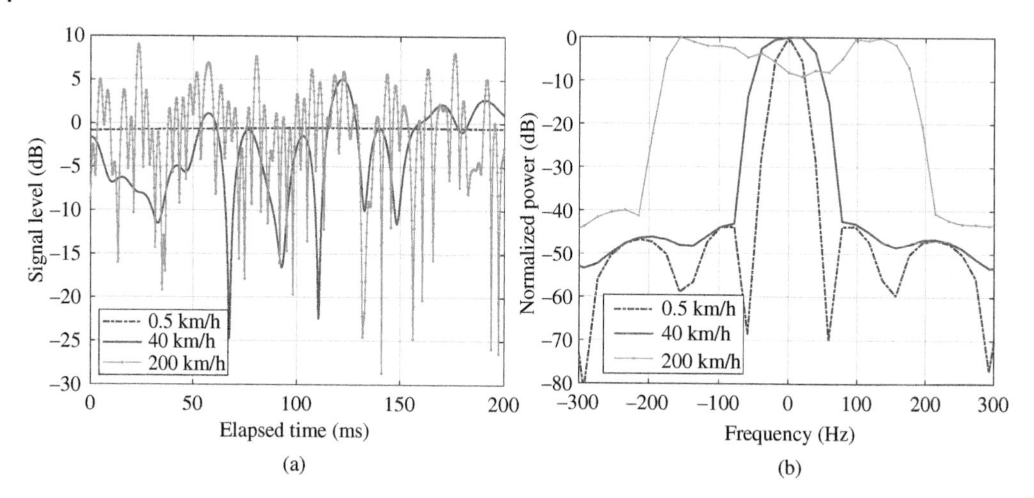

Figure 10.14 Illustration of time selectivity of a channel in time and frequency domains. (a) Time variation of the channel's envelope. (b) Spreading in frequency.

```
1  % Demonstration of Time selectivity for different
2  % relative speed
3  % The same code can be used to observe the effect of the
4  % carrier frequency (fc) on the Time selectivity.
5  fs = 5000; % sampling frequency
6  fc = 900e6; % Carrier frequency
7  n = (0:1:1000-1)';
8  time = n./fs;
9  Tx_Signal = 1i*ones(length(time),1);
10 c  = 3e8; % speed of light
11 Speed    = [0.5,40,200]; % km/h
12 % Open figures
13 fig1 = figure(1); ax1 = axes; hold(ax1,'on');
14 xlim([-300 300]); ylim([-80 0]); grid on; box on;
15 xlabel('Frequency (Hz)'); ylabel('Normalized Power (dB)')
16 fig2 = figure(2) ;ax2 = axes ; hold(ax2,'on');
17 xlabel('Elapsed time (ms)'); ylabel('Signal level (dB)')
18 grid on; box on;
19 for i=1:length(Speed)
20 f_Dmax    = (Speed(i)*1000/3600)*fc/c; % Max. Doppler spread
21 Channel = comm.RayleighChannel('SampleRate',fs,...
22     'MaximumDopplerShift',f_Dmax); % Create Channel object
23 Rx_Signal = Channel(Tx_Signal); % Pass the signal
24                                 % through the channel
25 [Pxx,F] = pwelch(Rx_Signal,[],[],[],fs,'center');
26 % Figures
27 plot(ax1,F,10*log10(Pxx/max(Pxx)),'linewidth',1.5)
28 plot(ax2,time.*1000,20*log10(abs(Rx_Signal)))
29 end
```

10.2.4 Novel Channel Characteristics in the 5G Technology

In an effort to address the challenges of the explosively growing mobile data demand, the fifth generation (5G) of the wireless technology identifies communication at the mmWave frequency band as one of its key enablers. The mmWave frequency band offers larger bandwidths that are necessary for the wideband transmissions (i.e. high data rate) envisioned by 5G. However, as we have noted in Section 10.1, the physical propagation phenomena are highly frequency-dependent. Consequently, mmWave signal are prone to severe pathloss due to excessive atmospheric absorption as well as poor scattering and diffraction phenomena. It also suffers from blockage effect from surrounding objects due to its small wavelength and poor penetration ability [6]. Compared to the traditional

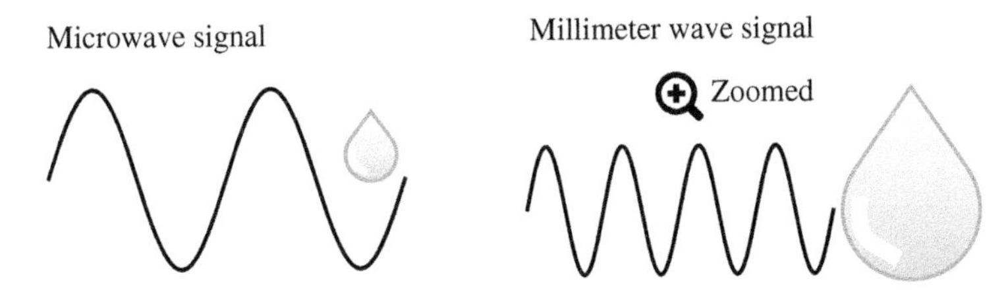

Figure 10.15 Perception of the microwave and mmWave signals in front of an obstacle. At mmWave frequencies, even a rain drop may appear significantly big.

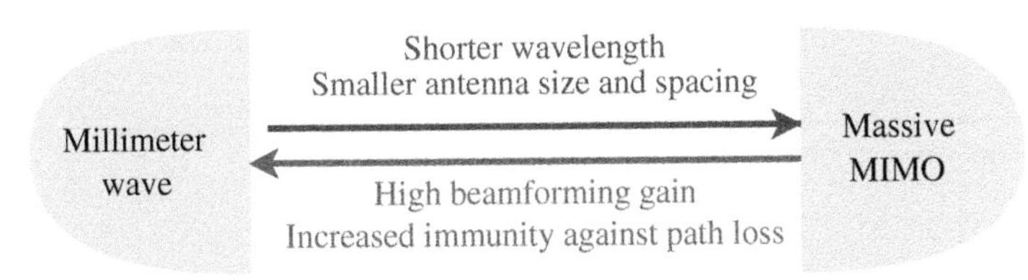

Figure 10.16 Inter-dependence of the mmWave and massive MIMO technologies.

microwave signal, even a small object appears relatively big to the mmWave signal and thus can significantly attenuate the signal. Perception of this phenomenon is illustrated in Figure 10.15.

Quantitatively, the level of attenuation changes drastically from about 10^{-3} dB/km at microwave band to about 20 dB/km at 60 GHz. As a result, effective communication at mmWave band can be achieved only with directional transmission where emitted power is focused to the direction of the receiver such that the aforementioned power loss is compensated for.

To this end, 5G technology identifies massive multiple-input multiple-output (mMIMO) as another key enabler in which both transmitter and receiver are equipped with large number of antennas. Apart from its ability of improving system capacity through spatial multiplexing and reliability through spatial diversity techniques, mMIMO, through beamforming techniques, inherently overcomes the elevated path loss problem in mmWave frequencies. In fact, mmWave and mMIMO technologies mutually depend on each other. The benefit of mMIMO can be enjoyed only when different transmit-receive antenna pairs experience uncorrelated channels. This can be realized when the spacing between antenna elements is at least half of the signal wavelength (λ). The extremely small wavelength of mmWave signals facilitate practical implementation of the mMIMO concept, as large number of antennas can be packed together with adequate inter-element spacing even on a small mobile device. Such symbiotic relationship is summarized in Figure 10.16.

The use of directional transmission via mMIMO beamforming sparks some new channel features that have not been significant in the earlier wireless generations. Due to the directional transmission, the number of scatters interacting with the transmitted signal is significantly reduced. Therefore, only few resolvable paths can be observed in the angular domain, leading to the so called channel sparsity in spatial domain. Likewise, if we consider a wideband transmission, the channel taps or clusters become highly resolvable in the time domain, making the channel sparse in the time domain as well. As a result, mmWave channel profiles usually exhibit only a few channel taps or clusters when observed in angle-delay domain as illustrated in Figure 10.17. These characteristics are of paramount importance from mMIMO channel estimation and user scheduling perspectives [7].

In the traditional single-input single-output (SISO) systems, channel is usually characterized in two fundamental domains, time and frequency. However, in the multi antenna systems with directional transmissions, polarization and spatial domain of the channel are also significant. Therefore,

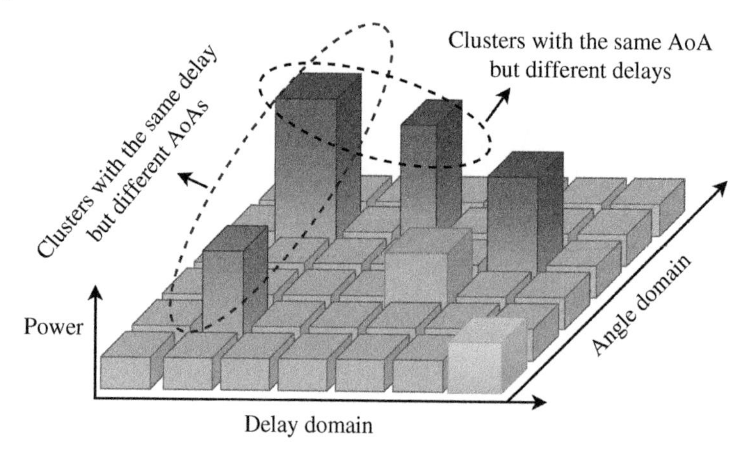

Figure 10.17 Illustration of a sparse channel in angle-delay domains.

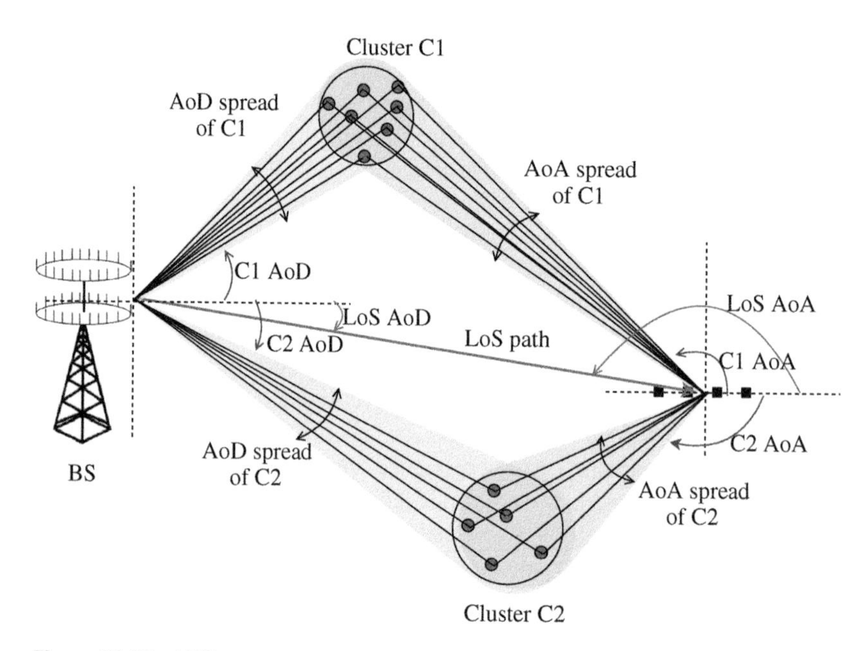

Figure 10.18 Millimeter wave massive MIMO spatial channel model.

for the mmWave mMIMO systems, the spatial representation of the channel is commonly considered, where antenna field patterns featuring the AoD and AoA are also taken into account. This is called spatial channel model and it is depicted in Figure 10.18. A simplified transfer function $\mathbf{H}_{tr,rx}$ of such a spatial channel for NLOS scenario is given in Figure 10.19, and the definitions of the used parameter are given in Table 10.2. The calculation of the parameters and generation of the spatial channel coefficients based on the 3rd Generation Partnership Project (3GPP) standardized link level simulation models are given in Section 10.4.3.

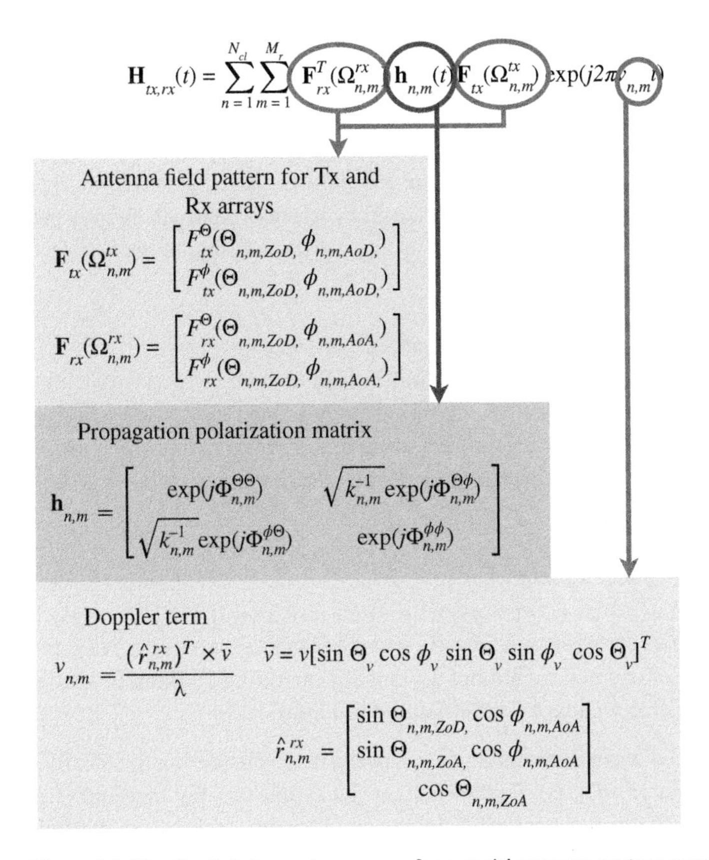

Figure 10.19 Spatial channel response for a multi antenna system considering NLOS scenario.

Table 10.2 Definition of parameters in Figure 10.19.

Parameter	Definition
N_{cl}	Number of clusters ($n = 1, 2, \ldots N_{cl}$)
M_r	Number of rays in a cluster ($m = 1, 2, \ldots M_r$)
ϕ	Azimuth angle
Θ	Zenith angle
v	Mobile speed
ϕ_v	Travel azimuth angle
Θ_v	Travel zenith angle
κ	Cross polarization power ratio (XPR)
$\phi_{n,m,ZoA}$	Azimuth angle of arrival of mth ray in nth cluster
$\phi_{n,m,ZoD}$	Azimuth angle of departure of mth ray in nth cluster
$\Theta_{n,m,ZoA}$	Zenith angle of arrival of mth ray in nth cluster
$\Theta_{n,m,ZoD}$	Zenith angle of departure of mth ray in nth cluster
$\phi_{n,m,AoA}$	Azimuth angle of arrival of mth ray in nth cluster
$\phi_{n,m,AoD}$	Azimuth angle of departure of mth ray in nth cluster
$\Theta_{n,m,AoA}$	Zenith angle of arrival of mth ray in nth cluster
$\Theta_{n,m,AoD}$	Zenith angle of departure of mth ray in nth cluster
$\left[\Phi^{\Theta\Theta}_{n,m} \quad \Phi^{\Theta\phi}_{n,m} \quad \Phi^{\phi\Theta}_{n,m} \quad \Phi^{\phi\phi}_{n,m}\right]$	Random initial phases for each ray mth ray in nth cluster for the cross zenith–azimuth polarizations

10.3 Channel as a Source of Interference

Interference is a classical drawback that limits the capacity, coverage, and effectiveness of wireless communication technologies. In literature, interference is defined as any unwanted signal power causing disruption of communication including the use of phone, television, machine type communications, etc. Mainly, the features of interference caused by wireless channel are grouped considering the fading characteristics of the medium, such as *large-scale* and *small-scale* fading.

10.3.1 Interference due to Large-Scale Fading

In the early times of wireless systems, support of wireless communication in large areas with a high user density could not be feasible by just a single BS. Actually, the number of UE was exceeding the number of channels, leading to an unjustifiable queuing time for communication. Integration of *cellular technology* was a headway in solving the problem of this spectrum scarcity as discussed in the following sections.

10.3.1.1 Cellular Systems and CoChannel Interference

In cellular systems, a single high-power BS is replaced with much low power BSs, where each provides coverage to a smaller portion of the service area. As a matter of fact, the reuse of the same channel frequencies for multiple users named cochannel users while operating in different spatial regions is considered to enhance network capacity, as illustrated in Figure 10.20.

- *Spatial guard:* Spatial separation between cochannel users is performed considering the fact that the average received signal power ($P_r(d)$), i.e. free space (Eq. 10.1) and two-ray medium (Eq. 10.2), by any user at any location (d) away from the BS decays as a function of distance. In other words, the large area is split into several small regions named *cells*, which are represented with the dotted lines for the ideal case in Figure 10.20. In case the cells are totally isolated, the same radio channel can be utilized within each cell and zero energy leakage occurs. Therefore, the capacity of the system improves with the increase of the number of cells within a fixed area. For instance, if the service providers offer 100 duplex channels for communications, no more than

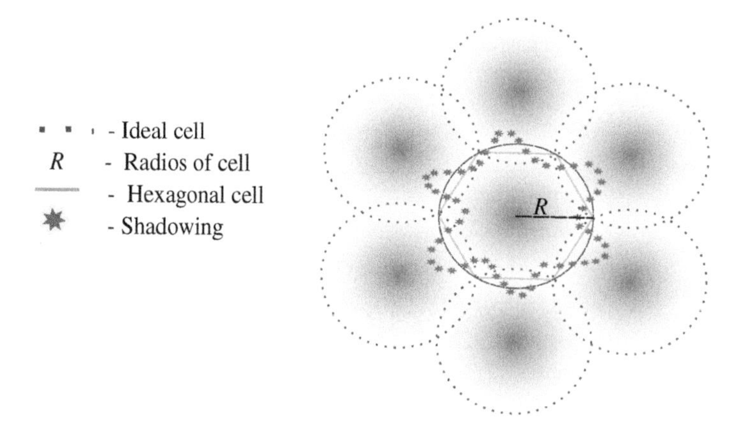

Figure 10.20 Cochannel interference in cellular systems.

100 connections can be established within a fixed area. Consequently, if the given area is split into two cells, the system capacity is doubled under the assumption of zero energy leakage between adjacent cells, utilizing the same radio resources.

The cell parameters, such as radius and distance between nearest cochannel users get smaller and control of transmitted power from adjacent cells becomes a challenging task since a given area cannot be fully covered via the utilization of circular shaped cells, which represent an ideal wave propagation scenario. Hence, the adjacent cells need to intersect with each other, as illustrated by cross regions in Figure 10.20. For this reason, energy leakage exists between the adjacent cells, limiting the performance of wireless systems. Additionally, the received signal strength by the desired UE at any point not only depends on displacement between BSs and receiver but also needs to consider the characteristics of shadowing, which is shown by the starred line in Figure 10.20.

- *Cochannel interference* (CCI): It occurs due to large-scale characteristics of the channel, i.e. distance and shadowing in cellular systems, where the user of interest is exposed to a substantial energy strength leaking from the cochannel users that seriously affects system performance, and resulting in lower data rates, higher dropouts, poor communication quality, and even complete loss of connection link.

The worst signal-to-interference (SIR) scenario occurs when the area is split into multiple cells and all the channels are used within each cell. In other words, there always exists a user from each cell causing interference on the user of interest. Since the average received signal power at any distance (d) away from the BS decays as a function of the separation, a general mathematical evaluation of SIR is given as follows

$$SIR = \frac{P_{r,d}(d)}{\sum_{i=0}^{I-1} P_{r,i}(d_i)}, \tag{10.26}$$

where $P_{r,d}(d)$ and $P_{r,i}(d_i)$ denote the strength of received signal from the desired user and interferes, respectively, with respect to their corresponding distances. I is the number of interferes. Regarding Eq. (10.26), not only the number but also the distances between the cochannelled UEs affect the SIR permanence.

In this sense, appropriate cellular system parameters need to be used in order to maximize the system capacity $C = B\log_2(1 + SINR)$ while treating interference as noise, as given in [8]. CCI is more severe in urban areas, due to a large number of BSs and mobile users. Unlike thermal noise/AWGN problem, CCI can not be mitigated by increasing transmission power. In fact, higher transmission power elevates the intensity of the interference suffered by the nearby cochannel user. Hence, new techniques such as cell sectorization, splitting, fractional frequency reuse, interference cancellation, and interference reduction are used in practice to manage CCI in cellular systems.

10.3.1.2 Cochannel Interference Control via Resource Assignment

In the literature, several channel assignment techniques have been proposed to compensate for the spectrum scarcity and achieve the objective of high capacity via proper control of interference for efficient utilization of the radio spectrum listed as follows. Note that the hexagonal cell shape is used in this chapter as it has a near perimeter and area to the ideal cell with a circular shape.

- *Cluster Size* (N_c): This defines the number of cells in which all the channels are used for data communication. As an example, in Figure 10.21a, each group of cells with the same color represents a cluster. It is assumed that the channels are equally split between the cells within the cluster and reused for all the other clusters. In case all the cells have the same radios and each BS emits the

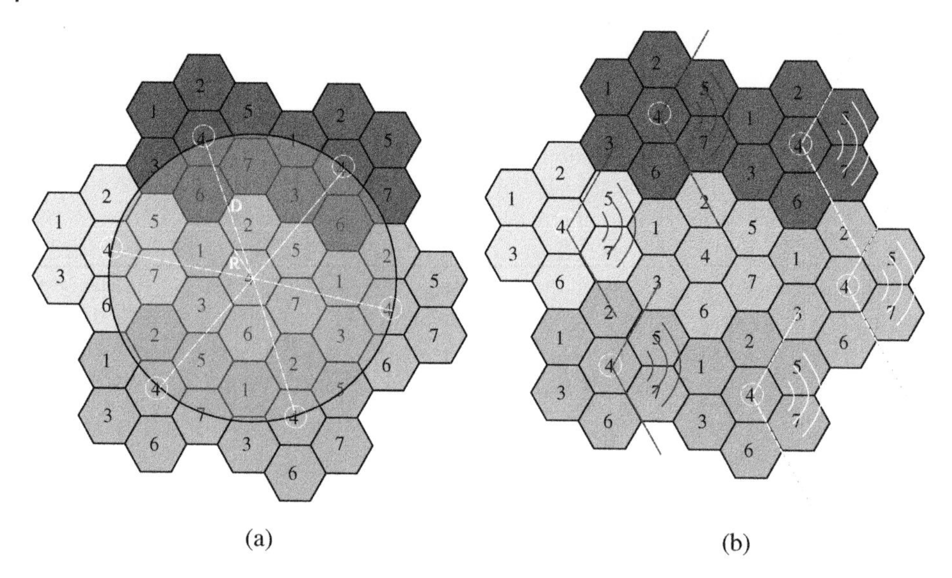

Figure 10.21 Hexagonal cell with reuse factor $N_c = 7$. (a) Conventional case. (b) 120° sectoring.

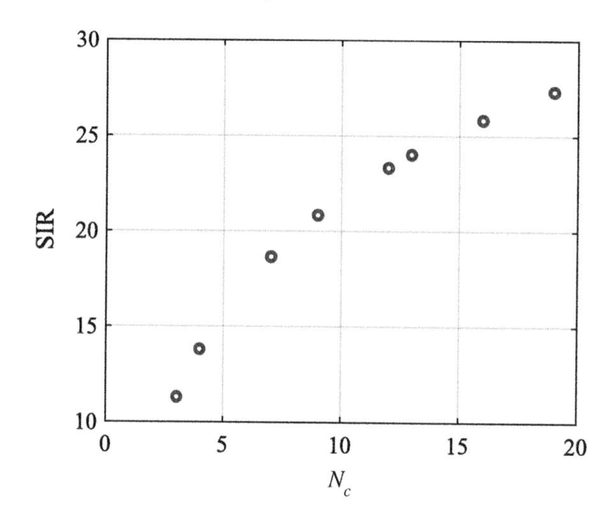

Figure 10.22 SIR vs N_c.

same energy, CCI becomes a function of the displacement between cochannel cells. The *Cochannel reuse ratio* (Q) parameter is defined to show the relation between CCI and displacement, and it is given as

$$Q = \frac{D}{R} = \sqrt{3N_c}, \tag{10.27}$$

where D is the distance between the cochannel cells. The interference is reduced with the increase of the cluster size, resulting in higher SIR (Eq. 10.26). In Figure 10.22, the SIR performance with respect to the cluster size is given, assuming that only first-tier cochannel UEs affect the desired signal, where $SIR = Q^n/6$. On the other hand, the number of accessing UEs per cell linearly decreases with the increase of N_c, which limits the system capacity. In the following Matlab code, we show how the SIR varies with respect to the cluster size. As can be seen in the Figure 10.22, the interference decreases with the increase of the cluster size as the displacement between cochannel UEs becomes significant, acting as an real guard.

```
1   % Calculation of cochannel interference (CCI) with
2   % respect to reuse factor Nc.
3   I = 6; % First tire interfer
4   eta = 4; % Path loss factor
5   Nc = [3,4,7,9,12,13,16,19];
6   for ii = 1:length(Nc);
7       Q  = sqrt(3*Nc(ii));
8       SIR(ii) =(Q^eta)/I;
9   end
10  figure
11  plot(Nc,10*log10(SIR),'ob','LineWidth',2);
12  xlabel('N'); ylabel('SIR');
13  title('SIR vs N'); grid on
```

- *Sectorization:* It is performed to split the area into sections in order to mitigate interference in Eq. (10.26) and enhance capacity by utilizing directional antennas, as illustrated in Figure 10.21b. In case 120° and 60° sectorization are applied at each BS, the number of the interfering cochannel cells can be reduced from 6 to 2 and 1, respectively, considering only the first tier of a cellular system in Figure 10.21a. Therefore, the CCI is controlled at the cost of higher transceiver complexity and energy consummation compared with omnidirectional transmission designs. *Beamforming* can be considered as an extreme case of sectorization, where beamshaping is performed to serve UEs in a specific direction.
- *Cell splitting*: Refers to the process of subdividing a congested cell into smaller cells (microcells) in order to improve the system capacity and serve more UEs without elevating the CCI problem. The system capacity is improved since, by introducing the microcells, the number of times that the channels are reused is increased. Heights and transmission powers of the microcells' BSs must be appropriately reduced so that the minimum co-channel reuse ratio (Q) given in Eq. (10.27) and the SIR performances given by Eq. (10.26) are maintained.
- *Fractional frequency reuse* (FFR): This groups the UEs into cell-center UEs and cell-edge UEs while partitioning the spectrum in order to control CCI. For instance, the cell area is split into two regions, including the area near to the BS and the border region of the cell, as illustrated in Figure 10.23. Higher power is allocated to channels for the edge region compared with the inner one[9], which is expected to be isolated within the inner region and not to pass the cell border. In this way, a more solid space guard is guaranteed in case the inner region is utilized by cochannel UEs and a higher number of cochannel UEs can enhance the system capacity with respect to the conventional channel reuse methods. This scheme is known as *strict* FFR, which is illustrated in Figure 10.23a. *Soft* FFR not only employs the same spectrum partitioning strategy as *strict* FFR, but also allows the inner UEs to share the spectrum with the other cells, as shown in Figure 10.23b, aiming to increase the number of available channels at the cost of lower SIR.

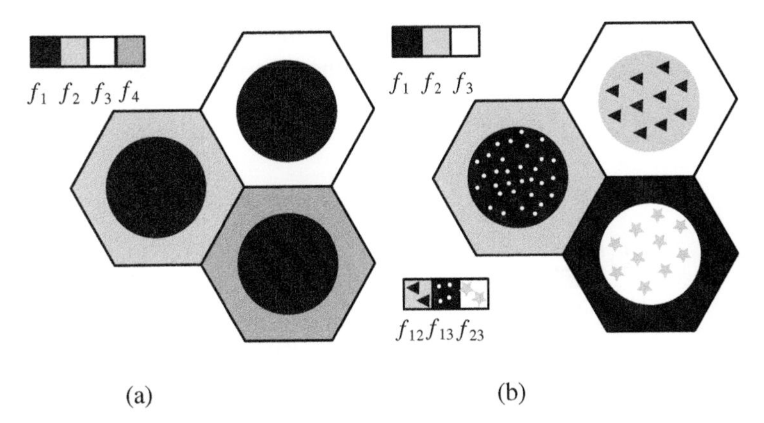

Figure 10.23 Hexagonal cell with three cell-edge reuse factor. (a) Strict FFR. (b) Soft SFR.

10.3.2 Interference due to Small-Scale Fading

Different from *large-scale fading*, where CCI is caused due to aerial spreading of the signal, mainly time spreading and frequency dispersion of the signal due to the wireless channel lead to new interference types in *small-scale fading*. These interference types are given as follows.

- *Inter-symbol interference*: This occurs due to the time spreading of transmitted signal pulses because of multipath propagation, causing the adjacent pulses to overlap in time, as illustrated in Figure 10.24. This interference type leads to substantial degradation of communication reliability. Therefore, in order to solve and control ISI, several techniques have been proposed by academia and industry. Mainly, parameters for *frequency-flat fading* are explored to control the ISI problem, where it can be avoided by leaving enough space in between consecutively transmitted symbols ($T_s >> \tau_{max}$). T_s and τ_{max} denote the symbol duration and maximum excess delay, respectively. However, *frequency-flat fading* criteria represents any medium with limited channel bandwidth, resulting in a challenge for performing transmission with high data rates [10]. Not only the fact that the symbol duration of the transmitted signal is limited to τ_{max}, but also the time domain resolvability of the channel multipath is limited since the channel behaves as a single tap. Therefore, multitap equalization cannot be performed to take advantage of the time diversity of the time spread signal. The smart use of an equalizer is like an "anti-channel" at the cost of high system complexity.
- *Inter-carrier interference*: It is a manifestation of the *time selectivity* of the channel in multicarrier systems, i.e. orthogonal frequency division multiplexing (OFDM). The transmitted subcarriers of a classical OFDM signal experiences some frequency modulation, corresponding to the loss of orthogonality among the subcarriers, as shown in Figure 10.25. The carrier frequency shift or modulation is denoted by f_{D1} for the first subcarrier and so on for the others. Different from the ISI of *frequency-selective fading* and the constraints on data rate (T_s), a shorter symbol duration than coherence time ($T_s << \tau_c$) is needed to mitigate the ICI impact in a *fast fading* medium. Moreover, the short time duration of the symbol increases the time domain resolution of the channel multipath. Full knowledge of channel parameters can help to mitigate the ICI effect at the cost of high system complexity. The reader is advised to visit Chapter 7 in order to gain more insights regarding the impact of *fast fading* in OFDM-based multicarrier systems. For instance, since *time selectivity* (Doppler) causes loss of orthogonality in the frequency domain, its effect on the system performance depends on the carrier frequency of the transmitted signal and subcarrier spacing (SCS). The latter is the main driving factor of the 5G wireless technology on the multinumerology concept, where different SCS values are agreed by academia and industry to be utilized for various use cases and applications.

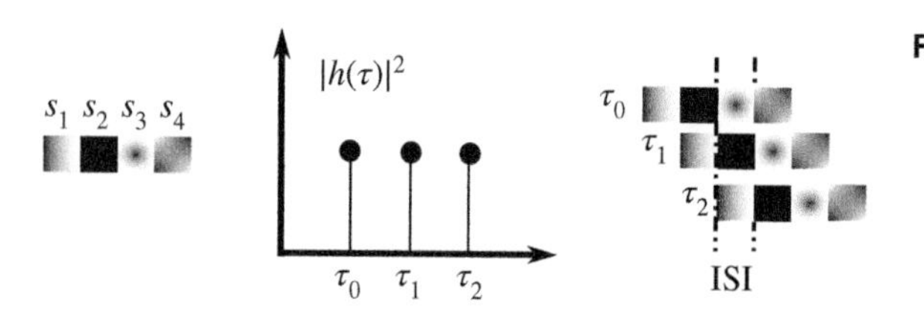

Figure 10.24 Illustration of ISI.

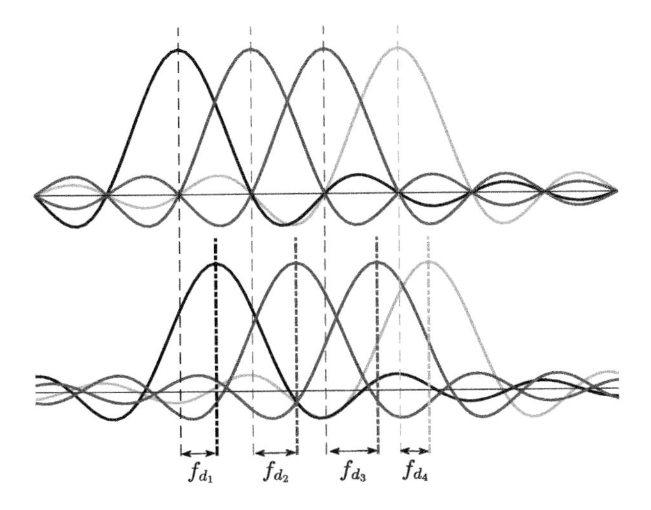

Figure 10.25 Illustration of ICI in OFDM.

10.4 Channel Modeling

Characterization of the wireless propagation phenomena and their related impairments on the communication signal is a stepping stone toward designing a well-optimized wireless system. However, a rigorous characterization of the propagation effects in a form that is useful for designing and optimizing a wireless communication system is a difficult task. In order to simplify the process, the concept of channel modeling is generally used. The basic objective of channel modeling is to identify and analytically represent the most essential and dominant propagation effects relevant to the considered communication scenario. In other words, a channel model captures only the portion of the propagation phenomena that has significant impact on the system of interest while ignoring the rest.

Channel modeling is a continuous practice whose importance and necessity grow with the advancement in wireless technology itself. For example, the paradigm shift from microwave to mmWave frequencies in 5G necessitated the extension of the existing channel models or development of new models that capture propagation effect which might have been insignificant in microwave frequencies. This is mainly because of the fact that wireless propagation phenomena are highly frequency dependant. Likewise, other factors such as environment settings in which communication system is deployed, i.e. rural, urban, mountainous, underground mine, etc., and signaling schemes may affect the choice of the channel model.

In the development of a new channel model the first and foremost thing that needs to be considered is the decision on the aspects of channel characteristics that need to be captured and the ones that can be ignored with respect to the intended system. This is important for balancing between complexity of the model and its adequacy in expressing the essential propagation impairment. In general, a channel model faces trade-offs in terms of its *accuracy*, *generality*, and *simplicity*. By accuracy, we refer to how close the channel effects realized by the model are to the reality. Generality, on the other hand, refers to the applicability of the model to different scenarios. Simplicity of a model is quantified by its computation and storage complexity level. Majority of the channel models available in the literature can be categorized into *analytical* and *physical* models. Further classification under these two categories are as summarized in Figure 10.26 [11].

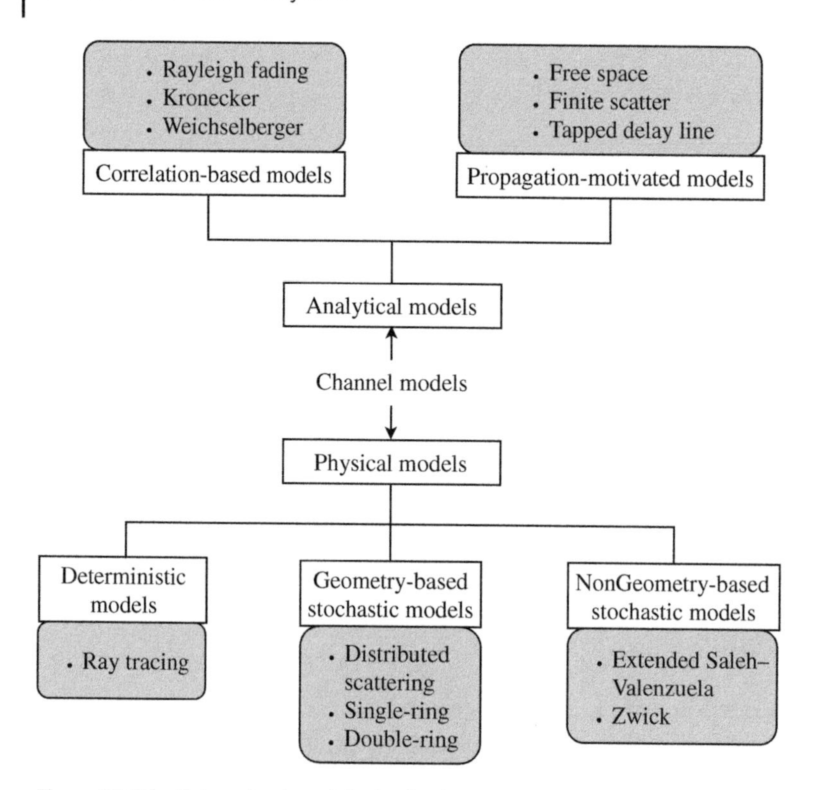

Figure 10.26 Categorization of the basic channel models.

10.4.1 Analytical Channel Models

Analytical channel models characterize channel effects mathematically without explicitly considering any physical EM propagation mechanisms. This category of channel models can be further divided into correlation-based and propagation-motivated models.

10.4.1.1 Correlation-based Models

Correlation-based analytical models characterize rich scattering wireless channels by considering the existing correlation between multipath channel components. These models are generally less complex and they are commonly used for system performance evaluation. This class of channel models is also appropriate for characterizing wireless channels in multi-antennas systems such as MIMO systems, in which channel correlations between each transmitter–receiver antenna pair is of paramount importance. Examples of the channel models that fall under this class includes Rayleigh/Rician fading channel models, Kronecker, and Weichselberger models as shown in Figure 10.26.

10.4.1.2 Propagation-Motivated Models

Propagation-motivated models, on the other hand, represent wireless channels in terms of propagation parameters. Free space model, for example, serves as the simplest propagation model which considers atmospheric absorption effects undergone by the transmitted signal in the propagation medium (free space). However, when a more realistic scenario is considered, more comprehensive models which encompass other propagation effects such as reflections, diffraction, and scattering become necessary. Finite-scatter, tapped delay line, and the virtual channel representation models are examples of the propagation models that take into account the aforementioned propagation mechanisms and their resulting multipath phenomenon.

10.4.2 Physical Models

Contrary to the analytical channel models described just above, *physical models* characterize wireless channel by considering the physical EM propagation phenomena taking place between transmitter and receiver antennas. In this case, propagation parameters such as complex amplitude, angles of departure and arrival of the EM wave, and delay effects experienced by multipath replicas of the communication signal are explicitly modeled. Physical channel models are generally independent of antenna configuration and operational bandwidth of the system. Further subclasses that fall under physical channel models include geometry-based stochastic, non geometry-based stochastic, and deterministic models as summarized in Figure 10.26.

10.4.2.1 Deterministic Model

Deterministic channel modeling approach aims to reproduce the actual propagation process undergone by the EM wave in a given communication scenario. This modeling approach is suitable for scenarios in which the scatterers in the environment are nearly static. For example, it can be used to develop a channel model for dense urban environment where there are a significant number of large buildings and other static infrastructures that determine paths taken by the EM signal on its way to the receiver. A three dimension map of the communication site containing all major structures along with the EM properties of their construction materials (i.e. permittivity and permeability coefficients) is first constructed and then used to extract the corresponding propagation process via computer simulation. A common example of the deterministic physical channel model is the RT technique. RT employs fundamental physical principles such as Maxwell's equations, geometrical optic theory, and the uniform theory of diffraction to trace propagation of the EM wave between transceiver nodes.

Owing to the great accuracy achieved by deterministic models, deterministic models are often used as alternatives to practical channel measurements. However, these models lack generality, as they are usually specific to a given communication scenario (i.e. site-specific). Additionally, prohibitively long simulation times may be required to generate a model, especially when complex environment settings are considered.

10.4.2.2 Geometry-based Stochastic Model

Geometry-based stochastic modeling approach follows similar assumption like in the deterministic modeling that channel response is determined by geometrical locations of the scatterers. However, in this case, location of the scatterers are not prescribed in the database beforehand, they are rather assumed to be random, following a certain probability distribution. When the scatterers are stochastically identified, RT technique is then used to derive the required channel model.

Depending on the location of scatters in the scenario, geometry-based stochastic models may assume single- or multiple-bounce scattering patterns. In the single-bounce or single-ring scattering model, a BS is assumed to be in an elevated location such that no scatterer exists in its vicinity, and the user equipment (UE) is assumed to be surrounded by a significant number of scatterers. Such a scenario in which a single ring of scatter around the UE is illustrated in Figure 10.27a. Generally, the scatterers in the ring of radius R are assumed to be uniformly distributed over $[-\pi, \pi)$. In some scenarios, such as communication in microcells, BSs are located below rooftop heights such that they are also surrounded by a significant number of scatterers. Such scenarios are incorporated

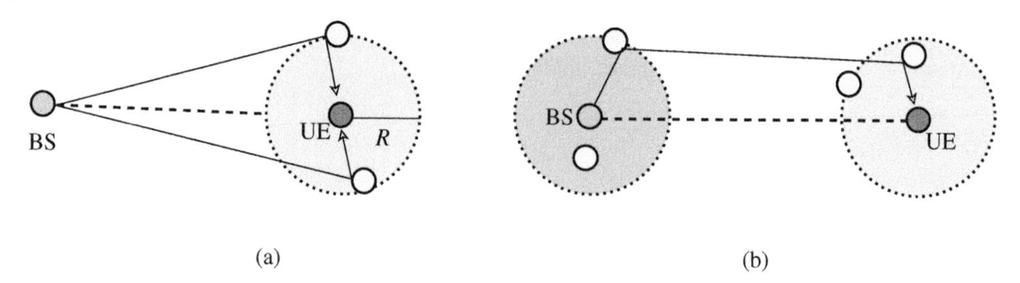

Figure 10.27 (a) Single-ring scatter. (b) Double-ring scatter.

in the multiple-bounce or double-ring scattering models in which both BS and UE are assumed to be surrounded by rings of scatters as illustrated in Figure 10.27b.

10.4.2.3 Nongeometry-based Stochastic Models

The last sub-category of the physical channel models is the nongeometry-based stochastic models. This particular modeling approach represents EM propagation between transceiver ends using stochastical parameters only without considering geometry of the physical environment. Examples of the channel models under this subclass include the extended Saleh-Valenzuela and Zwick models. The Saleh-Valenzuela model proposes to model a wireless channel as clusters of MPCs in the delay domain [12]. It uses a doubly-exponential decay process to simultaneously model channel's PDP and characterize the profiles of the MPCs within individual clusters. The scope of the Saleh-Valenzuela modeling approach can be easily extended into the space domain by considering AoA and AoD statistics. In this case, spatial clusters are generated based on the mean cluster angle and the clusters' angular spread. The illustration of the Saleh-Valenzuela modeling approaches in time and space domains is given in Figure 10.28. Note that when signal bandwidth is large enough, all MPCs become resolvable in time domain, which inhibits the formation of clusters in the channel's PDP. In order to account for such typical scenarios, the Zwick model was proposed [13]. In the Zwick approach, the MPCs are generated and treated independently (i.e. no cluster). Amplitude fading of the MPCs is ignored, but their phase changes are incorporated into the model via geometric modeling of the transmitter, receiver, and scatters motions.

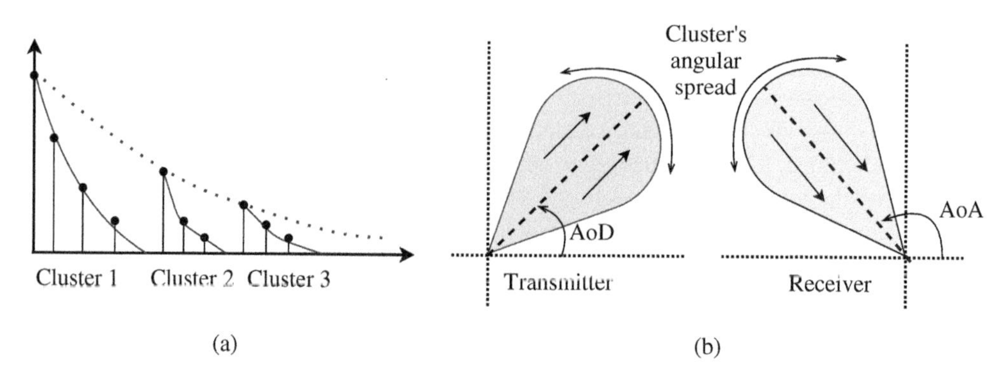

Figure 10.28 Illustration of the Saleh-Valenzuela model. (a) Delay domain. (b) Space domain (single cluster).

10.4.3 3GPP 5G Channel Models

In order to facilitate proper modeling and performance evaluation of the 5G physical layer (PHY) techniques, 3GPP has provided two channel models for link level simulation. These models include tapped delay line (TDL) and clustered delay line (CDL) models. Both models are valid for 0.5–100 GHz frequency range with maximum bandwidth of 2 GHz. This range is compatible with 5G technology which is standardized to operate at both microwave and mmWave frequencies.

10.4.3.1 Tapped Delay Line (TDL) Model

The TDL model is considered for evaluation of the simplified, non-MIMO systems, i.e. SISO based systems. CIR of the TDL model is similar to the one given in Eq. (10.10). Five different PDPs are considered under this model based on the characteristics of the environment and the presence or absence of the LOS component as shown in Table 10.3. The power and delay values of each tap are given in [14].

Doppler spectrum of each tap is characterized by the classical Jake's model with the maximum Doppler shift f_{Dmax} obtained as described in Section 10.2.3. For TDL_D and TDL_E models in which LOS path is considered (see Table 10.3), the Doppler spectra contain a peak at $f_D = 0.7 \times f_{Dmax}$ with an amplitude such that the resulting distributions have the specified Rician K-factor.

Note that delay spread values of the taps can be modified/scaled by a user in order to achieve a desired RMS delay spread. For instance, a specified normalized delay value for an n-th tap, $\tau_{n,model}$, can be scaled to obtain a new delay value $\tau_{n,scaled}$ by

$$\tau_{n,scaled} = \tau_{n,model} \times \text{DS}_{desired}, \tag{10.28}$$

all in [ns], where $\text{DS}_{desired}$ is the desired delay spread. Similarly, the model also allows the user to adjust the values of the K-factors specified for TDL_D and TDL_E models. For example, if the specified K-factor, K_{model}, is to be changed to $K_{desired}$, powers of all NLOS taps (i.e. the Rayleigh fading taps) must be scaled by

$$P_{n,scaled} = P_{n,models} - K_{desired} + K_{model}, \tag{10.29}$$

all in [dB], where $P_{n,scaled}$ and $P_{n,models}$ are the scaled and model path power (i.e. predefined by the model) of the nth tap. After the power scaling process, the delay spread must be normalized. The normalization is performed by calculating the actual RMS delay spread after the K-factor adjustment and then dividing the delay of each tap by the calculated RMS delay value.

An example Matlab code for generating and visualizing TDL channel model is given below. In this code, we use `nrTDLChannel` object from 5G toolbox. Channel profile visualizations are given in Figure 10.29 for TDL_C (without LOS component) and TDL_E (with LOS component).

Table 10.3 Summary of the TDL model's specified PDP models.

PDP model	Number of taps	LoS tap dist.	NLoS taps dist.
TDL_A	23	—	Rayleigh
TDL_B	23	—	Rayleigh
TDL_C	24	—	Rayleigh
TDL_D	13	Rician ($K = 13.3$ dB)	Rayleigh
TDL_E	14	Rician ($K = 22$ dB)	Rayleigh

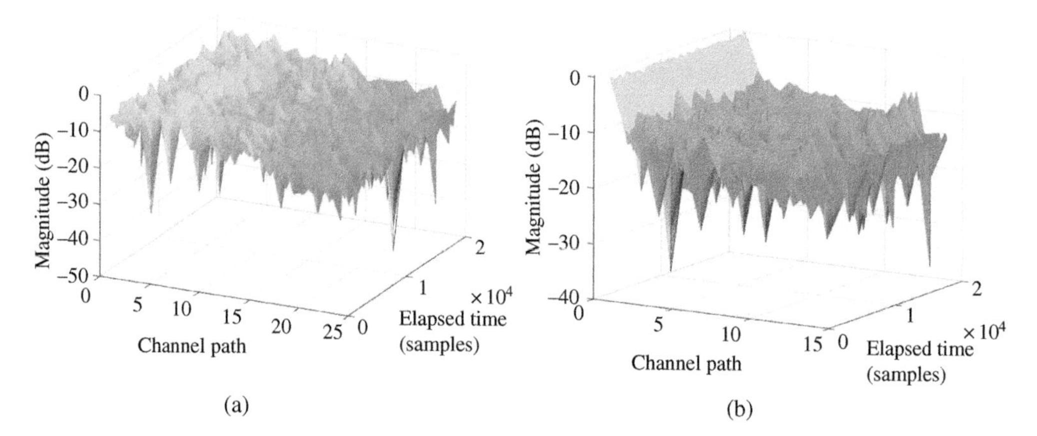

Figure 10.29 Profiles of the TDL channel. Observe the presence of a LOS component in TDL_E. (a) TDL_C. (b) TDL_E.

```
1  % Generation and visualization of TDL channel coefficient
2  % via Matlab toolBox.
3
4  % Specify channel model and the related parameters
5  tdl = nrTDLChannel;
6  tdl.DelayProfile = 'TDL-C';         % 'TDL-E'
7  tdl.SampleRate = 500e3;             % samples/second
8  tdl.MaximumDopplerShift = 300; % Hz
9  % Setup a SISO system
10 tdl.NumTransmitAntennas = 1;
11 tdl.NumReceiveAntennas = 1;
12 % Generate input Tx signal
13 Tx_signal = ones(20000,tdl.NumTransmitAntennas);
14 % Get channel's path gaind
15 [Rx_signal, pathGains] = tdl(Tx_signal);
16 % Figure
17 figure()
18 mesh(10*log10(abs(pathGains)));
19 view(26,17); xlabel('Channel Path');
20 ylabel('Elapsed time (Samples)'); zlabel('Magnitude (dB)');
```

10.4.3.2 Clustered Delay Line (CDL) Model

The CDL channel model is essentially an extension of the above discussed TDL model developed specifically for the spatial channels such as those pertaining to mmWave massive mMIMO systems. Like in the case of the TDL model, five PDPs are defined for the CDL model as well, three for NLOS scenarios and two for LOS scenarios. Note that, in this case, channel clusters are considered instead of taps and the power angular spread within a cluster is considered to follow Laplacian distribution [14].

Scaling of the predefined cluster parameters is also possible in this model. Delays of the clusters as well as K-factors of the PDP models with LOS component can be adjusted to the desired values through similar procedures given above for the TDL model. Predefined values of the angles of arrivals and departure (in both zenith and azimuth directions) of the clusters can be changed to the desired values via angular translation and scaling process. By using angular translation, the tabulated mean angle, $\mu_{\vartheta,model}$ is changed to the desired mean angle $\mu_{\vartheta,desired}$. The scaling process is used to adjust angular spread of the clusters. Angular translation and scaling in CDL model is obtained as

$$\vartheta_{n,scaled} = \frac{AS_{desired}}{AS_{model}} \left(\vartheta_{n,model} - \mu_{\vartheta,model}\right) + \mu_{\vartheta,desired}, \tag{10.30}$$

where $\vartheta_{n,model}$ is the tabulated angle of n-th cluster, $\vartheta_{n,scaled}$ is the scaled angle, AS_{model} is the tabulated RMS angular spread, and $AS_{desired}$ is the desired RMS angular spread. Note that, the angle parameter ϑ used here can be zenith or azimuth angle.

10.4.3.3 Generating Channel Coefficients Using CDL Model

In general, generation of the channel coefficients begins with identification of the environment type, network layout, and antenna array parameters. The propagation condition, LOS or NLOS, needs to be specified as well. Based on this information, path loss is calculated and large-scale parameters, such as delay spread, shadowing fading, angular spread, and K-factor, are generated. Afterwards, small-scale parameters, such as cluster power, angles of arrival and departure, and XPR, are generated. Once all these parameters are obtained, the channel coefficients can be generated using the equation given in Figure 10.19. The full roadmap for generating the CDL channel coefficients is given in Figure 10.30.

Note that, for link level simulations, cluster powers, delays, and other large-scale parameters shown in Figure 10.30, are already defined in [14] for different propagation conditions and scenarios. Procedures for generating AoDs and AoAs for each ray within the clusters, random coupling of rays and XPRs are summarized as follows:

- Generating departure and arrival angle for each ray in each cluster: An azimuth AoDs for the m-th ray in the nth cluster is given by

$$\phi_{n,m,AoD} = \phi_{n,AoD} + c_{ASD}\alpha_m, \tag{10.31}$$

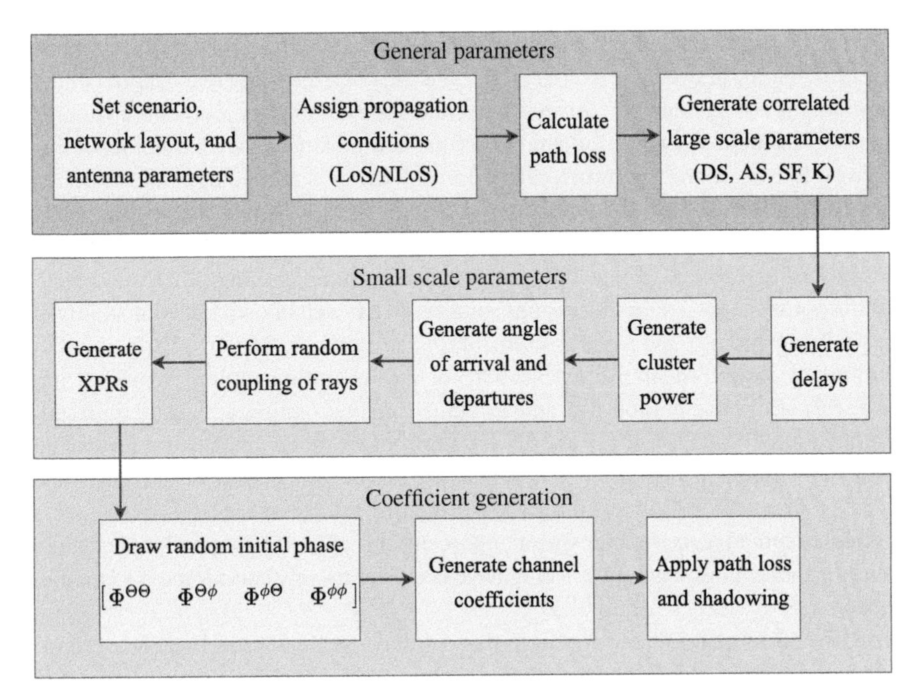

Figure 10.30 CDL channel coefficient generation steps. Source: 3GPP Radio Access Network Working Group [14]. ETSI.

where $\phi_{n,AoD}$ is the n-th cluster's azimuth AoD, c_{ASD} is the cluster-wise RMS azimuth spread of departure angle, and α_m is the ray offset angle within the cluster. numerical values of $\phi_{n,AoD}$, c_{ASD}, and α_m are predefined and tabulated in [14]. In a similar manner, $\phi_{n,m,AoA}$, $\Theta_{n,m,ZoD}$, and $\Theta_{n,m,ZoA}$ can be obtained as

$$\phi_{n,m,AoA} = \phi_{n,AoA} + c_{ASA}\alpha_m, \tag{10.32}$$

$$\Theta_{n,m,ZoD} = \phi_{n,ZoD} + c_{ZSD}\alpha_m, \tag{10.33}$$

$$\Theta_{n,m,ZoA} = \phi_{n,ZoA} + c_{ZSA}\alpha_m. \tag{10.34}$$

- Coupling of the rays within a cluster for both azimuth and zenith directions: Randomly couple AoD angles $\phi_{n,m,AoD}$ to AoA angles $\phi_{n,m,AoA}$ within each cluster. Similarly, couple ZoD angles $\Theta_{n,m,ZoD}$ to the ZoA angles $\Theta_{n,m,ZoA}$ in a random fashion. Again, couple randomly AoD angles $\phi_{n,m,AoD}$ to the ZoD angles $\Theta_{n,m,ZoD}$ within each cluster.
- Generation of the XPR values:

$$\kappa_{n,m} = 10^{X/10}, \tag{10.35}$$

with X being the per-cluster XPR (in dB) given in [14].

10.4.4 Role of Artificial Intelligence (AI) in Channel Modeling

The above discussed approaches of developing channel models are quite complex and time-consuming due to the required exhaustive processing and analysis of the measured data. Additionally, the developed models are usually given in terms of over-simplified mathematical relationships which in turn degrade their degrees of fidelity. Due to the critical nature of the envisioned futuristic use cases and applications, more reliable channel models will be strictly required and any discrepancy between a channel model and the corresponding real channel that it is ought to present might not be tolerable. Furthermore, since the complexity level of the channel evolves with the wireless technology, it is apparent that the complexity of the channel modeling process will increase drastically in the future generations of wireless technology. Therefore, an alternative approach of modeling the wireless channel more efficiently with reasonable complexity must be sought.

To this end, Artificial Intelligence (AI) techniques, such as machine learning (ML), are considered to be potential approaches of generating channel models. ML is well known for its capability of enabling a system to learn, predict, and make assessments of a situation without involving a human intelligence. Most importantly, the astonishing capability of ML in handling multidimensional and multivariety data with affordable complexity, and its ability of constructing a realistic representation of a phenomenon without requiring a strict definition of its model, makes it a promising approach for channel modeling in the future wireless networks. In fact, there has already been some growing effort of exploiting various ML techniques for estimating the channel's fading parameters such as path loss, delay spread, and Doppler spread. ML techniques based on supervised learning, unsupervised learning, and reinforcement learning have been extensively considered for channel estimation recently.

Deep learning (DL) is another ML based approach that exhibit a great potential in terms of channel modeling. With the help of its hidden layers based architecture, DL possess a great capability of tackling problems that do not have precise numerical models. The multiple hidden layers are capable of suppressing noise while appropriately preserving the intrinsic distinctive features that

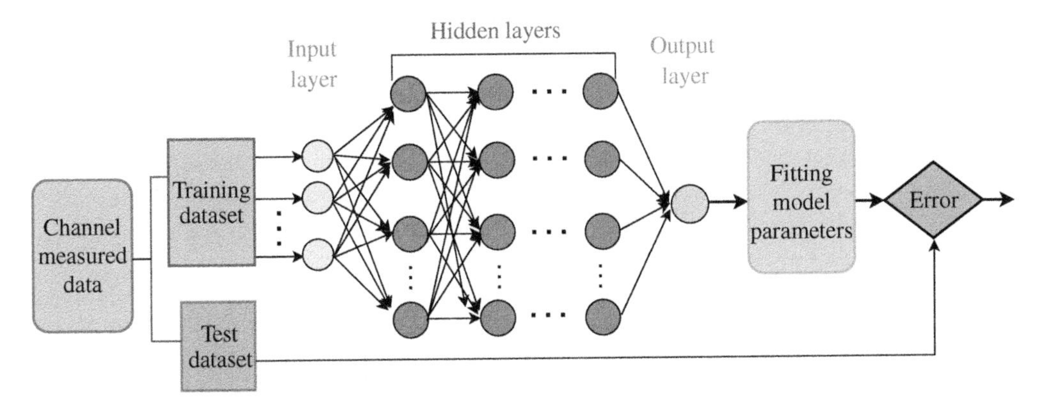

Figure 10.31 Illustration of the channel modeling steps with DL based neuro-network.

define the characteristics of the propagation phenomena. Generally, DL takes raw data as input, thus, it does not necessitate preprocessing of the data for feature extraction.

An example of modeling a channel by using a feedforward deep neural network (DNN) approach is shown in Figure 10.31 [11]. The collected real channel measurement data are first used for training by using back propagation or Newton algorithm, followed by the DNN algorithm that estimates the weights parameters to build the model. Test data are then used to examine performance of the developed model. The error calculated at the end serves as a measure of the accuracy of the developed channel model as well as efficiency of the used algorithm.

In general, the research on the applicability of ML techniques on wireless channel modeling is still in its infancy and there are a number of challenges that need to be addressed. Here are some of these challenges:

- Since the performance of the ML approaches is determined by the nature of the data set used for training process, the sufficiency of training data set in terms of quality and quantity from a channel modeling perspective needs to be investigated.
- Traditional ML problems, such as underfitting and overfitting of the models during the training process, pose a threat on the accuracy of the developed models.
- Suitable criteria that govern the selection of a certain ML technique for a particular scenario are yet to be investigated.
- ML have not yet been able to solve the problem of generality. That is, when a new environment is considered, the model need to be retrained using new, appropriate data set.
- The idea of using AI techniques in channel modeling and estimation brings a new physical layer security (PLS) threat.

10.5 Channel Measurement

Measurement of the propagation characteristics of wireless channel is a complex task. However, it is very crucial ingredient for designing a radio system and network planning. For instance, path loss and shadowing fading characteristics of a channel need to be accurately measured in order to determine the network's coverage area. Furthermore, channel measurement plays an important role in developing channel models, which are also essential tools for designing a wireless system and validating new algorithms. Note that it is also possible to obtain channel models via simulation-based approaches, such as RT, which are advantageous in terms of cost. However, plenty of prior information regarding propagation environment details are necessary to develop an acceptable simulation-based model, leading to high computational complexity which in turn limits the general applicability of such approaches. Consequently, channel measurement and empirical channel modeling are inevitable tasks. In a nutshell, channel measurement can be essential for (i) observing and understanding behavior of the channel in the relatively new environmental setups, (ii) developing more accurate channel models, and (iii) testing and validating new communication systems.

Generally, the channel measurement process is conducted through well-designed techniques known as channel sounding techniques. Conceptually, a typical sounding technique involves sending a sounding signal that excites the channel and observing its response at the receiver. Design of the sounding signal depends on the sounding technique used. Based on the transmit bandwidth of the sounding signal, the sounding techniques can be categorized into narrowband and wideband channel sounders [15].

Narrowband channel sounding: Narrowband simply refers to the transmission where the inverse of the signal bandwidth is much greater than the multipath delays of the channel. Statistical characteristics pertaining to such transmission are usually determined from measurements carried out at a single frequency. Consequently, narrowband sounding techniques employ an unmodulated carrier wave (single tone) signal as a sounding signal. The carrier signal is transmitted to excite the channel and the response is captured by a mobile or stationary receiver. Multipath effect of the channel is observed as the variation of the amplitude and phase of the received signal due to the random phases addition of the signal arriving over many scattered paths.

Narrowband measurement mainly captures time domain fading characteristics of the channel, which can be easily observed from the fading envelope of the received signal. Since this measurement captures channel response at a single frequency, it has infinite temporal resolution which makes it impossible to distinguish replicas of the signal arriving with different delays. Therefore, narrowband channel sounders are unable to unveil some important channel parameters such as maximum excess delay, RMS excess delay and coherence bandwidth, which are very crucial in understanding the small-scale characteristics of the channel. Such parameters are more related to the wideband channels.

Wideband channel sounding: Modern wireless systems such as cellular and wireless fidelity (Wi-Fi) normally operate in wideband channels. In order to capture the aforementioned wideband-related channel parameters, a sounding signal occupying a wide bandwidth is required. Several wideband channel sounding techniques based on periodic pulse and pulse compression techniques are available in the literature. Details about these techniques will be given in the subsequent subsections.

In general, wideband channel sounders can be categorized into time and frequency domain sounders. Time domain sounders focus on capturing temporal characteristics, whereas frequency domain sounders are intended to capture spectral characteristics of the channel. Nevertheless, spectral and temporal behaviors of the channel are like two sides of the same coin, that is, their corresponding parameters are usually interderivable. For instance, a fast Fourier transform (FFT) operation on the CIR of a given channel gives frequency correlation behavior of the same channel. However, a decision on which technique to use can be made based on the targeted application for the transmitted data as well as complexity in processing the measurement result.

10.5.1 Frequency Domain Channel Sounder

The narrowband channel sounding approach described previously can be extended and applied for wideband measurement as well, simply by stepping the single tone signal across discrete frequencies within the desired band. In other words, a transmitter is set to sequentially send a series of single tone signals with different carrier frequencies over the desired channel bandwidth with a constant transmission power. The effective received signal captured by a spectrum analyzer constitutes a transfer function that is equivalent to the low-pass transfer function of the measured channel. An inverse fast Fourier transform (IFFT) process can then be used to obtain the CIR. This approach serves as a straightforward, simple and inexpensive way of measuring a wideband channel. Furthermore, it has relatively high degree of accuracy, as the transmitter is tuned and phase-locked to each of the discrete center frequencies within the intended band. On the other hand, this technique has a number of drawbacks that limit its applicability. For instance, it becomes highly time-consuming when frequency steps are too small (i.e. large number of discrete frequencies need to be stepped on) or when the bandwidth to be measured is too large. Additionally, the frequency stepping technique produces meaningful results only when the channel remains constant during one complete measurement across the band. As a result this approach might not be suitable for measuring a rapidly varying channel due to mobility in the communication environment, i.e. no Doppler effect that can be accurately measured. Furthermore, the technique can face hardware limitation, as most of the equipment, antennas in particular, are optimized for a specific center frequency and thus may not give a reliable measurement result at other frequencies.

10.5.1.1 Swept Frequency/Chirp Sounder

As an alternative to the frequency stepping channel sounder, the frequency sweep technique or chirp sounder, which is relatively faster, is commonly used. With the chirp sounder technique, the desired channel bandwidth, ΔB, is swept continuously from the lower to higher frequency and the received signal is sampled without stopping the sweep. Time domain chirp signal or transmission waveform is given by

$$p(t) = \exp\left[j2\pi\left(f_o t + \Delta B\frac{t^2}{2T_c}\right)\right], \quad 0 \le t \le T_c, \tag{10.36}$$

where the term $f_o + \Delta B \cdot t/2T_c$ represents instantaneous frequency, f_{inst}, which increases linearly with time. T_c is the chirp duration.

Similar to the frequency stepping approach, this technique also measures the complex frequency response of a channel and then makes use of the Fourier relationship to obtain CIR, $h(t)$. Temporal resolution of the obtained CIR depends on ΔB and the frequency domain windowing applied to improve the dynamic range in time domain. A vector network analyzer (VNA) is usually employed for the measurement which captures the channels frequency domain response in terms of S-parameters, $S_{21}(f_{inst})$. Measurement setup and procedure for this method is summarized in Figure 10.32.

Although this method is relatively faster compared to the frequency stepping technique, it is still not suitable enough for a varying channel measurement, as a one complete sweep can take up to few seconds in some cases. Also, cable losses and bulkiness of the VNA limit this technique to the low range, indoor channel measurement.

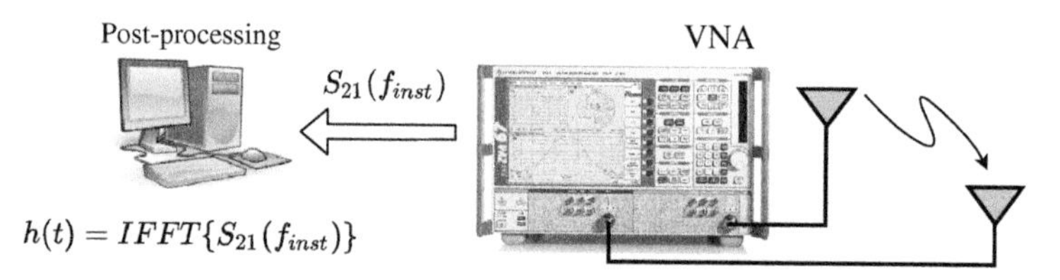

Figure 10.32 Setup of frequency sweeping technique with VNA.

10.5.2 Time Domain Channel Sounder

Time domain wideband channel measurements techniques are used to directly obtain the impulse response of the channel. These techniques are mainly based on either the periodic narrow pulse approach or the pulse compression method. Depending on the receiver structure, the pulse compression sounder can be realized either by matched filter or cross-correlation approaches. Generally, the transmitted signal $s(t)$ is given by

$$s(t) = \sum_{i=0}^{N_p-1} a_i g(t - iT_{Rep}), \tag{10.37}$$

where N_p is the number of pulses to be transmitted during measurement, a_i is the amplitude of the ith pulse, and $g(t)$ is the periodically repeated pulse in every fixed interval T_{Rep}. The difference between the aforementioned two types of time domain sounders is based on the choice of the $g(t)$ waveform.

10.5.2.1 Periodic Pulse/Impulse Sounder

The principle behind this technique is that a narrow (short duration) pseudo-impulse is periodically transmitted to excite the channel. The impulse must be sufficiently narrow to ensure that the signal bandwidth is larger than the coherence bandwidth of the channel being measured in order to capture all the echoes. That is, the pulse width T_w of the transmitted pseudo-impulse signal determines the minimum identifiable delay path resolution. The pulse repetition period T_{Rep} has to be sufficiently rapid to allow observation of the time-varying response of individual propagation paths, but also long enough to ensure that all multipath components decay between successive impulses. On the receiver side, the received signal is firstly filtered by a band pass filter (BPF) with a bandwidth $B = 2/T_w$. The signal is then amplified and measured by an envelope detector to get the attenuation of each received multipath component. The detected CIR is eventually displayed and stored on a high speed oscilloscope. Basically, each transmitted short pulse provides a "snapshot" of the multipath channel at a certain time instance. Average PDP is obtained by averaging the CIRs given by each snapshot over the measurement duration. The transmitter and receiver setups of this sounding technique is illustrated in Figure 10.33.

One important thing about this technique is that, accuracy of the measurement highly relies on the ability to trigger the oscilloscope on the first arriving signal. Therefore, proper synchronization method is of paramount importance for this particular sounding technique. In the case of low-range measurements in the indoor environment, synchronization between equipments can be achieved via direct cable connection. Generally, coaxial or fiber-optic cables are used. In case a long range measurement in an outdoor scenario is desired, global positioning system (GPS) can be used to establish synchronization through a reference signal. An added advantage of this approach is that the measurement locations can be automatically recorded. However, this approach requires both Tx and Rx equipments to have LOS connection with the GPS satellite. Alternative to the GPS approach,

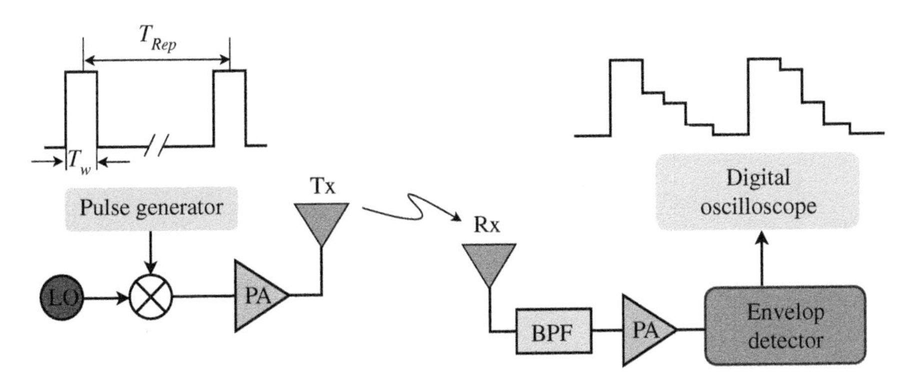

Figure 10.33 Channel measurement setup with periodic pulse technique.

presynchronized rubidium clocks can be employed to facilitate synchronization between equipments. Rubidium clocks are very stable and can retain the synchronization up to several hours. It is also possible to achieve synchronization through the wireless link itself. This is referred to as self triggering. In this case, the recording at the Rx is triggered by the received signal exceeding a certain threshold. However, if the first arriving signal is blocked or experiences a severe fading effect, the system may not trigger properly.

The pulse sounding technique uses an envelope detection technique which captures only the amplitude variation, whereas phase information of the channel, which contain details about the AoAs of the MPCs, is discarded. However, if coherent sources are available at the Tx and Rx, coherent demodulation of the received pulse can be employed to recover phase information, and thus obtaining the complex impulse response of the channel as well as Doppler shift information.

Theoretically, the periodic pulse sounder is an ideal technique if T_w approaches zero, which means that delay resolution is infinitely small. However, such pulse with very small T_w requires high transmit power in order to be able to detect weak multipath components, since all the energy is contained within a narrow pulse. Consequently, peak-to-average power ratio (PAPR) emerges as a major drawback of this technique due to a required large dynamic range of the transmit power amplifier in order to ensure accurate measurement. Another major limitation of the impulse sounder is that it is highly susceptible to interference and noise due to the wide BPF employed at the receiver to facilitate time-domain resolution of the MPCs.

10.5.2.2 Correlative/Pulse Compression Sounders

Correlative sounders are based on the theory of linear system. They assume that the channel and the transceiver equipments constitute a linear system whose output $y(t)$ is the convolution of the transmitted signal $s(t)$ and impulse response $h(t)$ of the channel, summarized as

$$y(t) = h(t) * s(t) = \sum h[\tau]s[t-\tau]. \tag{10.38}$$

It is well established in the theory of linear system that if white noise $w(t)$ is fed into the input of a linear system (i.e. $s(t) = w(t)$) and the obtained output is cross-correlated with a delayed version of the input (i.e. $w(t-\tau)$) then the resulting coefficients are proportional to $h(t)$ evaluated at $t = \tau$:

$$\mathbb{E}\{y(t) \cdot w^*(t-\tau)\} = \mathbb{E}\{h(t) * w(t) \cdot w^*(t-\tau)\}, \tag{10.39}$$

$$= \mathbb{E}\{h(t) * N_0\delta(\tau)\}, \tag{10.40}$$

$$= N_0 h(\tau), \tag{10.41}$$

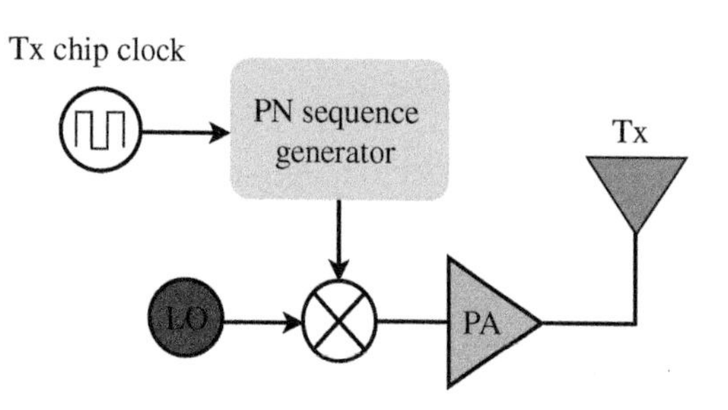

Figure 10.34 Correlative sounder transmitter system.

where $\mathbb{E}\{\cdot\}$ is the expectation notation, N_o is the noise power spectral density, and $\delta(\cdot)$ is the Dirac delta function. The process described just above shows that the channel response $h(t)$ can be obtained with the aid of a white noise signal and cross-correlation processing. However, generating a perfectly white noise is not feasible in practical scenarios. Alternatively, due to its noise-like characteristics, a pseudo-random binary sequence (PRBS), or simply pseudo-noise (PN) sequence, such as the maximum length PN sequence (m-sequence) can be used to achieve similar result. Generation of m-sequence is relatively easy and it is practically realizable by using linear feedback shift registers (LFSRs). The LFSRs with N_b shift registers and XOR gates generates an m-sequence signal of length N_{PN} where

$$N_{PN} = 2^{N_{\text{bit}}} - 1. \tag{10.42}$$

Figure 10.34 shows a block diagram of the essential components that constitute a transmitter setup of this technique. In simple terms, a carrier signal is spread over a large bandwidth by mixing it with a PN sequence with chip duration T_{Tx} and chip rate R_{Tx} (i.e. $R_{Tx} = 1/T_{Tx}$) which is then amplified and broadcasted to the channel. Principally, based on the implementation at the receiver, correlative sounders can be categorized into two types, a convolution matched filter and a swept time-delay cross correlation (STDCC) sounders, which are discussed below.

- **Swept Time-delay Cross-Correlation (STDCC) Sounder** STDCC is also known as spreading spectrum sliding correlator channel sounder and is the most widely used wideband channel measurement technique. With this receiver implementation approach of the correlative sounder, the received spread spectrum signal is despread by cross-correlating it with a locally generated PN sequence, identical to the one used at the transmitter but with slightly lower chip rate, R_{Rx}. Such practice of correlating the PN sequence signals with different chip rates (i.e. $R_{Tx} \neq R_{Rx}$) inherently implements a sliding correlator. The difference in the chip rates results in different time bases between the received and the locally generated sequences, such that the two signal are aligned with each other to give a maximum correlation after a duration $T_{slip} = 1/(R_{Tx} - R_{Rx})$. Additionally, since the received signal has gone through a multipath channel, individual path components produce maximum correlation with the local PN sequence at different times, depending on their respective time delays.
  The sliding correlation operation provides a time dilation of the measured CIR, thereby compressing the effective measurement bandwidth and thus easing hardware requirement. That is, the correlation process results in a narrowband signal, the bandwidth of which is determined by the relative rate (i.e. $R_{Tx} - R_{Rx}$) of the two correlated sequences. This allows narrowband processing of the correlator's output signal, which provides immunity to the passband noise and interference. The time dilation is quantified by the so called time-scaling factor γ given by

$$\gamma = \frac{R_{Tx}}{R_{Tx} - R_{Rx}}. \tag{10.43}$$

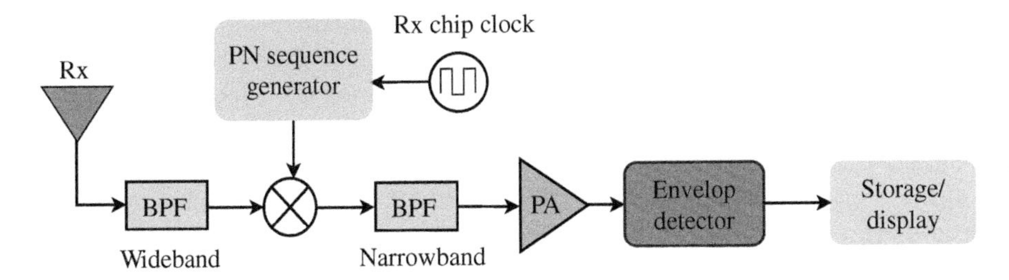

Figure 10.35 STDCC based CIR measurement system.

Due to the time dilation effect, the complete correlation over the whole PN sequence length takes γ times the actual propagation time. Such lengthened measurement time makes it possible to collect the data using slow data storage methods. A simplified setup of the STDCC receiver is shown in Figure 10.35.

Transmitter chip rate, R_{Tx}, scaling factor, γ, and length N_{PN} of the m-sequence have a wide ranging impact on the sounder's measurement capabilities, and thus their optimum values highly rely on the aspect of the channel desired to be measured. Here, relationships between these parameters and some of the channel aspects are highlighted:

1. *Multipath resolution:* The maximum delay spread (τ_{max}) of the channel that can be measured by the sounder is determined by N_{PN}. Specifically, period of the PN sequence, given by N_{PN}/R_{Tx}, has to be greater than the channel's maximum delay in order to ensure that all multipath echoes of the channel are captured. On the other hand, the sounder's ability to resolve consecutive multipath echoes is determined by the used chip rate R_{Tx}.
2. *Maximum Doppler resolution:* The maximum Doppler shift, $f_{D\max}$ that can be measured by STDCC sounder depends on R_{Tx}, γ, and N_{PN}. The relationship is given by

$$f_{D\,max} = \frac{R_{Tx}}{2\gamma N_{PN}}. \tag{10.44}$$

   Note that, for the fixed R_{Tx} and $\gamma, f_{Dmax} \propto 1/N_{PN}$, which poses a trade-off in selecting proper N_{PN} for τ_{max} and f_{Dmax} resolutions. That is, while longer PN sequence is beneficial for capturing the maximum delay spread of the measured channel, it, however, sacrifices sounder's ability of measuring the channel's Doppler spread accurately.
3. *Dynamic range:* In the context of STDCC technique, ignoring the effect of system noise, dynamic range (in dB) can be simply defined as the ratio of the magnitude of the correlation peak to the magnitude of the maximum instance of correlation noise, and it is purely a function of the PN sequence length:

$$\textit{Dynamic Range} = 20\log_{10} N_{PN}. \tag{10.45}$$

   If an additive white Gaussian noise (AWGN) channel is considered, dynamic range is given as a ratio of the peak power to the level of the noise power, given as

$$\textit{Dynamic Range} = 20\log_{10}\left(\frac{N_{PN}}{N_o B_{PN}}\right), \tag{10.46}$$

   where, B_{PN} is the bandwidth of the PN sequence.

   In essence, dynamic range determines the ability of the sounder to detect weak multipath echoes. However, STDCC sounder's dynamic range is limited by a number of factors such as noise, nonlinearities of the power amplifier, and the difference between the chip rates.

- **Convolution Matched filter Sounder** Matched filtering based implementation of the correlative sounder does not require the presence of the identical PN sequence at the receiver in order to recover the desired channel response. Instead, convolution processing of the received PN signal with a matched filter is employed. A commonly used matched filter is the surface acoustic wave (SAW) filter. Since the SAW filter is matched to the waveform of the sounding signal used at the transmitter, the need for the local generation of the identical PN sequence is removed, which in turn reduces hardware cost and complexity at the receiver.

10.5.3 Challenges of Practical Channel Measurement

Field measurement of wireless channel faces several challenges including [16]:

- *Time consuming*: Real measurement often require considerable amount of time and effort to conduct as a statistically enough amount of data needs to be collected.
- *Lack of repeatability*: Behavior of the measured channel is tightly coupled to the physical environment over which we do not have control. In case a wireless experimentation necessitates precise experimental conditions, the uncontrollable mobility of physical objects and people in the environment makes the required conditions impossible to reproduce.
- *Less flexibility*: Bulkiness of the measurement equipment often makes the whole process less flexible.
- *Hardware effects*: It is often difficult to separate distortions introduces by the channel and those introduced by the equipment.
- *Synchronization*: For long range channel measurements in particular, substantial amount of effort is required to synchronize the devices in order to reduce frequency drift and phase noise.
- *Cost and complexity*: Field measurements are generally expensive and complex. Their cost and complexity increase swiftly as communication systems become more sophisticated.

These challenges motivate the necessity of the channel emulation techniques which are discussed in the next section.

10.6 Channel Emulation

So far in this chapter and in the previous chapters, computer simulations have been adopted as a means for analyzing and understanding wireless systems. Under this section, we will shed light on other approaches that have been utilized in the literature as reliable testbeds for performance evaluation of the newly developed wireless devices and algorithms.

Prior to their mass production, newly developed wireless devices or algorithms need to be carefully tested and their performance thoroughly analyzed under realistic propagation conditions. In this regard, a direct and more reliable way of conducting such testings is through experimentation with real hardware and software in the real-world environment. Although such on-site tests provide a desired level of realism in the measurement and the performance analysis obtained therefrom is a highly reliable one, they are, however, expensive, time-consuming, and difficult to repeat. Ideally, a good experimentation method should, in addition to achieving realism and fidelity of the measurement, be able to provide controllable and repeatable experimental conditions. Due to the random nature of the RF propagation environment, it is difficult to meet such demands with on-site measurement. Consequently, researchers have resorted to computer simulations for the testing and verification process. Although computer simulations can overcome the problem of repeatability and reconfigurability, they are, however, hampered by the lack of realism and fidelity, as well as the required excessive run time. In order to make the measurement process tractable, simulation setups are usually over-simplified, which makes the conducted analysis less reliable from a practical perspective. Nevertheless, simulation-based analyses are still beneficial in the early stages of developing a wireless system [17].

In order to retain the advantages of both on-site measurement and simulation, wireless channel emulators are widely used by researchers and engineers. Channel emulators refers to the instrument or setup capable of reproducing the actual radio wave propagation effects in a controllable manner in the laboratory environment. In essence, emulation is a midway between realistic field measurement and computer simulation, serving as an efficient, reliable, and less expensive testing platform.

10.6.1 Baseband and RF Domain Channel Emulators

Conventionally, channel emulators generate the desired wireless channel effects based on the predetermined channel models. The generated channel coefficients are then introduced on the intended communication signal in baseband or radio frequency (RF) domain. Generally, baseband emulators are more flexible as they are capable of working with signals designated for any frequency band. However, they necessitate that the input RF signal is first down converted and sampled before being digitally processed to introduce the desired channel effect. Once the channel effects are added, the output signal needs to be up converted back to its original RF band. Consequently, baseband channel emulators introduce high processing delays in addition to their complex structure and high cost due to the required up and down conversion circuitry elements. Furthermore, the RF-baseband-RF conversion stages introduce unwanted hardware distortions to the signal besides the desired channel effects. Such distortions include analog-to-digital converter (ADC)/digital-to-analog converter (DAC) quantization noise, phase noise, and I/Q impairments [18].

On the other hand, RF domain channel emulators introduce the desired channel effects directly to the input RF signal without going through digitization and baseband conversion processes, which significantly overcomes the aforementioned challenges associated with the baseband emulators. State-of-the art RF domain emulators can be roughly categorized into two groups: as a pure RF circuit based, or as reverberation chamber (RVC) based emulators. Pure RF circuit based emulators are only suitable for non over-the-air (OTA) tests whereas RVC emulators demonstrate remarkable capabilities in emulating RF propagation environment for OTA tests. In simple terms, OTA test refers to the experimental methodology used to predict performance and reliability of a wireless device in the emulated/pseudo-realistic propagation environment. Ideally, a device under test (DUT) is placed inside a test chamber (with controllable propagation environment) and is subjected to different propagation conditions to check how it performs.

10.6.2 Reverberation Chambers as Channel Emulator

10.6.2.1 General Principles

RVCs are electrically large metal cavities with a high quality factor (or simply Q-factor) that permit the excitation of a large number of modes with closely proximate resonant frequencies. Q-factor of the RVC is defined as the ratio of the steady state energy retained within the chamber to the dissipated power per each RF cycle. Proper functioning of RVC as a reliable measurement facility is achieved when it is operated at high enough frequency (above the lowest usable frequency) such that its Q-factor is above the required threshold, Q_{thr}, given by

$$Q_{thr} = \left(\frac{4\pi}{3}\right)^{2/3} \frac{V^{1/3}}{2\lambda}, \tag{10.47}$$

where V is the volume of the chamber and λ is the wavelength of the EM signal at a given operational frequency. In general, it is suggested that operational frequency is selected such that electrical dimensions of the chamber are at least 8λ to 10λ [19].

The reason for on operating the RVC at high Q-factor is to ensure that a very high mode density is achieved inside the chamber. Mode density is defined as the number of modes (resonant frequencies) that can be excited in the chamber per the given frequency bandwidth. Field distribution of the excited modes create locations of high and low field magnitudes, also referred to as hot and cold spots, respectively. Measurement with RVC requires that these hot and cold spots are uniformly distributed within the chamber. This is in order to ensure that the measurement samples obtained from the chamber are statistically independent (i.e. uncorrelated). That is, the average of the power measured at any location within the chamber is constant, within some standard deviation. The smaller the standard deviation the higher the measurement accuracy. At this stage, RVC is said to be spatially uniform. Spatial uniformity of RVC is achieved by continuously relocating the hot and cold spots through the so called mode stirring techniques. There are two basic approaches of performing mode stirring, *mechanical stirring* and *frequency stirring*.

Mechanical stirring, as its name implies, involves physically moving something within the chamber in order to relocate the positioning of the hot and cold spots inside the chamber. Common examples of mechanical stirring are spatial/position stirring and paddle stirring. Spatial stirring is achieved by displacing or reorienting the DUT. That is, measurement is repeated with the DUT placed at different locations within the chamber. In the paddle stirring approach, a large, turnable, metallic paddle, preferably irregularly shaped, positioned somewhere inside the chamber is turned either continuously or in discrete steps in order to reorient the locations of the hot and cold spots while measurements are taken. The measurements collected at different paddle position are then averaged over all measured paddle positions. Here it is pertinent to note that step size at which the paddle is turned has very strong relevance to the effectiveness of this stirring mechanism. Too short paddle steps may be highly correlated, degrading efficiency of the stirring process. Similarly, the effectiveness of this mode stirring technique can be enhanced by placing more than one paddle stirrer inside the chamber.

In the case of *frequency stirring*, spatial uniformity of the field inside the chamber is realized by sweeping the excitation frequency through a given window of frequencies. Changing the center frequency during the sweeping process changes the electrical size of the chamber which in turn, causes the excited modes to change constantly. This dynamicity of the excited modes causes continuous relocation of the hot and cold spots inside the chamber. Samples measured at each of the discrete frequency points within the given window are averaged over the number of frequency points. The computed average measurement is attributed to the center frequency in the corresponding window of frequencies. Similarly, effectiveness of the frequency technique is determined by the selection of frequency step sizes for the frequency sweeping. The frequency step size has to be larger than coherence bandwidth B_c of the chamber in order to ensure that the samples measured between consecutive frequencies are totally uncorrelated. B_c is related to the Q-factor, Q_f of the chamber and the frequency f_c at which the chamber is operated by relationship $B_c = f_c/Q_f$ [20].

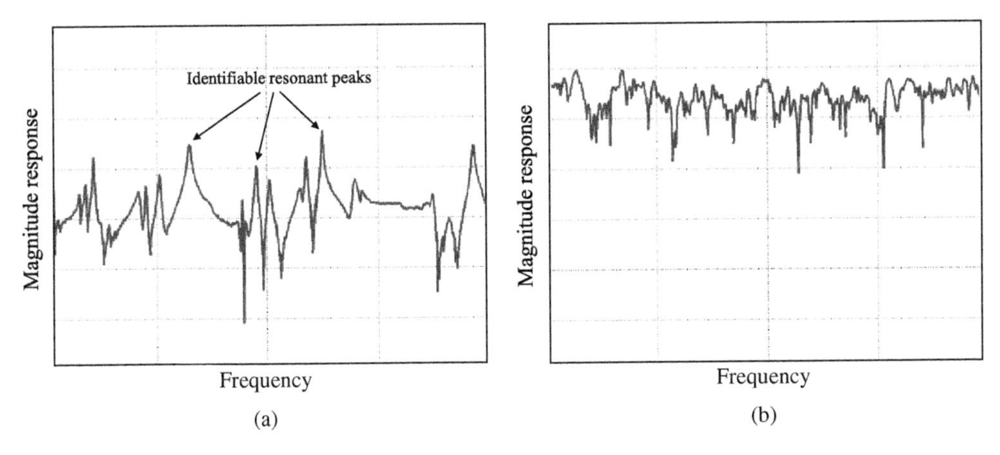

Figure 10.36 Frequency response for (a) undermoded RVC (b) overmoded RVC.

However, it is important to note that no stirring mechanism is perfect practically. There is always an unavoidable degree of spatial dependency of the field inside the RVC, especially when it is operated at low frequencies or when the size of DUT is relatively large. Therefore, using more than one stirring mechanism is highly recommended.

There are several ways of counterchecking whether spatial uniformity is achieved inside the chamber or not. But before characterizing the spatial uniformity, it is important to first examine density of the excited modes in the chamber. This can be done by simply observing frequency response of the measured data samples at a single spatial position. Frequency response of the undermoded chamber is characterized by separated individual resonant peaks, whereas such peaks are not identifiable for an overmoded chamber due to the overlapping of many significant modes at any one of the given frequencies, as shown in Figure 10.36 [21]. Once mode density is confirmed to be high enough, accuracy of the stirring mechanisms can be assessed by plotting the distributions of the measured data and comparing it with the PDF predicted for an ideal chamber. However, this approach requires that the measurement from multiple locations inside the chamber are compared with the corresponding theoretical PDF, as well as with one another, in order to ensure that they are not only of the correct type, but also there is only a little variation in their respective magnitudes and shapes. Another alternative, which is more straightforward, is just by looking at the standard deviation of the measured samples. For an ideal chamber with an infinite number of measurement samples, standard deviation is zero. Therefore, standard deviation of a well-stirred chamber is expected to be relatively small.

The above discussed approaches of double checking chamber performance are visual based approaches. A more reliable way of ascertaining the performance of the chamber is by conducting a goodness-of-fit (GoF) test such as Kolmogorov-Smirnov and Anderson–Darling tests [22]. GoF tests statistically measure the confidence level with which the samples measured from the chamber follow a targeted probability distribution. For more detail about GoF tests, reader is referred to references [22, 23].

10.6.2.2 Emulating Multipath Effects Using RVC

The RF signal emitted inside RVC naturally experiences multipath dispersion due to multiple bouncing off the highly reflective walls enclosing the chamber. This dispersion characterizes the typical multipath effect observed in the realistic propagation environment. It is already established in the literature that the multipath effect inside a well-stirred chamber inherently emulates

Rayleigh fading characteristics of the wireless channel. However, with some more sophisticated measurement setups, the capability of RVC can be extended further to emulate other varieties of fading processes experienced in the realistic channels.

Similar metrics used to quantify dispersion in the realistic environment can also be applied to the enclosed space in the chamber to measure and adjust the dispersion characteristics of the emulated channel. In the case of delay spread, for example, dispersion is quantified by using maximum excess delay (τ_{max}), mean excess delay (τ_{mean}), and RMS excess delay (τ_{rms}) derived from the PDP. Theoretically, PDP of the mode stirred chamber follows an exponentially decaying function with its time constant $\tau_c = Q_f/\omega$, where ω is the radian frequency (i.e. $\omega = 2\pi f_c$, with f_c being carrier frequency) [24]. Mathematically,

$$PDP(\tau) = P_{max}e^{-\frac{\tau}{\tau_c}} = P_{max}e^{-\frac{\omega}{Q_f}\tau}, \tag{10.48}$$

where P_{max} is the tap's maximum power. Empirically, PDP is found directly from the measured CIR, $\hat{h}(t,\tau)$ averaged over $t = 1, 2, \ldots, t_N$ stirring instances as [25]

$$PDP(\tau) = \langle|\hat{h}(t,\tau)|^2\rangle_t. \tag{10.49}$$

In order to control time dispersion of the emulated channel, the RF absorber loading technique is normally used. In this technique, materials capable of absorbing RF energy are introduced into the chamber to reduce the ring-down duration of the signal reverberating in the chamber. Appropriate number of RF absorber materials can be loaded to reduce the emulated delay spread to a required level. Below is the experiment demonstrating delay spread measurement with RVC.

- **DEMONSTRATION: Delay Spread Emulation Using RVC**
  Let us consider the experimental setup given in Figure 10.37. The setup is comprised of a typical RVC with 120 cm × 68 cm × 55 cm dimension. The chamber is equipped with all necessary components: stirrer with controllable speed, pieces of RF absorber made from foam materials, and Tx and Rx antennas. In this particular experiment, a Rhode–Schwarz ZVA67 VNA was used for collecting measurement samples.
  During the experiment, a combination of frequency, paddle position, and Rx antenna position stirring mechanisms is used. The VNA is set to sweep a window of 500 MHz bandwidth centered at 5 GHz operational frequency. Note that at this operational frequency, dimensions of the used RVC correspond to $20\lambda \times 11.3\lambda \times 9.2\lambda$ electrical dimensions, well above the least acceptable dimensions suggested by Corona [19] (i.e. 8λ to 10λ). 2001 frequency points are swept at each of

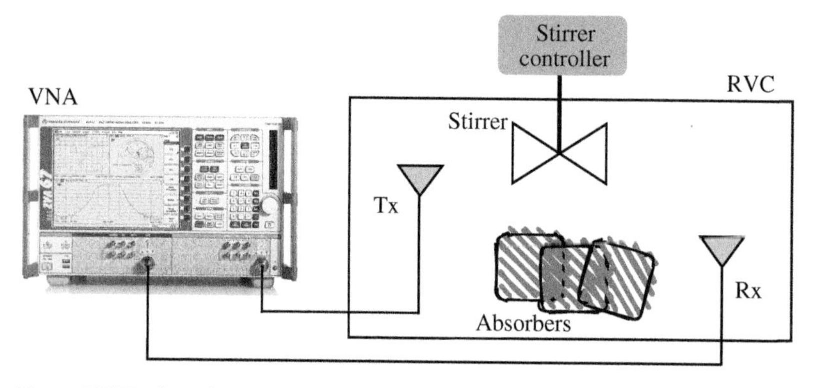

Figure 10.37 Experimental setup for delay spread analysis.

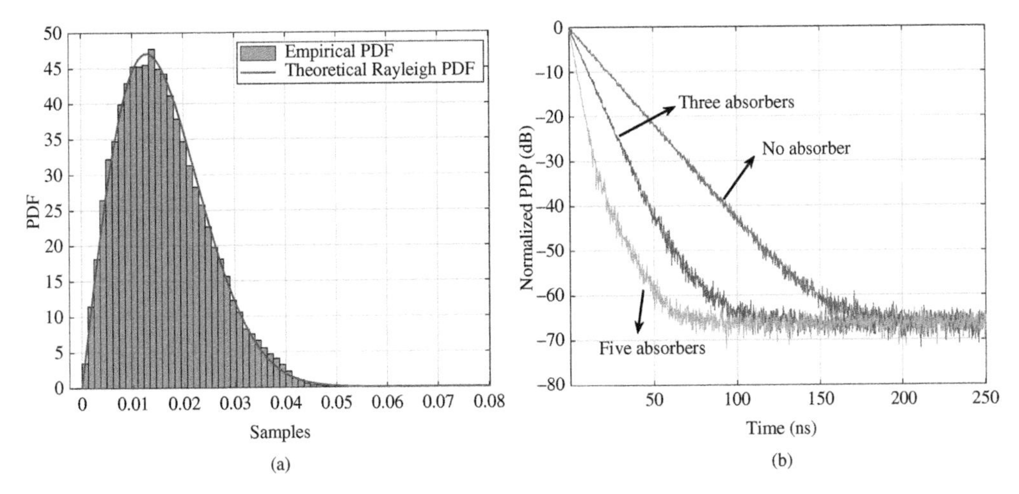

Figure 10.38 Measurement results. (a) PDF of the measured samples compared with theory. (b) PDP obtained with different number of absorbers loading.

the 5 different paddle positions and 4 antenna locations inside the chamber, making a total of 40 020 samples for PDF estimation.

As we have discussed earlier, it is important to first check if spatial uniformity is achieved with the selected stirring mechanisms. To this end, PDF of the magnitudes of the collected samples is computed and then compared with the theoretical PDF of the Rayleigh distribution. A snippet of Matlab code for this is given below and the PDFs are plotted in Figure 10.38a. The figure shows visually that the two PDFs are best fit of each other, suggesting that the chamber operates efficiently at the selected frequency.

```
1   % Code computing PDF of the collected measurement
2   % samples and compare it with Theoretical Rayleigh PDF
3   % Parameter Setup
4   FrequencyPoints = 2001; % Frequency stirring points
5   MechStirPoints  = 20; % Mechanical stirrind instances
6   TotalMeasSamples = FrequencyPoints*MechStirPoints;
7   bins = 40; % for Empirical PDF estimation
8   % Load measurement data (S-parameters)
9   load('Measurement.mat');
10  MeasSamples = S21;
11  MeasSamples = reshape(MeasSamples,TotalMeasSamples,1);
12  % Computing the Theoretical PDF
13  TheorySamples = linspace(0,0.5,TotalMeasSamples/bins);
14  [phat,¬]=raylfit(abs(MeasSamples));
15          % estimate Rayleigh parameter.
16  TheoryPDF=raylpdf(TheorySamples,phat); % Theoretical PDF
17  % Figure
18  figure; hold on; grid on; box on
19  histogram(abs(MeasSamples),bins,'Normalization','pdf')
20                                   % Empirical PDF
21  plot(TheorySamples,TheoryPDF,'linewidth',1.5)
22  xlabel('Samples');ylabel('PDF');
23  legend('Empirical PDF','Theoretical Rayleigh PDF')
```

Delay spread of the emulated channel is then observed for three scenarios defined by number of absorbers loaded into the chamber. In the first scenario, no absorber is used, and the second and third scenarios correspond to three and five pieces of loaded absorbers, respectively. With similar measurement processes described above, the VNA measures frequency response, $H(f)$ of the emulated channel over the 500 MHz bandwidth in terms of S-parameters. Empirical CIR, $\hat{h}(t,\tau)$, is computed through inverse discrete Fourier transform (IDFT) processing of the measured $H(f)$ as

$$\hat{h}(t,\tau) = IDFT\{H(f)\}. \tag{10.50}$$

The relationship given in Eq. (10.49) is then employed to compute the PDP for each scenario. Matlab code for the PDP computation is given below. The computed PDPs are shown in Figure 10.38b, which show clearly that RF absorber loading is an effective way of controlling delay spread in the RVCs.

```
1  % Computing PDP of the Three considered Scenarios.
2  % Parameter setup
3  FrequencyPoints = 2001; % Frequency stirring points
4  MechStirPoints = 20; % Mechanical stirrind instances
5  Bandwidth = 500e6; % Swept frequency window
6  Tau = (0:1:FrequencyPoints-1)./Bandwidth;
7  % Load measurement data (S-parameters)
8  load('Measurement1.mat'); H_Scenario1 = S21; % No absorbers
9  load('Measurement2.mat'); H_Scenario2 = S21; % 3 absorbers
10 load('Measurement3.mat'); H_Scenario3 = S21; % 5 absorbers
11 % Compute CIR at each mechanical stirring instance
12 % and then take their average to find PDP
13 for n=1:MechStirPoints
14     h_Scenario1(n,:)=ifft(H_Scenario1(n,:));
15     h_Scenario2(n,:)=ifft(H_Scenario2(n,:));
16     h_Scenario3(n,:)=ifft(H_Scenario3(n,:));
17 end
18 PDP_Scenario1=mean(abs(h_Scenario1).^2);
19 PDP_Scenario2=mean(abs(h_Scenario2).^2);
20 PDP_Scenario3=mean(abs(h_Scenario3).^2);
21 % Figure
22 figure;hold on; grid on; box on
23 plot(Tau./10e-9,10*log10(PDP_Scenario1/max(PDP_Scenario1)))
24 plot(Tau./10e-9,10*log10(PDP_Scenario2/max(PDP_Scenario2)))
25 plot(Tau./10e-9,10*log10(PDP_Scenario3/max(PDP_Scenario3)))
26 xlabel('Time (ns)'); ylabel('Normalized PDP (dB)')
```

RVC is also a reliable tool for emulating the time selectivity/Doppler spread effects of a wireless channel [25, 26]. The rotating stirrer employed in the chamber for mode stirring can also serve as a source of mobility inside the chamber, which induces Doppler effect to the reverberating signal. This particular way of creating Doppler effect corresponds to the scenario in which the mobility is due to the motion of surrounding objects while transmitter and receiver antennas are fixed. In order to control bandwidth and shape of the emulated Doppler spectrum, the factors influencing intensity of Doppler spread in the realistic environment need to be mapped into stimuli conditions for RVC experiment. For example:

- Effect of mobility speed can be easily emulated by using a stirrer with controllable speed. Here, it is important to note that due to the rotational motion of the stirring paddles, different points on the paddle experience different linear speed depending on their distance from axis of rotation. Therefore, the speed inside RVC should be treated as random variable with particular distribution rather than a single value. Furthermore, due to the resonant nature of the RVCs, an MPC can interact with the stirrer several times, undergoing a Doppler shift each time, before being captured by Rx. Consequently, the effective Doppler spread observed in the chamber is usually larger than its theoretically calculated value.
- Changing frequency of the RF signal from the signal generator can be done to study effect of operational frequency.

- Effect of motion intensity on the emulated Doppler spectrum can be reproduced by introducing more than one stirrer into the chamber. Motion intensity implies a ratio between the percentage of MPCs coming from stationary and moving objects. Therefore, introducing more than one stirrer into the chamber increases the probability that the MPCs interact with a moving object within the RVC, which increases motion intensity.
- RF absorbers can be used to characterize the effect of AoA on the Doppler spectrum. When absorbers are placed properly in the chamber they can prevent MPCs from a particular direction from interacting with the stirrer, affecting symmetricity of the resulting Doppler spectrum.

The following is demonstration of emulating Doppler spread and controlling Doppler spectrum by using different stimuli factors.

- **DEMONSTRATION: Doppler Spread Emulation Using RVC**
  The Doppler spread emulation setup is given in Figure 10.39. The key components in this setup are the RVC equipped with a speed controllable stirrer, vector signal generator (VSG), vector signal analyser (VSA), and antennas. Experimental investigation of Doppler spread involves transmitting a tone from the VSG at a desired operation frequency so that the spreading due to Doppler effect can be easily observed in frequency domain of the received signal. VSA is used to record the signal capture by Rx antenna in the chamber for the postprocessing stage on Matlab. Note that it is important to record enough number of samples in order to obtain sufficient statistics. In this experiment, data are captured and recorded for 20 seconds.
  Before starting the data recording, the span of the VSA must be appropriately adjusted in order to avoid an aliasing problem. In this regard, it is very important to know the relationship between span ΔW and sampling frequency f_s of the used VSA model. In this particular experiment, N5172B VSG and N9010A VSA models from Keysight Technologies are used, where f_s and ΔW are related by $f_s = \Delta W \times 1.28$. Span is set to $\Delta W = 8$ kHz, which corresponds to a sampling rate of 10.24 kHz. The Doppler spectrum is revealed by postprocessing the recorded I/Q data samples on Matlab. The postprocessing steps are as follows:
  1. *Frequency offset correction*: Frequency offset due to mismatch between local oscillators (LOs) of the VSA and VSG must be estimated and corrected before handling the data. However, this step can be avoided by synchronizing the devices through their appropriate synchronization ports by using a wired connection.
  2. *Computation of channel correlation function*: This is done by calculating autocorrelation of the captured I/Q data samples.

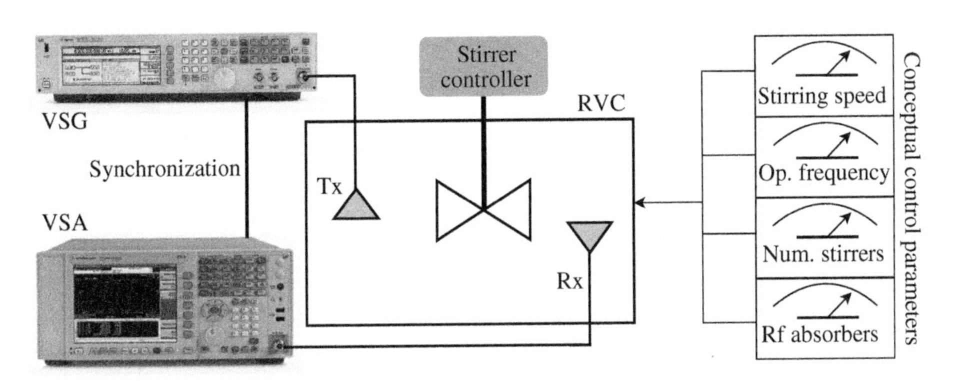

Figure 10.39 Experimental setup for Doppler spread emulation.

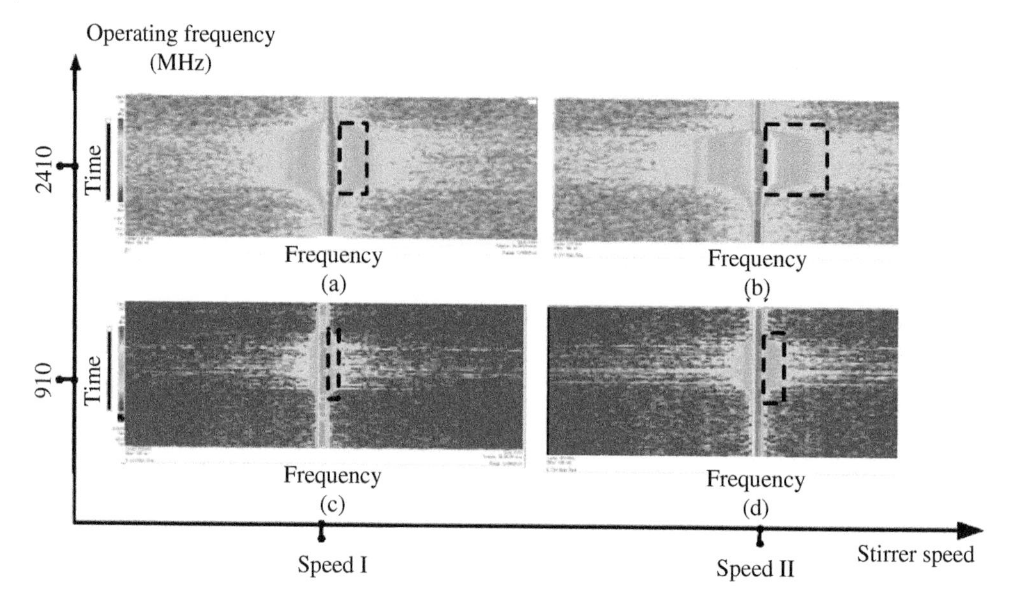

Figure 10.40 Doppler spectrogram of the measurements: (a) 2410 MHz stirrer speed low. (b) 2410 MHz stirrer speed high. (c) 910 MHz stirrer speed low. (d) 910 MHz stirrer speed high. Dashed boxes emphasize the extent of spread [26].

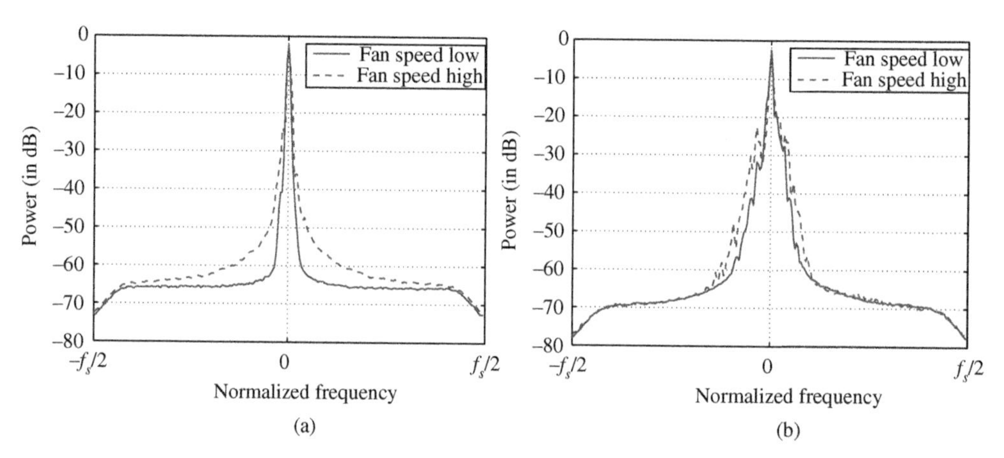

Figure 10.41 Impact of operating frequency and speed on Doppler spectrum. (a) $f = 910$ MHz. (b) $f = 2410$ MHz [26].

3. *Obtaining Doppler spectrum*: Finally, the Doppler spectrum is obtained by applying FFT process on the channel correlation function obtained in the previous step.

It is also possible to observe the spreading directly from the VSA. One way of doing this is by observing spectrogram of the received signal, as shown in Figure 10.40. In this figure, effects of speed and operation frequency are shown. By observing the Figure 10.40a–d along horizontal and vertical axes, it is clear that amount of spreading increases with the increase in speed of the stirrer and operational frequency, as expected. Similar result can be observed on the Doppler spectra, shown in Figure 10.41, obtained through postprocessing of the captured data on Matlab. As pointed out earlier, the effect of motion intensity is investigated by introducing another identical stirrer into the chamber. In this experiment, the speed of the stirrers is set their highest

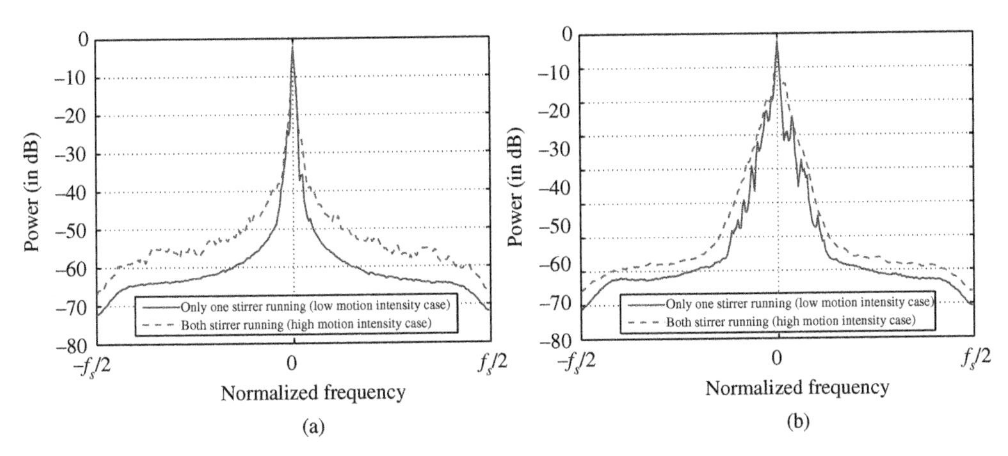

Figure 10.42 Impact of motion intensity on Doppler spectrum [26]. (a) $f = 910$ MHz. (b) $f = 2410$ MHz.

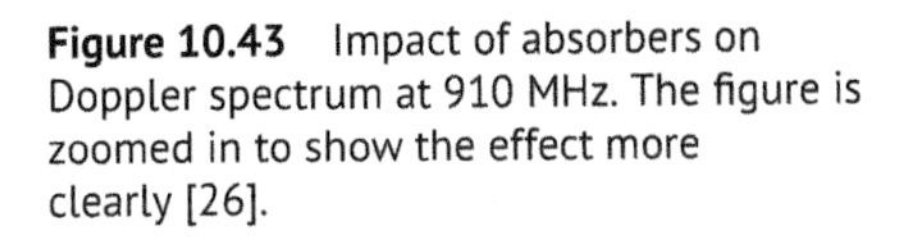

Figure 10.43 Impact of absorbers on Doppler spectrum at 910 MHz. The figure is zoomed in to show the effect more clearly [26].

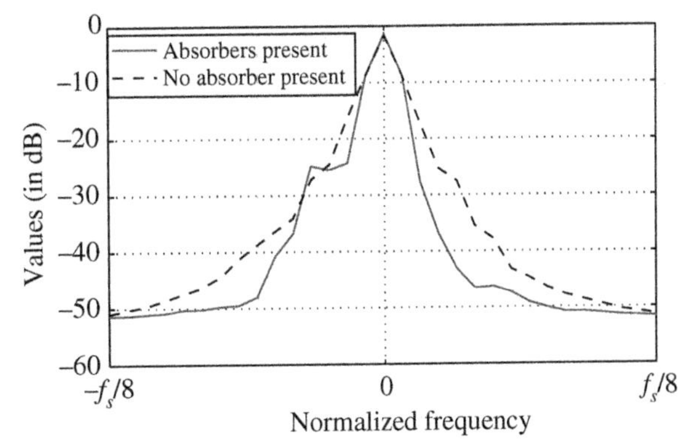

values and the Doppler spectra are observed for two different operation frequencies. The Doppler spreading is observed to increase with the increase in motion intensity for both frequencies, as shown in Figure 10.42.

The last part of the experiment investigates the impact of angular spread on the shape of the Doppler spectrum by using RF absorbers. A study in [25] suggests that an effective way of manipulating the shape of Doppler spectrum is by placing the absorbers around the stirrer, rather than the Rx antenna. The result is shown in Figure 10.43, in which the shape of the Doppler spectrum appears to be more symmetric prior to the introduction of the absorbers into the chamber. However, it should be understood that using RF absorbers is not the only way of controlling the shape of the emulated Doppler spectrum, judiciously changing the location of the antennas and stirrer inside the chamber can produce a similar effect.

The experimental demonstrations given above feature very basic channel emulation setups. In the recent literature of RVC studies, there are many advanced setups developed to facilitate OTA testings of more complex propagation characteristics and practical evaluation of the state-of-the art technologies such as MIMO systems [27]. Most of these advanced emulators are available commercially.

10.6.3 Commercial Wireless Channel Emulators

Commercial companies provide more comprehensive and standard compliant channel emulators. Commercial channel emulators are usually designed to emulated real-world channel, as specified by the standardization bodies such as 3GPP and International Telecommunication Union (ITU). Despite their high cost, most of the researchers and wireless systems developers prefer these commercial emulators because of their capability of emulating the channel while considering various channel-related concepts, such as scheduling, diversity, smart antennas, MIMO systems, and beamforming. Here, we give a brief overview of some of the interesting commercial channel emulators that target some state-of-the-art communication scenarios:

- *Air-to-ground channel emulator:* The **Propsim F8** channel emulator [28] from Keysight Technologies offers a platform for testing communication systems that incorporate aerospace, unmanned aerial vehicless (UAVs), satellites, and other airborne radio systems in the laboratory environment. This emulator enables realistic generation of all radio channel characteristics relevant to the aeronautical channels, such as high Doppler shifts and long delays. The desired channel effects can be emulated via built-in models or customer-specific data imported from third party scenario tools.
- *Drive test emulator:* A drive test is a performance test methodology that helps mobile operators analyze and benchmark interoperability of the targeted wireless devices and the real network. The test usually involves an exhaustive field measurements. It is, however, very important as it enables wireless engineers identify and resolve different practical issues prior to the network deployment process. The **Anite Virtual Drive Test toolset** [29] developed by Keysight Technologies is a customized emulator for conducting such tests in the laboratory environment. The emulator uses real data captured in the field to build tests that replay drive or indoor test routes by emulating real-world RF network conditions in a controllable manner. This replay can be performed with a real network infrastructure or a simulated network.
- *Ad Hoc network channel emulator:* The **Propsim Mobile Ad-Hoc Network (MANET)** channel emulator offers an efficient way of evaluating end-to-end performance of radio systems in Ad Hoc networks [30]. Mobile Ad Hoc networks are comprised of radio nodes that dynamically self-organize into random network topologies, making them difficult to characterize. With the MANET channel emulation solution, the radio locations and movements are defined in the Propsim test scenario file, which controls the time-varying dynamic link conditions such as network topology, path loss, multipath, Doppler and propagation delay, during the test run. The link conditions can also be controlled by a customer via a local area network (LAN) interfaced external computer.
- *5G massive MIMO emulator:* **NYU 5G emulator** [31] is developed to specifically reflect 5G related features, such as massive bandwidth and hundreds of antenna elements operating at mmWave frequencies, which determine the channel conditions. In this emulator, a different approach of emulating wireless channel involving multiple antennas is used in order to reduce hardware cost and complexity due to the usage of large number of antennas. Specifically, the NYU emulator emulates not only the wireless channel, but also the beamformer (phased-array antenna elements) at both Tx and Rx DUTs. In the traditional emulation paradigm, the beamforming operation are performed by the DUTs themselves. In this case, each antenna element needs a cable connection to the channel emulating filters. Note that the phased-array antenna elements cannot be connected via cables. Figure 10.44 shows the differences between traditional emulators and the NYU 5G emulator.

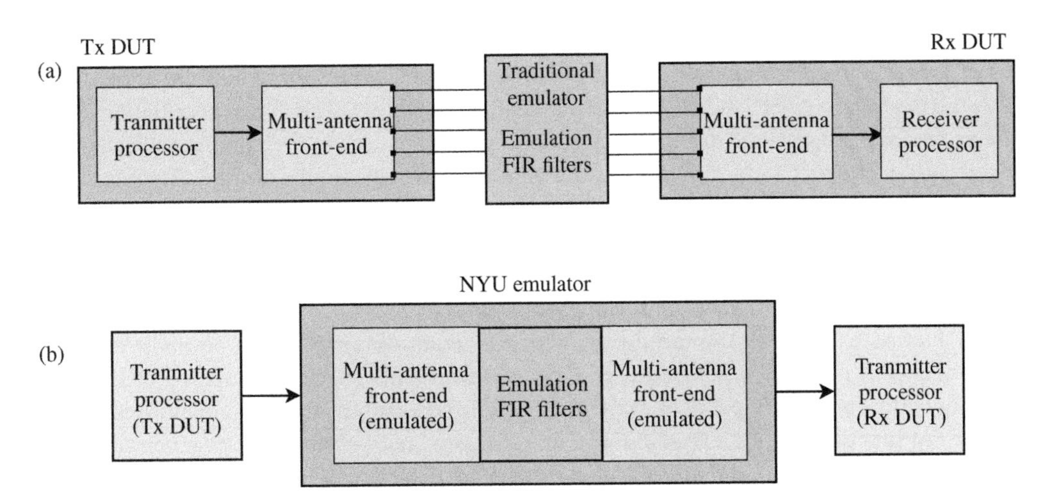

Figure 10.44 (a) Traditional emulation paradigm with beamforming operations integrated into the DUTs themselves. (b) NYU emulator with emulated beamforming operations.

10.7 Wireless Channel Control

Wireless channel has been perceived as an uncontrollable entity in the traditional wireless technologies, including the imminent 5G. This perception is generally attributed to the random nature of the propagation environment. Consequently, the designing process of wireless systems has always excluded propagation environment in the formulation of the system optimization problem. Such system design and optimization practice gradually increased the signal processing and hardware complexities on the transceivers. Additionally, this traditional approach relies on techniques that trade-off system's resources, such as time, spectrum and power, to mitigate the so called uncontrollable channel-related effects. Considering the explosive growth in the number of wireless users and mobile devices, accompanied with the tremendous increase in the amount of data volume that needs to be handled by wireless networks, the aforementioned system resources have become too precious to be traded-off. In addition to this, the day-by-day increase in the societal needs has led to the emergency of numerous applications and use cases with diverse and stringent performance requirements. In this regard, the traditional paradigm of designing wireless systems has proven to be very slow, if not unable, to catch up with such fast growing demands. To this end, researchers have resolved to start considering the optimization of the "problematic" wireless medium itself during the system design in the future generations of wireless technology. However, this requires the ability of controlling the wireless propagation environment.

In order to enable the control of wireless environment, several approaches have been proposed and discussed in the literature. Here we will shed light on two promising techniques: *reconfigurable antenna (RA)* and *reconfigurable intelligent surface (RIS)*.

RA is an emerging technology that equips antennas with the ability to dynamically modify their radiation characteristics, such as operating frequency, radiation pattern, and polarization. The concept of RA is based on the fact that the way antennas radiate energy is determined by the antenna's geometry and surface current distribution. Therefore, controlling the distribution of current in the antenna leads to the controlling of its radiation pattern [32]. Such capability of RAs offers some additional degrees of freedom for optimizing wireless systems. From the channel perspective, RAs are able to control how the transmitted signal interacts with the environment it travels through just by adjusting its radiation pattern. This is due to the fact that the channel response observed at the receiver, is determined by both the distribution of the scatterers in the propagation environment as well as the interplay between MPCs and antenna's radiation pattern. The spatial distribution of the scatterers varies relatively with distance, height and radiation pattern of the antenna, which in turn determines the temporal and angular/spatial domain characteristics of the channel. Therefore, RA can be used to deliberately create favorable channel conditions which enhances systems throughput and reliability. Furthermore, the fact that different radiation states lead to the realization of different channel signatures can be exploited for extra data transmission under the media-based modulation concept [33], thereby, improving system capacity.

RIS is another example of the technology which is under development in an effort to facilitate channel control capability. Essentially, RISs are man-made surfaces composed of low cost and energy efficient (nearly passive) elements that can be electronically controlled to reconfigure their EM functionalities, such as absorption, reflection, and polarization. Based on the same concept as RAs, i.e. the EM emissions from a surface are determined by the distribution of electrical current over it, RIS aims to control and modify the current distribution over its reflective elements in a way that enables exotic EM functionalities such as absorption, anomalous reflections, and reflection phase modification. The RISs are envisioned to be integrated into the propagation environment so that they act as artificial intelligent scatters that can alter the traditional wireless propagation characteristics. In this way, the propagation environment can be characterized by controllable scattering, reflection, and absorption phenomena, thereby, providing a way of overcoming their negative effects [34].

A simplified structure of the RIS is shown in Figure 10.45, along with the visualization of some of its interesting EM functionalities. A simple scenario of an RIS aided communication in the presence of a blockage is illustrated in Figure 10.46. In this case, RIS placed on the facade of a big building can be used to establish seamless connectivity with good quality of services for the users behind the blockage.

Although these approaches of controlling the propagation environment are quite promising, their developments are still in their infancy and there are many challenges yet to be solved in order to guarantee their practical viability. In the case of RIS, for example, the proper adjustment of the elements' EM functionalities requires the knowledge of the channel. The question of how

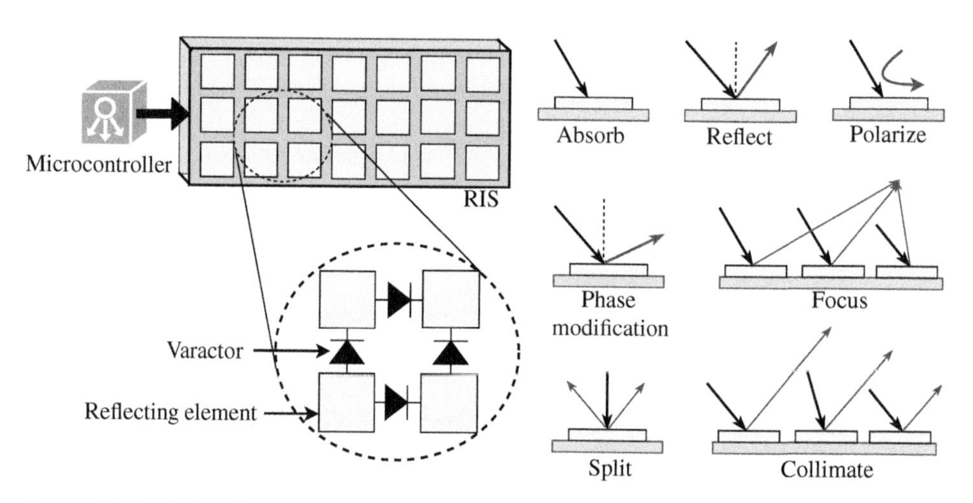

Figure 10.45 A simplified structure of RIS and some of its peculiar EM functionalities.

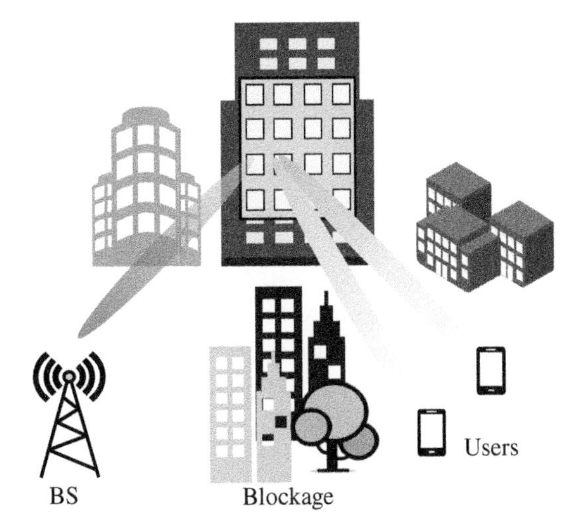

Figure 10.46 An example of the RIS aided communication scenario.

to perform channel estimation with the passive RIS elements needs to be answered. If we resort to performing channel estimation at the BS, the best way of feeding back the estimated channel coefficients to the RIS needs to be investigated.

10.8 Conclusion

This chapter summarizes the concept of wireless channel. It explains various aspects of the channel from its fundamental propagation principles and the associated effects, followed by the techniques that are used for measuring, modeling, and emulating these effects. The chapter also shed light on recent perceptions of wireless channel, where there is an increased need for achieving a capability of controlling its effects. For the sake of brevity and complexity of the discussion, some concepts are only briefly explained, an interested reader can refer to the given references for further discussions and insights on those concepts.

References

1 T. S. Rappaport, *Wireless Communications: Principles and Practice*. Prentice-Hall PTR New Jersey, 1996, vol. 2.

2 S. Tomasin, *Use of Millimeter Wave Carrier Frequencies in 5G*. Padova: University of Padova, 2018.

3 S. Y. Seidel, T. S. Rappaport, S. Jain, M. L. Lord, and R. Singh, "Path loss, scattering and multipath delay statistics in four European cities for digital cellular and microcellular radiotelephone," *IEEE Transactions on Vehicular Technology*, vol. 40, no. 4, pp. 721–730, 1991.

4 J. D. Parsons, *The M5obile Radio Propagation Channel*. Wiley, 2000.

5 A. Doukas and G. Kalivas, "Rician K factor estimation for wireless communication systems," in *2006 International Conference on Wireless and Mobile Communications (ICWMC'06)*. IEEE, Bucharest, 2006, pp. 69–69.

6 S. Doğan, M. Karabacak, and H. Arslan, "Optimization of antenna beamwidth under blockage impact in millimeter wave bands," in *2018 IEEE 29th Annual International Symposium on Personal, Indoor and Mobile Radio Communications (PIMRC)*. IEEE, Bologna, 2018, pp. 1–5.
7 B. Wang, F. Gao, S. Jin, H. Lin, and G. Y. Li, "Spatial-and frequency-wideband effects in millimeter wave massive MIMO systems," *IEEE Transactions on Signal Processing*, vol. 66, no. 13, pp. 3393–3406, 2018.
8 C. E. Shannon, "A mathematical theory of communication," *The Bell System Technical Journal*, vol. 27, no. 3, pp. 379–423, 1948.
9 C. Kosta, B. Hunt, A. U. Quddus, and R. Tafazolli, "On interference avoidance through inter-cell interference coordination (ICIC) based on OFDMA mobile systems," *IEEE Communications Surveys and Tutorials*, vol. 15, no. 3, pp. 973–995, 2012.
10 H. Zhang, Y. Shi, and A. S. Mehr, "Robust equalization for inter symbol interference communication channels," *IET Signal Processing*, vol. 6, no. 2, pp. 73–78, 2012.
11 M. Nazzal, M. A. Aygül, and H. Arslan, "Channel modeling in 5G and beyond," in *Flexible and Cognitive Radio Access Technologies for 5G and Beyond*. UK: The Institution of Engineering and Technology (IET), 2020, ch. 11.
12 Saleh, A. A. and Valenzuela, R., "A statistical model for indoor multipath propagation," *IEEE Journal of Selected Areas Communication*, vol. 5, no. 2, pp. 128–137, 1987.
13 Zwick, T. and Fischer, C. and Wiesbeck, W., "A stochastic multipath channel model including path directions for indoor environments," *IEEE Journal of Selected Areas Communication*, vol. 20, no. 6, pp. 1178–1192, 2002.
14 3GPP Radio Access Network Working Group, "Study on channel model for frequencies from 0.5 to 100 GHz (Release 15)," 3GPP TR 38.901, Tech. Rep., 2018.
15 D. Demery, J. Parsons, and A. Turkmani, "Sounding techniques for wideband mobile radio channels: a review," *IEE Proceedings I (Communications, Speech and Vision)*, vol. 138, no. 5, pp. 437–446, 1991.
16 V. Tarokh, *New Directions in Wireless Communications Research*. Springer, 2009.
17 G. Judd and P. Steenkiste, "Using emulation to understand and improve wireless networks and applications," in *Proceedings of the 2nd Conference on Symposium on Networked Systems Design and Implementation*, vol. 2. USENIX Association, California, 2005, pp. 203–216.
18 M. Karabacak, A. H. Mohammed, M. K. Özdemir, and H. Arslan, "RF circuit implementation of a real-time frequency spread emulator," *IEEE Transactions on Instrumentation and Measurement*, vol. 67, no. 1, pp. 241–243, 2017.
19 P. Corona, J. Ladbury, and G. Latmiral, "Reverberation-chamber research-then and now: a review of early work and comparison with current understanding," *IEEE Transactions on Electromagnetic Compatibility*, vol. 44, no. 1, pp. 87–94, 2002.
20 K. Madsen, P. Hallbjorner, and C. Orlenius, "Models for the number of independent samples in reverberation chamber measurements with mechanical, frequency, and combined stirring," *IEEE Antennas and Wireless Propagation Letters*, vol. 3, no. 1, pp. 48–51, 2004.
21 J. P. Hof, "Modeling the Dispersion and Gain of RF Wireless Channels inside Reverberent Enclosures," Ph.D. dissertation, Carnegie Mellon University, 2005.
22 C. Lemoine, P. Besnier, and M. Drissi, "Investigation of reverberation chamber measurements through high-power goodness-of-fit tests," *IEEE Transactions on Electromagnetic Compatibility*, vol. 49, no. 4, pp. 745–755, 2007.
23 F. J. Massey Jr, "The Kolmogorov-Smirnov test for goodness of fit," *Journal of the American Statistical Association*, vol. 46, no. 253, pp. 68–78, 1951.

24 D. A. Hill, M. T. Ma, A. R. Ondrejka, B. F. Riddle, M. L. Crawford, and R. T. Johnk, "Aperture excitation of electrically large, lossy cavities," *IEEE Transactions on Electromagnetic Compatibility*, vol. 36, no. 3, pp. 169–178, 1994.

25 A. B. Kihero, M. Karabacak, and H. Arslan, "Emulation techniques for small-scale fading aspects by using reverberation chamber," *IEEE Transactions on Antennas and Propagation*, vol. 67, no. 2, pp. 1246–1258, 2018.

26 S. Güzelgöz, S. Yarkan, and H. Arslan, "Investigation of time selectivity of wireless channels through the use of RVC," *Measurement*, vol. 43, no. 10, pp. 1532–1541, 2010.

27 X. Chen, J. Tang, T. Li, S. Zhu, Y. Ren, Z. Zhang, and A. Zhang, "Reverberation chambers for over-the-air tests: an overview of two decades of research," *IEEE Access*, vol. 6, pp. 49 129–49 143, 2018.

28 K. Technologies, *Propsim Channel Emulation Aerospace, Satellite and Airborne Radio System Testing*, 2016 (accessed August 24, 2020), https://www.mrc-gigacomp.com/pdfs/Keysight-Propsim-F8-aerospace-satellite-brochure.pdf.

29 ——, *Anite Virtual Drive Testing Toolset*, 2016 (accessed August 24, 2020), https://www.mrc-gigacomp.com/pdfs/Keysight-Propsim-Virtual-Drive-Testing-Brochure.pdf.

30 ——, *Propsim Channel Emulation Mobile Ad-Hoc Network Testing*, 2016 (accessed August 24, 2020), https://www.mrc-gigacomp.com/pdfs/Keysight-Propsim-{MANET}-radio-channel-emulation-brochure.pdf.

31 N. Wireless, *First Wireless Emulator Suitable for 5G Massive MIMO Systems*, 2017 (accessed August 24, 2020), https://www.microwavejournal.com/blogs/25-5g/post/28899-first-wireless-emulator-suitable-for-5g-massive-mimo-systems.

32 Z. Bouida, H. El-Sallabi, M. Abdallah, A. Ghrayeb, and K. A. Qaraqe, "Reconfigurable antenna-based space-shift keying for spectrum sharing systems under Rician fading," *IEEE Transactions on Communications*, vol. 64, no. 9, pp. 3970–3980, 2016.

33 E. Seifi, M. Atamanesh, and A. K. Khandani, "Media-based MIMO: outperforming known limits in wireless," in *2016 IEEE International Conference on Communications (ICC)*. IEEE, 2016, pp. 1–7.

34 E. Basar, M. Di Renzo, J. De Rosny, M. Debbah, M.-S. Alouini, and R. Zhang, "Wireless communications through reconfigurable intelligent surfaces," *IEEE Access*, vol. 7, pp. 116 753–116 773, 2019.

11

Carrier and Time Synchronization

Musab Alayasra[1] *and Hüseyin Arslan*[1,2]

[1] *Department of Electrical and Electronics Engineering, Istanbul Medipol University, Istanbul, Turkey*
[2] *Department of Electrical Engineering, University of South Florida, Tampa, FL, USA*

Received signal degradation is coming not only from the channel in the form of noise and fading but also from equipment, which might be thought of as part of the channel. In additive white Gaussian noise (AWGN) and fading channel models, no impairments from hardware are considered and the received signal is implicitly assumed to be synchronized with what was transmitted. In other words, the optimal sampling time instant was assumed to be known, and the signal was assumed to have been demodulated perfectly with no frequency mismatch between oscillators at transmitter and receiver sides. However, practically, mismatch in both time and frequency between communication sides is inevitable, and the receiver must be synchronized with the transmitter; otherwise, the performance of the system is degraded severely if it operates at all.

In this chapter, carrier and time synchronizations, which were briefly discussed in Chapter 8, are addressed and the conventional receiver structure is further advanced to include those functionalities. Signal modeling and the effects of impairments are first discussed in the next section. More detail on radio frequency (RF) impairments are provided in Chapter 5. In later sections, carrier and time synchronization approaches are detailed and implemented. Developed methods are presented for single-carrier signals and using linear modulation schemes. Synchronization for orthogonal frequency division multiplexing (OFDM) systems is discussed in Chapter 7, and synchronization as part of a blind receiver is provided in Chapter 12.

11.1 Signal Modeling

The main target of this chapter is synchronization for lab equipment, i.e. software defined radio (SDR) equipment, and since the processed signals given to and returned by these devices are baseband signals, the focus will be on the baseband signal modeling. We consider two signal representations at different rates, namely: sequence of samples, where the time index n is used, and sequence of symbols, where the time index k is used. For instance, the received signal can be either expressed at the sample rate and denoted by $r(nT_n)$ (or r_n) or at the symbol rate and denoted by $r(kT)$ (or r_k), where T and T_n are symbol and sample periods, respectively. The two periods are related by $T = N_s T_n$, where N_s is the oversampling ratio (i.e. number of samples per symbol).

In the AWGN channel, only one imperfection was assumed, which is the addition of white noise. In this chapter, a more practical channel with more imperfections coming from the propagation path and hardware is studied. The newly added signal errors are time error (or delay) and phase error.

Wireless Communication Signals: A Laboratory-based Approach, First Edition. Hüseyin Arslan.

At the transmitter side, before transmitting a frame, its training and data symbols are upsampled and filtered by a shaping filter. The resulted sequence of samples represents the wave carried by a carrier signal over the channel. At the receiver side, the same frame as a continuous wave is received and sampled. Notice that the same wave is represented as a sequence of samples at both sides, so it is required to have matching between samples in time to obtain the optimum downsampling point. Also, the location of the optimum downsampling point for symbol extraction within each N_s samples will not be the same at both sides. The total difference between optimum sampling points at transmitter and receiver sides is referred to as symbol timing error, which is due to signal propagation and sampling clock mismatch and impairments. Downsampling with timing error results in inter-symbol interference (ISI) that can severely degrade the system performance.

The symbol timing error is modeled as a fractional number, denoted by τ, and added to the ideal downsampling time index. The downsampling operation is a resampling operation where the sampling rate is reduced by a factor of N_s. Assume that the output of the matched filter is $r^F(nT_n)$, then the downsampler output with time error is given as:

$$\begin{aligned} r(kT;\tau_k) &= r^F(kN_sT_n - \tau_k N_s T_n) + z(kN_sT_n) \\ &= r^F((k-\tau_k)T) + z(kT), \end{aligned} \tag{11.1}$$

where $z(kT)$ represents the filtered noise samples that are modeled as complex Gaussian random variables. The problem of estimating and compensating for τ is referred to as optimum downsampling in this chapter. Symbol timing synchronization and frame edge detection are addressed in Section 11.4.

The phase error is divided into two parts: the first one grows linearly with time and it is a result of the carrier frequency offset f_o, and the second one is fixed over time and it is a result of the carrier phase offset θ_o. Oscillators at transmitter and receiver drift slightly from the carrier frequency to right or left and the frequency difference between them results in f_o. This offset causes rotation of points in the signal space over time, which results in a cross-talk between real and imaginary parts of the received signal. Another source of frequency offset is the Doppler shift due mobility. However, oscillators mismatch might be much larger than the offset due Doppler effect. Oscillators mismatch is not only in frequency but also occurs in time resulting in θ_o. Given that the transmitted signal through the channel is $x(nT_n)$, the received signal with phase error is given as:

$$y(nT_n) = x(nT_n)e^{j(2\pi f_o nT_n + \theta_o)} + w(nT_n), \tag{11.2}$$

where $w(nT_n)$ is a white Gaussian noise following $\mathcal{CN}(0,\sigma^2)$. Estimation and compensation for the carrier phase error, including both offsets f_o and θ_o, is addressed in Section 11.3.

Figure 11.1 shows the main stages related to the synchronization problem, which the signal goes through. Note that, in addition to the noise, f_o and θ_o are introduced in the channel since the RF circuit is assumed to be part of the channel. The timing error is partially due to propagation and

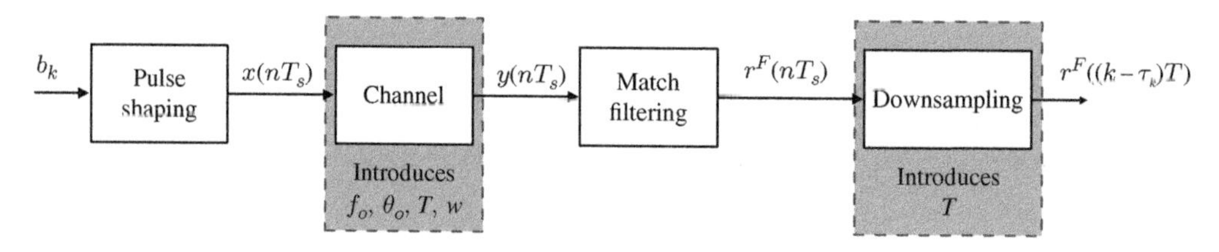

Figure 11.1 AWGN channel with time and phase errors modeling.

it is also introduced by the downsampling unit. The effect of these impairments on the system performance is discussed in Chapter 8.

For numerical analysis, the introduced impairments above are modeled as shown in the following code snippet. Note that the added delay is fixed for all symbols and it takes a finite number of values equal to multiples of T_n. To have a variable delay over time, *resample()* function might be used as shown later in Section 11.4.

```
1  % Simple implementation of a channel including the discussed impairments.
2
3  %% Signal Generation
4  % Symbols Generation
5  N = 1e3; % Number of symbols
6  M = 4; % Modulation order
7  data = randi([0 M-1], N, 1);
8  symbs = complex(qammod(data, M));
9  % Pulse Shaping
10 alpha = 0.5; % Rolloff Factor
11 sps = 8; % Oversampling Ratio
12 fltr = rcosdesign(alpha, 16, sps, 'normal'); % RC Filter
13 symbs_up = upsample(symbs, 8);
14 sgnl_in = conv(symbs_up, fltr);
15
16 %% Channel
17 % Frequency Offset
18 fs = 1e6; % Sampling Rate (Hz)
19 fo = 350; % Frequency Offset (Hz)
20 t = (0:length(sgnl_in)-1)'/fs; % Time Vector
21 sgnl_off = exp(j*2*pi*fo*t); % Frequency Offset Scaling Vector
22 sgnl_out = sgnl_in .* sgnl_off;
23 % Phase Offset
24 po = 0.5*pi; % Phase offset
25 sgnl_out = sgnl_out*exp(j*po);
26 % AWGN Channel
27 SNR = 20; % Signal to Noise Ratio (dB)
28 sgnl_out = awgn(sgnl_out, SNR, 'measured');
29 % Time Offset
30 dly = 3; % Delay (Samples)
31 sgnl_out = sgnl_out(dly:end); % Delay by Skipping Some Samples
```

11.2 Synchronization Approaches

In communication systems, an important step at the beginning of the communication process is called initial acquisition or initial synchronization. It is required whenever a device is turned on searching for connection or in the case of handover between cells. In this step, as the name suggests, initial operations necessary to establish a successful communication are performed. In addition, information required for reliable communication is obtained. The main goal in this process is to detect the existence of a signal corresponding to a given source in the network. Besides, part of the synchronization process is done at this stage.

Transmitters, like base stations, send known signals periodically to perspective terminals to initiate a communication link. The coordination between base stations, to avoid interference or to enable services like high-accuracy positions, is achieved traditionally based on the received global positioning system (GPS) signal. However, due to its limitations, like poor indoor support, dedicated synchronization networks are deployed. Synchronization networks are also needed for phase synchronization, which is important in applications like multi-cast services where multiple cells are transmitting to the same terminal at the same time.

For a terminal to connect to a given base station, it has to detect the known signal transmitted by that base station in the acquisition process. This is achieved by means of correlation between what is received and a locally generated replica of the expected signal. Based on hypothesis testing, a decision on whether this signal exists or not is made. In a nutshell, the detection steps are as follows. The correlation operation is performed between consecutive bursts of whatever is received by the terminal. The results at different lags are compared to a given threshold value; if the correlation value at a given lag is above this threshold level, then the receiver decides on the existence of the expected signal at that time. The threshold value is selected based on different parameters like the expected received signal power. For good results, the used sequence in this process should have good correlation properties, so that better detection of the signals is possible. Examples of practically used sequences are Zadoff-Chu and the maximum length pseudorandom sequence (m-sequence).

Given that the transmitted sequence of symbols is known at the receiver side, it can be used for purposes other than communication initialization. Practically, it is used for channel estimation, adjusting the automatic gain controller (AGC), and some parts of the synchronization process. As shown in later sections, channel estimation and some synchronization steps can be achieved jointly jointly, especially coarse synchronization which is imporatnt at this stage. Mobile devices usually do not have high-end oscillators that are well centered at the carrier frequency with minor frequency offsets; instead their oscillators can have large offset values that need to be compensated using the received known signal. Even base stations in small 5G network cells, like pico and nano cells, are simpler than those deployed in larger cells and might be equipped with lower cost components including the oscillators. This makes the frequency offset compensation problem more serious with larger values. Coarse frequency offset compensation algorithms are applied at the initial synchronization stage to eliminate or reduce the frequency offset between oscillators, and later during data transmission, finer compensation algorithms might be applied depending on the waveform used at that stage. As shown later, it is critical to at least reduce the offset value, if it is relatively large, so that match filtering and downsampling are possible.

Waveform design plays an important role in the synchronization process, since one impairment can have different effects on different waveforms. For example, the effect of frequency offset on signal-carrier signal is a cross talk between real and imaginary parts of the received sequence of symbols (i.e. rotation in the signal space). However, its effect on OFDM is different. Assuming the offset value is not an integer multiple of the subcarrier spacing, inter carrier interference (ICI) occurs due to the shift of subcarrier signals in the frequency domain. Also by using different waveforms, a synchronization problem might be shifted to other parts of the receiver like equalization or channel estimation. For instance, if the OFDM signal is adopted, frequency offset compensation is handled during channel estimation. The waveform selection can also give advantages at some parts of the synchronization process.

As seen above, synchronization should not be thought of separately; instead, processing at the receiver side, which includes mainly channel estimation and synchronization, in addition to the waveform selection, should be considered jointly in the receiver design. Also, it should be pointed out that some applications might have restrictions on the waveform selection for different reasons, so they all should be considered when studying synchronization. In this chapter, the focus is on signal-carrier signals. Synchronization for OFDM is discussed in Chapter 7. In terms of what inputs are given to a synchronization algorithm, the algorithms are classified into two categories: data-directed or decision-directed. In the former, the processed sequence of samples or symbols is known in advance by the receiver. On the other hand, no knowledge of the received signal is available in decision-directed methods; instead received symbols are detected first and the detected symbols are used for synchronization. In some algorithms, both received and detected symbols are used as shown later in time synchronization.

For both methods above, in most practical implementations, some degree of knowledge about the system is available. This includes what type of waveform is used, modulation order, sampling rate, etc. However, blind receiver designs emerged recently based on machine learning (ML), where the level of knowledge of system parameters varies to a point where no knowledge is assumed. The implementation of such receivers is treated in Chapter 12.

In terms of how the received samples are processed, synchronization algorithms are divided into two types: feedforward and feedback approaches. In feedforward methods, each vector of samples, forming a burst, is processed all at once. The input to a feedforward algorithm is a vector of samples with some error to be estimated, and the output is the same sequence of samples but with the estimated error compensated. In feedback approaches, the samples are processed sequentially, and in each iteration, some information is fed back to be used in next iterations. Feedback approaches are widely used in practice since the early days of wireless communications systems and are simpler to implement compared to feedforward approaches. They are tracking approaches, where the error value to be compensated is tracked over time and fed back. On the other hand, feedforward approaches are based on estimation theory, where the error value is treated as an unknown characteristic of the received vector of samples that has to be estimated and removed. Techniques based on this approach are asymptotically optimum in the sense that with more samples as an input, the estimate will be more accurate. For infinite number of samples, the estimate will be exactly the actual error value.

For single carrier signals, it is better to compensate for phase errors after downsampling for less complexity and better performance. However, if the frequency offset value is relatively large compared to the symbol rate, there should be a mechanism to at least reduce the offset severity so that match filtering and downsampling are possible. In Section 11.3 this problem is addressed in detail and the synchronization ordering is discussed.

The rest of this chapter has two main sections as follows. In Section 11.3, carrier phase error is addressed where feedback and feedforward compensation techniques are presented. Time synchronization is considered in Section 11.4, where both frame edge detection and downsampling problems are addressed. For downsampling, both feedback and feedforward methods are shown. The developed methods, in general, are either maximum likelihood estimation (MLE) or heuristic-based methods.

11.3 Carrier Synchronization

Estimation and compensation for carrier impairments including both frequency offset, f_o, and phase offset, θ_o, are studied in this section. Developed methods show that phase and frequency offsets can be jointly compensated.

In frequency offset compensation, two cases based on the offset severity are considered. The first one is for large frequency offset value relative to the symbol rate $1/T$. In this case, frequency compensation must be applied before timing recovery. Otherwise, in the second case where $f_oT \ll 1$, it is better in terms of performance to recover symbols and then compensate for frequency offset. Another advantage of operating at the symbol rate is less complexity, since the number of processed samples is the smallest possible number for detection. Consider the next signal model to show why it is important to compensate first for large offsets.

Let the transmitted signal be given as:

$$x(nT_n) = \sum_k b_k g(nT_n - kT); \quad n = 0, 1, \dots, N_n - 1, \tag{11.3}$$

where N_n is the number of samples and b_k's are the transmitted symbols, which are assumed to be independent identically distributed random variables with zero mean value, and $g(nT_n)$ is the pulse-shaping filter. The above equation represents the convolution sum of upsampled sequence of symbols and sampled filter g_n [1, 2]. However, since the upsampled sequence of symbols is all zeros except at lN_s for $l = 0, 1, 2, \dots$, the convolution sum reduces to Eq. (11.3) shown above. Pulse shaping is covered in Chapter 6.

Assuming an AWGN channel with a flat frequency response over all frequencies (i.e. no bandwidth limitation), the received signal y_n with phase error is given as:

$$y_n = x_n e^{j(2\pi f_o T_n n + \theta_o)} + w_n, \tag{11.4}$$

where w_n is an additive white Gaussian noise following $\mathcal{CN}(0, \sigma^2)$. Since θ_o is constant over time and has no effect on the developed algorithms for frequency offset compensation, it is omitted for now (i.e. $\theta_o = 0$). Later in this section, we consider it back along with any phase ambiguity resulting from frequency offset compensation techniques.

At the receiver side, y_n is filtered by a filter matched to the effective filter corresponding to transmitter shaping filter and channel response. Since the channel is assumed to have an ideal response, this filter is just matched to g_n. Then, the output of this matched filter, r_n^F, is given as (assuming a symmetric g_n):

$$\begin{aligned} r_n^F &= y_n * g_n \\ &= \sum_l g_l s_{n-l} e^{j2\pi f_o T_n (n-l)} + z_n \\ &= e^{j2\pi f_o T_n n}(s_n * g_n') + z_n, \end{aligned} \tag{11.5}$$

where $(*)$ denotes the convolution operator. z_n is the filtered noise and g_n' is given as:

$$g_n' = \begin{cases} g_n; & f_o T \ll 1 \\ g_n e^{-j2\pi f_o T_n n}; & otherwise, \end{cases} \tag{11.6}$$

where for a relatively small frequency offset compared to the symbol rate (and not the sampling rate), the filtering mismatch due to frequency offset is assumed to be negligible. However, for larger offsets, the effect of offset on g_n' shape within one symbol duration is not neglected. In this case, the frequency offset should be compensated before filtering and time synchronization.

Note that for $f_o T \ll 1$, the effect of frequency offset on r_n^F is similar to its effect on y_n. In other words, the filtering operation is not affected by frequency offset and the filter output is just multiplied by a complex exponential representing the offset effect. In this case, time synchronization is conducted before frequency synchronization for the reasons mentioned earlier. Frequency offset compensation techniques performed after time synchronization are called time-aided.

In this section, compensation methods for large frequency offsets before time recovery are called coarse frequency compensation techniques. On the other hand, time-aided methods are called fine frequency compensation techniques. This naming convention is used to reflect the superior performance of time-aided compensation techniques. Figure 11.2 shows frequency offset compensation steps and their dependence or relation to symbol time recovery where they are applied at two stages, i.e. operating at rates $1/T_n$ and $1/T$, for improved accuracy. Another reason for applying both coarse and fine compensation techniques is to support low-end equipment suffering from high carrier frequency offset values.

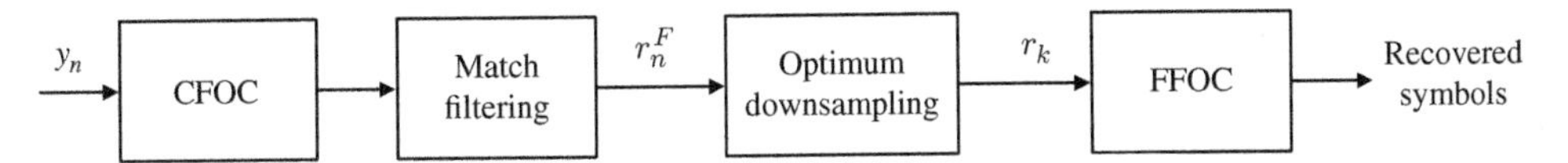

Figure 11.2 Carrier and time synchronizations ordering. Coarse frequency offset compensation (CFOC) is first applied to compensate for large frequency offset. Then, after match filtering and optimum downsampling, fine frequency offset compensation (FFOC) is applied to compensate for the small offset values.

In the case of coarse frequency offset compensation, two methods are shown; the first one is based on the discrete Fourier transform (DFT) of received signal and the second is based on phase increment between samples. For fine frequency offset compensation, feedforward and feedback-based methods are developed where maximum likelihood estimation (MLE) and heuristic-based approaches are involved.

11.3.1 Coarse Frequency Offset Compensation

For coarse frequency estimation, two methods are shown. The first one is based on the DFT of received signal raised to a given power. This method gives a peak in the frequency domain at a frequency value equals to the offset frequency multiplied by the power value. The second method is based on phase increments between samples where the speed of this increment is related directly to the frequency offset value.

11.3.1.1 DFT-based Coarse Frequency Offset Compensation

Recall that any linearly modulated symbol, b_k, can be written as

$$b_k = A_k e^{jc_k\theta_r}, \tag{11.7}$$

where $A_k \in \mathbb{N}$ and $c_k \in \mathbb{Z}$, and $\theta_r = 2\pi/D$ is an angular resolution of which the phase of any point in the signal space is an integer multiple and D is a fixed integer depending on the modulation order. For M-ary phase shift keying (PSK) modulation, D is equal to the modulation order M.

By raising symbols to power D, where the multiplication is defined without conjugation, the resulted $(b_n)^D$ will be real-valued regardless of its information content. In addition, if a constant-envelop modulation is used (i.e. M-ary PSK), $(b_n)^D$ will be a constant-valued signal where all its information content is freed up. This property is to be exploited next in detecting the frequency offset as proposed in [3].

For high signal-to-noise ratio (SNR) value, the noise term in Eq. (11.4) can be neglected and the received signal raised to power D is given as:

$$\begin{aligned}
(y_n)^D &\approx (x_n)^D e^{j2\pi D f_o T_n n} \\
&= e^{j2\pi D f_o T_n n} \sum_{k_1} \cdots \sum_{k_D} b_{k_1} g_{n-k_1N_s} \cdots b_{k_D} g_{n-k_DN_s} \\
&\overset{(a)}{=} e^{j2\pi D f_o T_n n} \left(\sum_k (b_k)^D (g_{n-kN_s})^D + \sum_{k_1} \cdots \sum_{\substack{k_D \\ k_1 \neq k_2 \neq \cdots \neq k_D}} b_{k_1} g_{n-k_1N_s} \cdots b_{k_D} g_{n-k_DN_s} \right) \\
&\overset{(b)}{=} c e^{j2\pi D f_o T_n n} + w'_n,
\end{aligned} \tag{11.8}$$

where in (a), coefficients with same indices are taken out of the multiplication and represented by the single summation term. Assuming constant-envelop modulation is used, b_k^D is constant for

Algorithm 1 DFT-Based Coarse Frequency Offset Estimation.

Input: **y**: vector of received signal samples, N_{DFT}: DFT size, f_s: sampling rate.
Output: $\hat{f}_o$: coarse frequency estimate.

Execution :

1: Find $D = 2\pi/\theta_r$.
2: $\mathbf{p} = \mathbf{y}^{\circ D}$, where ($^\circ$) denotes the element-wise power operator.
3: $P_\kappa = DFT\{\mathbf{p}\}$ with DFT size N_{DFT}.
4: $\hat{\kappa} = \arg\max_\kappa P_\kappa$.
5: $\hat{f}_o = \hat{\kappa} \times \frac{f_s}{DN_{DFT}}$.

all k and this summation is constant for all n. The second term corresponding to multiple summations in (a) is considered as a random disturbance with zero mean value since b_k's are independent random variables each with zero mean value. In (b), c is given as $c = \sum_k (b_k)^D (g_{n-k})^D$ and w'_n represents the random disturbance term in (a) multiplied by the complex exponential.

Based on Eq. (11.8b), the magnitude of the discrete-time Fourier transform (DTFT) of $(y_n)^D$, DTFT$\{(y_n)^D\}$, will have a peak value at a frequency equals to Df_o. This peak corresponds to the constant-envelop term $ce^{j2\pi Df_o T_n n}$. However, we don't deal with DTFT of the signal but its sampled version; the DFT. Consequently, the index, $\hat{\kappa}$, of the DFT sample having maximum value is used to obtain the offset estimate given as $\hat{f}_o = \hat{\kappa} f_s/(DN)$, where $f_s = 1/T_n$ is the sampling frequency and N_{DFT} is the DFT size. The offset frequency does not always match with the frequency spectrum samples; however, the larger the N_{DFT} is, the better its resolution will be and, as a result, more accurate estimate is obtained. A pseudocode of the developed algorithm above is shown in Algorithm 1.

Since the DFT of received signal has a resolution of f_s/N_{DFT}, the mean value of estimated frequency offset might not be exactly the real value. Figure 11.3 shows simulation results for the mean value of frequency estimate error using different N_{DFT}. The simulation setup is shown in

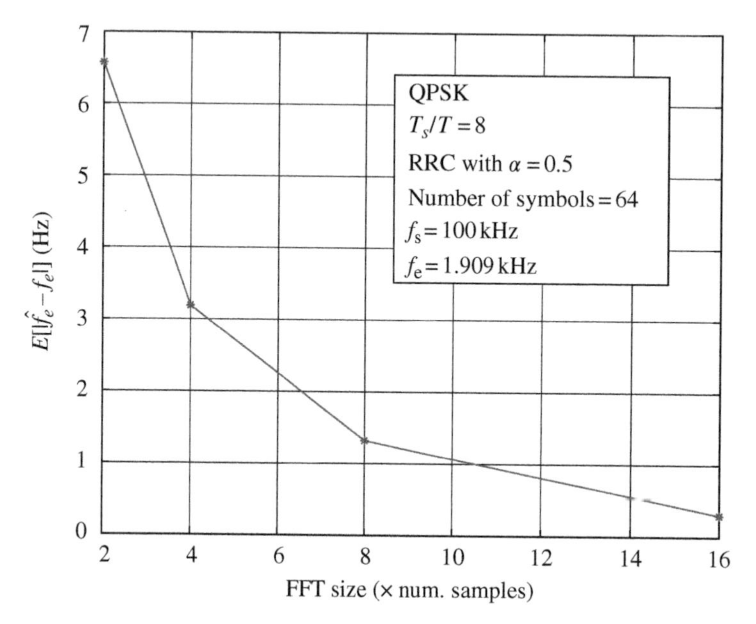

Figure 11.3 Simulation results showing the mean value of the frequency offset estimate error for different N_{DFT}. By increasing the DFT size (higher resolution of the spectrum), f_o will be closer to the samples and the error mean value will be less.

the figure. If f_o is happened to have a value that is integer multiples of f_s/N_{DFT}, then the mean error value will be zeros; however, if it is not the case, then the estimator is said to be biased. When the N_{DFT} is increased, f_o will be closer and closer to the spectrum samples and less error mean value is obtained as shown in the figure.

In this simulation, f_o and N_{DFT} were selected carefully so that f_o is always located in the middle between frequency spectrum samples even when their number is increased. This was done to show the overall behavior of the estimator as N_{DFT} increases. However, changing the N_{DFT} might result in very close samples to f_o at smaller sizes or even give samples that match exactly with f_o. Generally, increasing N_{DFT} gives less error mean value.

The estimation based on DFT is also used for fine offset compensation in Section 11.3.2 and this issue of finding the estimate between samples is treated in detail. Finally, the estimate variance of this estimator is discussed in the next section and compared to the phase-based estimator.

11.3.1.2 Phase-based Coarse Frequency Offset Compensation

The previously developed algorithm was built under the assumption of constant-envelop modulated signals. In this section, another method for coarse frequency estimation is provided for any linear modulation scheme [4], which is based on the phase increment between samples. Since the frequency offset is directly related to the signal phase, this increment can be exploited to estimate its value.

Given x_n and f_o, it can be easily seen that each sample y_n, in Eq. (11.4), is normally distributed with the probability density function (PDF)

$$f(y_n; x_n, f_o) = \frac{1}{\sqrt{2\pi\sigma^2}} \exp\left[-\frac{|y_n - \mu_{y_n}|^2}{2\sigma^2}\right], \tag{11.9}$$

where $\mu_{y_n} = x_n e^{j2\pi f_o T_n n}$ is the mean value of sample y_n. Assuming independent identically distributed samples y_n's, the PDF of $\mathbf{y} = [y_0, y_1, \dots, y_{N_n-1}]$, representing the sequence of received samples is their joint PDF given as:

$$f(\mathbf{y}; \gamma) = \left(\frac{1}{2\pi\sigma^2}\right)^{N_n/2} \exp\left[-\frac{1}{2\sigma^2}\sum_{n=0}^{N_n-1} |y_n - \mu_{y_n}|^2\right], \tag{11.10}$$

where $\gamma = \{\mathbf{x}, f_o\}$ is the estimation vector given that $\mathbf{x} = [x_0, x_1, \dots, x_{N_n-1}]$. The maximum likelihood estimate of γ, denoted by $\hat{\gamma} = \{\hat{\mathbf{x}}, \hat{f}_o\}$, is the one that maximizes the log-likelihood function of $f(\mathbf{y}; \gamma)$ as follows:

$$\begin{aligned}
\hat{\gamma} &= \arg\max_{\gamma} L(f(\mathbf{y}; \gamma)) \\
&\overset{(a)}{=} \arg\max_{\gamma} \left\{\frac{N_n}{2}\ln\left(\frac{1}{2\pi\sigma^2}\right) - \frac{1}{2\sigma^2}\sum_{n=0}^{N_n-1} |y_n - \mu_{y_n}|^2\right\} \\
&\overset{(b)}{=} \arg\min_{\gamma} \left\{\sum_{n=0}^{N_n-1} y_n^2 - \sum_{n=0}^{N_n-1} 2y_n\mu_{y_n}^* + \sum_{n=0}^{N_n-1} |\mu_{y_n}|^2\right\} \\
&\overset{(c)}{=} \arg\max_{\gamma} \left\{\frac{1}{N_n}\sum_{n=0}^{N_n-1} 2y_n x_n^* e^{-j2\pi f_o T_n n} - \frac{1}{N_n}\sum_{n=0}^{N_n-1} |x_n|^2\right\},
\end{aligned} \tag{11.11}$$

where in (a) the left term is independent of γ so it is omitted. Also, the multiplication term $1/2\sigma^2$ is removed for the same reason. In (b), maximization is replaced by minimization since we omitted the negative sign. The first left term in (b) is the summation of observations which are fixed and not affected by $\hat{\gamma}$ selection, so it is omitted as shown in (c). The introduced coefficient $1/N_n$ has no effect on the estimation process, but it is added as it is required next.

As a consequence of the law of large numbers, for a sufficiently large N_n, the sample average $\frac{1}{N_n}\sum_{n=0}^{N_n-1}|x_n|^2$ is approximated by $E[|x_n|^2]$. As a result, for sufficiently large N_n, it can be considered as a constant value regardless of $\hat{\gamma}$ selection, so it is omitted. Therefore, Eq. (11.11) is reduced to

$$\hat{\gamma} = \arg\max_{\gamma} \sum_{n=0}^{N_n-1} y_n x_n^* e^{-j2\pi f_o T_n n}. \tag{11.12}$$

Next, we assume high SNR value such that $y_n \approx x_n e^{j2\pi f_o T_n n}$. Another assumption that we would make is $x_n \approx x_{n-1}$ and it implies that the change in any two adjacent received sample values (i.e. y_n and y_{n-1}) is mainly due to frequency offset. Based on these two approximations, the likelihood function is written as:

$$\hat{\gamma} = \arg\max_{\gamma} \sum_{n=1}^{N_n-1} y_n y_{n-1}^* e^{-j2\pi f_o T_n}. \tag{11.13}$$

The complex exponential is independent of the summation index, so for any given $\mathbf{x}$, $\hat{f}_o$ that maximizes the likelihood function is given as:

$$\hat{f}_o = \frac{1}{2\pi T_n} \angle \left\{ \sum_{n=1}^{N_n-1} y_n y_{n-1}^* \right\}. \tag{11.14}$$

As seen in the above equation, the implementation of this technique is straightforward as shown in Algorithm 2.

The performance in terms of the estimate variance of this estimator is shown in Figure 11.4. In addition, a comparison with the DFT-based method discussed in Section 11.3.1 is provided.

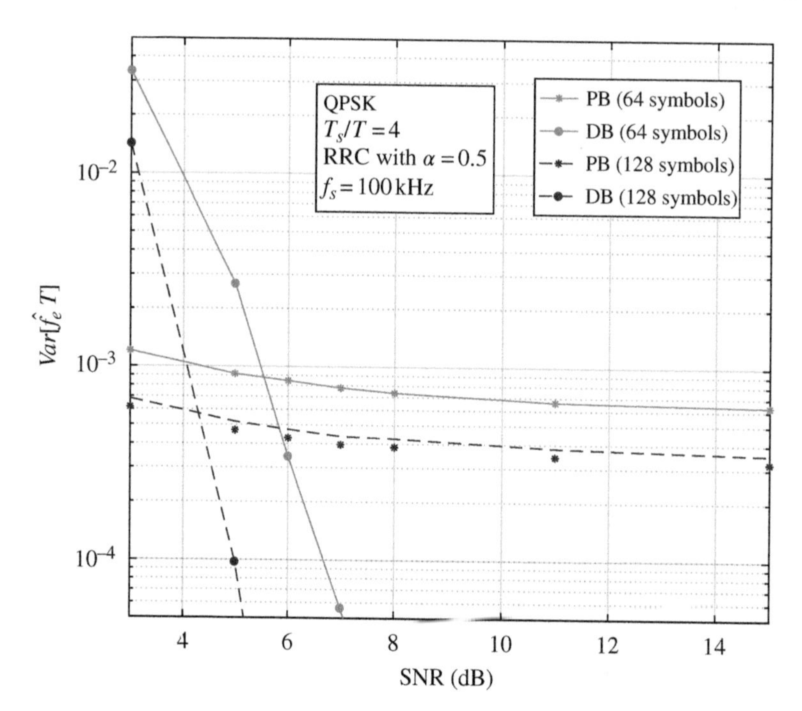

Figure 11.4 Performance comparison between two coarse frequency estimation methods: DFT (DB) and phase (PB)-based approaches. DFT-based algorithm is shown to have variance going to zero for higher SNR values, unlike the phase-based algorithm which has variance floor.

Algorithm 2 Phase-Based Coarse Frequency Offset Estimation

Input: y_n: received signal samples, f_s: sampling rate.
Output: $\hat{f}_o$: coarse frequency estimate.

Execution :

1: Find $d_n = y_n y_{n-1}^*$ where $n = 1, 2, \dots, N_n - 1$

2: $\hat{f}_o = \frac{f_s}{2\pi} \angle(\sum_n d_n)$.

It can be easily seen that the DFT-based technique performs better; however, it was shown only for constant envelop modulation schemes. On the other hand, the phase-based technique has no limitations on the received signal envelope and it is not biased for any f_o value. As discussed in the previous section, to have an unbiased estimate in the DFT-based approach, f_o must match one of the frequency spectrum samples, otherwise the closest sample is selected.

The two methods discussed above operate at the sampling rate and require no time recovery. In the next section, time-aided estimation methods operating at the symbol rate are provided. As mentioned earlier, two frequency estimation methods might be applied at different stages for more accuracy and support for devices suffering from high offset values like low-end SDR equipment or aged high-end equipment that might have this issue if not calibrated.

11.3.2 Fine Frequency Offset Compensation

Before starting with the estimation techniques, we first give the signal model and assumptions used throughout this section. The frequency offset is assumed to be relatively small (i.e. $f_o T \ll 1$), and presented algorithms operate at the symbol rate $1/T$ using recovered symbols after time synchronization. After filtering and downsampling, the received symbols are given as:

$$r_k = b_k e^{j2\pi f_o Tk} + z_k, \tag{11.15}$$

where $k = 0, 1, \dots, N - 1$ and N is the number of symbols and T is the symbol duration. z_k's are noise samples that are assumed to be independent and identically distributed random variables where each is modeled as a complex Gaussian random variable following $\mathcal{CN}(0, \sigma)$. Without loss of generality, we assume that $T = 1$.

In the next two sections, two methods are presented. The first one is MLE-based and it is shown to be related to the DFT of received symbols in a similar manner to the developed method in Section 11.3.1. This method is said to be feedforward, which, like all developed methods until now, has no feedback and operate on the buffered sequences of received symbols at once. The second one is a heuristic-based method that is based on phase-locked loop (PLL) theory.

11.3.2.1 Feedforward MLE-Based Frequency Offset Compensation

All received symbols b_k's in Eq. (11.15) are assumed to have fixed amplitude. Therefore, r_k is given as

$$r_k = Ae^{j\theta} e^{j2\pi f_o Tk} + z_k. \tag{11.16}$$

This is a valid assumption when s_k's are training symbols. Another case where this assumption is valid is for symbols based on a constant-envelop modulation and relatively high SNR. In this case, raising r_n to the modulation order, M, as explained in Section 11.3.1 will give constant valued s_n's regardless of their information content multiplied by $e^{j2\pi M f_o Tk}$. Therefore, after estimating the offset, it is divided by M to get the actual offset value.

The value $Ae^{j\theta}$ of received symbols is fixed but assumed to be unknown. In the developed algorithm shown next, A and θ are considered as estimation parameters among f_o and their estimates are given in terms of f_o estimate.

Let $\mathbf{r} = [r_0, r_1, \dots, r_{N_k-1}]$ be a vector representing the received symbols and $\gamma = \{A, f_o, \theta\}$ be an estimation vector containing unknown but fixed parameters to be estimated. Then, similar to the discussion around Eq. (11.10), the PDF of $\mathbf{r}$ for a given γ is given as:

$$f(\mathbf{r};\gamma) = \left(\frac{1}{2\pi\sigma^2}\right)^{N/2} \exp\left[-\frac{1}{2\sigma^2}\sum_{k=0}^{N-1}|r_k - \mu_{r_k}|^2\right], \tag{11.17}$$

where $\mu_{r_k} = Ae^{j(2\pi f_o k+\theta)}$ is the mean value of symbol r_k.

Similar to derivation in Eq. (11.11), the maximum likelihood estimate of γ, denoted by $\hat{\gamma} = \{\hat{A}, \hat{f}_o, \hat{\theta}\}$, is the one that maximizes the log-likelihood function of $f(\mathbf{r};\gamma)$ as follows:

$$\begin{aligned}\hat{\gamma} &= \arg\max_{\gamma} L(f(\mathbf{r};\gamma)) \\ &= \arg\min_{\gamma} \left\{\sum_{k=0}^{N-1} r_k^2 - \sum_{k=0}^{N-1} 2r_k\mu_{r_k}^* + \sum_{k=0}^{N-1}|\mu_{r_k}|^2\right\} \\ &\overset{(a)}{=} \arg\max_{\gamma}\left\{2(Ae^{-j\theta})\sum_{k=0}^{N-1} r_k e^{-j2\pi f_o k} - \sum_{k=0}^{N-1} A^2\right\}.\end{aligned} \tag{11.18}$$

The summation in the left term in (a) represents the DTFT of the signal r_k, and it is denoted by $R(e^{j\omega})$ where $\omega = 2\pi f_o$. Let $b = Ae^{j\theta}$, then the problem will be in terms of the estimation vector $\boldsymbol{\beta} = \{\omega, b\}$. Consequently, the maximum likelihood estimate in Eq. (11.18) is represented as follows:

$$\begin{aligned}\hat{\boldsymbol{\beta}} &= \arg\max_{\boldsymbol{\beta}} L(f(\mathbf{r};\boldsymbol{\beta})) \\ &= \arg\max_{\boldsymbol{\beta}} \left\{2b^* R(e^{j\omega}) - N_s|b|^2\right\},\end{aligned} \tag{11.19}$$

where $\hat{\boldsymbol{\beta}} = \{\hat{\omega}, \hat{b}\}$. For a given ω, it can be found that the value of b that maximizes $L(f(\mathbf{r};\boldsymbol{\beta}))$ by ensuring $\partial L(f(\mathbf{r};\boldsymbol{\beta}))/\partial b = 0$ is given as:

$$b = \frac{R(e^{j\omega})}{N}. \tag{11.20}$$

By substituting Eq. (11.20) in Eq. (11.19), we find that

$$\hat{\omega} = \arg\max_{\omega} P(e^{j\omega}), \tag{11.21}$$

where

$$P(e^{j\omega}) = \frac{1}{N}|R(e^{jw})|^2, \tag{11.22}$$

and it is called the periodogram of signal r_k, which is an estimate of the signal power spectrum [2]. Finally, the values $\hat{\omega}$ and $\hat{b} = R(e^{j\hat{\omega}})/N$ are the maximum likelihood estimates of ω and b, respectively.

Based on $P(w)$ definition, it can be seen that the maximum likelihood estimator of signal parameters is related to the Fourier transform of observed samples, which is a relationship that was proven in [5]. The Cramér–Rao bound of the estimator, $\sigma_{\hat{f}_o}^2$, which gives a lower bound on the estimate error variance, was also derived in [5] and it is given as:

$$\sigma_{\hat{f}_o}^2 \geq \frac{12 f_s^2 \sigma^2}{(2\pi)^2 A^2 N(N^2-1)}. \tag{11.23}$$

Note that for $N \gg 1$, the estimate variance is proportional to $1/N^3$ and for higher sampling rate more samples are required to maintain lower deviation.

The algorithm used to get the signal frequency would be simply to find the DTFT of $\mathbf{r}$, then the estimated frequency value is the one at which $|R(e^{jw})|$ is maximum. However, as mentioned in Section 11.3.1, we don't deal with the DTFT of the signal, which is a function of continuous variable, but its DFT. In addition, finding the global maximum of $|R(e^{jw})|$ is computationally demanding since it might have many local maxima. The DFT of signal r_k is given as:

$$\begin{aligned} R(\kappa) &= R(e^{jw})\Big|_{w=2\pi\kappa/N} \\ &= \sum_{l=0}^{N-1} r_l e^{j\frac{2\pi}{N}l\kappa}. \end{aligned} \tag{11.24}$$

As a sampled representation of the actual signal, using $R(\kappa)$ to estimate the frequency by searching for κ at which it is maximum would not give an accurate estimate. Figure 11.5 shows both DTFT and DFT for a given signal. Note that the maximum value is located between two DFT samples; the two samples with maximum $|R(\kappa)|$ values. Therefore, even if the DFT size is increased to get higher resolution and closer estimate (as discussed in Section 11.3.1), finding just the maximum value of $|R(\kappa)|$ is not enough, and there should be a mechanism to detect the frequency value between samples at which $|R(\kappa)|$ is maximum.

Since the accurate frequency estimate is expected to be in the vicinity of $\hat{\kappa}$, an accurate estimate of frequency offset value is given as:

$$\hat{f}_o = \frac{\hat{\kappa} + \delta}{N}\frac{1}{T}, \tag{11.25}$$

where $\delta \in (-1, 1)$ is a fractional value added to the estimate $\hat{\kappa}$. Searching for $\hat{\kappa}$, such that

$$\hat{\kappa} = \arg\max_{\kappa} |R(\kappa)| \tag{11.26}$$

is called a coarse search. The next step of finding δ is referred to as a fine search. Several algorithms have been reported in the literature for estimating δ based on DFT samples interpolation. The goal of these algorithms is to use samples around $\hat{\kappa}$, i.e. $\hat{\kappa} \pm d$ where $d = 1, 2, \ldots$, to find the estimate $\hat{\delta}$. For instance, in [6, 7], the ratios of DFT samples at $\hat{\kappa}$ and $\hat{\kappa} \pm 1$ are exploited for estimation, and the use of five samples centered around $\hat{\kappa}$ for estimation was shown in [8]. Another approach to

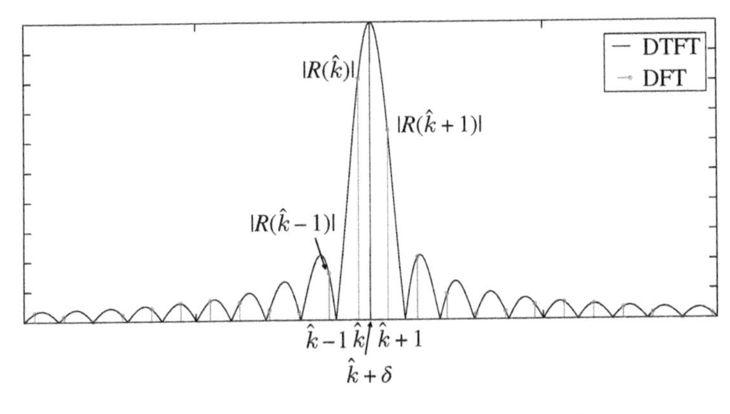

Figure 11.5 DFT and DTFT for the same arbitrary signal. Finding the global maximum of $P(e^{j\omega})$ requires two searching steps. The first one is finding $\hat{\kappa}$ at which $P(\kappa)$ is maximum. Next, samples around the maximum value are used in interpolation to locate $\hat{\kappa} + \delta$.

enhance the performance of the estimator is based on iterative implementation as shown in [9]. The estimated frequency residual in an iteration is used as a bias for the next one to give more accurate estimate. Since the goal in this section is to provide a general understanding of fine DFT-based frequency offset compensation approaches and for the sake of simplicity, the approach in [6] is adopted next.

Consider the three DFT samples including $\hat{k}$ and the two samples around it given as follows:

$$C_p = \sum_{l=0}^{N-1} r_l e^{-j2\pi l \frac{\hat{k}+p}{N}}; \quad p \in \{0, \pm 1\}. \tag{11.27}$$

By substituting Eq. (11.16) into Eq. (11.27) and considering only the signal part of r_k, those coefficients are given as:

$$\begin{aligned} C_p &= Ae^{j\theta} \sum_{l=0}^{N-1} e^{j2\pi l \frac{(\hat{k}+\delta)}{N}} e^{-j2\pi l \frac{\hat{k}+p}{N}} \\ &= Ae^{j\theta} \sum_{l=0}^{N-1} e^{j2\pi l \frac{\delta-p}{N}} \\ &\overset{(a)}{=} Ae^{j\theta} \frac{1+e^{j2\pi(\delta-p)}}{1-e^{j2\pi\frac{(\delta-p)}{N}}} \\ &\overset{(b)}{\approx} Ae^{j\theta} \frac{1+e^{j2\pi\delta}}{j2\pi(\delta-p)/N}. \end{aligned} \tag{11.28}$$

In (a), the power series identity $\sum_{n=0}^{N-1} z^n = (1+z^N)/(1-z)$ is applied. In (b), since $(\delta - p)/N \ll 1$, $e^{j2\pi(\delta-p)/N}$ is approximated by the first two terms of its Taylor series expansion; $e^{j2\pi(\delta-p)/N} \approx 1 - j2\pi(\delta-p)/N$. In addition, $e^{j2\pi(\delta-p)} = e^{j2\pi\delta}$ for any p.

Next, the following two ratios are defined:

$$\rho_1 = \Re\left\{\frac{C_1}{C_0}\right\} = \frac{\delta}{\delta-1}, \tag{11.29}$$

$$\rho_2 = \Re\left\{\frac{C_{-1}}{C_0}\right\} = \frac{\delta}{\delta+1}. \tag{11.30}$$

Real value is taken because, in case of having additive noise, the ratios will be complex-valued and only their real parts are taken as a noisy estimate of δ [6]. Note that the ρ_1 and ρ_{-1} are directly related to the residual frequency δ and two estimates, as follows, are possible:

$$\delta_1 = \frac{\rho_1}{\rho_1 - 1}, \tag{11.31}$$

$$\delta_2 = \frac{\rho_2}{1-\rho_2}. \tag{11.32}$$

Given a sufficient number of symbols, both estimates would give the same value in the ideal case with no noise. However, estimate errors based on δ_1 and δ_2 were derived in [6] and shown to follow normal distributions $\mathcal{CN}(0, \sigma_1)$ and $\mathcal{CN}(0, \sigma_2)$, respectively. σ_1 and σ_2 depend on both SNR and the offset sample location, i.e. $\hat{k} + \delta$, with respect to $\hat{k}$. As it might be expected, the variance increases for lower SNR values. In terms of sample location, it was shown that the estimate based on the ratio of closest samples to $\hat{k} + \delta$ has less variance, so it is selected as an estimate. This is illustrated in Figure 11.6, where the offset value at $\hat{k} + \delta$ can be located either between the interpolation samples C_0 at $\hat{k}$ and C_1 at $\hat{k} + 1$ (case 1), or C_{-1} at $\hat{k} - 1$ and C_0 at $\hat{k}$ (case 2). In case 1, it can be easily seen

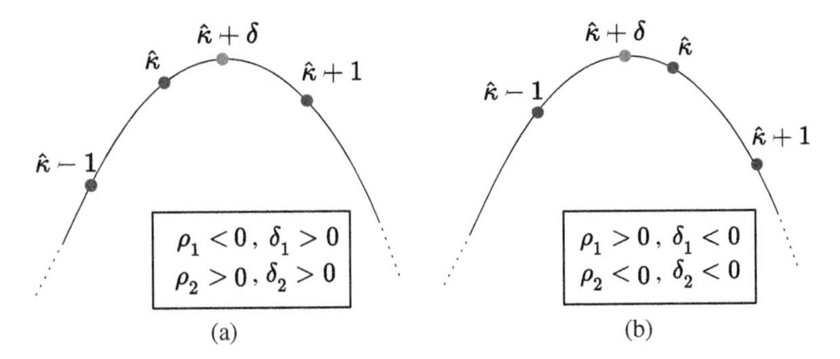

Figure 11.6 Two cases for the location of the estimate ($\hat{\kappa} + \delta$) between DFT samples. (a) Case 1: the fine estimate is located between the coarse estimate and next sample. (b) Case 2: the fine estimate is located between the coarse estimate and previous sample.

based on their definition that both δ_1 and δ_2 are positive fractional values; otherwise, they are less than zero. Therefore, the estimate of δ in Eq. (11.25) is δ_1 if $\delta_1 > 0$ and $\delta_2 > 0$, else it is δ_2.

Based on discussion and derivations provided above, the MLE-based fine frequency estimation algorithm is given in Algorithm 3.

The performance of this algorithm is measured in terms of its estimate error variance. Figure 11.7 shows the performance of this method for different sequence lengths over a range of SNR values. The variance decays exponentially by increasing the SNR value. It can be read from results that doubling the size of the training sequence enhances the performance by approximately 10 dB.

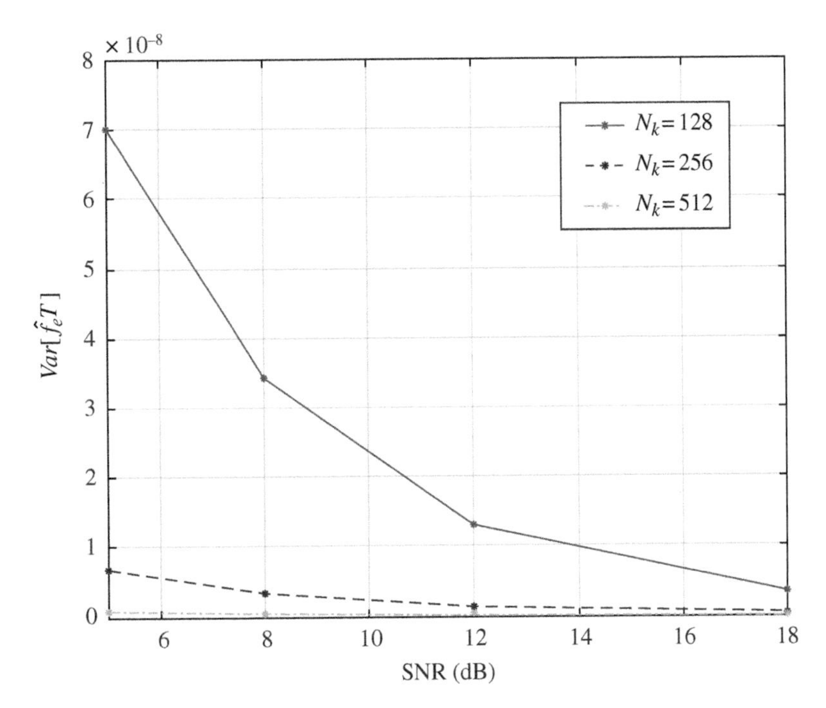

Figure 11.7 Maximum-Likelihood-based fine frequency compensation algorithm performance for different sequence lengths and over different SNR values. An offset value of f_o = 350 Hz and symbol rate of 1 MHz are selected in this simulation.

Algorithm 3 MLE-Based Fine Frequency Offset Estimation.

Input: $\mathbf{r}$: received symbols vector with length N.

Output: $\hat{f}_o$: fine frequency estimate.

Execution:

1: Find $\hat{\kappa}$ based on the DFT of $\mathbf{r}$ as defined in Eq. (11.26).

2: $C_p = R(\hat{\kappa} + p)$, where $p \in \{0, \pm 1\}$.

3: $\rho_l = \Re\left\{\frac{C_l}{C_0}\right\}$, where $l \in \{\pm 1\}$.

4: Find $\delta_1 = \frac{\rho_1}{\rho_1 - 1}$ and $\delta_2 = \frac{\rho_2}{1 - \rho_2}$.

5: **if** (δ_1 and δ_2) > 0 **then**

6: $\quad\hat{\delta} = \delta_1$.

7: **else**

8: $\quad\hat{\delta} = \delta_1$.

9: **end if**

10: Find $\hat{f}_o$ as defined in Eq. (11.25) where δ is replaced by its estimate $\hat{\delta}$.

11.3.2.2 Feedback Heuristic-Based Frequency Offset Compensation

In this section, frequency offset compensation is performed based on the drifting speed of symbols in the signal space.

Unlike phase offset, which is a fixed phase value-added to all symbols, frequency offset produces a time-varying phase. Consider Figure 11.8 that shows different time snapshots for the constellation diagram of received quadrature phase shift keying (QPSK) symbols, all having the same value $(1 + j)$. All symbols are expected to come at the same space point; however, due to frequency offset, they are drifting away from that point at a speed that depends on the offset value as follows:

$$\begin{aligned}\Delta_k(2\pi f_o k) &= 2\pi f_o(k+1) - 2\pi f_o k \\ &= 2\pi f_o,\end{aligned} \tag{11.33}$$

where $\Delta_k()$ is the difference operator over the time index k, which is an approximate of derivation in the continuous time domain. We note that the phase difference between two adjacent symbols is directly proportional to the frequency offset value. $(2\pi f_o)$ is referred to as the angular frequency or angular speed. This property is exploited in the algorithm to be developed next.

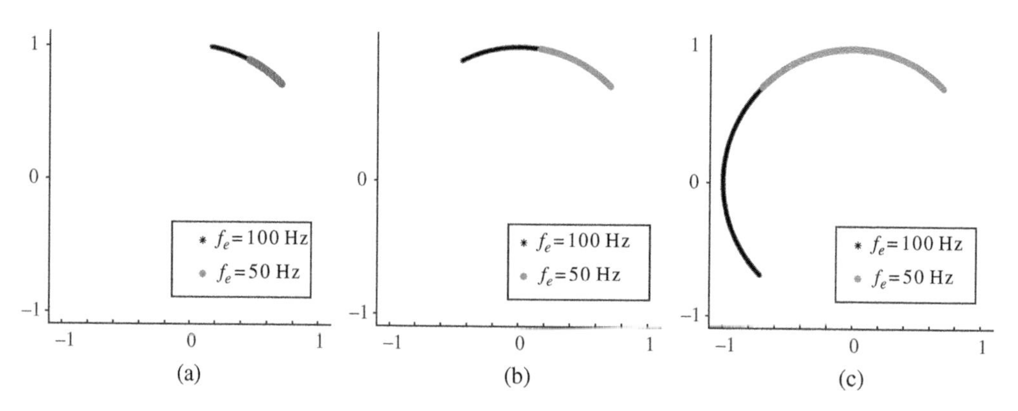

Figure 11.8 Snapshots of the signal space of fixed valued QPSK symbols with frequency offset at different times. Symbol rate and frequency offset value are 50 kHz and 100 Hz, respectively. (a) First 50 received symbols. (b) First 100 received symbols. (c) First 250 received symbols.

The method shown in this section is said to be feedback-based method, which means the estimate in each iteration is fed back to the next one in a controlled manner. It is also called heuristic-based and that means it is based on our understanding of the frequency offset effect on received symbols. Finally, it will be called decision-assisted, which means, in each iteration, the detected transmitted symbol is exploited in the frequency offset estimation.

The idea behind this method is to simply remove the linearly growing phase from each received symbol based on accumulated estimates from previously received symbols. Figure 11.9a shows the block diagram for an implementation of this method. Each received symbol is phase-corrected based on the current phase error estimate, $\phi_k^{(e)}$. Then, the phase-corrected symbol, r'_k, is fed to an MLE-based detector to detect what symbol was transmitted. The detected symbol, $\hat{b}_k$, along with r'_k are sent to a phase error tracking unit, which is responsible of providing a phase error estimate for the next iteration, $\phi_{k+1}^{(e)}$, based on their phase difference.

The phase error tracking unit is the main part of the shown receiver part, and its parts are shown in Figure 11.9b. Its inputs at time instance k are r'_k and $\hat{b}_k$, which are used to find the angle difference $\theta_k = \angle r'_k - \angle \hat{b}_k$. The output is $\phi_k^{(e)}$ and it is used to correct the growing phase error in the received symbol r_k before being used for detection.

The three main components of the error tracking unit are: phase error detector, loop filter, and phase controller. The ultimate goal is to track the time growing phase so it can be removed gradually from received symbols over time. Ideally, this can be achieved using an integrator, which accumulates estimated phase errors over time and provides for each symbol the right phase error value to be removed. This integrator is referred to as a phase controller in Figure 11.9b. It has control over the decision on the error estimate value at each iteration. This decision is made based on the stored phase error estimate from previous iteration, $\phi_{k-1}^{(e)}$, and the filtered phase difference $\Delta\phi_k^{(f)}$. This value is given as $\Delta\phi_k^{(f)} = g_0(\Delta\phi_k) + g_1(\Delta\phi_k + \Delta\phi_{k-1})$ where g_0 and g_1 are the loop filter gains. Filtering in the loop is critical to ensure stable and working PLL in the presence of noise.

The PLL response or behavior is controlled by the loop filter. The design of this filter determines PLL capabilities of tracking phase errors that can be one of two types of interest. The first one is called unit step phase error where the phase changes from one value to another at a given time instance and remains fixed for all symbols. This type represents the phase offset θ_o. The second one is called ramp phase error in which the phase changes linearly as a function of time. This change is due to frequency offset f_o.

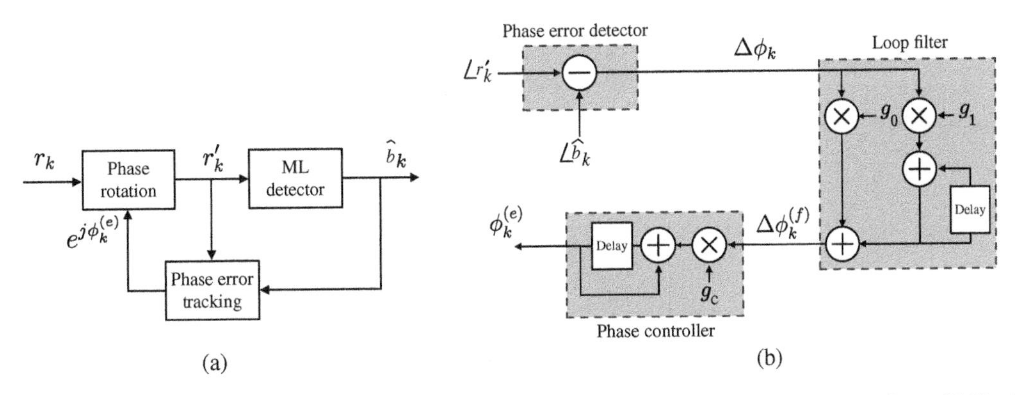

Figure 11.9 (a) Block diagram for a feedback-based carrier phase error compensation technique. (b) Block diagram for a second-order PLL used to track phase changes due to frequency and phase offsets.

As shown in [10], in case of having fixed phase error, to have $\phi_k^{(e)}$ locked to θ_0 in the steady state, the loop filter must have a nonzero fixed gain g_0. In case of having linearly time-changing phase error (i.e. frequency offset), the filter must have an integrator with gain g_1 in addition to this fixed gain filtering. As shown in Figure 11.9b, the loop filter output is the summation of two terms coming from those two parts of the filter. The first one is $g_0(\Delta\phi_k)$, which is alone a fixed gain filter. The second one is $g_1(\Delta\phi_k + \Delta\phi_{k-1})$, which is an integrator. Together, they form a loop filter that, with proper gains selection, guarantees $\Delta\phi_k$ locked to zero in the steady state for both cases of phase errors.

PLL gains $\{g_c, g_0, g_1\}$ are selected based on its desired behavior as follows [10]

$$g_c g_0 = \frac{4\zeta\lambda_o}{1 + 2\zeta\lambda_o + \lambda_o^2}, \tag{11.34}$$

$$g_c g_1 = \frac{4\lambda_o^2}{1 + 2\zeta\lambda_o + \lambda_o^2}, \tag{11.35}$$

where $\lambda_o = B_n T/(\zeta + 1/4\zeta)$. Coefficients B_n and ζ are called noise bandwidth and damping factor, respectively. $B_n T$ is the normalized loop noise bandwidth. Noise bandwidth is the bandwidth of an ideal rectangular filter that would give the same white noise power for a given noise input as if it would pass through the loop filter. The selection of ζ affects the PLL transition behavior from one state to another as shown in Figure 11.10, where the graphs show the scaled derivative of $\phi_k^{(e)}$ with respect to k. The acquisition time, which is defined as the time required to lock to a zero-phase error, is less for smaller ζ values. This is shown in the figure for locking to the frequency offset value. In addition, the selection of ζ affects the loop bandwidth, where for larger ζ values the loop

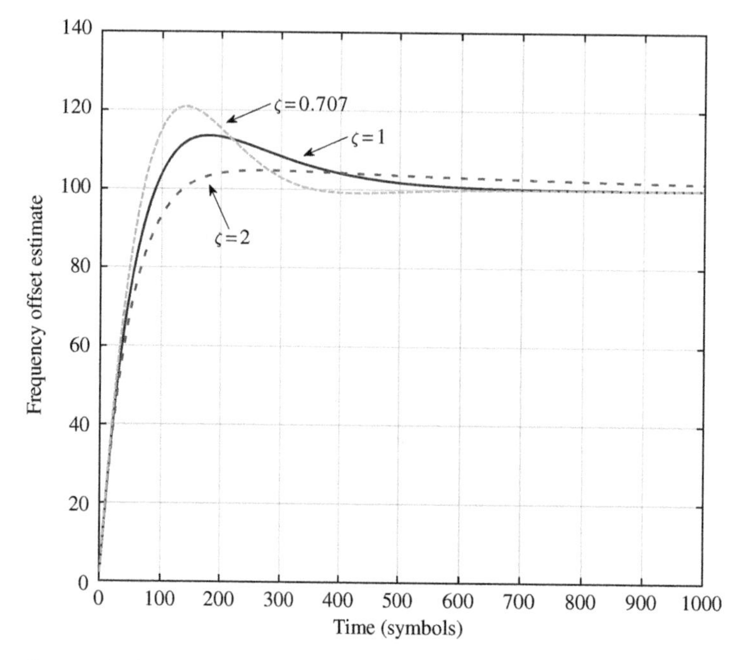

Figure 11.10 Second order PLL response to phase error due to frequency offset 100 Hz.

will have more bandwidth. For detailed analysis of PLL applied for frequency offset compensation both in digital and analog forms, the reader is referred to [10, 11].

The following code snippet shows an implementation of the discussed PLL-based method above. Loop gains are selected based on the selection of damping factor ζ and normalized noise bandwidth B_nT, which are selected to be $\zeta = 0.5$ and $B_nT = 0.1$. In addition, the phase controller gain is set to be $g_c = 1$. Based on those values, g_0 and g_1 are calculated as in Eqs. (11.34) and (11.35). Symbols are modulated using a QPSK modulator and passed through an AWGN channel with $SNR = 25$ dB. Frequency and phase offsets are introduced with the values: $f_o = 800$ Hz and $\theta_o = 0.16\pi$. Vectors definitions are omitted here since the focus is on the working principle of PLL.

```
1  % Receiver implementation for joint frequency and phase offset compensation
2  % based on PLL.
3
4  N=800; % Number of Symbols
5  M = 4; % Modulation Order
6  fk = 1e6; % Symbol rate (Hz)
7  fe = 800; % Frequency Offset (Hz)
8  th_e = 0.16*pi; % Phase Offset (rad)
9
10 %% Data Generation
11 data = randi([0 M-1], N, 1);
12 symbs = complex(qammod(data, M));
13
14 %% Channel
15 sgnl_off = exp(j*2*pi*fe/fk*(0:N-1)).';
16 symbs_off = symbs .* sgnl_off .* exp(j*th_e);
17 symbs_off = awgn(symbs_off, 25, 'measured');
18
19 %% Detection and Syncronization
20 dm = 0.5; % Damping Factor
21 BnT = 0.1; % Normalized Noise Bandwidth
22 lmd = BnT/(dm+1/(4*dm)); % Lambda Constant
23 gc = 1; % Phase Controller Gain
24 g0 = 4*dm*lmd/(1+2*dm*lmd+lmd^2); % Loop filter parameter 1
25 g1 = 4*lmd^2/(1+2*dm*lmd+lmd^2); % Loop filter parameter 2
26 for i=1:N
27     sn(i) = symbs_off(i)*exp(-j*ph(i)); % Phase Rotation
28     b(i) = sign(real(sn(i)))+j*sign(imag(sn(i))); % Max. Likelihood Detec.
29     ph_d(i) = angle2(sn(i)) - angle2(b(i)); % Phase Error Detection
30     fl2(i) = g1*ph_d(i) + fl2_st(i); % Current Filter Integrator Output
31     fl2_st(i+1) = fl2(i); % Store The Current Integrator Status
32     ph_d_f(i) = g0*ph_d(i)+fl2(i); % Filter Output
33     ph(i+1) = kc*ph_d_f(i) + ph(i); % Update Next Phase Error Value
34 end
```

Figure 11.11 shows the results after running the code snippet above two times for different B_nT values, namely: 0.1 and 0.01. It can be seen that higher B_nT value results in less transition time. The capability of second-order PLL to compensate for both phase and frequency offsets is shown in Figure 11.11b. In addition to stopping symbols from drifting over time away from a given location, the constellation itself is rotated back to the right position. This is shown as a tail of symbols, where, before complete phase offset correction, earlier symbols still have phase offsets that gradually go to zero as time passes.

Other phase error detection methods can be used in PLL, and it could be MLE-based as shown in [10]. A comprehensive treatment of PLL can be found in several titles like [4, 10–12]. Depending on the extent of used digital parts, PLLs are called digital PLL or all digital PLL. Digital PLL with different implementations for phase error compensation is presented in [11].

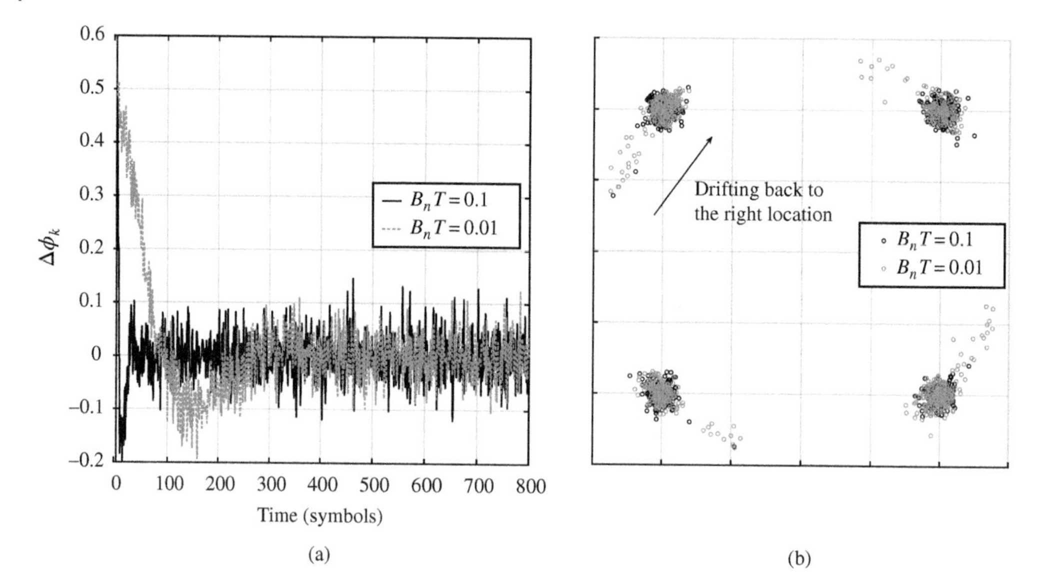

Figure 11.11 Outputs of second-order PLL for 800 QPSK symbols passed through an AWGN channel with $SNR = 25$ dB, $f_o = 800$ Hz, and $\theta_o = 0.16\pi$. Two B_nT values of 0.1 and 0.01 are considered in this simulation. (a) $\Delta\phi_k$ over times shows the locking speed of PLL. (b) The constellation diagram of corrected received symbols showing capabilities of this PLL to correct both phase and frequency offsets.

11.3.3 Carrier Phase Offset Compensation

As it has the same effect over all received symbols, estimation of phase offset, θ_o, might be achieved as part of the channel estimation. Presented methods in previous section were shown to provide joint frequency and phase offset estimation and compensation. Recall that, in Section 11.3.2, the maximum likelihood estimate was found for unknown parameters vector γ which includes a phase θ. The built algorithm was based on the assumption of having a known sequence and A and θ are unknown but fixed values. θ represents the phase offset θ_o. Based on Eq. (11.20), it was shown that the estimate of θ_o is given as $\hat{\theta}_o = \angle\left(R(e^{j\hat{\omega}})/N\right)$, where $R(e^{j\hat{\omega}})$ is calculated based on the estimated frequency offset value.

Another method to directly estimate θ_o based on the sequence of frequency offset compensated symbols is as follows. In Eq. (11.18a), the exponent $e^{-j2\pi f_o k}$ is neglected for low or zero f_o value. Therefore, for a given A, the phase offset estimate $\hat{\theta}_o$ is found based on the following:

$$\hat{\theta}_o = \arg\max_{\theta_o} \left\{ e^{-j\theta_o} \sum_{k=0}^{N-1} r_k \right\}, \tag{11.36}$$

and it is given as

$$\hat{\theta}_o = \angle\left(\sum_{k=0}^{N-1} r_k\right). \tag{11.37}$$

For PLL-based approaches, in Section 11.3.2, a second order PLL was used to compensate for both frequency and phase offsets. However, for signals with only phase offset, first-order PLL is enough to lock to a zero-phase error. The implementation of this method is the same as the one shown in Figure 11.9 but with $g_1 = 0$. In this case, the loop filter is simply a constant gain g_0, and the resulted PLL is first order, which is the simplest implementation of PLL.

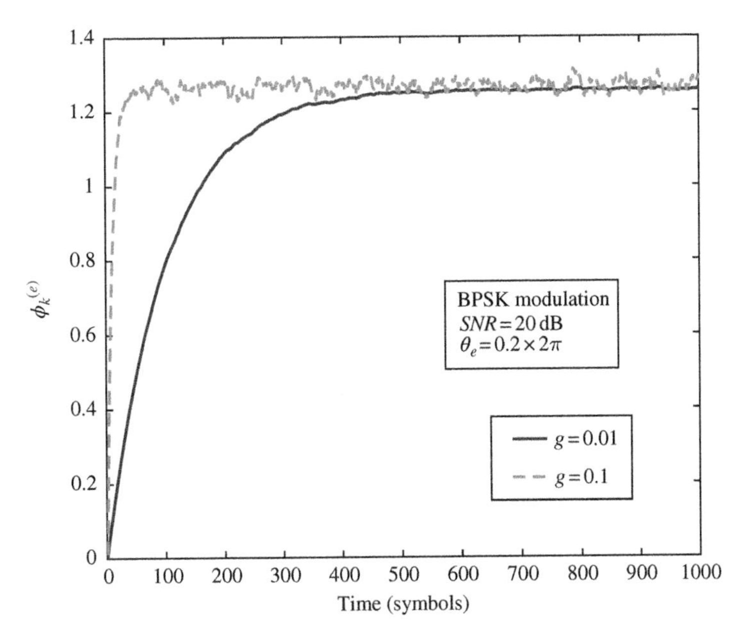

Figure 11.12 PLL response for different gain values, $g_0 \in \{0.01, 0.1\}$.

The transition behavior of the first-order PLL is governed by the loop filter response as follows [10]:

$$\phi_k^{(e)} = \theta_o(1 - e^{-g_0 k}). \tag{11.38}$$

We note that $\phi_k^{(e)}$ converges (or locks) to θ_o at a rate exponentially related to g_0. Figure 11.12 shows the PLL response for different g_0 values. It shows the value of $\phi_k^{(e)}$ over time and how fast it locks to θ_o. For larger values of g_0, the PLL converges faster to the phase offset value, but the estimate is more sensitive to noise. g_0 can be thought of as the rate at which the PLL locks to the phase error, and it also gives a weight to the contribution of each estimate to the total phase error estimate. Therefore, g_0 should be selected based on SNR, where for higher SNR values, larger g_0 value is chosen.

Carrier phase error compensation algorithms might lock to a wrong phase due to the unknown constellation rotation. For example, in binary phase shift keying (BPSK), if $|\theta_o| > \pi/2$, then PLL will lock to a wrong angle that is 180° shifted. This is called phase ambiguity, and one of the methods to solve this problem is using a sequence of known symbols sent among the data symbols, which might be the frame preamble. Once the phase error is detected, it is removed from this known sequence of symbols and the resulted compensated sequence is compared to a locally generated replica. If they match, then the estimated phase is correct, otherwise, the constellation should be rotated back by a specific angle that ensures matching between received and locally generated sequences.

11.4 Time Synchronization

Timing problems are addressed in this section and they include frame synchronization and optimum downsampling. Frame synchronization is the problem of finding the frame edge and, in this section, it is achieved by means of the correlation operation. Optimum downsampling is the problem of searching for the best sampling point in every symbol duration to detect that transmitted symbol. Feedforward and feedback methods are shown for this problem. The received signal is assumed to have no frequency offset or to have it but with very low value compared to the symbol rate (refer to the introduction in Section 11.3 for more detail on this assumption).

11.4.1 Frame Synchronization

One of the very first steps in synchronization is to detect the beginning of the received frame of data. The beginning of a frame corresponds to the sudden change in received signal level moving from noise to information signal as shown in Figure 11.13. This sudden change is called the frame edge and it is detected by energy or preamble detection methods. The energy detection approach, as the name implies, is based on the detection of changes in the energy level. The time instance at which the energy level exceeds a given threshold is detected as the frame edge. For low SNR values, this method might not give accurate estimate since the noise level is higher. In the preamble-based approach, the frame preamble sequence is correlated with a locally generated replica to detect the frame edge. Since the correlation can be thought of as an averaging operator, the preamble-based approach gives better performance compared to that of energy-based approach. However, in case of having large frequency offset values, the received preamble might get distorted so the cross-correlation will not give correct result. This problem can be tackled by using the energy detection method or the auto-correlation operator. In the latter case, the frame preamble is designed to have two identical sequences, which are used in the correlation instead of using a locally generated signal. In this section, the preamble detection method based on cross-correlation is demonstrated. The sequence of symbols used as a preamble should have good correlation properties to have more accurate estimate of the edge. Barker and Zadoffchu sequences are examples of

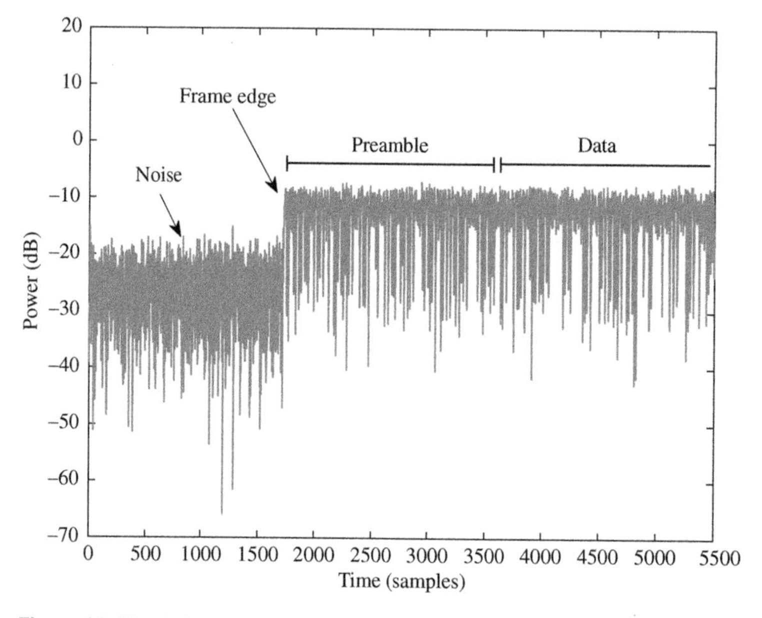

Figure 11.13 A time snapshot showing the beginning of a frame having BPSK symbols filtered by a root-raised cosine filter. The frame is divided into two parts, namely preamble and data.

sequences used practically in communication systems [13, 14]. In general, if the training sequence is repeated, it is required to have good periodic correlation properties. On the other hand, if it is only one sequence in the preamble, it is required to have good aperiodic correlation properties. Another type of sequences that is used in this section is called the maximum length sequence (m-sequence), which is a pseudo-noise sequence [15].

To detect the edge, a locally generated sequence is first filtered by the same filter used to shape the transmitted sequence. Then, the cross-correlation between locally generated and received symbols is calculated. Correlation results will give a peak value that is used to detect the frame edge. The decision on the peak level is made based on a given threshold. If the correlation value at a given lag index is above this threshold, then the index at that value is used to detect the frame edge.

The following code snippet shows a simple implementation of the preamble-based frame synchronization, and Figure 11.14 shows the results of this implementation. The lag index with maximum correlation value is used to obtain the frame edge estimate.

```
1  % Frame edge detection based on preamble detection.
2
3  % Data Generation
4  M = 4; % QPSK
5  data = randi([0 M-1], Nd, 1); % Nd: Number of symbols
6  symbs = pskmod(data, M);
7  msq = mseq(2,6);
8
9  % Filtering
10 sps = 8; % Oversampling Ratio
11 span = 16; % Filter Span
12 fltr = rcosdesign(0.5, span, sps, 'normal');
13 msq_up = upsample(msq, sps);
14 symbs_up = upsample(symbs, sps);
15
16 msq_fltr = conv(msq_up, fltr);
17 symbs_fltr = conv(symbs_up, fltr);
18
19 % Building the Frame and Attaching a Guard
20 f1 = [msq_fltr; zeros(length(symbs_fltr)-sps*span,1)];
21 f2 = [zeros(length(msq_fltr)-sps*span,1); symbs_fltr];
22 frame = f1+f2;
23 frame = [zeros(Ng,1); frame]; % Ng Zero Samples as a Guard Addition
24
25 % Channel
26 frm_rc = awgn(frame, 15, 'measured');
27
28 % Cross-correlation
29 [cr, lg] = xcorr(frm_rc, msq_fltr);
30 plot(lg, abs(cr));
31 pt = lg(max(cr) == cr)+sps*span/2+1; % First Symbol Index
```

11.4.2 Symbol Timing Synchronization

Assuming no frequency offset, which was treated in Section 11.3, the received signal is filtered using a filter matched to the overall transmission filter that includes both transmitter shaping filter and channel response, and the sampled output is represented by a sequence of samples $r(nT_n)$ for $n = 1, 2, \dots, N_n$. One sampling location has to be chosen within one symbol duration, and this location might or might not match the currently available samples. The problem of finding that optimum sampling point is to be considered in this section.

The received symbols after downsampling are given as:

$$r(kT; \tau_k) = r^F(kT - \tau_k T) + z(kT), \tag{11.39}$$

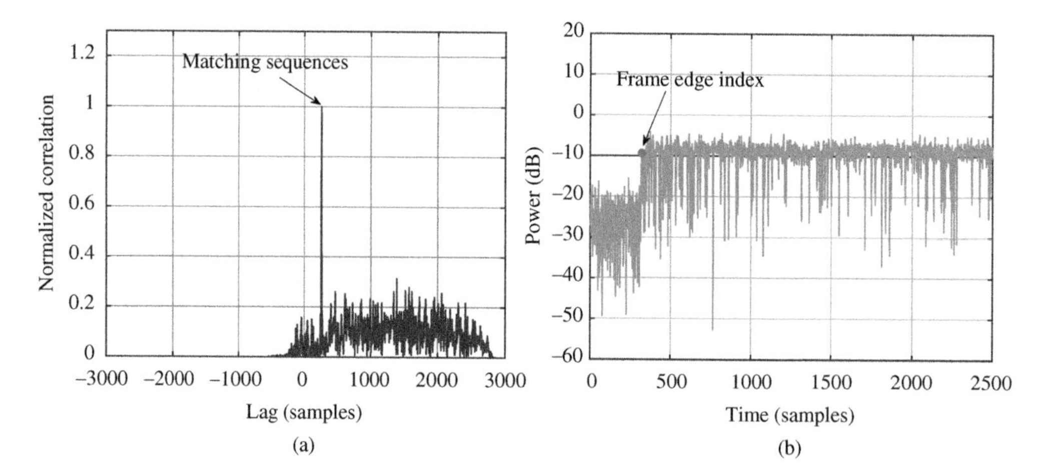

Figure 11.14 Frame edge index detection. (a) The result of correlating the locally generated m-sequence with the received frame. (b) Frame edge index detection based on the estimated lag index.

where $z(kT)$, for $k = 1, 2, \dots, N$, are noise samples that are normally distributed following $\mathcal{CN}(0, \sigma)$. $\tau_k \in [0, 1)$ is a fractional number representing the downsampling error. Note that it is subscripted by the symbols index k and that means it might change from one symbol to another. This is due to the imperfection of the sampling clock that can have deviation from its ideal periodicity.

Based on the model provided above, the problem of optimum downsampling can be defined as estimating the variable τ_k for each received symbol. Two approaches are provided next. The first one is a feedforward technique based on MLE, where buffered received symbols are processed all at once. In this method, τ_k is assumed to be fixed or having negligible variation over time. The second approach is based on PLL theory where received symbols are processed in order and the time offset for each of them is compensated based on accumulated estimates from previously received symbols.

11.4.2.1 Feedforward MLE-based Symbol Timing Synchronization

Similar to derivations of MLE-based carrier offsets estimation approaches in Section 11.3, a simple MLE-based optimum downsampling algorithm is shown here. The downsampling error is assumed to be constant for all symbols and given by $\tau_k = \tau$. In addition, the number of samples per symbol, N_s, is assumed to be large enough so that τ is approximated by one of the samples; $\tau \approx nT_n$ for $n = 1, \dots, N_s$.

Given a vector $\mathbf{r} = [r(0), r(T), \dots, r((N-1)T)]$ representing the N independent identically distributed received symbols, their joint PDF is given as:

$$f(\mathbf{r}; \tau) = \left(\frac{1}{2\pi\sigma^2}\right)^{N/2} \exp\left[-\frac{1}{2\sigma^2}\sum_{k=0}^{N-1} |r(kT; \tau) - \mu_{r_k}|^2\right], \tag{11.40}$$

where $\mu_{r_k} = r^F((k-\tau)T)$. The τ estimate, $\hat{\tau}$, is the value that maximizes the log-likelihood function of $f(\mathbf{r}; \tau)$ given as:

$$\begin{aligned}\hat{\tau} &= \arg\max_{\tau} L(f(\mathbf{r}; \tau)) \\ &= \arg\min_{\tau} \left\{\sum_{k=0}^{N-1} r(kT; \tau)^2 - \sum_{k=0}^{N-1} 2r(kT; \tau)\mu_{r_k}^* + \sum_{k=0}^{N-1} |\mu_{r_k}|^2\right\}\end{aligned}$$

Algorithm 4 Phase-based Coarse Frequency Offset Estimation.

Input: r_n: received signal samples with length N_n. N_s: oversampling ratio.
Output: $\hat{\tau}$: symbol timing offset.

Execution :

1: **for** i = 1 to N_s **do**
2: r_k = downsample(r_n, N_s)
3: LK(i) = sum($|r_k|^2$)
4: **end for**
5: LK_M = max(LK(1),LK(2),…, LK(N_s))
6: $\hat{\tau}$ = i where LK(i) == LK_M

$$= \arg\max_{\tau} \left\{ \sum_{k=0}^{N-1} 2r(kT;\tau)(r^F((k-\tau)T))^* - \sum_{k=0}^{N_s-1} |r^F((k-\tau)T)|^2 \right\}$$

$$\overset{(a)}{=} \arg\max_{\tau} \sum_{k=0}^{N-1} |r(kT;\tau)|^2, \tag{11.41}$$

where in (a), SNR is assumed to be high enough so that $r(kT) \approx r^F((k-\tau)T)$. The derivation is similar to that in Section 11.3.1. Based on this result, the MLE-based estimation of the optimum downsampling sample is shown in Algorithm 4.

Since the estimate is based on the absolute value of received samples, this method is robust against frequency offsets. However, as discussed in Section 11.3, relatively large frequency offset has to be compensated first since match filtering will not be matched anymore.

11.4.2.2 Feedback Heuristic-based Symbol Timing Synchronization

Similar to the approach shown in Section 11.3.2, a heuristic-based approach is presented here for time offset compensation. Figure 11.15 shows the block diagram of the system. The received signal is first match-filtered to get the signal $r^F(nT_n)$ for all n. Then, the samples $r^F(nT_n)$ are fed to a downsampling unit that is responsible of selecting the best sample among N_s samples, where N_s is the number of samples per symbol, based on the estimate $\hat{\tau}_k$. In addition, this unit might give another samples that are required for time error detection, as shown later. The resulted symbol $r((k-\tau_k+\hat{\tau}_k)T)$ is sent to an MLE-based detector to detect what symbol was transmitted. Both detected symbol $\hat{b}_k$ and $r((k-\tau_k+\hat{\tau}_k)T)$ and, depending on the error detection method, other samples are sent to a time error tracking unit to detect the time offset for the next symbol.

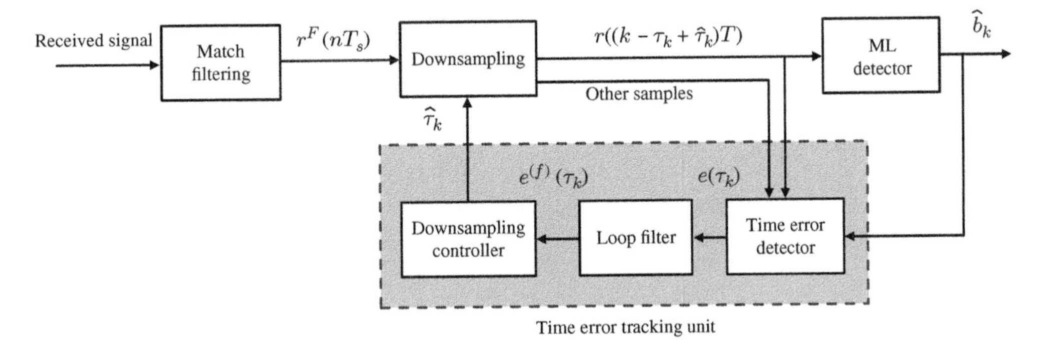

Figure 11.15 Block diagram for a PLL-based symbol timing synchronization technique.

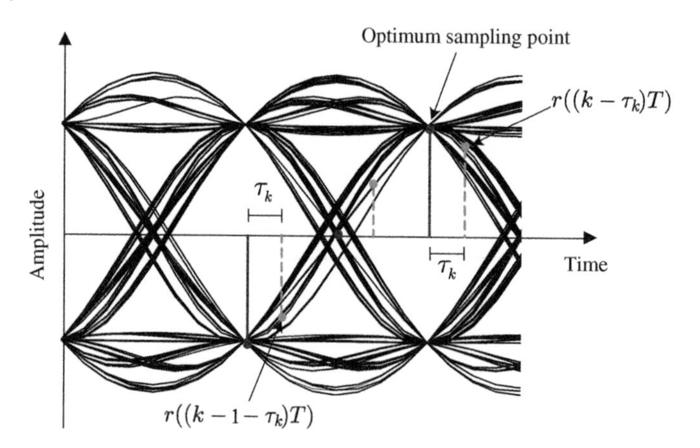

Figure 11.16 Eye diagram of received signal sampled at different places. Samples shown using solid lines are the optimum sampling and those with dashed lines show drifted samples that result in ISI.

The time error tracking unit has three components: time error detector, loop filter, and downsampling controller. The time error detector gives an error value $e(\tau_k)$ that is function of $\hat{b}_k$ and $r((k-\tau_k+\hat{\tau}_k)T)$ and other samples in its vicinity. The loop filter is used to control the system response as discussed in Section 11.3.2 where it has two gains g_0 and g_1, which are adjusted based on the system requirements (i.e. damping factor, ζ, and noise bandwidth, B_n, requirements). The filtered error value $e^{(e)}(\tau_k)$ is used by a downsampling controller to decide on what sample to use as a representation of the current symbol. The operation of those units is detailed again when their software implementation is provided, but first the design of time error detector is discussed.

A method called zero crossing for error detection is shown in this section [10]. As the name implies, it is based on the zero-crossing point in the eye diagram as shown in Figure 11.16 for a BPSK signal. Ideally, when downsampling with no time error, only one sample is taken per symbol. Here, the case is different where two samples are considered per one symbol. The first one is that used to detect what was transmitted as shown in the figure for $r((k-\tau_k)T)$. The second one is a sample in the middle between $r((k-\tau_k)T)$ and $r((k-1-\tau_k)T)$. Assuming an even number of samples per symbol, this sample is $r((k-\frac{1}{2}-\tau_k)T)$. Given that $r(kT) \neq r((k-1)T)$, if $\tau_k = 0$, then $r((k-\frac{1}{2}-\tau_k)T) = 0$, otherwise, $r((k-\frac{1}{2}-\tau_k)T)$ will have a value that depends on τ_k and amplitudes of $r((k-\tau_k)T)$ and $r((k-1-\tau_k)T)$.

Based on above observations, we note that $r((k-\frac{1}{2}-\tau_k)T)$ can be used to detect if the downsampling is optimal or not, where it should be as close as possible to the zero value. However, if $r((k-\tau_k)T)$ and $r((k-1-\tau_k)T)$ carry equivalent symbols, that means the signal will not cross the zero between the times $k-1-\tau_k$ and $k-\tau_k$, and the error function $e(\tau_k)$ should give zero, which means there is no information of the error value. This error function is given as follows:

$$e(\tau_k) = r\left(\left(k-\frac{1}{2}-\tau_k\right)T\right)(\hat{b}_{k-1}-\hat{b}_k), \tag{11.42}$$

where $\hat{b}_k$ and $\hat{b}_{k-1}$ are the current and previously detected symbols. To understand the response of time error detector, we consider the possible cases for $e(\tau_k)$. If $\hat{b}_k = \hat{b}_{k-1}$, then $e(\tau_k) = 0$, which is a desired result as mentioned previously. For $\hat{b}_{k-1} < \hat{b}_k$, $e(\tau_k)$ is a positive value if $\tau_k < 0$, else it is a negative value. The same is true if $\hat{b}_{k-1} > \hat{b}_k$.

$e(\tau_k)$ depends on the received signal power and pulse shaping. For example, if a raised cosine filter is used, τ_k will depend also on the roll-off factor of the filter. The plot of $e(\tau_k)$ mean value versus τ_k is a useful to understand the dependence of $e(\tau_k)$ on τ_k under different cases. This function is called the S-curve of the time error detector. It is desirable to have a linear S-curve, so that $e(\tau_k)$ is a linear function of τ_k given as $e(\tau_k) = g_d\tau_k$ where g_d is detector gain. This gain affects loop filter gains selection in a same manner as shown in Section 11.3.2, where

$$g_d g_0 = \frac{4\zeta\lambda_o}{1 + 2\zeta\lambda_o + \lambda_o^2}, \tag{11.43}$$

$$g_d g_1 = \frac{4\lambda_o^2}{1 + 2\zeta\lambda_o + \lambda_o^2}, \tag{11.44}$$

for $\lambda_o = B_n T/2(\zeta + 1/4\zeta)$.

An implementation of the receiver block in Figure 11.16 is shown in the next code segment. To have a fixed delay added to all symbols, instead of variable one using *resample()* function, a number of samples representing the delay might be removed from the sequence of samples or zeros might be appended to the beginning. The resolution of τ_k estimate in this implementation is equal to the sample duration T_n. However, practically, digital resamplers are used for interpolation between samples. For detailed treatment of this topic, refer to [11]. Vectors definitions are omitted for the sake of brevity.

```
1   % Symbol time synchronization based on PLL and using zero-crossing time
2   % error detection method.
3
4   %% Transmitter
5   % Data Generation
6   N = 1e3;
7   data = (2*(randn(N,1)>0)-1);
8
9   % Filtering
10  sps = 8;
11  span = 16;
12  alpha = 1;
13  fltr = rcosdesign(alpha, span, sps, 'normal');
14  data_up = upsample(data, sps);
15  sgnl = conv(data_up, fltr);
16
17  %% Channel
18  sgnl = awgn(complex(sgnl), 25, 'measured'); % AWGN Channel
19  sgnl_err = resample(sgnl, 1e3, 10+1e3); % Time Offset
20
21  %% Receiver
22  sgnl_err = sgnl_err/rms(sgnl_err); % Normalization
23
24  gd = 1; % Detector Gain
25  dm = 0.5; % Damping Factor
26  BnT = 0.1; % Normalized Noise Bandwidth
27  lmd = BnT/2/(dm+1/(4*dm)); % Lambda Constant
28  g0 = 4*dm*lmd/(1+2*dm*lmd+lmd^2)/gd; % Loop filter parameter 1
29  g1 = 4*lmd^2/(1+2*dm*lmd+lmd^2)/gd; % Loop filter parameter 2
30  while k=Ns
31      % Symbols Selection
32      s_p = real(sgnl_err(k-sps)); % Current Symbol
33      s_z = real(sgnl_err(k-sps/2)); % Zero-Crossing Symbol
34      s_n = real(sgnl_err(k)); % Previous Symbol
35      % Maximum Likelihood Detection
36      a_p = sign(s_p);
37      a_n = sign(s_n);
38      % Time Error Detection
39      e(i) = gd*s_z*(a_p-a_n);
40      % Loop Filtering
41      fl2(i) = g1*e(i) + fl2_st(i);
42      fl2_st(i+1) = fl2(i);
43      e_f(i) = g0*e(i)+fl2(i);
44      % Downsampling Control
45      k = k+sps+round(e_f(i));
46      i= i+1;
47  end
```

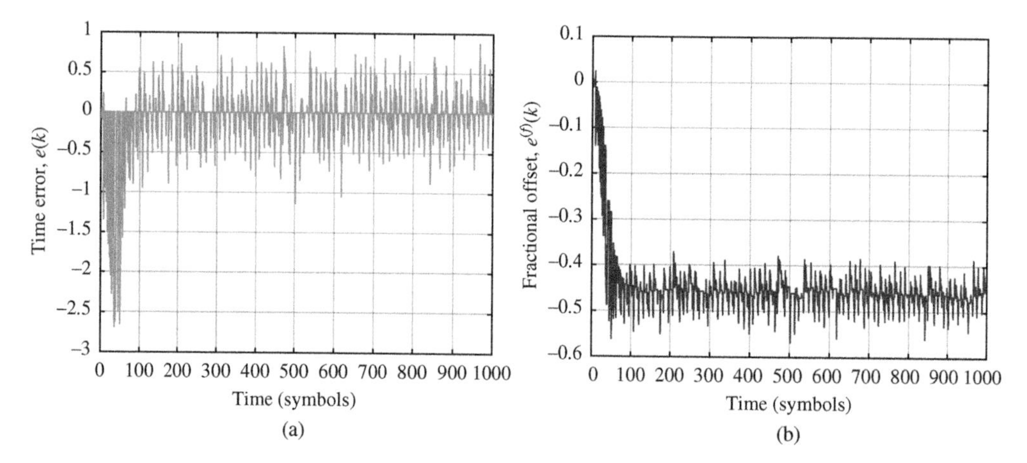

Figure 11.17 Time response of the zero-crossing approach shown with $B_nT = 0.1$ and $\zeta = 0.5$. (a) The time error $e(k)$. (b) The filtered time error $e^{(f)}(k)$ used to control the downsampling unit.

Figure 11.17a and b show the values of $e(k)$ and $e^{(f)}(k)$ as results of the implementation provided in the code above. $e(k)$ locks to zero error and $e^{f}(k)$ locks to the error value that should be added to symbols. The speed at which PLL moves to the steady state is governed mainly by the damping factor ζ.

The provided example above was given for BPSK signal where only the real part is considered. In case of using quadrature amplitude modulation (QAM) modulation, where data are received over both real and imaginary parts, the error function might be given as:

$$
\begin{aligned}
e(\tau_k) = {} & \Re\left\{ r((k-\frac{1}{2}-\tau_k)T)\right\}(\hat{b}_{k-1}^{(r)}-\hat{b}_k^{(r)}) \\
& +\Im\left\{ r((k-\frac{1}{2}-\tau_k)T)\right\}(\hat{b}_{k-1}^{(i)}-\hat{b}_k^{(i)}),
\end{aligned}
\tag{11.45}
$$

where $\hat{b}_k^{(r)}$ and $\hat{b}_k^{(i)}$ are real and imaginary parts of the detected symbol $\hat{b}_k$, respectively. However, this can result in a different gain g_d of the time error detector. Another option for $e(\tau_k)$ is to keep using the same one shown for BPSK since the same concept applies.

The zero-crossing method shown here is said to be decision-aided since detected symbols are used in the time error detection. For better system performance, a known sequence of symbols might be used. In this case, $\hat{b}_k$'s in the error function would be known and there is no need for detection. The approach in this case is called data-assisted.

Other time error detection functions are developed in the literature. The most popular and widely used ones in practice are Gardner [16] and Mueller and Müller [17] time error detection functions. Gardner method is very similar to zero-crossing method shown above, but it is not decision aided. Among those methods, Mueller–Muller method requires the minimum possible number of samples for synchronization which is one sample per symbol. For more information on error detection functions, the reader is referred to [10].

11.5 Conclusion

In this chapter, the problems of time and carrier synchronizations were addressed separately for single-carrier systems with linear modulation schemes. It was shown that the developed compensation algorithms should be conducted in a given order based on the frequency offset severity. In all cases, if the frequency offset value is relatively large, coarse frequency offset estimation and compensation must be applied first. On the other hand, it is always preferable to recover symbols first by optimally downsampling the received signal, then carrier synchronization is performed.

For full implementation, either feedforward or feedback approaches might be adopted for both time and carrier synchronizations. When working with SDR equipment, the received signal is sampled and buffered into bursts that are returned to the processing unit, which is computer software, and each of these bursts or multiple of them are processed at once. With the increasing power of computers, it might be better to process data using the asymptotically optimal approaches for both time and frequency, which are based on MLE.

References

1 J. Proakis and M. Salehi, *Digital Communications*. McGraw-Hill, 2008.

2 A. V. Oppenheim, *Discrete-Time Signal Processing*. Pearson Education India, 1999.

3 A. M. Wyglinski, R. Getz, T. Collins, and D. Pu, *Software-Defined Radio for Engineers*. Artech House, 2018.

4 H. Meyr, M. Moeneclaey, and S. Fechtel, *Digital Communication Receivers, Volume 2: Synchronization, Channel Estimation, and Signal Processing*, ser. Wiley Series in Telecommunications and Signal Processing. Wiley, 1997.

5 D. C. Rife and R. R. Boorstyn, "Single-tone parameter estimation from discrete-time observations," *IEEE Transactions on Information Theory*, vol. 20, no. 5, pp. 591–598, 1974.

6 B. G. Quinn, "Estimating frequency by interpolation using fourier coefficients," *IEEE Transactions on Signal Processing*, vol. 42, no. 5, pp. 1264–1268, 1994.

7 ——, "Estimation of frequency, amplitude, and phase from the DFT of a time series," *IEEE Transactions on Signal Processing*, vol. 45, no. 3, pp. 814–817, 1997.

8 M. D. Macleod, "Fast nearly ML estimation of the parameters of real or complex single tones or resolved multiple tones," *IEEE Transactions on Signal Processing*, vol. 46, no. 1, pp. 141–148, 1998.

9 E. Aboutanios and B. Mulgrew, "Iterative frequency estimation by interpolation on fourier coefficients," *IEEE Transactions on Signal Processing*, vol. 53, no. 4, pp. 1237–1242, 2005.

10 M. Rice, *Digital Communications: A Discrete-time Approach*. Pearson/Prentice Hall, 2009.

11 F. Ling, *Synchronization in Digital Communication Systems*. Cambridge University Press, 2017.

12 U. Mengali, *Synchronization Techniques for Digital Receivers*, ser. Applications of Communications Theory. Springer US, 1997.

13 R. Barker, "Group sysnchronizing of binary digital systems," *Communication Theory*, pp. 273–287, 1953.

14 R. Frank, S. Zadoff, and R. Heimiller, "Phase shift pulse codes with good periodic correlation properties (corresp.)," *IRE Transactions on Information Theory*, vol. 8, no. 6, pp. 381–382, 1962.

15 S. W. Golomb and G. Gong, *Signal Design for Good Correlation: For Wireless Communication, Cryptography, and Radar*. Cambridge University Press, 2005.

16 F. Gardner, "A BPSK/QPSK timing-error detector for sampled receivers," *IEEE Transactions on Communications*, vol. 34, no. 5, pp. 423–429, May 1986.

17 K. Mueller and M. Müller, "Timing recovery in digital synchronous data receivers," *IEEE Transactions on Communications*, vol. 24, no. 5, pp. 516–531, 1976.

12

Blind Signal Analysis

Mehmet Ali Aygül[1], Ahmed Naeem[1], and Hüseyin Arslan[1,2]

[1] *Department of Electrical and Electronics Engineering, Istanbul Medipol University, Istanbul, Turkey*
[2] *Department of Electrical Engineering, University of South Florida, Tampa, FL, USA*

Blind signal analysis (BSA) plays an essential role in wireless communication when the receiver does not know most or all of the received signal parameters. This lack of information may relate to signal characteristics such as carrier frequency, symbol rate, occupied bandwidth (BW). This chapter aims to provide an in-depth understanding of BSA with laboratory implementation for different applications. Besides, a blind receiver is given as a case study, which would help to understand different model-based estimation and identification techniques. Furthermore, the use of machine learning (ML)-based methods is motivated in cases where the unknown features are too many or the relationship between them is too complex for model-based methods. Along this line, the chapter also explains ML-based methods with their future directions and challenges.

12.1 What is Blind Signal Analysis?

The word "blind" in BSA refers to the concept that there is no knowledge about the received signal. However, the standard-based receivers are designed with a basic knowledge of what is transmitted at a specific frequency band group. Besides, the receiver knows particular waveform and signaling formats that are well defined in the standards. Therefore, in BSA, some of the signal specifications and parameters can be assumed to be known.

BSA is a framework with which future wireless communication systems can achieve learning, reasoning, and adaptation. Particularly, it consists of identifying transmitted waveforms and their specific features/parameters, analyzing the radio signals in every possible dimension of the multi-dimensional electro-space, identifying interference, and other impairments coupled to the received signal in any of these domains, and even extracting channel characteristics from the received signal.

12.2 Applications of Blind Signal Analysis

BSA has a long and rich history in the context of military communication, where the reception and decoding of enemy signals for gathering information is the primary goal. Besides, it is also used for commercial applications including cognitive radio (CR) and software defined radio (SDR), public safety radio, spectral efficiency improvement for next-generation cellular networks, optimization of the operation for various networks in unlicensed bands, testing, and measurement of the communication devices.

The usage of BSA varies depending on the applications and its model. For example, identifying the signal of interest is a very important concept for CR, which is done through spectrum sensing.

Wireless Communication Signals: A Laboratory-based Approach, First Edition. Hüseyin Arslan.

On the other hand, identifying who is out there, determining what type of network resources and communication devices are available to interact with, is an important aspect for public safety communication. This section reviews spectrum sensing, parameter estimation and signal identification, radio environment map (REM), equalization, modulation identification, and multi-carrier parameter estimation in the context of BSA. However, BSA is the vast domain and the applications given below are just a few selected ones aimed at illustrating the concept. Also, most of the techniques that have been discussed in this chapter are relevant to some specific waveform and signaling.

12.2.1 Spectrum Sensing

Spectrum sensing is a commonly used technique for obtaining awareness about the radio environment and spectrum usage. Arguably, the first step in blind signal identification is to determine the presence of a signal in the multidimensional electro-space. Several approaches are available for spectrum sensing. One of the commonly used technique is based on energy detection (ED).

The ED-based method, which is also known as radiometry/periodogram, is the most common way for sensing the spectrum, due to its low computational and implementation complexity [1]. Besides, it is more generic since receivers do not need to have any knowledge about the signal of interest. In this method, the output of the energy detector is compared with a specific threshold, which depends on the noise floor, to learn the existence of the signal. This method is detailed below.

Let the received signal, $y(n)$, be denoted by the following simple form:

$$y(n) = x(n) + w(n), \tag{12.1}$$

where n is the sample index, $x(n)$ is the signal to be detected, and $w(n)$ is the additive white Gaussian noise (AWGN) sample. Note that when there is no transmission by the primary user (PU), $x(n) = 0$. The decision metric for the ED-based method can be written as follows:

$$D = \sum_{n=0}^{N-1} |y(n)|^2, \tag{12.2}$$

where N denotes the size of the observation vector. The occupancy decision of a band is acquired by comparing the decision metric (D) against a threshold. This can be represented by two following hypotheses:

$$y(n) = \begin{cases} w(n), & \mathcal{H}_0 : \text{ there is no PU,} \\ x(n) + w(n), & \mathcal{H}_1 : \text{ a PU is present.} \end{cases} \tag{12.3}$$

Despite the success of ED-based methods in several applications, it still has some challenges; the selection of a threshold for detecting the existence of signal, poor performance under low signal-to-noise ratio (SNR) values, inability to differentiate interference from desired signal and noise, and sensitivity for security attacks.

The detection method performance can be measured with the probability of detection (p_D) and the probability of false alarm (p_F). p_D refers to the signal detection probability on the considered frequency when it is truly present. p_F indicates the wrong decision probability of the test when it considers that the frequency is occupied but actually it is not. Therefore, detection performance highly depends on the threshold value.

The threshold defined in the ED-based method depends on the noise variance. Therefore, even a small noise power estimation error can cause significant performance loss. To overcome this problem, the noise level is estimated dynamically by separating signal subspaces and the noise using multiple signal classification methods [2]. The noise variance is obtained as the smallest eigenvalue of the incoming signal's autocorrelation. Furthermore, the threshold is chosen using the estimated value for satisfying a constant p_F. An iterative algorithm is proposed to find the decision threshold in [3]. The threshold is found iteratively to satisfy a given confidence level, i.e. p_F. In [4], the performance of ED-based sensing is investigated over different fading channels. The probability of detection under AWGN and fading (Rayleigh, Nakagami, and Ricean) channels are derived using closed-form expressions. Similarly, the average probability for the detection of ED-based sensing methods under Rayleigh fading channels is derived in [5]. More specifically, the effect of log-normal shadowing is obtained using numerical evaluations and it is observed that the performance of the ED-based method degrades considerably under Rayleigh fading.

12.2.2 Parameter Estimation and Signal Identification

12.2.2.1 Parameter Estimation

Once the ED is done in the band of interest, which may indicate the existence of a signal, the sampled signal proceeds to the feature extraction unit for the detection of the signal features. Signal features might reside in various dimensions of the signal. Therefore, multidimensional feature extraction and signal analysis techniques are needed. Some of the parameters, which can be used to identify signals, are as below: (Note that these parameters can also be used for other applications.)

- Time domain features: received signal strength indicator (RSSI), peak-to-average power ratio (PAPR), complementary cumulative distribution function (CCDF), duty cycle, frame/burst length
- Frequency domain features: BW, carrier frequency (f_c)
- Cyclostationarity-based features of the signal: spectral correlation, cyclic features
- Statistical properties: autocorrelation function properties, variance, mean, cumulants, and moments (second, third, etc.)
- Sampling related features: chip rates, symbol rates
- Angle of arrival
- Distinguishing between single-carrier or multi-carrier
- Spread spectrum or narrowband
- Hopping sequence
- Multi-carrier parameters of the signal: time domain (cyclic prefix [CP] duration, symbol duration) and frequency domain (number of subcarriers, subcarrier spacing)
- Type of modulation and its order

The number of the feature set is a design parameter. Some or all of the features can be extracted from the multidimensions and can be used for classification. When the dimensionality of the feature set is increased, it can provide better results. However, this improvement comes at the cost of added complexity. The features to be extracted can be selected by considering the characteristics of a potential system. These features can be represented with a vector, which is a point in multidimensional feature space. Then, this vector can be used for classifying the detected transmission into one of the K candidate transmissions using a classifier (like ML, which is explained later).

12.2.2.2 Signal Identification

Blind signal identification includes identifying the technology of the received signal (like whether the signal fits any known technologies such as worldwide interoperability for microwave access [WiMAX], global system for mobile communications [GSM], long-term evolution [LTE], etc.) or identifying the type of signaling technique (if the signal is a spread spectrum signal or orthogonal frequency division multiplexing [OFDM] signal, etc.). The identification of different parameters is required at the initial block of the receiver chain so that appropriate receiver blocks can be used later on for signal processing if needed.

Identification should be performed under the effects of different channels and radio impairments including, but not limited to, frequency and phase offsets, I/Q imbalance. It cannot be assumed that time, frequency, and phase synchronization is being achieved or that the equalization and other channel compensations are performed before identifying the blind signal and its type.

Due to practical and real-time assumptions, blind signal identification is a challenging task. Also, with the vast digital communication technologies and concepts like SDR, the possible waveforms that can be designed are limitless. Therefore, in many of the studies, the focus is on identification from a finite set of wireless technologies. Specifically, these studies assume that *a priori* standard information corresponding to these technologies is available at the receiver.

In blind signal identification, different features/characteristics are extracted from the received signal, and they are used for selecting the most probable PU technology by employing various classification methods. In [6], classification is done using the features obtained by ED-based methods. These features include the amount of energy detected and its distribution across the spectrum. In [7], the reference features used are channel BW and its shape. In [8], the center frequency and operation BW of the received signal are extracted using ED-based methods. Furthermore, in [9] for detection and signal classification, the cyclic frequencies of the incoming signal are used. These features can provide insight into the signal characteristics and the hidden periodicities in both frequency and time domains. For time-domain cyclostationary features, the cyclic autocorrelation function is used while taking the fast Fourier transform, spectral correlation function (SCF) is obtained for frequency-domain features.

Some knowledge of signal features can be obtained by identifying the transmission technology of the signal in a standard-based transmission. This identification enables the receiver to extract further information about the signal with higher accuracy [8]. For instance, if the technology of the transmitter is identified as a bluetooth signal, this information can be used by the receiver for extracting some useful information in the space dimension (e.g. the range of bluetooth signal is known to be around 10 meters).

12.2.3 Radio Environment Map

The concept of REM has grown into a very promising domain especially for fifth generation (5G) and beyond wireless communication networks. The concept is used for both military communication (e.g. signal intelligence, aircraft detection) and commercial purposes (e.g. measure and understand the environment for CR and SDR, channel estimation, signal detection). Unlike the statistical techniques, REM does not learn the environment but the effects of the environment on the characteristics are learned.

REM can be used for sensing and learning everything about the environment. The general process of REM is illustrated in Figure 12.1. The direct observations from the radio and the knowledge derived from network support can contribute to environmental awareness. Learning and reasoning help the CR to identify specific radio scenarios, learn from past experience and observation, and make decisions accordingly. To implement, populate, and exploit the REM, various technologies can be employed. For example, ML, detection and estimation, cross-layer optimization, network-based ontology, and site-specific propagation prediction (more examples are given in Figure 12.1).

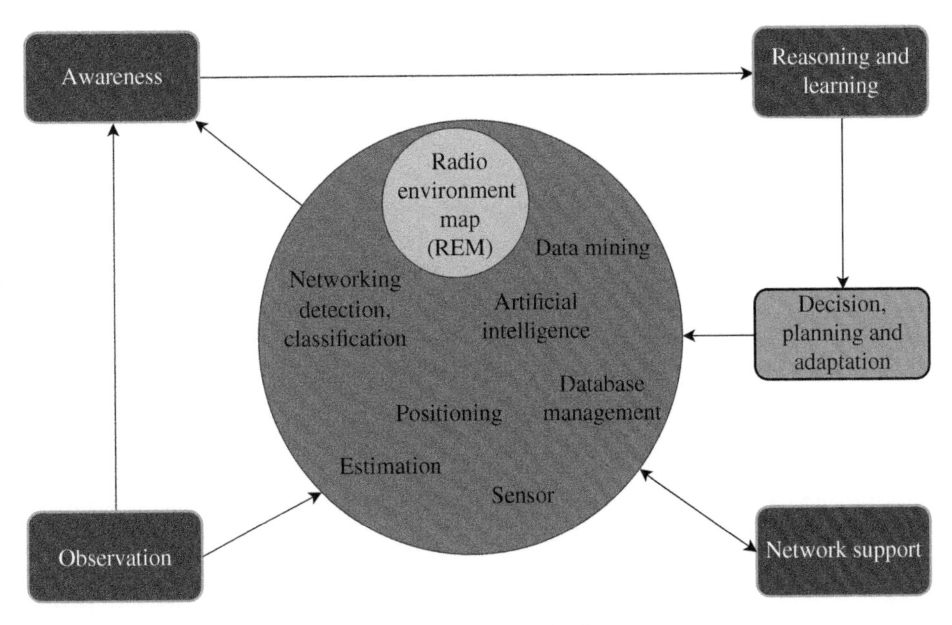

Figure 12.1 General REM process and some of its applications.

REM and BSA have a symbiotic relationship. The information in the REM, such as user profiles and mobility, can be used to aid in BSA applications. Similarly, BSA can also be used to build the REM, for example, by blindly detecting features and fingerprints of the received signals, enabling user equipment (UE) identification/authentication/activity tracking. These applications include channel modeling and estimation, physical layer security, and autonomous vehicle applications, which are enlisted in detail. More applications and discussions on REM can be found in Chapter 13.

- *REM for channel modeling and estimation:* REM can be used for blind channel modeling and estimation. Particularly, small-scale and large-scale characteristics (Chapter 10) of the channel can be learned. Besides, one can say that if the radio has the capability of learning the effects of the environment, it will help the radio to identify and compensate these effects. Some of the features that can be estimated by REM are as follows: angle of arrival estimation, line-of-sight (LOS) and non LOS (NLOS) identification, Doppler estimation, mobility estimation, sparsity-level estimation, and detection of whether the signal is sparse or not.
- *REM for physical layer security:* Due to the broadcast nature of the wireless medium, security is a critical issue for wireless networks. Traditional security threats include eavesdropping, spoofing, and jamming. Especially, authentication of the spoofing and jamming attacks is critical for BSA. Most of the authentication techniques depend on the environment characteristics. REM can learn any effects of the environment (location, angle, mobility, channel characteristics, etc.). Then, this information can be used to differentiate legitimate users, jammers, and spoofers. However, the same information can be used by malicious parties as well for security threats (eavesdropping, spoofing, and jamming).
- *REM for autonomous vehicle applications:* With the recent innovations in the automotive industry, considerable research is done on how to enhance communication and sensing between vehicles and infrastructure. REM is popularly used in this area to provide better knowledge of the

radio environment. The blind received signals at the vehicle can be identified and different sensing parameters such as the velocity of the vehicle and the distance between different vehicles can be extracted. Then, this information can be updated in the REM database.

12.2.4 Equalization

One of the major problems of wireless communication is inter-symbol interference (ISI). It causes distortion and interference in a given transmitted symbol by its neighboring transmitted symbols. The main cause of ISI being imposed on the transmitted signal is due to the multipath effect of the channel. Linear channel equalizers are commonly used to compensate channel effects. These equalizers are linear filters that approximate the inverse of the channel responses. On the other hand, the nature of the structure of an equalizer is adaptive because the channel characteristics are unknown and the channel is time-varying.

Standard techniques used for equalization, first, employ a preassigned time slot (periodic for the time-varying channel) during which a training signal that is known by the receiver in advance is being transmitted. At the receiver, the equalizer coefficients are either changed or adapted using some adaptive algorithms (e.g. least mean square, recursive least squares). This would ensure that the equalizer output accurately matches the training sequence. However, including these training sequences in addition to the transmitted information adds overhead and reduces the throughput of the system. Therefore, blind adaptation schemes, which do not require any training, can be preferred to reduce the overhead of the system.

Blind equalization is used to cancel the effects of a channel on the received signal when the transmission of a training sequence in a predefined time slot is not possible. With the help of this equalization, the instantaneous error can be determined by incorporating the signal properties. Then, this error is used for updating the adaptive filter coefficient vector. This process can be seen in Figure 12.2.

For blind channel equalization, a commonly used adaptive algorithm is a constant modulus algorithm (CMA). The CMA method assumes that input to the channel must be a modulated signal with a constant amplitude at every instant in time. If a deviation is observed in the received signal amplitude from a constant value, it is assumed that distortion is introduced by the channel. This distortion is mainly caused by multipath effects or band limitation. Moreover, CMA can also be used for quadrature amplitude modulation (QAM) signals, where the amplitude does not remain constant at every instant of the modulated signal. The instantaneous error can be calculated by

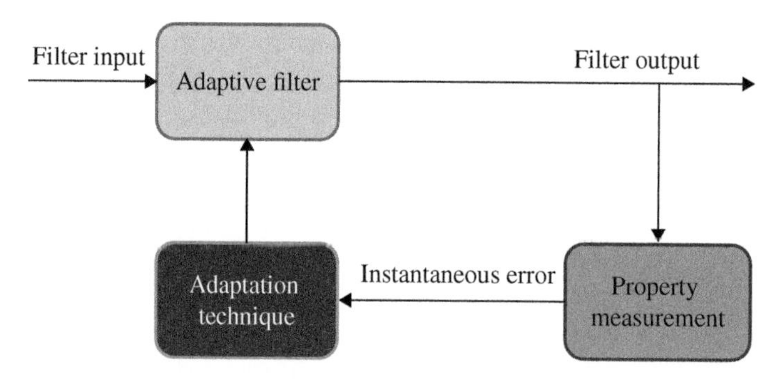

Figure 12.2 Adaptive filter model for blind equalization.

considering the nearest valid amplitude level of the modulated signal as the desired value. To measure the deviation of the output signal from the desired amplitude level, different types of error equations can be used.

For a time-varying channel, the performance of an equalizer depends also on the tracking property. This shows the ability of an equalizer to track down the changes in the channel. Using a recursive linear predictor at the output of the equalizer can improve the tracking property of a blind equalizer. Then, the instantaneous gain of the overall system can be adjusted by the predicted value of the amplitude.

12.2.5 Modulation Identification

Blind modulation identification is widely used in both military and civilian applications. Similar to the blind signal identification, some features are extracted to identify different types of modulation schemes. The design of a proper algorithm to detect the modulation scheme of any blind signal received has two steps. First is the preprocessing of the received signal and second is the selection of a proper technique or classifier to detect the modulation. The preprocessing stage may include reduction of noise, estimation of symbol rate, carrier frequency estimation, etc., which will be detailed in the next section.

There are generally two types of techniques for modulation identification; likelihood-based and feature-based methods. There are three approaches in likelihood-based methods: average likelihood-ratio test, hybrid likelihood-ratio test, and generalized likelihood-ratio test. On the other hand, the feature-based method uses different features that are extracted or estimated from the blind signal. These extracted features are related to modulation and would help distinguish different modulation schemes from each other. Some features may include instantaneous amplitude, frequency, and other statistics of the signal.

In the feature extraction, it is essential to extract the right features that serve to distinguish the signals efficiently in the given test environment. In this chapter, QAM, frequency shift keying (FSK), and phase shift keying (PSK) modulation identification is done through extracting the features. Along this line, first QAM modulation and then FSK and PSK modulations are identified as follows.

- *QAM modulation identification*: PSK and FSK are angle-modulated signals, and the instantaneous amplitude remains relatively constant for them. Therefore, the standard deviation of the amplitude sequence, which will either show small or large variations, can be taken into account to identify amplitude-modulated or non-amplitude-modulated. The instantaneous amplitude sequence is normalized and centered on measuring these variations. Large samples of data are tested with zero-mean and unit variance, which are used to develop thresholds for the standard deviation of the amplitudes. Accordingly, amplitude-modulated and non-amplitude-modulated signals can be separated. This will allow us to separate QAM from PSK and FSK as the first modulation classification.
- *FSK and PSK modulation identification*: The frequency is the change in phase over time. Since the data used is discrete, the difference between consecutive samples is taken as a means for estimating the derivative with a backward difference. FSK signals have instantaneous frequencies that stay within a small specific range and remain concentrated together. Therefore, its standard deviation will be relatively smaller than PSK. Accordingly, when the standard deviation of a signal is lower than the threshold, it is considered as FSK and anything above is PSK.

After identifying the modulation types, the next step is to identify the modulation levels. To determine the multiple-FSK levels, the instantaneous frequencies of the new frequency offset resistant FSK data are considered. This is calculated by taking the differences in instantaneous phases between consecutive samples. The standard deviation of this instantaneous frequency dataset is calculated for each FSK level, and a consistent pattern of values is obtained. Then, the modulation level can be identified according to the standard deviation value since they stay within a constant threshold for each different level. Once the modulation type of the signal has been identified as QAM or PSK, the appropriate moments and cumulants of the received data are compared against threshold values to further classify them into different levels.

12.2.6 Multi-carrier (OFDM) Parameters Estimation

Identification of a signal being multi-carrier or single-carrier is needed to further estimate their parameters. Multi-carrier, such as OFDM, signals are considered as a composition of a large number of independent random signals. Based on the central limit theorem, the sampled OFDM signal can be assumed to have a Gaussian distribution. This observation is used in [10] for the classification of any blind signal as multi-carrier or single-carrier by applying a normality test to the incoming signal. However, due to the time-varying channel, multipath-fading, and interference, the single-carrier signals at the receiver have a Gaussian distribution, especially in low SNR values. Furthermore, when power control is utilized in time-multiplexed OFDM, the Gaussian approximation is not true anymore, which results in a failure of the normality test for classification.

Once that it is identified whether the signal is a single-carrier or multi-carrier, the next step is to estimate its parameters. There are many methods used for finding the OFDM symbols and CP duration based on cyclostationarity of OFDM signaling due to CP extension. Distance between the peaks at correlation can help in the estimation of OFDM symbol duration (T_s) [11]. The time-frequency transform of the received blind signal and its entropy can be used to estimate the length of CP. Furthermore, using a cyclic correlation-based algorithm, the T_s can be estimated [10]. Then, CP duration can be estimated by performing a correlation test. Moreover, T_s can be estimated from the received samples by exploiting the CP. All the expected values should be tested as the receiver is assumed to have no prior information on the T_s. Discrete correlation of the received blind signal is written as:

$$R_y(\Delta) = \begin{cases} \sigma_s^2 + \sigma_w^2 & \Delta = 0 \\ \frac{T_g}{T_s+T_g}\sigma_s^2 e^{-j2\pi f_o \Delta t T_s} & \Delta = T_s \\ 0 & otherwise, \end{cases} \tag{12.4}$$

where f_o represents frequency offset due to inaccurate frequency synchronization. T_g denotes the length of CP. σ_w^2 and σ_s^2 correspond to the variance of the noise and signal, respectively. Hence, the autocorrelation-based estimation of T_s can be constructed as:

$$\hat{T}_s = \arg\max_{\Delta} |R_y(\Delta)|, \Delta > 0. \tag{12.5}$$

Once the $\hat{T}_s$ is found, estimation of the CP duration can be done by maximizing the autocorrelation of the received blind signal with a delayed version of itself over several hypotheses of cyclic duration intervals. Commonly, the CP length is chosen as a multiple of $\hat{T}_s$. The available sizes for many systems include 1/32, 1/16, 1/8, and 1/4. Thus, the CP size is limited to the mentioned ratios of the estimated T_s value from the previous step to reduce the complexity of the system.

Once the CP size and T_s are estimated, the repetition in the CP of OFDM systems can also be used to obtain time and frequency synchronization. Synchronization algorithms and methods based on the maximum correlation [12], minimum mean-square error (MMSE) [13], and maximum likelihood estimation (MLE) [14] can be used for this purpose. Furthermore, channel state information can also be estimated using the cyclostationarity features of the signal [15].

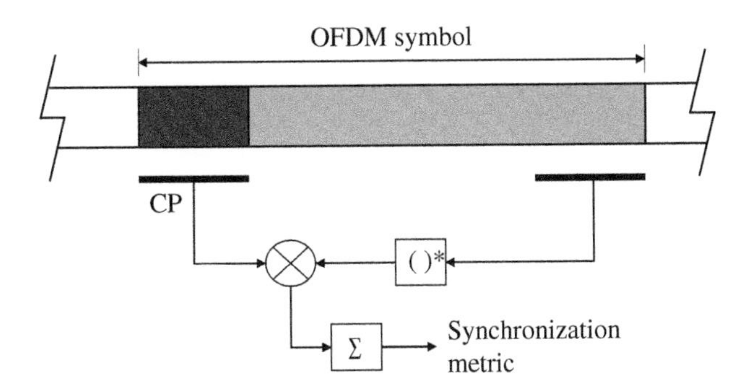

Figure 12.3 CP-based MLE for a single OFDM symbol.

The MLE method is one of the famous methods. It is shown for a single OFDM symbol in Figure 12.3. The synchronization metric obtained by the single OFDM symbol can be written as:

$$M(m) = \sum_{n=0}^{T_g-1} y(m-n)y^*(m-n+T_s), \tag{12.6}$$

where $y(n)$ is the received signal, T_g is the length of CP, and T_s is the length of the useful data part. Using Eq. (12.6), the timing position can be found as:

$$\hat{\theta}_\tau = \arg\max_m |M(m)|, \tag{12.7}$$

and frequency offset estimate is

$$\hat{f}_o = \frac{1}{2\pi}\angle M(\hat{\theta}_\tau). \tag{12.8}$$

The extraction of the synchronization parameters using the CP with one OFDM symbol is challenging in real-time applications since the CP is affected by the time-varying channel. Furthermore, the effect of frequency offset decreases the correlation between the repeated parts. Averaging over a number of OFDM symbols should be performed to overcome this problem. More details about OFDM can be found in Chapter 7.

12.3 Case Study: Blind Receiver

Waveform design is a critical aspect of wireless communication (readers are referred to Chapters 7 and 8 for discussion regarding multi- and single-carrier waveforms, respectively). It defines a framework about what to transmit, how to transmit, and where to transmit. Along this line, all of the applications in wireless communication systems depend on the type of waveform used. For example, if a frequency hopping waveform is used, the spectrum sensing is different than of spread spectrum. As another example, the roll-off factor (α) estimation may not be meaningful for OFDM, but some other estimation might be more important. Therefore, a priori information can help improve the estimation process by identifying which parameters are critical for a given scenario.

In the blind receivers, there is no or little known information about the transmitted signal. This lack of information makes the system more complicated. Thus, layered approaches which divide the problem into several layers are preferred in blind receivers. In this section, we discuss and implement some of the blind receiver layers including synchronization for single-carrier raised cosine pulse-shaped systems, as illustrated in Figure 12.4. Note that these layers are independent of

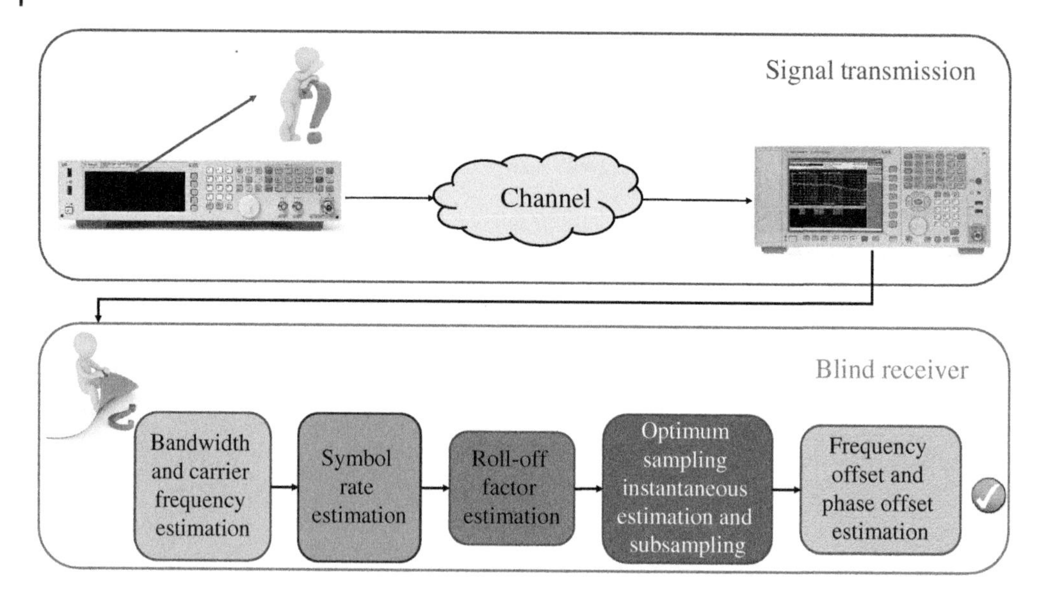

Figure 12.4 Illustration of basic steps in the blind receiver.

each other, but the output of the one layer is used in the input of the other layer. Also, the parameters given in the following are not a complete set and the list can be extended. Moreover, the blind receiver block diagrams (layers) might change for another waveform or signaling. Besides, more detailed information for synchronization including blind aspects is given in Chapter 11.

12.3.1 Bandwidth Estimation

BW signifies the area covered by the signal in the frequency spectrum. Various transmission BWs are used for different wireless standards (licensed or unlicensed) depending on the data rate requirements.

The center frequency value of the signal and the estimation of the BW for the occupied band are not known in the full blind signal processing. Therefore, the estimation of the occupied BW is the first step in the blind receiver. Besides, it is an essential parameter for system design, allowing for estimation of the carrier frequency.

The majority of baseband signal power is contained between the first two nulls of the spectrum surrounding the origin. Therefore, null-to-null BW has critical importance. It can either be estimated by ED-based methods, power spectrum analysis, or cyclostationary methods. These methods include single packet and packet pair techniques, which are two of the most popular BW estimation techniques.

In this chapter, null-to-null BW is estimated using a single-packet technique. In this technique, the start and end frequency of the null-to-null BW is calculated, and a threshold level is defined. Any point that is above the threshold is considered the desired signal, and all the remaining are assumed to be a part of the noise floor. The accurate threshold can be determined by first taking an average of the signal magnitude, while all the remaining frequencies above the magnitude are rejected. Then, the second average of the remaining signal magnitude is calculated, which is assumed to be the BW threshold level.

Table 12.1 Parameter settings of the transmitted signals.

Frequency	Amplitude	Modulation type	Filter	Symbol rate	IQ mod filter
915 MHz	−10 dBm	QAM	RRC	24.300 ksps	40 MHz

12.3.2 Carrier Frequency Estimation

Carrier frequency estimation is important for the blind signal receiver. Besides, it can also help differentiate between various types of signals depending on their operating frequencies. Knowledge of the carrier frequency is especially useful for systems with large BWs such as wireless local area network (WLAN). The carrier frequency can be calculated based on the BW of the signal. More specifically, the center point of the calculated BW indicates the carrier frequency. If the BW estimation is not correct due to the noise in the environment, the carrier frequency value will also be incorrect, and the frequency offset is caused. In this chapter, the null-to-null BW and carrier frequency estimations are simulated with the following MATLAB code for different roll-off factor values. Table 12.1 shows the parameters of transmitted signals for these simulations.

Simulation results (Table 12.2) show that the carrier frequency is estimated correctly for all α values. Besides, all of the estimated carrier frequencies are exactly the same. Null-to-null BW (B_{null}) is also estimated correctly and increases with increase in α value.

```
1  % Estimation of null-to-null BW and carrier frequency
2  load blind_rx.mat % Load your blind received signal
3  span=FreqValidMax-FreqValidMin;
4  sample_rate=1/XDelta; % Equivalent to 1/XDelta (sampling frequency)
5  % Power spectrum
6  [Y,X]=pwelch(symb_rx); % Assign the x values of psd(Y) to X
7  a=((X/max(X))*(1/XDelta));
8  plot(a,fftshift(10*log10(Y)));
9  grid
10 xlabel('Frequency (Hertz/sample)')
11 ylabel('Power/frequency (dB/Hz/sample)')
12 psd_freq=((X/max(X))*(1/XDelta));
13 psd_pwr=fftshift((Y));
14 psd_sgnl=[psd_freq psd_pwr];
15 norm_psd=psd_pwr./max(psd_pwr); % Normalized magnitude
16 temp_thresh=mean(norm_psd); % Temporary threshold is defined
17 % The peaks above the temporary threshold are removed,
18 % and a new threshold is defined from the remaining signal magnitudes
19 thresh_test=norm_psd<temp_thresh;
20 temp_sgnl=thresh_test.*norm_psd;
21 act_thresh=mean(temp_sgnl(temp_sgnl≠0));
22 % Determines the null-to-null BW start
23 BW_check=norm_psd>act_thresh; % Check from signal above threshold
24 BW_start=BW_check.*psd_freq;
25 % Choose the minimum frequency
26 null_BW_start=min(BW_start(BW_start≠0));
27 % Determines the null-to-null BW stop
28 rev_norm_psd=fliplr(norm_psd'); % Flip the signal from left to right
29 BW_check2=rev_norm_psd>act_thresh;
30 rev_BW_stop=BW_check2'.*psd_freq;
31 rev_null_BW_stop=min(rev_BW_stop(rev_BW_stop≠0));
32 % Choose the minimum frequency
33 null_BW_stop=max(psd_freq)-rev_null_BW_stop;
34 % Approximation of the null-to-null BW
35 null_null_BW=null_BW_stop-null_BW_start
36 % Approximation of the carrier frequency
37 f_c=mean(null_BW_start:null_BW_stop)+FreqValidMin
```

Table 12.2 Transmitted and estimated signal parameters for null-to-null BW and carrier frequency estimations.

Transmitted signal	Estimated signal	
α	B_{null} kHz	f_c MHz
0.01	25	915
0.35	31	915
0.99	45	915

12.3.3 Symbol Rate Estimation

The symbol rate indicates the number of symbols transmitted in a second. It is one of the most important parameters to make an accurate estimation in the blind receiver. This is because the other parts in the receiver block are estimated according to the symbol rate. Deciding the symbol rate also means deciding the oversampling rate (OSR), this step proceeds with the decision of the optimum sampling phase.

The symbol rate for the same bit rate will vary depending on the modulation type used. Therefore, symbol rate information is required in the blind receiver for the radio frequency (RF) signal to be demodulated. In the literature, different methods such as inverse Fourier transform, wavelet transform, and cyclostationarity are presented for the estimation of the symbol rate.

In this chapter, cyclostationary features of the transmitted signal are used to estimate the symbol rate. By cyclostationarity of a signal, it implies that a signal can be stationary within a cycle and that its autocorrelation function fluctuates in time. The symbol rate is estimated from the peaks of the spectral correlation density shown in Figure 12.5. In this figure, each peak indicates the value of the symbol rate of the transmitted signal. After BW and symbol rate estimation is done, the roll-off factor can be estimated.

12.3.4 Pulse-Shaping and Roll-off Factor Estimation

Symbols must be converted into the pulse form in order to transmit information over the air. The information is shaped with the help of a filter. This filter is the most significant parameter that determines the BW of the signal and ISI.

Information transmitted in the time plane has a response in the frequency plane. For that reason, the chosen pulse shape should be optimum for both time and frequency (and possibly space) dimensions. The type of filters to be used vary depending on the channel state where the signal is transmitted and they should be estimated in BSA.

The estimation of the filter includes the estimation of the roll-off factor that belongs to the filter. The roll-off factor will be estimated using BW of the signal and symbol rate information. More specifically, if the assumption is made that a Nyquist filter was used during modulation, roll-off factor (α) can be estimated using the previously found parameters (null-to-null BW and symbol rate) as follows:

$$\alpha = \frac{B_{null}}{R_s} - 1 \tag{12.9}$$

where R_s is the symbol rate.

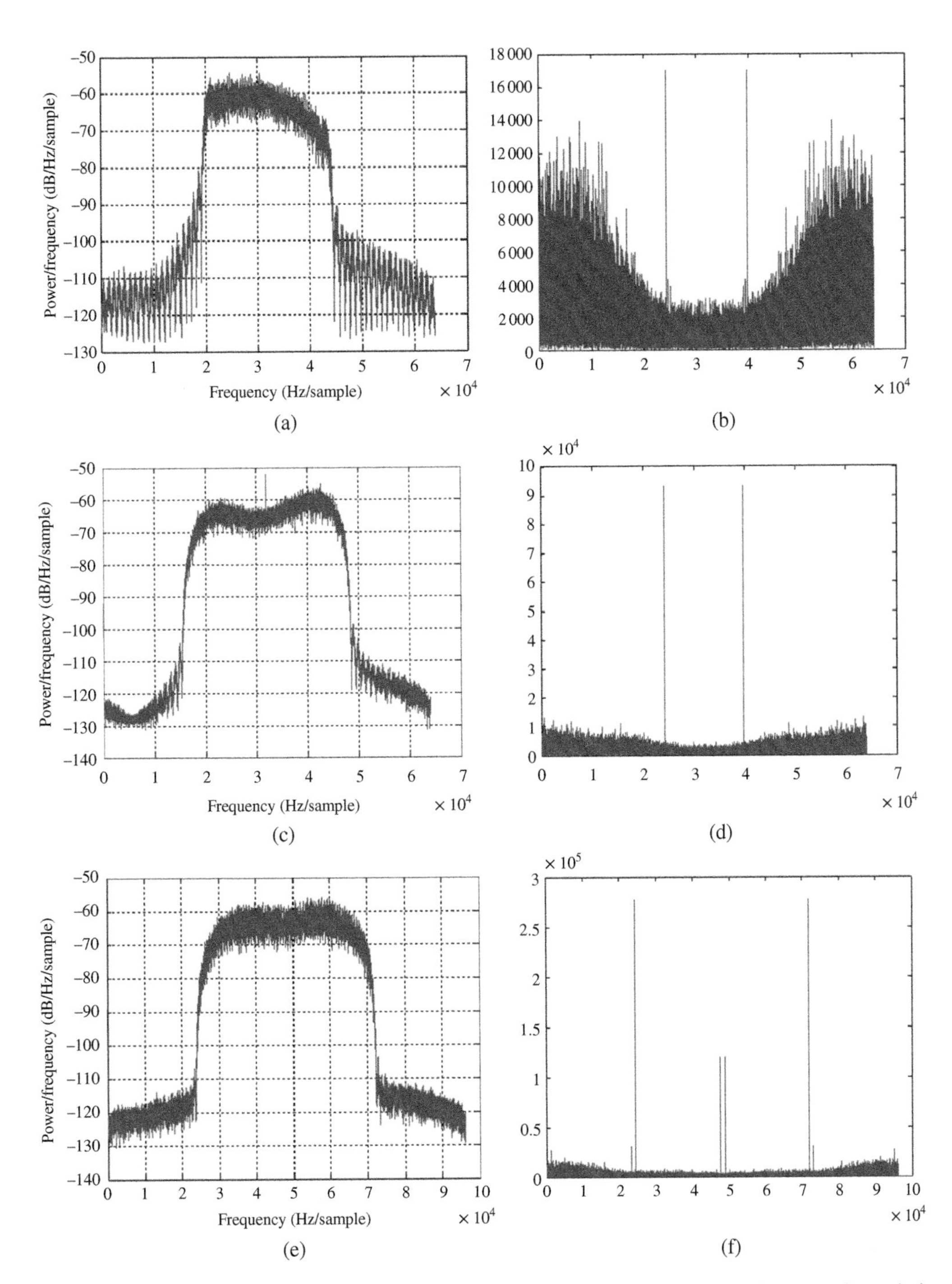

Figure 12.5 (a), (c), and (e) are estimated null-to-null BWs and (b), (d), and (f) show the spectral correlation density and their peaks indicate the symbol rates, when $\alpha = 0.01, 0.35$, and 0.99, respectively.

Table 12.3 Transmitted and estimated signal parameters for symbol rate and α estimations.

Transmitted signal	Estimated signal	
α	Symbol rate (ksps)	α
0.01	24.330	0.01
0.35	24.331	0.29
0.99	24.322	0.86

The symbol rate and roll-off factor are estimated in the code below. Again, the transmitted signal parameters, which are defined in Table 12.1, are used. Table 12.3 shows that symbol rates and α values are estimated with small errors.

```
1  % Symbol rate and Nyquist filter roll-off factor estimation
2  % Symbol rate approximation
3  sgnl_mag=abs(fft(abs(10*log10(symb_rx.^4))));
4  [Y1,X1]=max(sgnl_mag); % The first peak is undesired
5  sgnl_mag_1=sgnl_mag(sgnl_mag≠Y1);
6  plot(sgnl_mag_1)
7  % Cutoff the signal at the BW (since symbol rate is less than the BW)
8  adj=1;
9  temp_null_null_BW=null_null_BW;
10 while ceil(temp_null_null_BW)>length(sgnl_mag_1)
11     temp_null_null_BW=temp_null_null_BW*0.1; % Normalized BW
12     adj=adj*10;
13 end
14 sgnl_mag_2=sgnl_mag_1(1:ceil(temp_null_null_BW));
15 % Reverse the elements of the signal
16 % the first peak will be the rotated symbol rate
17 sgnl_mag_rev=fliplr(transpose(sgnl_mag_2));
18 sgnl_rev_srt=fliplr(sort(sgnl_mag_rev)); % the peaks are the first elements
19 % Give the percentage of the max values to determine the threshold
20 thresh=mean(sgnl_rev_srt(1:ceil(0.0007*length(sgnl_rev_srt))));
21 peak=sgnl_mag_rev≥thresh;
22 % Search for the first peak above the defined threshold
23 m=0;
24 while m<length(peak)
25     m=m+1;
26     if peak(m)==1
27         rev_sym_rate=m;
28     break
29     end
30 end
31 symbol_rate=(ceil(temp_null_null_BW)-rev_sym_rate)*adj
32 % Approximation of the roll-off factor
33 alpha=(null_null_BW/symbol_rate)-1
```

12.3.5 Optimum Sampling Phase Estimation

The upsampled signal that has been sent by the transmitter goes through the process of downsampling at the receiver. The downsampling process is carried out using OSR that is decided along with the symbol rate. Downsampling is done by selecting the optimum sample phase to keep ISI at the minimum level.

The ISI caused by the estimation of the wrong sample phase will be observed in the eye and constellation diagrams. If the signal is downsampled at the time of optimum sampling, the location of the symbols can be distinguished from each other in the constellation diagram. Otherwise, all symbol levels are mixed with each other. We here note that the effects of different pulse-shaping and different modulation types also can be examined in an eye diagram. For instance, a higher level of eye diagram is obtained when the modulation type is high, while for lower modulation types the eye diagram is of low level.

Clustering methods are used for optimum sampling phase estimation. For example, some predefined algorithms in MATLAB (e.g. "fcm" and "kmeans") can be used for this purpose. On the other hand, an optimum sampling phase can also be estimated using an ED-based method. In this method, which is less computationally complex compared to the eye diagram approach, the total energy is calculated at each sampling phase.

12.3.6 Timing Recovery

In wireless communication systems, the coherent receiver must know the exact symbol timing to decode the transmitted data accurately. The main purpose of timing recovery is to recover a clock

```
1  % Timing correction using band edge timing recovery method
2  bf=0;settle_time=30;OSR_rate=16;OSR_time=1/OSR_rate;
3  % Symbol rate information extracted from the peaks obtained from SCF
4  k1=1/64;k2=1/1024;k3=1/32;k4=1/16;k5=1/1024;k6=1/2048;k7=4/pi;k8=-1/1024;
5  % Peaks given at: F=0,2Fc,-2Fc,Fs,-Fs,2Fc+Fs,2Fc-Fs,-(2Fc+Fs),-(2Fc-Fs)
6  f_c=0;f_s=1/XDelta;beta=0; % Carrier and sampling frequency of Tx signal
7  for ii=4:length(symb_rx)-10 % Interpolation section
8      if ii>settle_time
9          if beta>OSR_time, bf=bf+1;beta=0;int_rx(ii)=symb_rx(ii+bf);
10         elseif beta<-OSR_time, bf=bf-1;beta=0;rx_int(ii)=symb_rx(ii+bf);
11         else
12             if beta≥0
13                 int_rx(ii)=symb_rx(ii+bf)*(OSR_time-beta)/OSR_time+...
14                     symb_rx(ii+1+bf)*beta/OSR_time;
15             else
16                 int_rx(ii)=symb_rx(ii+bf-1)*abs(beta)/OSR_time+...
17                     symb_rx(ii+bf)*(OSR_time-abs(beta))/OSR_time;
18             end
19         end
20     else
21         int_rx(ii)=symb_rx(ii);
22     end
23     % Upper and lower band-edges
24     upper_rx=int_rx(ii)*exp(j*2*pi*(f_c-Symb_rate/2)*ii/f_s);
25     lower_rx=int_rx(ii)*exp(j*2*pi*(f_c+Symb_rate/2)*ii/f_s);
26     % Low pass filter
27     upper_rx_lpf=k1*upper_rx+(1-k1)*upper_rx_lpf;
28     lower_rx_lpf=k1*lower_rx+(1-k1)*lower_rx_lpf;
29     % Control loop
30     if mod(ii,OSR_rate)==0
31         rn=upper_rx_lpf.*conj(lower_rx_lpf);
32         rn_av=k2*rn+(1-k2)*exp(j*Delta_w)*rn_av;
33         rn_n=exp(j*k3*sign(imag(rn_av*conj(rn_n_f))));
34         rn_n_f=rn_n*rn_n_f*(1-k4*(abs(rn_n*rn_n_f)^2-1));
35         temp=rn_n_f*k5+(1-k5)*temp*exp(j*2*Delta_w);
36         Delta_w=k6*imag(rn_n_f*conj(temp));f=f+k7*Delta_w;
37         if f>limit, V=limit;
38         elseif f<-limit, V=-limit;
39         else V=f;
40         end
41         beta=beta+k8*V+imag(rn_n_f)/512;index1=index1+1;
42     end
43 end
```

at a specific symbol rate(s) from the modulated waveform, which enables the conversion of the given signal into a discrete-time sequence. In the beginning, since the receiver does not know the optimum sampling time of the signal, the whole signal is sampled and then later corrected sample by sample. In the literature, there are several methods that are used for timing recovery estimation. Two commonly used timing recovery methods are discussed in this chapter and implemented. These two methods are *band edge* timing recovery and *Gardner* timing recovery methods and their estimated results are shown in Figure 12.6. We here note that the quadrature phase shift keying (QPSK) modulation type is used for signal generation in these simulations.

Band edge timing recovery method: In the band edge timing recovery method, similarity of the upper and lower edges of the signal is observed. These edges are pushed to a control loop for obtaining the correction term. Then, this term is calculated for each sample and updated recursively.

Gardner timing recovery method: Gardner timing recovery method uses three samples for each symbol. The correcting term is obtained from three samples, which are a sample and $\pm\frac{T}{2}$ spaced samples. First, this term is used for interpolation to generate optimum samples. Then, it is recalculating for each sample recursively.

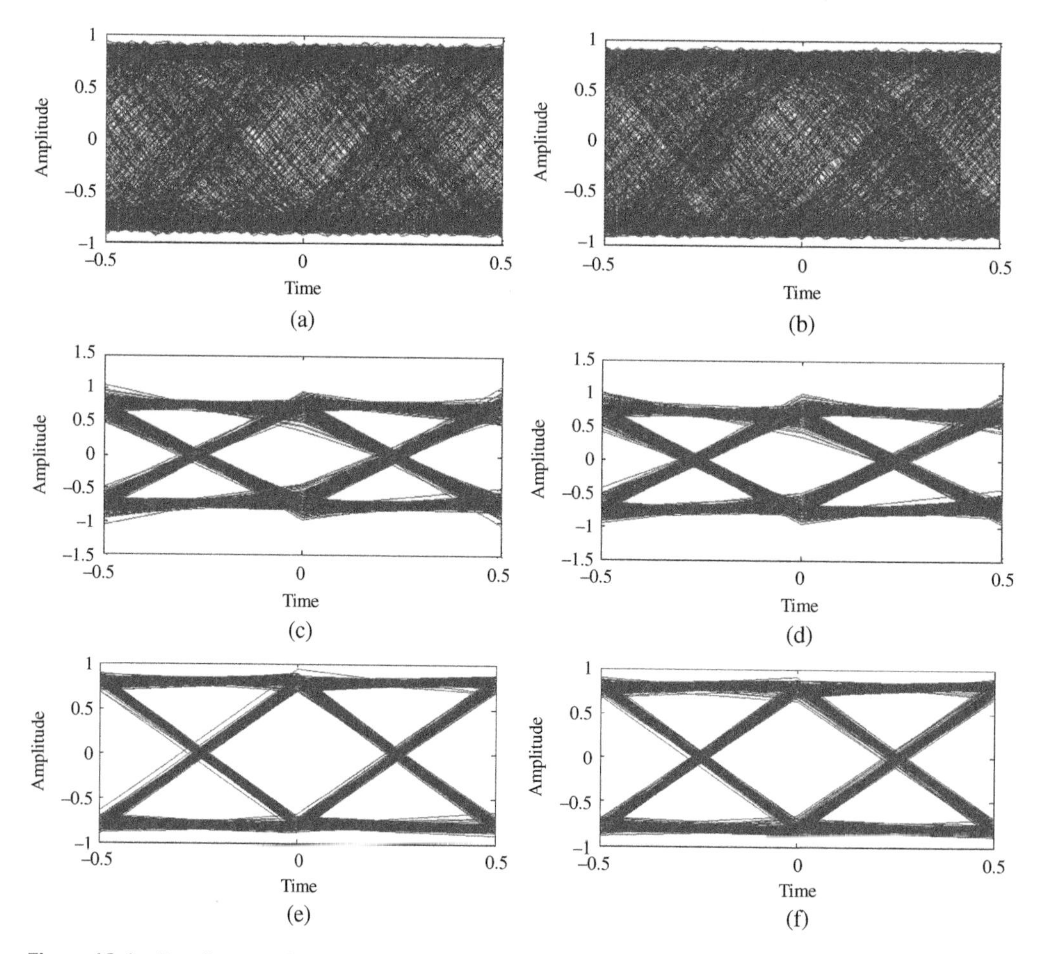

Figure 12.6 Eye diagrams for in-phase and quadrature signals; (a) and (b), before timing recovery; (c) and (d), recovered by the Gardner method; (e) and (f) recovered by the band recovery method.

```
1  % Timing Correction using Gardner timing recovery method
2  ind_x=20;mu=0;
3  for ii=ind_x:OSR_rate:length(symb_rx)-10
4      %Interpolation Block
5      if ii>settle_time
6          if mu>1
7              bf=bf+1;mu=0;
8          elseif mu<-1
9              bf=bf-1;mu=0;
10         end
11         down1_rx=symb_rx(ii+bf);
12     else
13         down1_rx=symb_rx(ii);
14     end
15     % Selecting three samples for each symbol correcting term
16     down2_rx=symb_rx(ii+bf+round(OSR_rate/2));
17     down3_rx=symb_rx(ii+bf-round(OSR_rate/2));
18     in=in+1;rx_int(in)=down1_rx;
19     e_gardner=real(conj(down1_rx)*(down2_rx-down3_rx)); % Optimum sample
20     tau=gamah*e_gardner;
21     mu=mu*alpha+(1-alpha)*tau;
22     mu_all(in)=mu;
23 end
```

12.3.7 Frequency Offset and Phase Offset Estimation

The difference between the local oscillators used at the transmitter and receiver side causes a frequency offset. The offset will always be in every system since there are no two perfectly identical oscillators. Frequency offset causes rotation in the symbol location, and the amount of rotation in each constellation point depends on the sample index. Due to this phase rotation, the constellation diagram forms a circular shape, which can be observed in Figure 12.7a.

Estimating the frequency offset value on the receiver side is one of the most challenging areas for the blind receiver. It can be done with feedforward and feedback algorithms. The feedforward algorithms are used to eliminate the higher frequency offsets, while the feedback algorithm removes the remaining residual errors.

The *polarity decision stop and go* algorithm is also a popular method to estimate the frequency offset. In this method, the polarity detection of the received symbol is performed, and an error signal

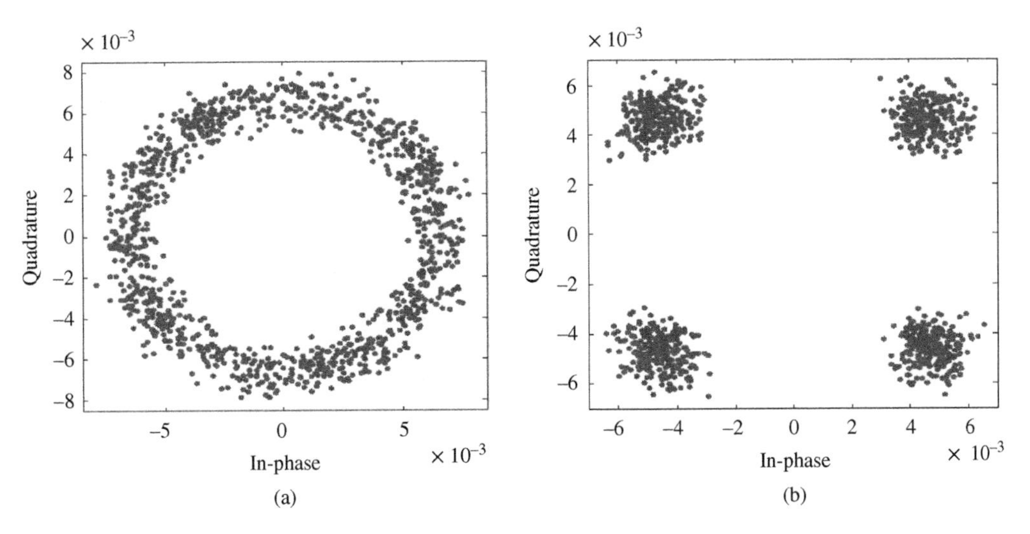

Figure 12.7 Constellation diagrams for frequency offset estimation and correction.

is generated. This error signal is fed into a stop and go block algorithm for estimation. A MATLAB code for frequency offset estimation is provided below. In this code, a threshold is set to capture the constellation diagrams, from which the frequency offset can be estimated accurately. After estimating the frequency offset, it is removed from the received signal. The effect of this compensation can be observed from Figure 12.7.

Phase offset is due to the effect of the channel, which causes a shift in the constellation diagram, and the entire constellation diagram rotates at the same amount. Commonly used estimation techniques include decision-directed phase estimation, maximum likelihood phase estimation, S-curve tool, etc. These methods work efficiently and give good accuracy when carrier frequency and timing have been acquired beforehand.

```
1   % Estimation of frequency offset
2   % Defining threshold for the down-sampled received signal
3   threshold=max(abs(down_rx))*.8;len=length(down_rx);
4   z_max=0.7071*.7;z=0;z_avg=0;z_est=0;k=1;track=0;p=0;
5   for index=1:len
6       down_rx(index)=down_rx(index)*exp(-j*2*pi*index*z_avg*Symbol_rate/FS);
7       if abs(down_rx(index))>threshold % Power threshold value
8           d=abs(down_rx(index))/sqrt(2)*(sign(real(down_rx(index)))+...
9               j*sign(imag(down_rx(index))));
10          p=imag(down_rx(index)/d);
11          if abs(p)<z_max, z=p;
12              if z<z_max
13              offset=z/index;z_avg=z_avg+offset;
14              end
15          end
16      else
17          if real(down_rx(index))<0&&imag(down_rx(index))>0
18              s(index)=down_rx(index)*exp(-j*pi/2);
19          elseif real(down_rx(index))<0 && imag(down_rx(index))<0
20              s(index)=down_rx(index)*exp(-j*pi);
21          elseif real(down_rx(index))>0 && imag(down_rx(index))<0
22              s(index)=down_rx(index)*exp(-j*3*pi/2);
23          else
24              s(index)=down_rx(index);
25          end
26          if abs(atan2(imag(s(index)),real(s(index)))- pi/4)<pi/20
27              d = abs(down_rx(index))/sqrt(2)*(sign(real(down_rx(index)))+...
28                  j*sign(imag(down_rx(index))));
29              p=imag(down_rx(index)/d);
30              if abs(p)<z_max, z=p;
31                  if z<z_max
32                  % Removal of offset after estimation
33                  offset=z/index;z_avg=z_avg+offset;
34                  end
35              end
36          end
37      end
38  end
```

12.4 Machine Learning for Blind Signal Analysis

Model-based methods are successfully applied for BSA, both in civil and military areas. However, these methods do not perform well when the models are unavailable, relationships between different features of the signal are complex, training information is scarce or the system is completely blind (turned into a black box), and it is necessary to form multiple nonlinear relations. Considering, the success of ML-based methods, especially in image and signal processing, data augmentation, and such areas, current literature proposes that ML-based methods can be successfully

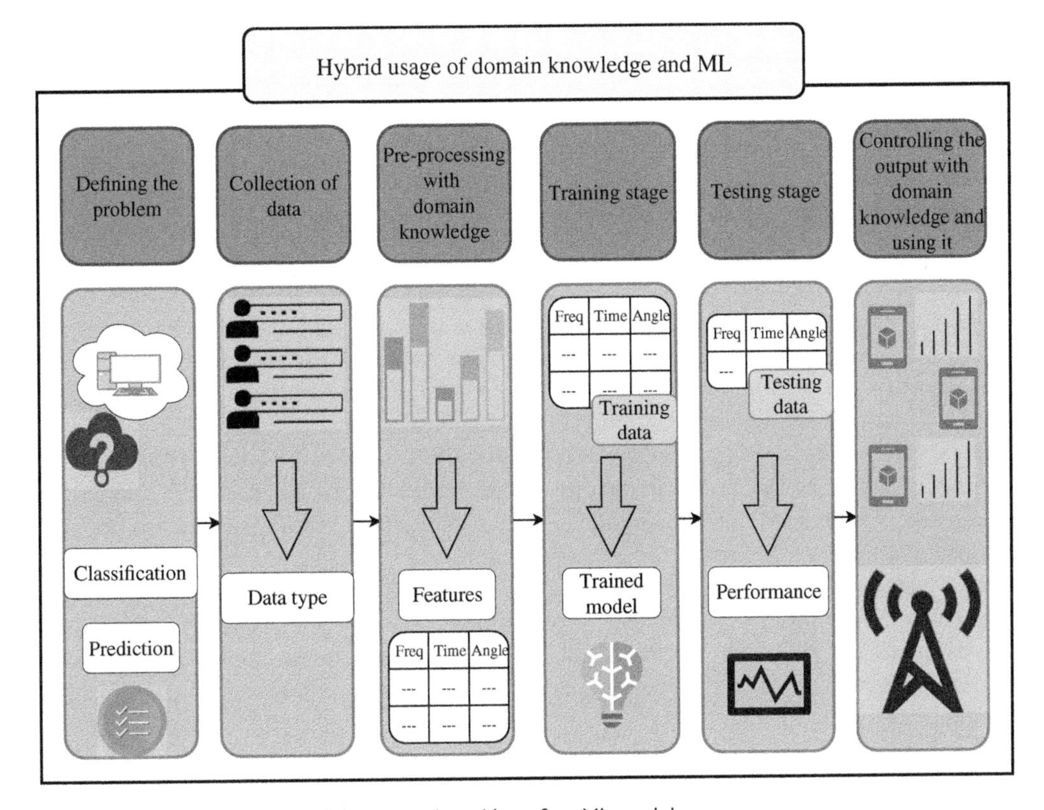

Figure 12.8 Block diagram of the general working of an ML model.

applied to the BSA. In this section, preliminary information for ML is presented and applications of the multidisciplinary domain (BSA and ML) including signal and interference identification, multi-RF impairments identification, channel modeling and estimation, and spectrum prediction are given with their future directions and challenges.

ML methods are data-dependent, which means they need a large amount of data to learn a specific problem. Therefore, after defining the problem, the first and foremost step is the collection of data. Once the data is ready, feature selection with domain knowledge is crucial because the model accuracy of the ML method is dependent on the suitability of the selected features for a given problem. The selected feature needs to be robust to real-time conditions, yet able to provide maximum (distinctive) information about the problem at hand. Once the distinct features are selected, a suitable model would be chosen for a specific problem. On the bases of these features and the data provided, the machine learns and trains itself for further evaluation and testing of either prediction or classification. Also, after the prediction, controlling the output with domain knowledge is preferable. The ML steps are summarized in Figure 12.8.

Domain knowledge is very important to fully utilize the ML-based methods. The features that are defined in the previous sections can be used with these methods. For instance, in [16], the standard deviation of the instantaneous frequency and the maximum duration of a signal are extracted using time-frequency analysis. Then, neural networks (NN) are used for the identification of active transmissions using these features. As another example, cyclostationary-based methods that use the outputs of SCF and autocorrelation function as salient features for the classifiers can be used for signal identification. In [17], these features are utilized for signal classification using NN.

The selection of the ML technique depends on what kind of data is used for a specific problem. There are three primary ML techniques; supervised learning, unsupervised learning, and reinforcement learning (RL). These techniques are explained in the following list and deep learning (DL) is motivated for more complex problems in the following section.

- *Supervised learning:* Supervised learning is a model that has known response (output) to labeled data (known input). It trains the model to either use classification algorithms or regression for its output. Some common methods used are K-nearest neighbor, support vector machine, NN, linear regression, and decision trees. Classification categorizes the data into specific classes, while regression is used for the prediction of continuous data.
- *Unsupervised learning:* The primary aim of unsupervised learning is to develop an understanding and learning of the hidden structure of the unlabeled data. In this learning, the unlabeled data is processed using different methods that divide the data into the clusters. The main aim is to find different patterns that can help distinguish between the unlabeled data. Common unsupervised methods include K-means, Gaussian mixture models, and principal component analysis.
- *Reinforcement learning:* RL tries to imitate the human behavior of learning new things from the interaction with the environment. The RL setup includes a kind of conversation between an agent and the environment. The environment gives a state to the agent in return and the agent takes action over that state. Then, the environment gives back a reward to the agent, and so on. The loop continues until the environment terminates this process. Therefore, the agent learns based on trial and error methods. RL methods include Q-learning, multi-armed bandit learning, deep RL, and actor-critic learning.

12.4.1 Deep Learning

As the amount of data increases and the problems become more complex, DL methods will be preferred instead of ML methods. The most significant difference that distinguishes ML from DL is, ML learns in one single hidden layer while DL learns in more than one hidden layer, as illustrated in Figure 12.9. In that case, it can allow learning with raw data without doing sophisticated feature engineering. Some of the DL architectures used in the literature are as follows; long short-term memory (LSTM) networks, convolutional neural networks (CNN), repetitive NN, feedforward deep network, deep belief networks, Boltzmann machine, and generative adversarial networks.

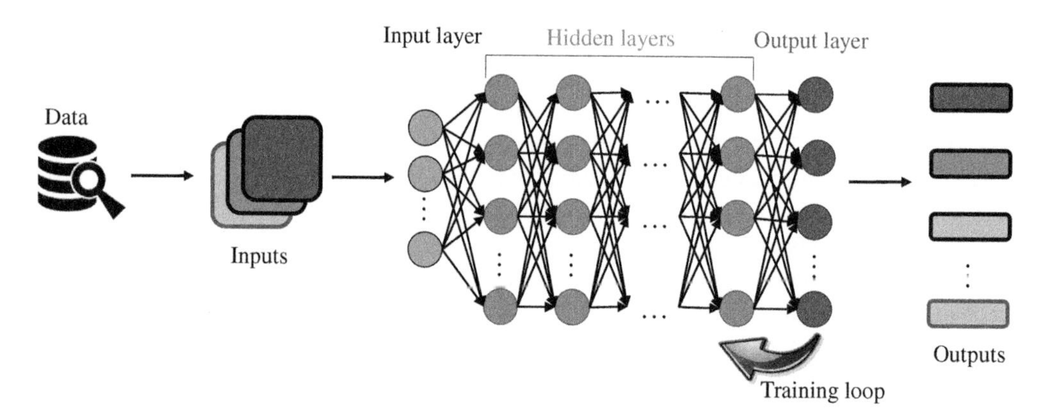

Figure 12.9 Illustration of a DL architecture.

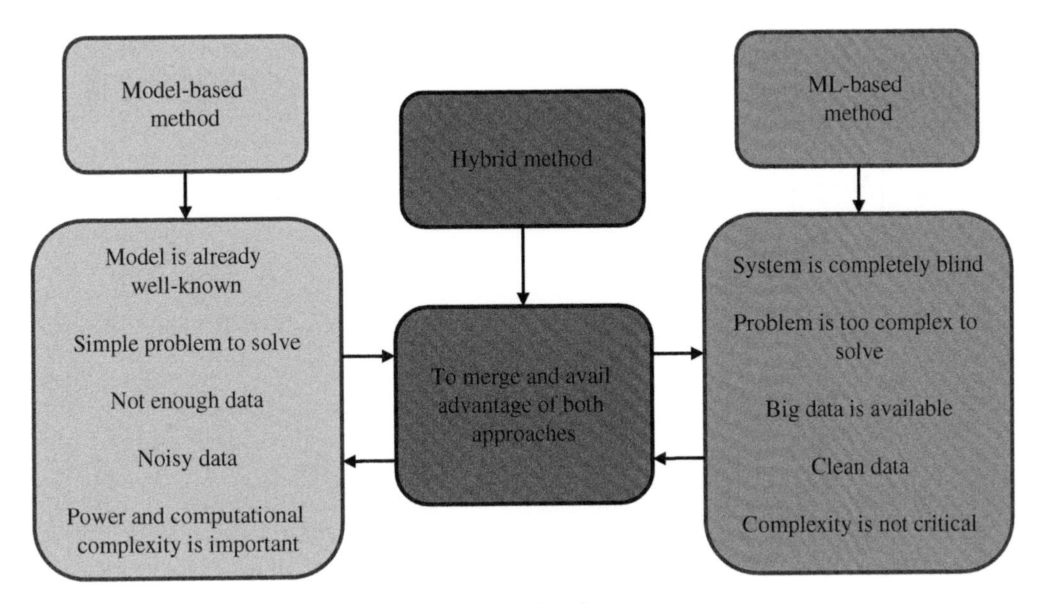

Figure 12.10 When to use model and ML-based methods?

12.4.2 Applications of Machine Learning

Model-based methods fail or they do not work efficiently in some of the BSA applications. The domain (wireless communication) knowledge and ML can be merged to solve these issues. Even ML methods can be used without sophisticated feature engineering in some scenarios. These scenarios are explained in Figure 12.10. Also, some of their specific applications are detailed below. Note that they are not a complete list and can be extended.

12.4.2.1 Signal and Interference Identification

Blind signal and interference identification have a great significance in radio surveillance, spectrum awareness, improving interference management, and electromagnetic environmental assessment. Signal processing (model-based methods) and ML methods can be considered two main categories of approaches for this problem. While model-based methods were discussed earlier in the chapter, this section highlights the ML-based approaches for the signal and interference identification problem.

Most works in the literature assume the presence of AWGN channel and/or prior information about the signals. However, practical wireless communication systems are significantly affected by multipath fading, path loss, time shift, and frequency selectivity, etc. Due to this complex nature of the real-world environment, it is difficult to represent the system by model-based methods. Thanks to the aforementioned advantages of the ML, these methods offer promising solutions.

Figure 12.11 shows an overview of how three different signals can be separated and identified using an ML model. The mixing matrix is the combined matrix of all the signals that have been superimposed together and sent to a receiver. In the ML models, the features that have been selected are either frequency domain, time domain, time-frequency domain, etc., as discussed in Chapter 3.

12.4.2.2 Multi-RF Impairments Identification, Separation, and Classification

Most of the signal processing in the transceiver is actualized in digital space to reduce the implementation cost. However, RF front-end components usually represent a significant part of power

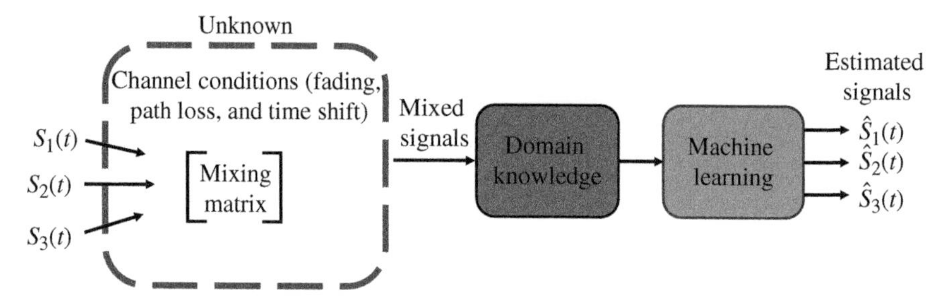

Figure 12.11 Blind signal identification with ML.

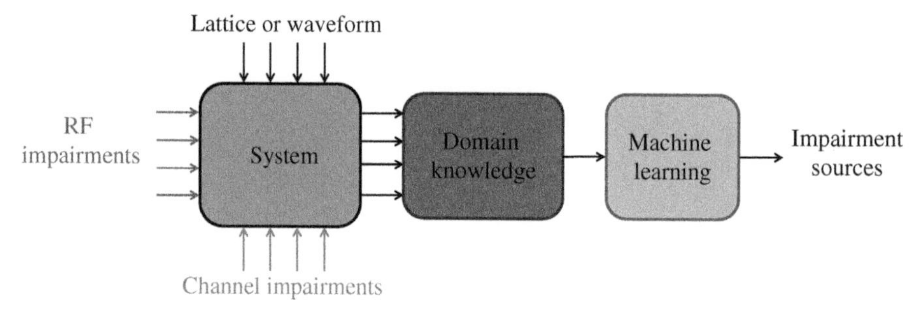

Figure 12.12 Multi-RF impairments detection with ML.

consumption and cost and determine the overall performance of the radio system, as discussed in Chapter 5.

In the literature, there are many methods for detecting single RF impairments using model-based approaches. However, studies on the detection of multiple RF impairments are scarce. The primary reason for this is the lack of models (or the knowledge) regarding the relationships between different impairments, leading to the failure of existing model-based methods. For these complex problems, ML-based methods are highly promising.

Figure 12.12 shows an overview of multi-RF impairments detection using an ML model. In this figure, RF impairments, lattice or waveform, channel impairments showed as a black box, then domain knowledge is used to measure these various impairments as feature extraction. Finally, ML is used to identify impairments.

12.4.2.3 Channel Modeling and Estimation

In wireless communication systems, channel modeling is essential. The channel effect can be acquired by conventional channel modeling that utilizes various mathematically simplified models. However, the correct channel cannot be represented precisely by these models. The reason for that is these models tend to have a simplifying (but imperfect) assumptions. Furthermore, conventional channel modeling generally requires a lot of time and computational power, which means complexity. The new additional requirements that beyond 5G on communication systems will bring as mentioned in [18], the complexity will be increased significantly. ML approaches developed for big data applications are suited to deal with such complexity. Besides, ML applied to wireless communication channel modeling is expected to improve channel modeling quality.

Channel estimation is as essential as channel modeling in the wireless communication systems. It is a mathematical prediction method of the natural propagation for the signal which supports the receiver for the approximation of the effected signal. However, massive multiple-input multiple-output (MIMO), quick handover, higher data rate, and channel modeling in 5G require more sophisticated methods than the traditional stochastic and deterministic ones. Along this line, researchers are focusing on finding an effective and accurate method with less complexity. As it is mentioned before, ML methods can be promising [19] for this complex problem.

12.4.2.4 Spectrum Occupancy Prediction

Spectrum prediction is important for capturing the information about the spectral evolution and identifying the holes in the spectrum. The channel predictions can be utilized by secondary users (SUs) to give the decision of when and where to implement a transmission without affecting PUs. It can also be used by SUs to minimize the fluctuation between channels, decrease energy usage, and increase the quality of service.

There are many model-based methods for the spectrum occupancy prediction, as discussed earlier. However, these methods are not capable of covering non-stationarity, especially in the case of 5G and beyond systems where different types of users and high mobility are expected [20]. Therefore, the recent approaches for the spectrum prediction include the application of ML-based methods. For instance, the non-stationarity problem is covered within the non-stationary hidden Markov [21] and DL models [22] as improved frameworks for the spectrum prediction.

To show the practical usage of the DL-based method, we refer to the experiment in [23]. In the experiment, the training and testing datasets are acquired from real-world spectrum measurements. In Figure 12.13, an image of the measurement setup used is given. The setup includes the UHALP 9108 receiving antenna that is connected to the Agilent EXA N9010A spectrum analyzer.

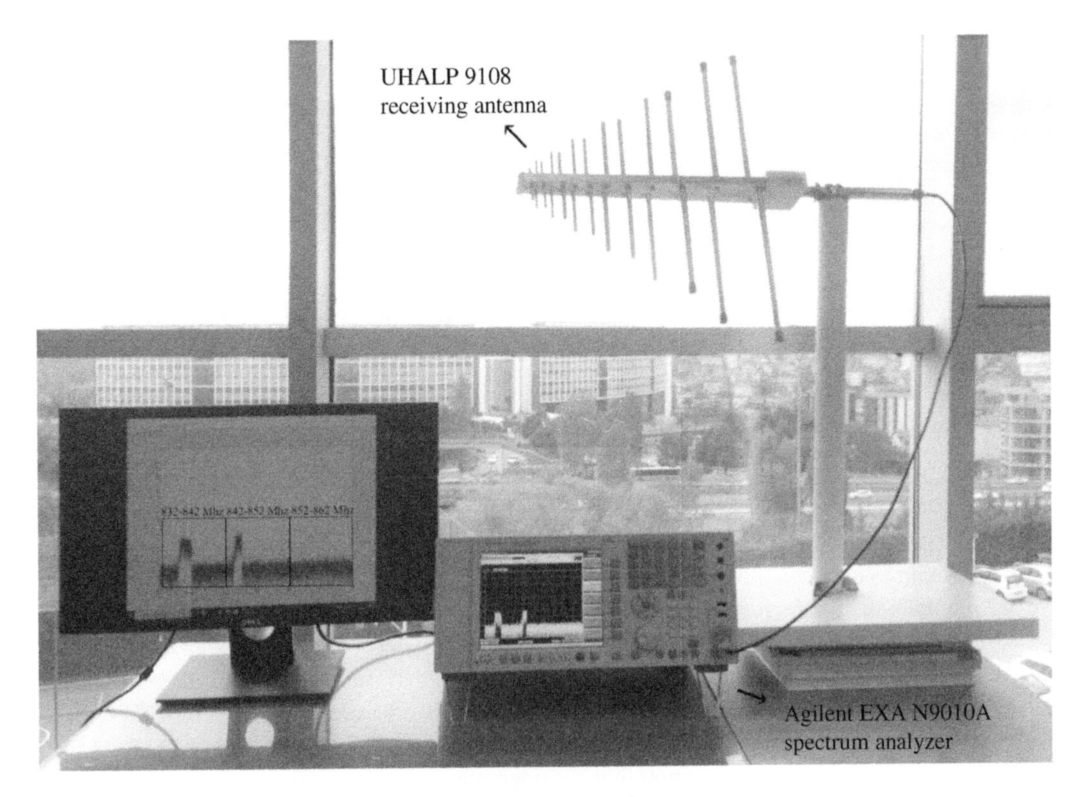

Figure 12.13 An illustration of the spectrum measurement setup.

Table 12.4 The *precision* (Π), *recall* (Ψ), and F_1-score spectrum occupancy predictions performance of the deep learning.

	832–842 MHz	842–852 MHz	852–862 MHz
Π	0.9192	0.9810	0.9864
Ψ	0.6726	0.6669	0.6708
F_1-score	0.7768	0.7940	0.7985

The measurements of received signals are between 832 and 862 MHz, each of these 10 MHz bands belong to a different operator in Turkey. LSTM is used as a DL-based classifier.

The setting of input and output data for LSTM training is made as follows. For the formation of a sample, a sliding window, which sweeps time axes, is used. For instance, when the length of the sliding window is selected as 7, then each input will have a 1×7 vectors of binary occupancies. Besides, the output data for this input data is the power spectral density (PSD) value of the following time interval. In the testing stage, which represents runtime operation, one uses spectrum measurements for time lags to predict the corresponding spectrum occupancy at the next time instant. This is achieved by feeding the binary occupancies as a grid to the DL model, which generates binary number occupancies.

A classifier model's performance metrics can be assigned as *precision* (Π), *recall* (Ψ), and F_1-score. The *precision* metric is the percentage of the actual positive values out of the total positive predictions, the *recall* is the quantification of true positives out of the total positives, and the F_1-score is the harmonic average of *precision* and *recall* that quantifies the accuracy of a classifier model. Table 12.4 shows that the DL-based classifier is able to predict spectrum occupancy with a success rate of 79% on average. Comparison between different DL methods including methods that utilize multi-dimensional correlations can be found in [24].

12.5 Challenges and Potential Study Items

12.5.1 Challenges

Although the ML paradigm wants to fulfill BSA requirements, there are still some major problems in applying this paradigm practically. A list of these challenges is given below.

- The ML methods are trained to solve a specific problem and may not show good performance when it is used to solve another problem in a similar area. Models generally need to be trained again based on the data of the problem. Although this is not a problem for some static networks, it is difficult to use such models in real-time applications.
- ML methods are hungry for data and need to learn from data gradually. It works best if the quality of the available data is high.
- Domain knowledge is essential to fully utilize the dataset for both complexity and performance. On the other hand, ML-based techniques should also be known very well. Therefore, experience in both disciplines is of significant importance for the use of ML in BSA.
- The butterfly effect is a phenomenon that considers the sensitivity of the system toward a small change in the input which can cause chaos in the output. This phenomenon can be used by

attackers deliberately to create a security attack and make the system unstable with changing the input in ML methods for wireless communication.
- Storage problem; because of the fact that ML needs a large amount of data for training the model, at times the storage of such big data can be a hurdle at situations where data storage is limited.
- ML may be vulnerable to malicious modifications and attacks for inputs. These modifications can be dangerous because they can lead to wrong outputs.
- The way of application of ML and DL methods to datasets and the level of privacy that ML and DL methods should protect are the subjects to be investigated.
- The working principle of deep neural networks (DNNs) is like black boxes. It is very difficult to know how any DL model manipulates parameters and inputs to draw conclusions. While it is impossible to know how the human brain achieves the result, it can be said to be the result of embedded neurons in complex interconnected layers. Also, it is impossible to know the complex path that the decision-making process moves from one layer to another in DNN. Therefore, it may not be appropriate to use DNN in applications where validation is important, as it will be very difficult to precisely estimate the layer that has errors.

12.5.2 Potential Study Items

Here, the list of potential study items is provided to the readers, which can be helpful for revisiting the chapter and giving some future directions. Previous chapters help the readers to get better hands-on laboratory equipment, which would help them to generate different signals. In this chapter, the generated signals can be used for different blind estimations enlisted below:

- *Blind filter roll-off detection*: From a given signal with an root raised cosine (RRC) filter, identify the roll-off factor.
- *Blind modulation analysis*: Identify the modulation type and order from the received signal.
- *Sparse or not identification and sparsity-level estimation*: From a given signal, identify whether the signal is sparse or not. If it is sparse, what is the sparsity level?
- *LOS and NLOS identification*: From a given signal, identify whether the signal is coming from LOS or NLOS environment.
- *Channel estimation*: Given datasets that are obtained from noisy channel measurements of the channel frequency response of a wireless channel, find the representation of these channels. Hint: First, estimate the number of taps. Then, find the channel coefficients.
- *Blind numerology identification*: Identify the number of numerology and numerology types of a blind received signal.
- *RF fingerprinting*: Identify which transmitter module is used from the received signal.
- *Blind signal analysis*: Identify how many different types of signals are existed in the time domain, frequency domain, time-frequency domain, and angle domain. Then, extract each symbol roughly. Finally, analyze the extracted signal.

12.6 Conclusions

BSA plays an important role when there is a lack of information about the transmitted signals. Along this line, this chapter reviewed the main efforts in BSA with model-based methods. Besides, some applications in BSA were given with codes and detailed explanations for laboratory implementations. Also, ML-based methods have been shown as a promising framework for BSA. This is especially the case when it is used with model-based approaches. Furthermore, this chapter is expected to shed some light on this important area for further exploration by students, researchers, and engineers. However, these methods are still in the early stages and face some challenges against their effective usage. Last but not least, potential study items were given for future work.

References

1 T. Yucek and H. Arslan, "A survey of spectrum sensing algorithms for cognitive radio applications," *IEEE Communications Surveys and Tutorials*, vol. 11, no. 1, pp. 116–130, First 2009.

2 M. P. Olivieri, G. Barnett, A. Lackpour, A. Davis, and P. Ngo, "A scalable dynamic spectrum allocation system with interference mitigation for teams of spectrally agile software defined radios," in *Proc. IEEE International Symposium on Dynamic Spectrum Access Networks (DySPAN)*. IEEE, Baltimore, MD, 8–11 Nov. 2005, pp. 170–179.

3 F. Weidling, D. Datla, V. Petty, P. Krishnan, and G. Minden, "A framework for RF spectrum measurements and analysis," in *Proc. IEEE International Symposium on Dynamic Spectrum Access Networks (DySPAN)*. IEEE, Baltimore, MD, 8–11 Nov. 2005, pp. 573–576.

4 F. F. Digham, M.-S. Alouini, and M. K. Simon, "On the energy detection of unknown signals over fading channels," in *Proc. IEEE International Conference on Communications (ICC)*, vol. 5. IEEE, Anchorage, AK, 11–15 May 2003, pp. 3575–3579.

5 A. Ghasemi and E. S. Sousa, "Collaborative spectrum sensing for opportunistic access in fading environments," in *Proc. IEEE International Symposium on Dynamic Spectrum Access Networks (DySPAN)*. IEEE, Baltimore, MD, 8–11 Nov. 2005, pp. 131–136.

6 M. Mehta, N. Drew, G. Vardoulias, N. Greco, and C. Niedermeier, "Reconfigurable terminals: an overview of architectural solutions," *IEEE Communications Magazine*, vol. 39, no. 8, pp. 82–89, 2001.

7 J. Palicot and C. Roland, "A new concept for wireless reconfigurable receivers," *IEEE Communications Magazine*, vol. 41, no. 7, pp. 124–132, 2003.

8 T. Yucek and H. Arslan, "Spectrum characterization for opportunistic cognitive radio systems," in *Proc. IEEE Military Communications (MILCOM)*. IEEE, Washington, DC, 23–25 Oct. 2006, pp. 1–6.

9 K. Kim, I. A. Akbar, K. K. Bae, J.-S. Um, C. M. Spooner, and J. H. Reed, "Cyclostationary approaches to signal detection and classification in cognitive radio," in *Proc. IEEE International Symposium on New Frontiers in Dynamic Spectrum Access Networks (DYSPAN)*. IEEE, Dublin, 17–20 Apr. 2007, pp. 212–215.

10 H. Li, Y. Bar-Ness, A. Abdi, O. S. Somekh, and W. Su, "OFDM modulation classification and parameters extraction," in *Proc. 1st International Conference on Cognitive Radio Oriented Wireless Networks and Communications (CROWNCOM)*.IEEE, Mykonos Island, 8–10 June 2006, pp. 1–6.

11 A. Walter, K. Eric, and Q. Andre, "OFDM parameters estimation a time approach," in *Proc. Asilomar Conference on Signals, Systems and Computers (ACSSC)*, vol. 1. IEEE, Pacific Grove, CA, 29 Oct.–1 Nov. 2000, pp. 142–146.

12 T. Keller, L. Piazzo, P. Mandarini, and L. Hanzo, "Orthogonal frequency division multiplex synchronization techniques for frequency-selective fading channels," *IEEE Journal on Selected Areas in Communications*, vol. 19, no. 6, pp. 999–1008, 2001.

13 M. Speth, F. Classen, and H. Meyr, "Frame synchronization of OFDM systems in frequency selective fading channels," in *Proc. IEEE 47th Vehicular Technology Conference (VTC)*, vol. 3. IEEE, Phoenix, AZ, 4–7 May 1997, pp. 1807–1811.

14 J.-J. Van de Beek, M. Sandell, and P. O. Borjesson, "ML estimation of time and frequency offset in OFDM systems," *IEEE Transactions on Signal Processing*, vol. 45, no. 7, pp. 1800–1805, 1997.

15 R. W. Heath and G. B. Giannakis, "Exploiting input cyclostationarity for blind channel identification in OFDM systems," *IEEE Transactions on Signal Processing*, vol. 47, no. 3, pp. 848–856, 1999.

16 M. Gandetto, M. Guainazzo, and C. S. Regazzoni, "Use of time-frequency analysis and neural networks for mode identification in a wireless software-defined radio approach," *EURASIP Journal on Advances in Signal Processing*, vol. 2004, no. 12, p. 863653, 2004.

17 M. Oner and F. Jondral, "Cyclostationarity based air interface recognition for software radio systems," in *Proc. IEEE Radio and Wireless Conference*. IEEE, Atlanta, GA, 22 Sept. 2004, pp. 263–266.

18 M. Nazzal, M. A. Aygül, and H. Arslan, "Channel modeling in 5G and beyond," in *Flexible and Cognitive Radio Access Technologies for 5G and Beyond*. UK: The Institution of Engineering and Technology (IET), 2020, ch. 11.

19 S. M. Aldossari and K.-C. Chen, "Machine learning for wireless communication channel modeling: an overview," *Wireless Personal Communications*, vol. 106, no. 1, pp. 41–70, 2019.

20 B. Khalfi, B. Hamdaoui, M. Guizani, and N. Zorba, "Efficient spectrum availability information recovery for wideband DSA networks: a weighted compressive sampling approach," *IEEE Transactions on Wireless Communications*, vol. 17, no. 4, pp. 2162–2172, Apr. 2018.

21 X. Chen, H. Zhang, A. B. MacKenzie, and M. Matinmikko, "Predicting spectrum occupancies using a non-stationary hidden Markov model," *IEEE Wireless Communications Letters*, vol. 3, no. 4, pp. 333–336, 2014.

22 O. Omotere, J. Fuller, L. Qian, and Z. Han, "Spectrum occupancy prediction in coexisting wireless systems using deep learning," in *Proc. IEEE 88th Vehicular Technology Conference (VTC-Fall)*. IEEE, Chicago, IL, 27–30 Aug. 2018, pp. 1–7.

23 M. A. Aygül, M. Nazzal, A. R. Ekti, A. Görçin, D. B. da Costa, H. F. Ateş, and H. Arslan, "Spectrum occupancy prediction exploiting time and frequency correlations through 2D-LSTM," in *Proc. IEEE 91st Vehicular Technology Conference (VTC-Spring)*. IEEE, 25–28 May 2020, pp. 1–5.

24 M. A. Aygül, M. Nazzal, M. İ. Sağlam, D. B. da Costa, H. F. Ateş, and H. Arslan "Efficient Spectrum Occupancy Prediction Exploiting Multidimensional Correlations through Composite 2D-LSTM Models," *Sensors*, vol. 21, no. 1, p. 135, 2021. Multidisciplinary Digital Publishing Institute.

13

Radio Environment Monitoring

Halise Türkmen[1], *Saira Rafique*[1], *and Hüseyin Arslan*[1,2]

[1] *Department of Electrical and Electronics Engineering, Istanbul Medipol University, Istanbul, Turkey*
[2] *Department of Electrical Engineering, University of South Florida, Tampa, FL, USA*

Today, there are millions of wireless devices serving various applications and use-cases, from mobile phones to remote sensors and machinery. The spectrum is, however, a finite resource. In order to use the radio resources more efficiently and meet the demands of all the user equipments (UEs), several wireless technologies and concepts have been developed by academia and industry. Some of these, such as multiple accessing systems, the orthogonal frequency division multiplexing (OFDM) waveform used in fourth generation (4G) long term evolution (LTE) and the multinumerology OFDM concept adopted in fifth generation (5G) standards are discussed in Chapters 7 and 9. Other methods involve intelligent scheduling mechanisms to utilize empty areas in the spectrum without interfering with other transmissions. Increasing spectral efficiency is a multidimensional optimization problem and the dimensions involved are related to user patterns, channel and the environment. For example, knowledge on the mobility of the environment and user can optimize modulation orders, handover management and beam tracking, thus improving performance for metrics such as data rate and reliability. Additionally, the high number of wireless devices available and the inherent relationship between electromagnetic wave propagation and the physical environment have made these devices and their network architectures ideal sources of information for noncommunication-related applications, such as home monitoring and heart beat detection.

The number and significance of wireless sensing use-cases highlight the importance of a framework to acquire information on the radio environment. This chapter will introduce such a framework, go over relevant technologies and provide a case study for environment sensing. But first, for clarity, some terms will be briefly defined for the remainder of this chapter:

- The **radio environment** refers any radio or network-related parameter that might impact the operation of radio and network spectrum. For example, spectrum, radio channel, interference, radio user location, individual/group mobility, user context/profile, available networks and devices, etc.
- The **physical environment** or refers to the physical domain, the objects and users in it, along with their states (mobile/stationary).
- The **physical information** refers to anything related to the physical signal and the effects acting on it, such as, power, location, angle, data, mobility, size, signal-to-noise ratio (SNR), radio frequency (RF) front-end properties, impairments, and modulation, etc.

As mentioned before, the radio environment and physical environment are related through the propagation characteristics of the electromagnetic waves. To make the distinction more clear, take

Wireless Communication Signals: A Laboratory-based Approach, First Edition. Hüseyin Arslan.

the example of beam tracking. Here, the beam width, angle of departure, and carrier frequency are related to the radio environment. Obstacle presence, object, and user (receiver) mobility are related to the physical environment. The actions in the physical environment, such as mobility, effect the physical signal by inducing Doppler shift and changing the angle of arrival or beam width.

13.1 Radio Environment Map

Initially envisioned as a cognitive radio (CR) [1] enabler for wireless regional area networks (WRANs) nodes utilizing the empty TV white spaces in the spectrum as a secondary user, the radio environment map (REM) is defined as a multidimensional database containing past, present and predicted information on the radio environment [2]. This information includes and is not limited to area coverage, network subscribers, policies and usage patterns. The REM database can be formed from information provided by network operators, online databases, such as geographical and terrain maps, and dedicated or opportunistic sensors. Although the REM database was originally considered for WRAN networks, a similar database could be used to enable flexibility in cellular networks. A simple REM architecture is shown in Figure 13.1. Here, data input from external sources is either sent to the processor, which contains data processing and modeling algorithms, to derive higher level information or to the storage unit. The CR interface acts as a communication portal between the CR and the REM manager, which is responsible for retrieving the requested data.

The most well-studied area of REM is perhaps the radio frequency map construction. We will give this as an example for the working of the REM architecture. There are measurement capable devices (MCDs) – devices capable of making measurements on the spectrum power – present in the network. These devices can be dedicated and specifically placed in the geographical environment or opportunistically used by the REM capable device and randomly scattered. Whichever the case, the MCDs measure the received signal strength indicator (RSSI) at different frequencies and relay this data to the REM capable device. In the REM architecture, this data are sent to the manager, which

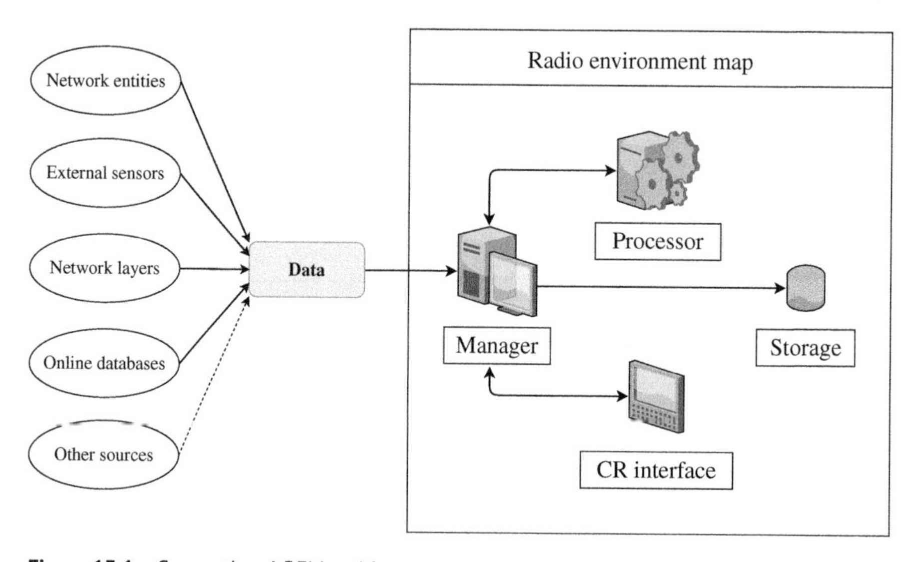

Figure 13.1 Conventional REM architecture.

forwards it to the processor and storage units. In the processor, specialized interpolation algorithms are used to fill the values between the taken measurements, thereby generating a coverage map.

While REM enabled networks will no doubt be able to improve spectral efficiency, restricting REM to a semi-intelligent database and CRs limits its potential. Forming and maintaining the database not only requires computational power, which may be significant if the REM has to be updated in short periods, but also introduces a spectral burden due to the amount of sensor data or information that needs to be collected by the REM enabled node. Ergo, the capabilities of REM enabled methods are also limited. Because of these reasons, more efficient data collection methods are required. In this regard, along with the current data sources, methods to extract information from the received signals are being developed in an effort to reduce the amount of data or information that has to be transmitted. Furthermore, the increase in number of devices, performance demands, and enabling technologies in 5G, sixth generation (6G), and beyond networks show a growing trend toward more flexible networks, which can only be possible through REM-like frameworks.

13.2 Generalized Radio Environment Monitoring

REM in the literature focuses on methods to store, process and utilize transmitted information, such as RSSI at a sensor node location, data from device sensors (e.g. camera), device or the network to enable some amount of cognition in the network. However, with the limitation of the REM scope to CR, the other information inherent in the signal structure or acquired by the signal during propagation is neglected and discarded. This is an unfortunate waste of valuable, readily available information, which then may need to be measured and transmitted separately. Exploiting this discarded information to the maximum extent to gain information and insight on the radio and physical environment can be vital to the future of cellular and wireless communication, not only to improve wireless capabilities, such as spectral efficiency, reliability, etc., but also for other noncommunication-related applications, such as mobility detection. While these applications are not directly related to wireless communication, the information extracted to enable them can be crucial in enabling old as well as novel cellular and wireless technologies. However, extracting this "high-level" information is not as straightforward as obtaining the conventional REM measurements – RSSI, for example. In this regard, the conventional REM framework can be generalized to satisfy the requirements of various devices and applications.

As stated, the effects of the propagation environment, RF front-ends in transmitter (Tx) and receiver (Rx), along with the mobility of the Tx and/or Rx on the transmitted signal can be used to extract, model or map the radio and physical environment. The effects of RF front-end impairments, wireless channel, interference, modulations, waveforms and spectrum accessing techniques on the received signal have been thoroughly explained and simulated in the previous chapters of this book. The effects of the environment on the electromagnetic wave propagation, such as mobility and objects in it, have also been explained in Chapter 10 and are reflected in the channel as frequency shift, multipath and gains. Consequently, it is accurate to say that the received signal, and by extension the physical layer (PHY) and medium access control (MAC) layer, are a large source of information on wireless devices, the radio environment and the physical environment. Taking this into account, a generalized definition for **Radio Environment Sensing/Monitoring** can be stated as a framework consisting of methods and protocols to extract data and high-level information from communication transmissions, sensing transmissions, external sensors, network entities and any other available entities in order to have the required awareness of the radio and physical environment. A depiction of the G-REM framework is shown in Figure 13.2.

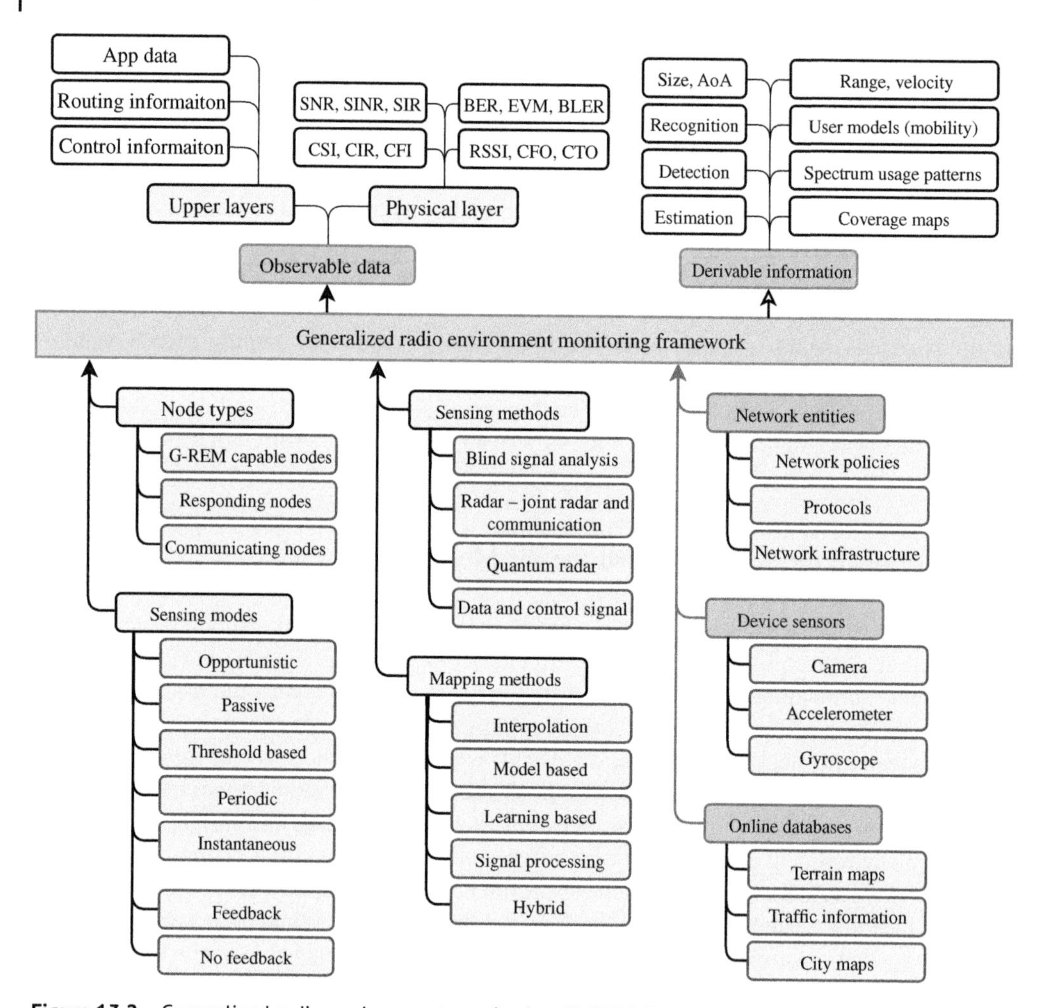

Figure 13.2 Generalized radio environment monitoring (G-REM) framework.

The G-REM framework consists of nine elements – the nodes participating in the monitoring process, sensing modes, sensing methods, mapping methods, the three external information sources, the observable data and the derivable information. Observable data and derivable information are the outputs of the framework and are requested by dependent applications. These applications will be discussed in Section 13.8. It is worth noting that the list of items for the elements in Figure 13.2 is not exhaustive and only contains the better known items.

Before discussing the G-REM elements in depth, a quick overview of how the G-REM framework operates will be beneficial. For reference, the G-REM process is given in Figure 13.3. The radio environment monitoring process is initiated by a data or information request from an application or device. More specifically, it is initiated by the absence of the requested information in the G-REM capable device. One important thing to note is that this is only for scenarios and devices where power efficiency is important and computational capabilities are limited. In more capable devices with no power constraints, the sensing process can continue indefinitely without the need for a request from an application.

In this context, rather than a semi-intelligent architecture, the generalized REM is now a stand-alone device capable of monitoring and learning the radio and physical environments, in order to support communication, as well as many other applications and use-cases.

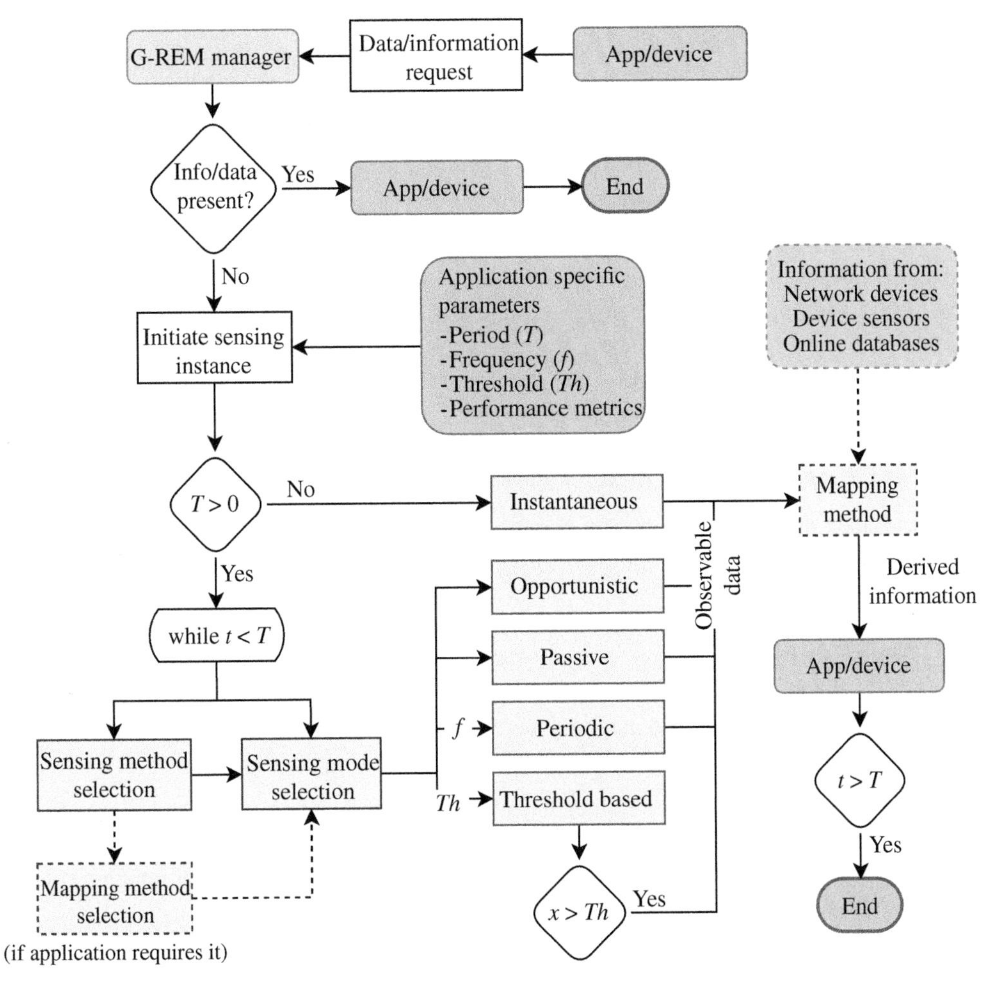

Figure 13.3 A simplistic schematic on the possible operation of the G-REM framework.

13.2.1 Radio Environment Monitoring with the G-REM Framework

Use-cases or scenarios determine the performance metrics used, which affects the number and values of the sensing parameters. These application-specific parameters are used as inputs to initiate the sensing process. Possible parameters are:

- Period (T_p): the duration of the sensing process
- Frequency (f_p): the rate of taking measurements
- Threshold (Th): the threshold below which measurements are neglected

The parameters and performance metrics also help determine the sensing method as well as the mapping method, if one is required. Accordingly, one of the five sensing modes are chosen and the sensed data or derived information is sent to the requesting application. Once the sensing period is finished, the instance is terminated. The sensing instance can have feedback or no feedback.

- Sensing Instance with Feedback: This is used in conventional radio environment mapping. Here, nodes in the network measure aspects of the signals transmitted by the G-REM capable node and transmit their measurements as a feedback.
- Sensing Instance without Feedback: In this scenario, the G-REM capable device takes measurements from the signals transmitted by the other nodes.

13.3 Node Types

In the G-REM framework, nodes are taken as devices with wireless capabilities. They do not necessarily have to be an element of a network, such as router or base station, and can be a user as well. Depending on the capabilities of a node and its function in the environment monitoring process, they can be classified under three headings:

- G-REM Capable Nodes: These nodes are able to initiate a sensing instance and have the hardware and computational power required to perform the sensing and mapping methods.
- Sensing/Responding Nodes: These nodes respond to the sensing instance initiated by the G-REM capable nodes by either sending sensing transmissions or taking measurements from the sensing transmissions sent by the G-REM capable nodes and transmitting these measurements to the G-REM capable node.
- Communicating Nodes: These nodes are communicating in the network and do not actively participate in the sensing instance. However, their transmissions may be utilized for sensing.

While these labels are not fixed for any node, a node cannot have more than one label in a specific sensing instance.

13.4 Sensing Modes

Different applications, sensing and mapping methods have different sensing requirements. Based on these requirements, the sensing instance can have the following modes:

- Instantaneous: This mode can be applied when there is already prior information and only verification or a check for changes is required. For example, if there is already some information about an object in the environment, a single transmission radio detection and ranging (radar) can be used to check whether or not the object is still present/stationary.
- Opportunistic: This mode utilizes the communication transmissions sent and received by the G-REM capable node for environment sensing without incurring any additional overhead to the spectrum.
- Passive: In this mode, the G-REM capable node simply utilizes the communication or sensing transmissions transmitted by and to other devices.
- Periodic: In this mode, sensing transmissions are sent with a specific frequency parameter continuously for the duration of the sensing instance. Here, of course, there is some spectral overhead depending on the frequency parameter.
- Threshold Based: This is generally a specific case of the periodic mode where the sensed data or information is only conveyed to the application if it passes a threshold. The threshold generally reflects the amount of change in the sensed measurement and therefore the amount of change in the environment.

The different sensing modes come with their own advantages and drawbacks regarding spectrum usage and accuracy or relevancy of the acquired data. The instantaneous, opportunistic and passive modes introduce minimal or no extra traffic to the spectrum, but the quality and frequency resolution of the data acquired is limited by the transmissions in the environment. Additional complexity is also present if prior information about the transmissions, such as waveforms and transmitting time, is not present. The periodic mode, on the other hand, can result in high quality information, but adds traffic and needs to be scheduled properly in order to prevent interference to the communication transmissions. This can be difficult in areas or times when the communication traffic is already high. The threshold-based sensing mode has the additional task of setting a suitable threshold.

Choosing the appropriate sensing mode and parameters may seem straightforward, and probably will be so if the communication transmissions are minimal to none, or are predetermined. However, in most real-world scenarios, communication transmissions are random. Additionally, the first and foremost task or wireless networks to this day has been communication. Therefore, any additional or extra transmissions must not deteriorate communication performance. Therefore, there is a trade-off between communication performance and sensing performance. This is further complicated in scenarios where the communication is aided by sensing, and therefore, the communication performance is dependent on the sensing performance.

13.5 Observable Data, Derivable Information and Other Sources

Now, what exactly is the information a base station or G-REM capable device can glean from the received signals? Or from the other sources shown in Figure 13.2? The transmitted signal is generated in the PHY and MAC layer of the open systems interconnection reference (OSI) model and passed through the channel before being received by the receiver. Therefore, the answer to the first question largely depends on what is added to the signal in each of these stages and the capabilities of the sensing and mapping methods used. There is a distinct difference between observable data and derivable information. Observable data are those acquired with minimal processing and are a direct reflection of the effective channel characteristics. Derivable information or high-level information is extracted from the observable data using one or more various data processing, machine learning and mapping algorithms as well as theoretical, empirical or statistical models. Specific examples of derivable information could be hand gestures and user patterns. Some generic examples for observable data and derivable information is given in Figure 13.2.

As for the second question, an exhaustive list for information from network entities can be known from referring to standardization documents and the network operators. However, there is no fixed list of data for the other sources. Information from device sensors is specific to the G-REM capable device's capabilities. Perhaps the most widely used wireless device, the mobile phone, nowadays is equipped with cameras, accelerometers, gyroscopes, pressure and fingerprint sensors. From these, data such as images, relative velocity, acceleration, and orientation can be obtained. However, utilizing these sensors continuously significantly affects the power efficiency; therefore, this information may not be available at all times. The last information source, online databases, refers to the enormous amount of free or cheap information databases available through the Internet. This information is accessible by all devices connected to the Internet, at the cost of (spectral) overhead and can provide information on the physical environment, models, communication-related trends and patterns, etc.

13.6 Sensing Methods

Other than Quantum Sensing and extracting information from the data and control signals methods, sensing methods using electromagnetic waves are dependent on the placement of the Txs and Rxs. These placements are called the sensing configurations. Before going into detail about the sensing methods themselves, these configurations will be explained.

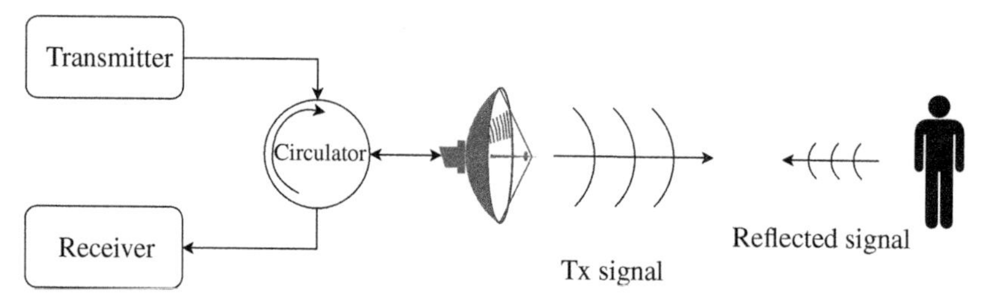

Figure 13.4 Monostatic system.

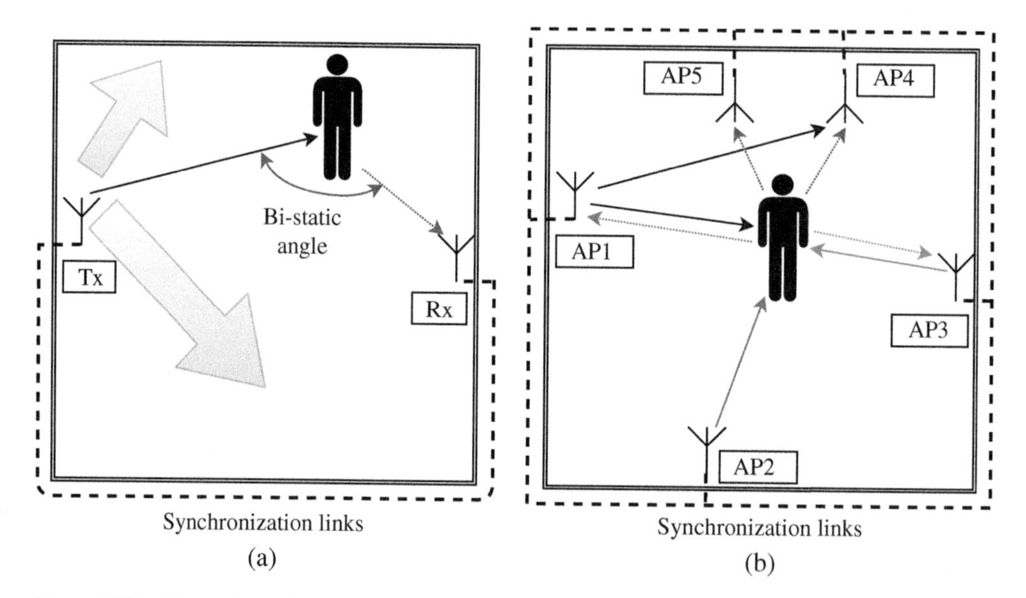

Figure 13.5 Illustration of the different modes of radio environment sensing. (a) Bistatic configuration. (b) Multistatic configuration.

13.6.1 Sensing Configurations

Depending on the number and placement of transmitter and receiver antennas, wireless systems can be classified as monostatic, bistatic, or multistatic. *Monostatic* systems have a single antenna or two colocated antennas, *bistatic* systems have two antennas separated by some distance and *multistatic* systems have multiple antennas, which may be monostatic or bistatic, for transmitting and receiving signals. The basic operation of these configurations are depicted in Figures 13.4 and 13.5.

Monostatic systems require a circulator or duplexer to isolate the received signal from the transmitted signal. This is because of the great power difference in the different signals, which can be on the order of 10^{12} and can cause the transmitted signal to eclipse the received signal. Compared to the other configurations, monostatic systems are more compact but more complex and costly.

For bistatic and multistatic systems, time synchronization between the transmitter and receiver is required and can be achieved through a wireless or wired synchronization link, shown in Figure 13.5a and b, or a reference oscillator.

13.6.2 Processing Data and Control Signal

For wireless communication systems, at the receiver the received signal is demodulated and the channel effects are removed to get the transmitted information. Several valuable parameters are calculated as a by-product of this process, some will be explained under this heading. In addition to this, the transmitted information itself can also give insight to UE requirements and patterns using data analysis and machine learning algorithms. The control signal contains information which enables the synchronization of the Tx and Rx, not only in time and frequency, but also in other domains, such as modulation. It may also contain some specifically requested information, such as global navigation satellite system (GNSS) signals or location of Tx/Rx. The exhaustive list of the data in the control signal, as defined in the standards for 5G – new radio (NR) can be found in [3] and will not be listed here.

13.6.2.1 Channel State Information (CSI)

Channel state information (CSI) describes how a signal propagates from transmitter to receiver. All the impairments that impact a transmitted signal are embedded in CSI. These impairments are caused due to environmental effects such as shadowing, multipath, scattering, etc. However, the environmental effects that act as an impairment for communication are a source of information with respect to sensing. Therefore, information that is embedded in CSI can be exploited for sensing.

CSI based sensing is being used in various macro and micro level activity recognition such as fall detection, presence detection, indoor positioning, gesture recognition, gait recognition and daily activity recognition. In the 802.11 protocols, two important techniques are being used: OFDM and multiple-input multiple-output (MIMO). In OFDM, the signal is sent via various orthogonal subcarriers at different carrier frequencies hence, the CSI is measured at the subcarrier level. Moreover, with the incorporation of the MIMO technique, spatial diversity is exploited and the signal is sent via multiple antennas. This results in enhanced multiplexing gain, array gain and diversity gain. The received signal is characterized by the following equation:

$$y_k = H_k x_k + w_k, \tag{13.1}$$

where k represents subcarrier index, w is the noise vector, $x_k \in \mathbb{R}^{N_T \times 1}$ and $y_k \in \mathbb{R}^{N_T \times 1}$ are transmitted and received signals respectively. N_T is the number of transmitter antennas and N_R is the number of receiver antennas. $H_k \in \mathbb{C}^{N_R * N_T \times 1}$ represents the CSI matrix of the subcarrier k

$$H_k = \begin{pmatrix} h_k^{11} & h_k^{12} & \dots & h_k^{N_T} \\ h_k^{21} & h_k^{22} & \dots & h_k^{2N_T} \\ \vdots & \vdots & \ddots & \vdots \\ h_k^{N_R 1} & h_k^{N_R 2} & \dots & h_k^{N_R N_T} \end{pmatrix}, \tag{13.2}$$

where h_k^{mn} is the CSI of the kth subcarrier between mth and nth receiver and transmitter antennas respectively. Each element h_k^{mn} in the CSI matrix H is a complex value and is defined as

$$h_k^{mn} = | h_k^{mn} | e^{j\phi}, \tag{13.3}$$

where, $| h_k^{mn} |$ is the amplitude and ϕ is the phase of the kth subcarrier between mth receiver antenna and nth transmitter antenna.

The code below calculates the CSI of transmitted OFDM symbols. Calculation of CSI is actually the measurement of channel response. After calculating the channel response via minimum mean-square error (MMSE) and least squares (LS) methods, the absolute value of the estimated channel response is derived to demonstrate CSI. The Figure 13.6 shows how CSI varies with the subcarrier index. This variation over longer periods can be used to detect desired actions, as done in wireless fidelity (Wi-Fi) sensing.

```
1  clear all; close all;
2  Nfft=32;  Ng=Nfft/8;  Nofdm=Nfft+Ng;  Nsym=100;
3  Nps=4; % Pilot Spacing
4  Np=Nfft/Nps; % Number of Pilots
5  Nd=Nfft-Np; % Data per OFDM Symbol
6  Nbps=2; M=2^Nbps; % Number of bits per (modulated) symbol
7  SNRs = [0:10:40];
8  for i=1:length(SNRs)
9     SNR = SNRs(i);
10    rand('seed',1); randn('seed',1);
11    MSE = zeros(1,6);
12    for nsym=1:Nsym
13       Xp = 2*(randn(1,Np)>0)-1;     % Pilot sequence generation
14       bits=randi(1,Nfft-Np,M);     % bit generation
15       mod_msg = qammod(bits,M,'UnitAveragePower',true);
16       ip = 0;     pilot_loc = [];
17       for k=1:Nfft
18          if mod(k,Nps)==1
19             X(k) = Xp(floor(k/Nps)+1); pilot_loc = [pilot_loc k];
20             ip = ip+1;
21          else X(k) = mod_msg(k-ip);
22          end
23       end
24       sgnl = ifft(X,Nfft); % IFFT
25       sgnl_cp = [sgnl(Nfft-Ng+1:Nfft) sgnl]; % Add CP
26       chnl = [(randn+j*randn) (randn+j*randn)/2]; % generates channel
27       chnl = fft(chnl,Nfft);
28       chnl_length = length(chnl); % channel and its time-domain length
29       chnl_power_dB = 10*log10(abs(chnl.*conj(chnl))); % channel power (dB)
30       y_chnl = conv(sgnl_cp, chnl); % channel path (convolution)
31       yt = awgn(y_chnl,SNR,'measured');
32       y = yt(Ng+1:Nofdm); % Remove CP
33       Y = fft(y); % FFT
34       for m=1:3
35          if m==1
36             chnl_est = LS_CE(Y,Xp,pilot_loc,Nfft,Nps,'linear');
37             method='LS-linear'; % LS estimation with linear interpolation
38          elseif m==2
39             chnl_est = LS_CE(Y,Xp,pilot_loc,Nfft,Nps,'spline');
40             method='LS-spline'; % LS estimation with spline interpolation
41          else
42             chnl_est = MMSE_CE(Y,Xp,pilot_loc,Nfft,Nps,chnl,SNR);
43             method='MMSE'; % MMSE estimation
44          end
45          chnl_est_power_dB = 10*log10(abs(chnl_est.*conj(chnl_est)));
46          chnl_est = ifft(chnl_est);
47          chnl_DFT = chnl_est(1:chnl_length);
48          chnl_DFT = fft(chnl_DFT,Nfft); % DFT-based channel estimation
49          chnl_DFT_power_dB = 10*log10(abs(chnl_DFT.*conj(chnl_DFT)));
50          MSE(m) = MSE(m) + (chnl-chnl_est)*(chnl-chnl_est)';
51          MSE(m+3) = MSE(m+3) + (chnl-chnl_DFT)*(chnl-chnl_DFT)';
52       end
53       Y_eq = Y./chnl_est;
54       ip = 0;
55       for k=1:Nfft
56          if mod(k,Nps)==1, ip=ip+1;
57          else  Data_extracted(k-ip)=Y_eq(k);
58          end
59       end
60       msg_detected=qamdemod(mod_msg,M);
61    end
62    MSEs(i,:) = MSE/(Nfft*Nsym);
63 end
64 figure
65 plot(abs(chnl_est)); xlabel('Number of Subcarriers'); ylabel ('CSI');
66 title('Amplitude Plot of Channel State Information')
```

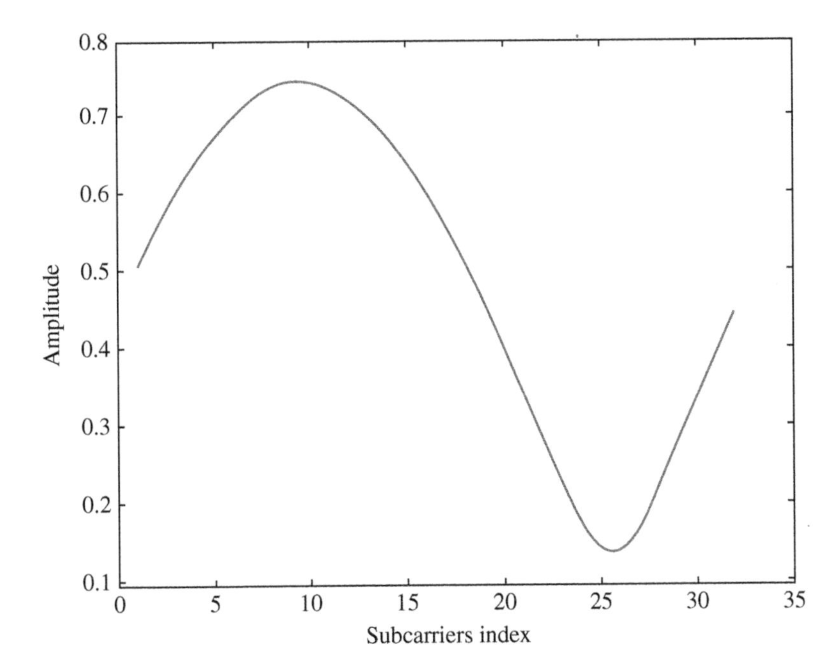

Figure 13.6 Amplitude of channel state information.

13.6.2.2 Channel Impulse Response (CIR)

RSSI-based sensing is limited by its inability to incorporate multipath effects. To completely characterize all the individual paths adopted by the transmitted signal, the wireless channel is modeled as temporal linear filter which is also termed as Channel Impulse Response (CIR). By assuming that the channel is time-invariant, channel impulse response (CIR) can be equated as

$$h(\tau) = \sum_{l=1}^{L} a_l e^{-j\theta_l} \delta(\tau - \tau_l), \tag{13.4}$$

where a_l, θ_l and τ_l are the amplitude, phase and time delay of the lth path. L is the total number of multipath. $\delta(\tau)$ is the Dirac delta function. Each impulse represents a multipath component with its own amplitude and phase.

13.6.2.3 Channel Frequency Response (CFR)

CFR is the frequency domain of CIR. CFR characterizes the effect of frequency selective fading which occurs due to constructive and destructive phases. It consists of amplitude-frequency response and phase frequency response.

CIR and CFR demonstrate small-scale multipath effects and are used for channel measurements. The channel response provides fine-grained frequency resolution and high time resolution to resolve different multipath components at the cost of slight hardware complexity.

13.6.3 Blind Signal Analysis

Blind signal analysis (BSA) is an emerging area where signal detection and analysis is aimed to be performed with limited or no information. Specifically, the Rx has no knowledge of the transmitted signal, but may have some predefined information on the communication parameters, like possible waveforms used, frame design, and operational radio resources. The predefined information is global and may be present in wireless standardization documents. Using this information and model and learning-based algorithms, BSA can be performed and various measurements and information can be collected about the signal, such as SNR, waveform, etc. As such, BSA can be especially useful in passive sensing instances, where prior information about the received signals are not known. A detailed explanation of BSA, along with a case study and various applications can be found in Chapter 11.

13.6.4 Radio Detection and Ranging

Radar technology was developed during the second world war to detect and intercept enemy aircraft. The underlying concept of radar is to transmit a pulse or continuous signal and measure the time elapsed between transmission and receiving the echoes. Radar sensing can give the following information: range, relative velocity, angle-of-arrival (AoA), time delay and Doppler frequency (through which the range and relative velocity are calculated). Radar configurations can be mono, bi or multistatic. Depending on the configuration, the range and velocity formulas need to be derived. The relations given in this section are for monostatic radar configurations. An important design parameter for radar systems is the resolution of the sensed information. Ideally, higher resolutions are desired in order to get measurements with less error. This section will briefly go over the basic concepts of radar, the relationships of radar design parameters and the quality of the observable information, and provide a simulation to visualize these relations. For more in-depth information, please refer to [4, 5].

The radar range equation given in Eq. (13.5) defines relationships between important design criteria for radar performance, the maximum transmit power, desired range and minimum receiver output SNR SNR_o, and through that, the minimum detectable received signal power.

$$SNR_o = \frac{P_{signal} G^2 \lambda^2 A_{RCA}}{(4\pi)^3 k_B T_o B N_F P_{loss} R^4} \tag{13.5}$$

Here, P_{signal} is the transmitted power, G is the directional gain of the antenna, λ is the wavelength, A_{RCA} is the effective cross-section of the object observed by the radar (radar cross-section area), k_B is the Boltzmann's constant, T_o is the real-world antenna temperature, B is the bandwidth of the transmitted signal, N_F is the Rx noise figure, P_{loss} contains the real-world radar losses and R is the desired range.

Range is the distance of the target from the radar. This is calculated using the speed equation and the time taken for the pulse to travel to the object and back to the receiver – or the time elapsed from transmitting the pulse and receiving its echo. The formula for range is given below

$$R = \frac{c\Delta t}{2}, \tag{13.6}$$

where Δt is the elapsed time and c is the speed of light.

Radar *range resolution* is the ability to detect and separate reflecting signals coming from targets or objects in close proximity to each other. The *down-range resolution* ΔR is the distance in meters to which the received signals from objects at same angles but different distances from the radar receiver can be resolved. Its expression is given below

$$\Delta R = \frac{c}{2B}. \tag{13.7}$$

The *cross-range resolution* ΔCR is the distance in meters to which the received signals from objects at the same distance but different angles to the radar receiver can be resolved.

The cross-range resolution is proportional to the object range and the antenna beamwidth at 3 dB $\theta_{3\,\text{dB}}$, as shown below

$$\Delta CR = \theta_{3\,\text{dB}} R. \tag{13.8}$$

Doppler Shift, the shifting of the signal frequency due to object mobility, is used to calculate the velocity of the object. Its effects can be seen on the power spectrum as shifting in frequency (Chapter 10). The relation between the Doppler shift and object velocity is given as below, where f_c is the carrier frequency and v_r is the relative velocity

$$f_D = \frac{2v_r}{\lambda} = \frac{2v_r f_c}{c}. \tag{13.9}$$

The velocity resolution is given by the formula

$$\Delta v_r = \frac{c}{N_{chrp} f_c T_{sweep}}, \tag{13.10}$$

where N_{chrp} is the number of transmitted chirps (or pulses, depending on the radar system) and T_{sweep} is the duration of a chirp/pulse.

Radar systems can be classified based on the utilized waveforms - pulses or continuous signals. A *pulsed radar* transmits and receives short bursts of high-power pulses. Figure 13.7 shows the pulse waveform and Figure 13.8 is a simplified schematic of a monostatic pulsed waveform radar system and its components. The transmitter generates the pulse-modulated signal which consists of a train of repetitive pulses. The Duplexer is a transmit-receive switch. It connects the antenna to the transmitter section during the transmission of pulses, and therefore, it protects the receiver from damaging with high-power transmitted pulses. In the receiving time, it forwards the echo signals to the receiver. The receiver is designed to separate the desired signal from noise and other inferences and amplifies the filtered signal in order to be displayed. The synchronizer controls and provides the timing information for the operation of whole radar system. The range can be calculated using Eq. (13.6). Doppler frequency can be extracted from the range rate, as given in Eq. (13.11), provided the range rate is not varying significantly over the time interval Δt.

$$\dot{R} = \Delta R/\Delta t \tag{13.11}$$

Unlike pulse radars, continuous wave (CW) radars generate a continuous signal and can also estimate the velocity of a mobile object using the Doppler effect and Doppler frequency – relative velocity relationship given in Eq. (13.9). Through Doppler shift estimation, CW radars can detect mobile, small objects in a large sensing area by filtering reflections from large, stationary objects (stronger reflections), leaving the weaker, shifted reflection signals. The main advantages of CW radars are that they are simpler systems, as the signals are not pulsed, and since the transmission

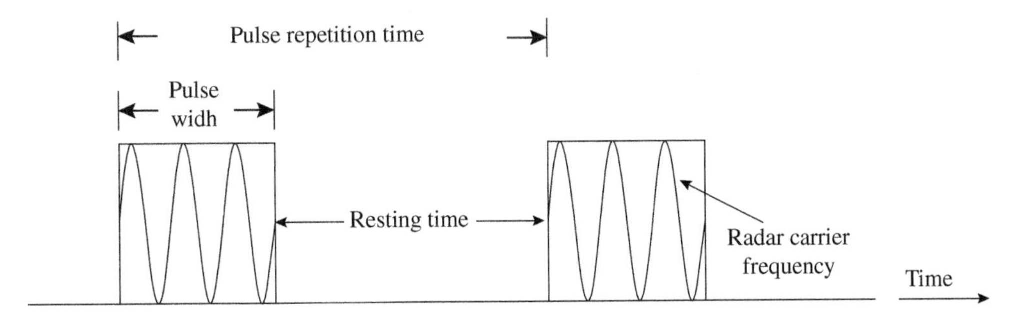

Figure 13.7 Pulse transmission.

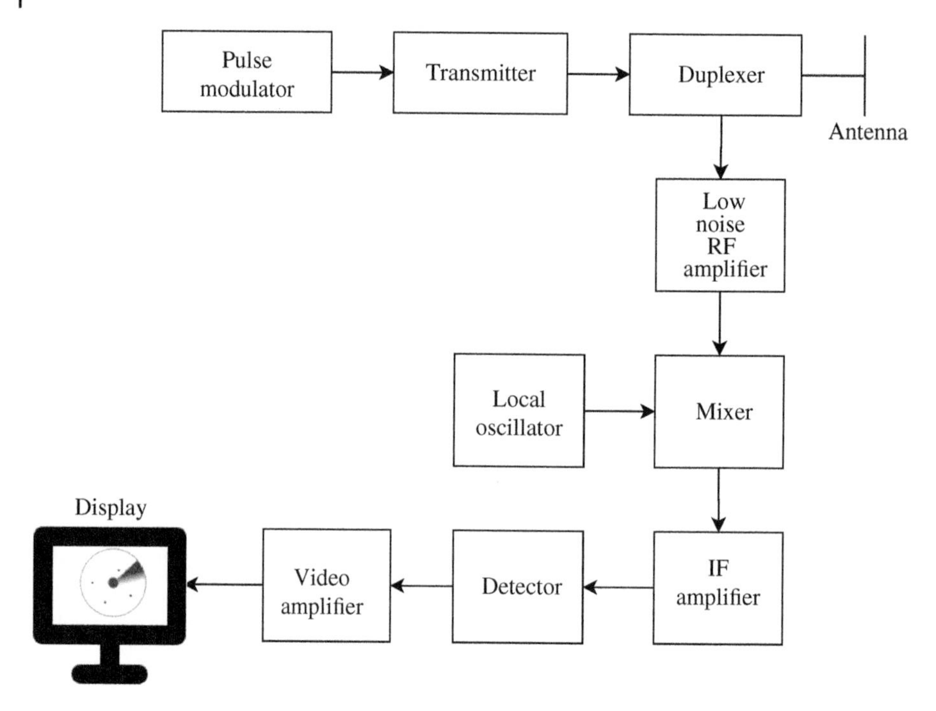

Figure 13.8 Block diagram of pulsed radar.

is continuous, more power reflects from the object, giving higher SNR than for pulsed radar systems. CW radar systems also can measure micro-Doppler shifts, giving very accurate target velocity results. Police/traffic radars, proximity sensors and missile detection systems are some application areas of CW radars.

Figure 13.9 shows a simplified block diagram of CW waveform radar system. The CW Generator transmits a wave at known carrier frequency, f_c. The reflection with Doppler shift f_D is received at

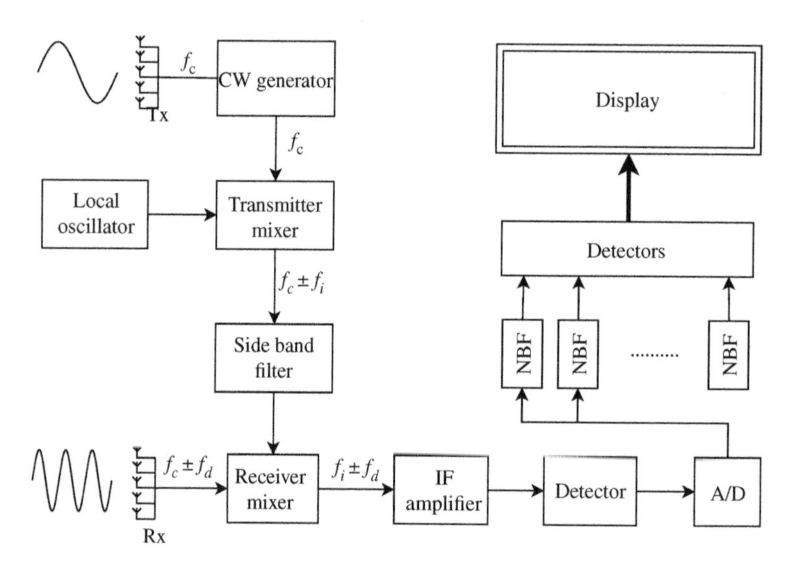

Figure 13.9 A simple block diagram of an CW waveform radar system.

the receiver and a mixer is used to remove the carrier frequency, leaving only the frequency from the stable local oscillator and f_D. This signal is then passed through narrow-band filters (NBFs) in order to separate it into different frequency components, which are then used to find the Doppler frequency using the relation $f_D = N\Delta f/2$, where N is the size of the FFT used to implement the NBF and Δf is the NBF bandwidth. The velocity can then be calculated using Eq. (13.9). It is obvious that to have a more accurate result, the NBFs should be as narrow as possible.

Frequency modulated continuous wave (FMCW) is the most common CW waveform and has periodically changing frequency. This allows synchronization to one cycle through estimating the phase or frequency offset, which is then used to get the time elapsed and calculate the object range. FMCW radars have good range resolution which is dependent on the transmitted wavelength and bandwidth, and can measure range and velocity simultaneously without extra steps. There are a few different frequency modulation techniques [4]. The most common radar waveform, *linear frequency modulation (LFM)* modulates the signal with linearly increasing and/or decreasing frequency with time. LFM can also be called *chirp modulation*, as LFM is a type of chirp signal where the chirpyness – the variation of frequency with time – is constant.

Among the different types of LFM radar chirp waveforms, *Saw-tooth Frequency Modulated Waveform* is the most common and is usually what is understood by the term "chirp." Figure 13.10 shows the transmitted and received signal for the saw-tooth frequency modulated chirp waveform, depicting the beat frequency f_{beat}, the chirp duration T_B, chirp bandwidth B and the delay Δt.

Initially, consider a stationary target and neglect the effects of the channel. The received reflected signal is the same as the transmitted signal, except with a delay Δt. The delay is the time elapsed between transmitting the signal and receiving the reflection. At a given time t, there will be a difference in frequency called the beat frequency, $f_{beat} = f_{received} - f_{transmitted}$. The beat frequency is proportional to the time delay and rate of change of frequency for the waveform and is used to find the range. When the target is mobile, there will be a Doppler shift along with the delay, which is added to the beat frequency.

For this example, the transmitted signal can be stated as

$$x(t) = A(t)e^{\frac{j2\pi t^2 B}{T_B}}, \tag{13.12}$$

where t is the time instant and $A(t)$ is the pulse envelope, which is rectangular in this case -

$$A(t) = \begin{cases} 1, & \text{if } 0 \leq t \leq T_B \\ 0, & \text{otherwise} \end{cases}. \tag{13.13}$$

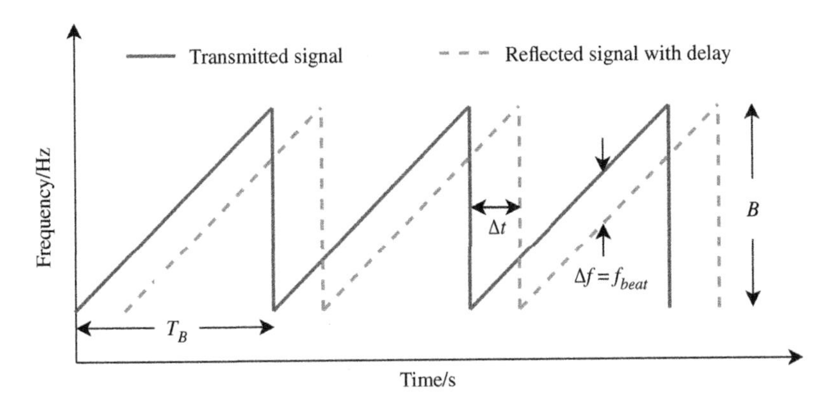

Figure 13.10 Saw-tooth frequency modulated wave.

Given that $f_{beat} = \frac{B\Delta t}{T_B}$, using Eq. (13.6), the range of the object can be found as

$$R = \frac{cT_B f_{beat}}{2B}. \tag{13.14}$$

Matlab code for generating a chirp signal with three chirps, its time plot and spectrogram is given below.

```
1  % Chirp Parameters
2  sweep_bandwidth = 320e6;
3  fsmpl = 320e6;   % sampling rate
4  N = 512;         % number of samples
5  Tchirp = N/fsmpl; % duration of one chirp
6  num_of_chirps = 3;
7
8  % For Spectrogram5
9  windowlength = 50;
10 noverlap = 49;
11 nfft = 512;
12
13 % Chirp generation
14 t = 1/fsmpl:1/fsmpl:Tchirp; % time instances
15 chrp = exp(1i*pi*(t.^2*sweep_bandwidth/(Tchirp))); % single chirp
16 chrp_tx = repmat(chrp, 1, num_of_chirps);            % cont. chirp
17
18 spectrogram(chrp_tx,windowlength,noverlap, nfft,fsmpl,'yaxis');
19
20 % To view the real part of the chirp in time:
21 fsmpl = 320e6*2;
22 Tchirp = N/fsmpl;
23
24 t = 1/fsmpl:1/fsmpl:Tchirp;
25 chrp = exp(1i*pi*(t.^2*sweep_bandwidth/Tchirp));   % single chirp
26 chrp_tx = repmat(chrp, 1, num_of_chirps);            % cont. chirp
27
28 plot(real(chrp_tx))
29 xlabel('Samples')
30 ylabel('Amplitude')
31 title('Saw-tooth Frequency Modulated Chirps')
```

Note that while the sampling rate satisfies the Nyquist criteria for complex signals, since the time plot shows only the real part of the signal, the sampling rate is doubled. This is only done for visual purposes, so that the frequency modulation with time can be clearly observed. The resulting plots are given in Figures 13.11 and 13.12a. The increase in frequency can be observed by the way the sinusoidal signal becomes compressed in time for the duration of one chirp.

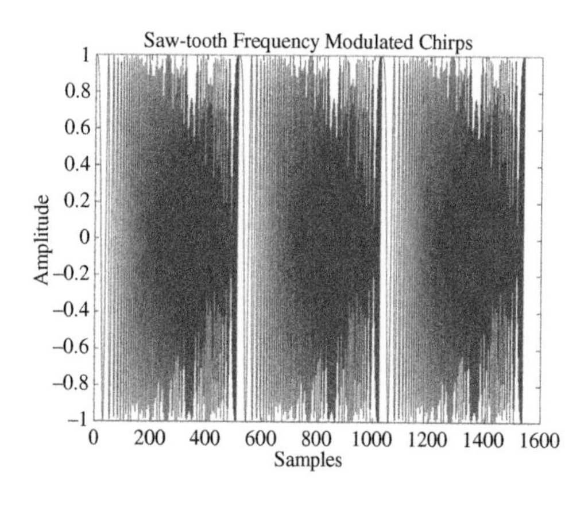

Figure 13.11 Time plot of three chirp saw-tooth frequency modulated chirp radar.

Considering a simple monostatic radar system, like the one given in Figure 13.4, the following effects need to be simulated – noise, delay and Doppler (mobility of the objects). Let there be one object in the environment, with range R and constant velocity v_r, respectively. The simulation parameters are taken as follows: $f_c = 30$ GHz, $B = 320$ MHz, $R = 100$ m, $SNR = 10$ dB, and $v_r = 50$ m/s.

Rayleigh channel model can be used for this simulation. From the SNR equation below

$$SNR = 10\log_{10}\frac{P_{chirp}}{P_{noise}}, \tag{13.15}$$

the power for the AWGN noise is calculated as

$$P_{noise} = \frac{P_{chirp}}{10^{SNR/10}}. \tag{13.16}$$

Using Eq. (13.6), the delays Δt is found as

$$\Delta t_1 = \frac{2R}{c} = \frac{2 * 100}{3^8} = 6.6667e^{-7}s = 0.6\ \mu\text{s}. \tag{13.17}$$

Similarly, using Eq. (13.9), the Doppler frequencies f_D can be found as 600 Hz. Then, the received signal becomes

$$y(t) = Hx(t) + w(t), \tag{13.18}$$

where w is the time varying noise an H is the channel matrix and contains the channel effects. The Matlab code implementation to add these effects is given below and the spectrogram of the received signal is shown in Figure 13.12b. A delay of 0.6 μs can be seen in the figure, along with the noise, while the frequency shift is not distinguishable, when compared to the transmitted signal in Figure 13.12a.

```
1  v = 50;              % velocity of object - m/s
2  R = 100;             % range of object - m
3  Fc = 30e9;           % carrier frequency - Hz
4  c = physconst('lightspeed');
5  SNR = 10             % dB
6
7  % Rayleigh channel gain
8  chnl_gain = sqrt(1/2)*(randn+1i*randn);
9  % Effect of doppler shift
10 f_Doppler = v*Fc/c;
11 doppler_shift = exp(1i*2*pi*f_Doppler/Tchirp);
12 % Effect of delay
13 delay = 2*R/c;
14 discrete_delay = round(delay*fsmpl);
15
16 % Create channel matrix
17 chnl = [zeros(discrete_delay, length(chrp_tx));
18 eye(length(chrp_tx)-discrete_delay,length(chrp_tx))*chnl_gain*doppler_shift];
19
20 % Create noise
21 noise_P = mean(abs(chrp_tx).^2)/10^(SNR/10);
22 noise = sqrt(noise_P/2).*(randn(length(chrp_tx),1) + ...
23 1i*randn(length(chrp_tx),1));
24
25 % Received signal
26 chrp_rx = chnl*chrp_tx' + noise;
27 spectrogram(chrp_rx',windowlength,noverlap,nfft,fsmpl,'yaxis');
```

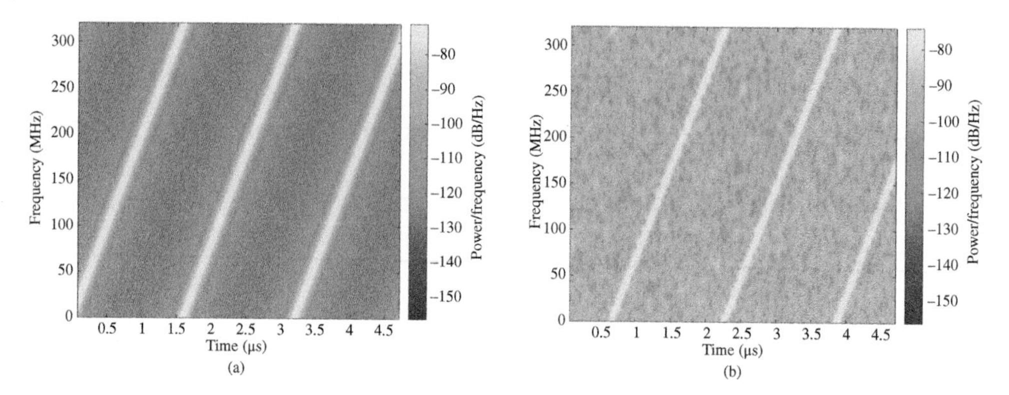

Figure 13.12 Spectrograms of the transmitted FMCW shirp signal and received FMCW chirp signal with channel effects. (a) Transmitted FMCW chirp. (b) Received FMCW chirp.

A Matlab implementation for a radar system with two targets with ranges $R_1 = 100$ m, $R_2 = 450$ m and velocities $v_1 = -400$ m/s, $v_2 = 1000$ m/s is given below.

```
1  dechrp = exp(-1i*pi*(t.^2*sweep_bandwidth/(T_chirp)));
2  dechrps = repmat(dechrp, 1, num_of_chirps);      % For dechirping
3  total_t = 1/fsmpl:1/fsmpl:Tchirp*num_of_chirps; % Time instances
4
5  % calculate delay and frequency offset to make the channel
6  % calculate channel gains (rayleigh)
7
8  % Create channel matrix
9  chnl=sparse(numel(total_t),numel(total_t));
10 for l = 1:2
11     chTime(i,:) = chnl_gain(i)*exp(1i*2*pi*f_Doppler(i)*total_t);
12     chTime(i,1:discrete_delay(i)) = 0;
13     chnl = chnl+sparse(1:numel(chTime(i,:)),[ones(1,...
14         discrete_delay(i)),1:numel(chTime(i,:))-discrete_delay(i)],...
15         chTime(i,:),numel(chTime(i,:)),numel(chTime(i,:)));
16 end
17
18 % Create noise
19 % Multiply chirps with channel and add noise
20
21 fbeat = chrp_rx.*dechrps.'; % Mix signals to get beat frequency
22 spectrogram(fbeat,windowlength,noverlap,nfft,fsmpl,'yaxis');
23 f_beats = reshape(fbeat',N,num_of_chirps);  % Reshape to separate chirps
24
25 f_beats_fft = abs(fft(f_beats,N));   % Take Nfft and normalize
26 f_beats_fft = f_beats_fft./max(f_beats_fft);
27 plot(t.*c/2, f_beats_fft);
28 title('Range Plot')
29 xlabel('Range (m)');
30 ylabel('Amplitude');
31
32 % 2D FFT using the FFT size for both dimensions
33 f_beats_2fft = fftshift(fft2(f_beats,N,num_of_chirps),2);
34
35 range_velocity_map = abs(f_beats_2fft);
36 range_velocity_map = 10*log10(range_velocity_map);
37
38 doppler_axis = [-(c/Fc)/(T_chirp*2):(c/Fc)/
39     (num_of_chirps*T_chirp):(c/Fc)/(T_chirp*2)-
40     (c/Fc)/(num_of_chirps*T_chirp)];
41 range_axis = t*c/2;
42
43 figure
44 surf(doppler_axis,range_axis,flipud(range_velocity_map),
45     'EdgeColor',"none");
46 view([90.0 90.0])
47 colorbar()
48 title('Range-Velocity Map')
49 ylabel('Distance (m)')
50 xlabel('Relative Velocity (m/s)')
```

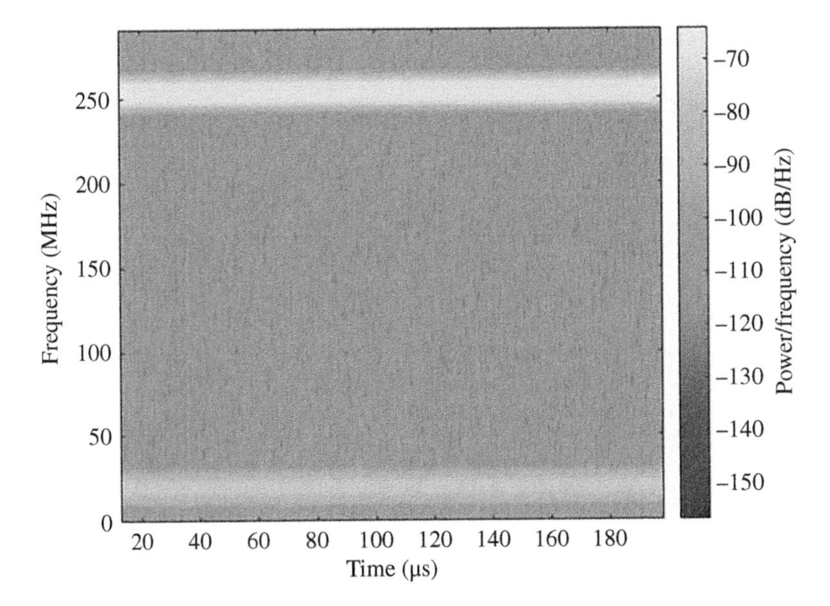

Figure 13.13 The beat frequencies corresponding to the two targets.

In order to increase the velocity resolution, the number or chirps is increased. As mentioned before and shown in Figure 13.9, to remove the carrier frequency, the received signal is mixed with the transmitted signal to give the beat signal, a process called dechirping. This corresponds to multiplication of the chirp signal with its Hermitian transpose. The beat frequencies for the two targets are shown in Figure 13.13.

Next, the fast Fourier transform (FFT) is applied on the beat signal and the output is mapped to distance. The resulting range plot has two peaks at the target ranges defined in the simulation, as shown in Figure 13.14.

Additionally, the 2D FFT can be applied to get the range-velocity map. Once again, this is mapped to the range and velocity using the sampling times and the resolution relationships in Eqs. (13.7) and (13.10), respectively. The outputs plotted on a surface plot is shown in Figure 13.15. As can be seen, there are two peaks at the ranges and velocities of the targets. Thresholding and other techniques can be used to detect these peaks.

13.6.4.1 Radar Test-bed

A radar system can be tested via software defined radios (SDRs) like universal software radio peripheral (USRP) devices. However, SDRs have certain limitations – they cannot coherently process the phase information of MIMO systems because it is complicated for them to synchronize the time and phase information of various radios. Therefore, a multiport network analyzer along with Simulink is deployed for MIMO radar systems. Aforementioned network analyzer is designed to measure the coherent phase scattering S parameters between its ports and hence it is suitable to be used as RF front-end for the MIMO radar systems. The network analyzer can transmit and receive stepped-frequency continuous wave signals with the required sweep bandwidth. A Simulink model is used to store the obtained S-parameters into an array to perform inverse fast Fourier transform (IFFT) operation to extract the range and angle information from the targets. The above mentioned set-up is expected to provide a range resolution of 33 cm and angle resolution of 19°.

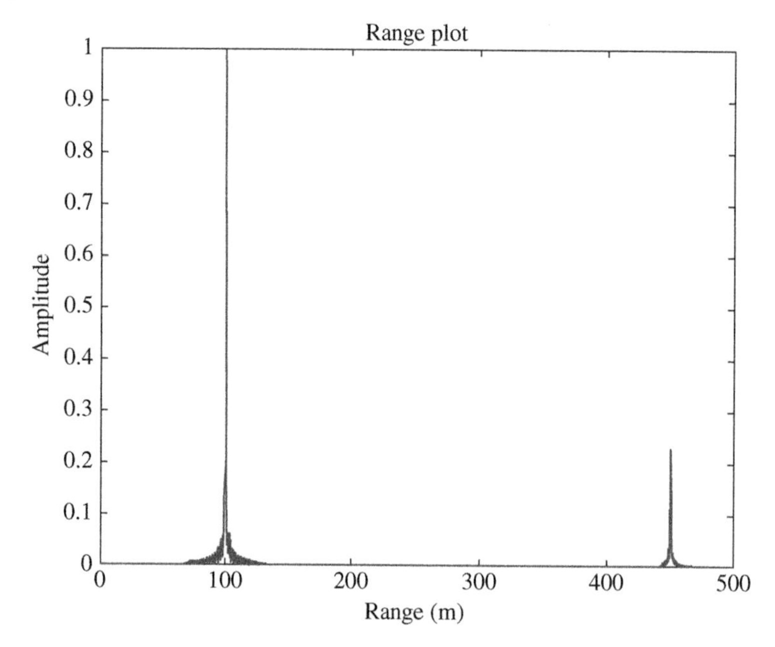

Figure 13.14 The range plot.

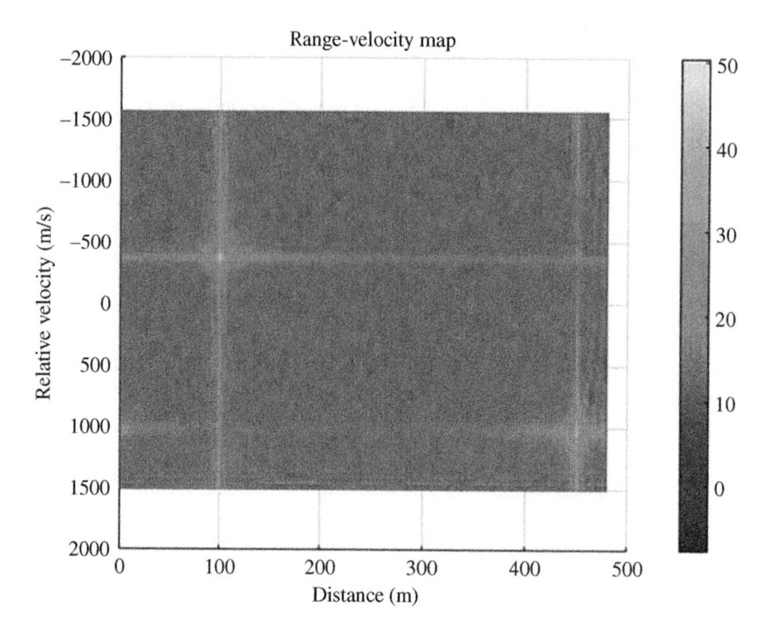

Figure 13.15 Range-velocity plot.

13.6.5 Joint Radar and Communication

Due to the increased hardware and computational capabilities of wireless communication device antennas, 5G and beyond communication systems also have radar functionality. Knowledge of mobile objects in the environment can improve the robustness of beam alignment and tracking techniques, aid in dynamic numerology assignment and, in general, support cognitive radio functionalities. As such, studies on developing radar enabled communication systems or joint radar and communication (JRC) systems, have gained momentum. This section discusses the different topologies for JRC systems and builds on the basic radar information, explaining how it can be used for JRC, and provides an OFDM-based example.

JRC can be deployed in four different topologies depending on the type of waveforms being used. These are Coexistence, Co-Design, RadComm and CommRad. RadComm and CommRad systems are a special case of Co-Design.

13.6.5.1 Coexistence

In *coexistence* approach, communication and radar systems operate independently as separate entities [6]. Both systems jointly access the same spectral resources; hence this approach is also termed as spectral coexistence. The key focus in this topology is the interference management. Interference can be reduced via sensing the state of the channel and then adjusting the Tx and Rx parameters accordingly. Consequently, the performance of individual systems is enhanced. The goal for communication system is to transmit information at high data rate with minimal error for a particular bandwidth, while the performance of a radar system is measured on the basis of its ability to detect a target. The most common performance metrics for the communication systems are bit-error-rate (BER), signal-to-interference-plus-noise ratio (SINR), mutual information (MI) and channel capacity. The target detection capability of the radar is measured by the probability of correct detection, missed-detection, and false alarm.

Figure 13.16 is the illustration of spectral coexistence, where radar and communication terminals are independently utilizing the same radio resources.

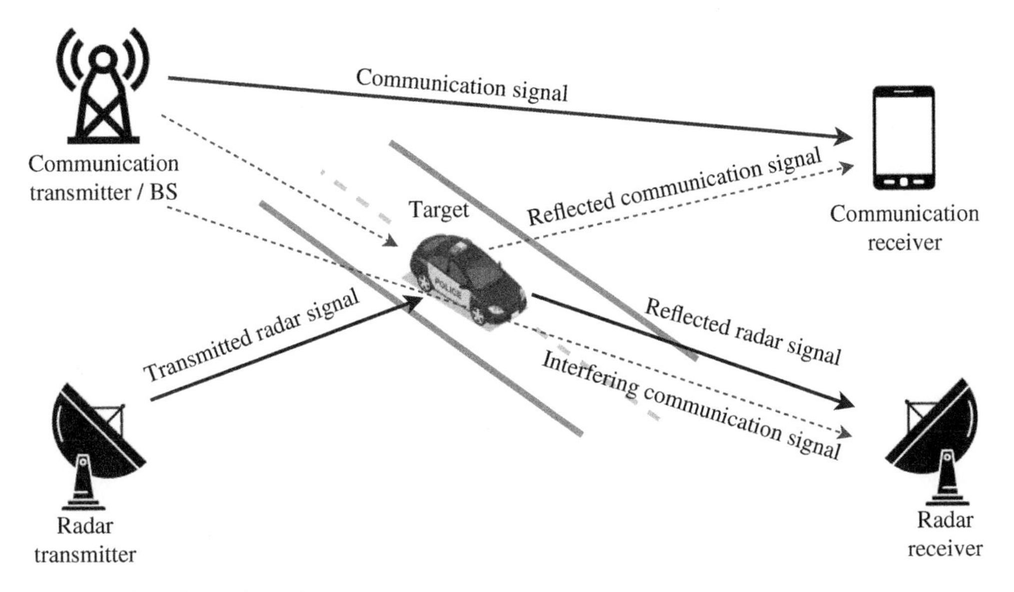

Figure 13.16 Spectral coexistence.

13.6.5.2 Co-Design

The codesign topology is actualized in three different ways based on how Tx and Rx are shared between the radar and communication systems.

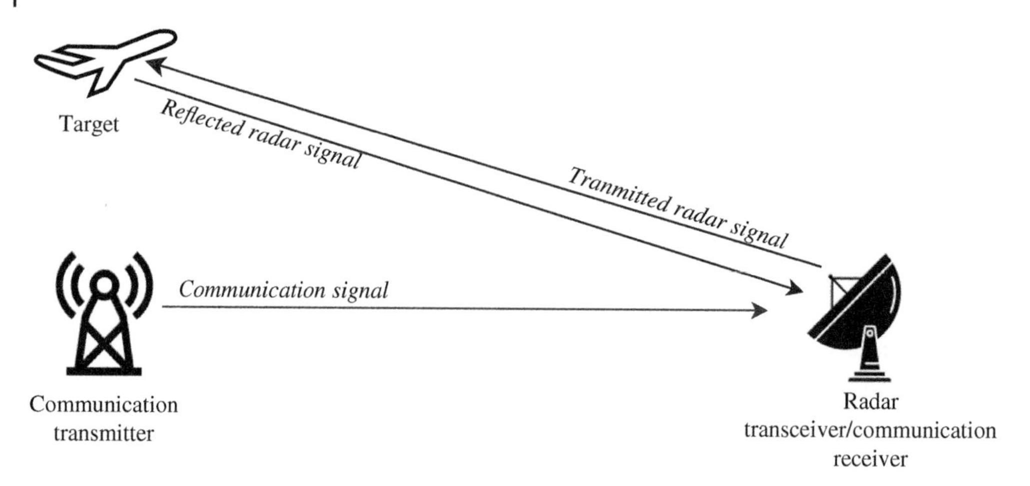

Figure 13.17 Joint multiple access topology.

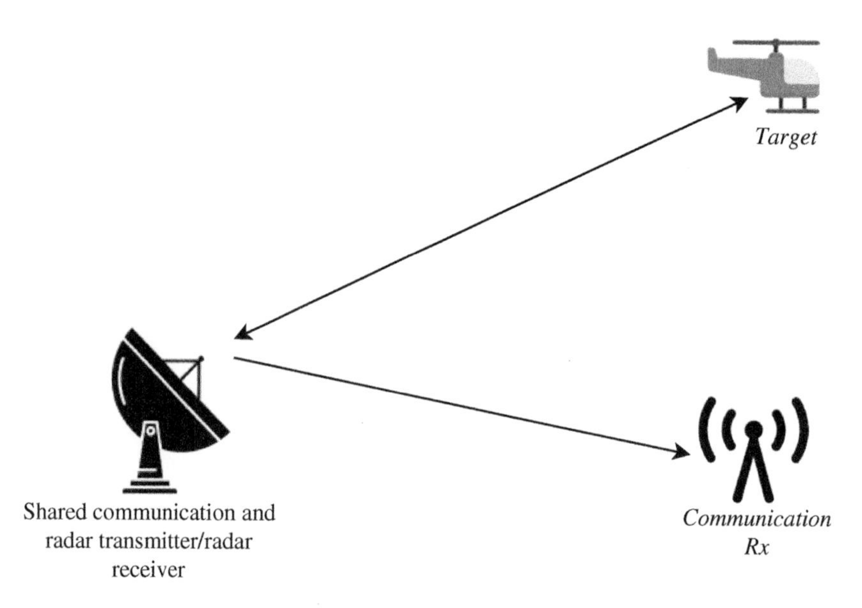

Figure 13.18 A DFRC system.

1. The communication and radar systems share a common receiver. This is actually a joint multiple-access channel, in which radar operates in the monostatic mode and both the systems transmit separate waveforms that are kept orthogonal using different degrees of freedom (DoF) such as frequency, time or code [7]. Figure 13.17 provides the illustration of this topology.
2. The monostatic radar also serves as a communication transmitter. Therefore, a single JRC waveform is transmitted in this Tx-shared codesign method. This form of codesign in which there is a common waveform for both radar and communication systems is referred as *dual function radar communication (DFRC) systems*. Proper waveform design that is suitable for both communications and radar systems is the fundamental concern in this topology. Figure 13.18 shows Tx-shared codesign topology.

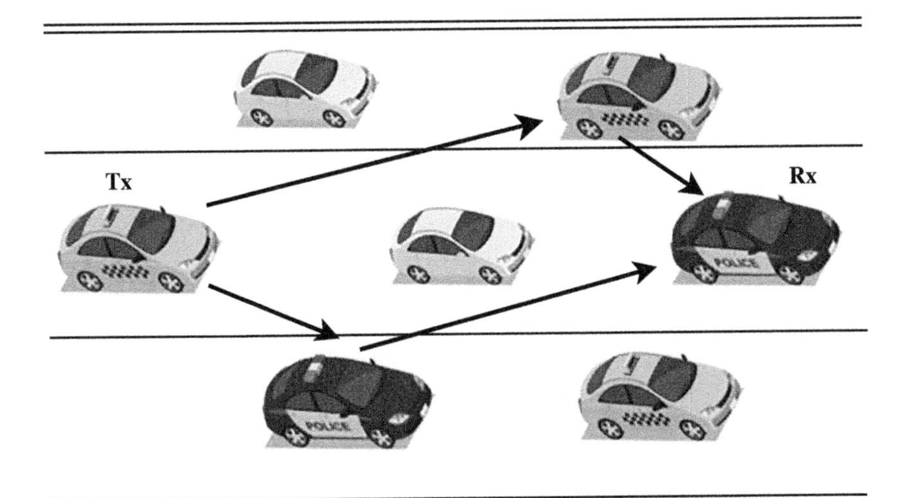

Figure 13.19 An example of shared Tx and Rx for communication and radar systems in vehicle-to-vehicle (V2V) networks.

3. Both Tx and Rx are shared between communication and radar systems. This design is also termed as bistatic broadcast codesign. An example of this design is vehicle-to-vehicle communication, where a JRC waveform is transmitted by the Tx vehicle. The waveform bounces off from different targets and is received by the Rx vehicle which is shown in Figure 13.19.

Recent research in JRC revolves around DFRC systems and the development of one emission for sensing the environment and sending data. A single waveform that is optimized for the use of communication and radar sensing is termed as a *hybrid waveform* or DFRC waveform. Hybrid waveforms are further divided into two types – radar centric (RadComm) and communication centric (CommRad).

13.6.5.3 RadComm

In radar-centric or RadComm approach, radar waveforms are used to send the data. The main goal in this approach is to improve data rate for communication purpose.

There are two techniques to achieve dual functional *RadComm waveform*.

1. **Information Embedded Waveform Diversity:** The traditional technique is to embed information into the radar chirp signal. In this method, the information can be transmitted toward a specific direction by emitting one waveform, from a group of specified waveforms during each radar pulse. Although this method is simple in terms of implementation, it degrades the visibility of target due to range sidelobe modulation (RSM), which occurs as a consequence of varying waveform during coherent processing interval (CPI). Moreover, numerous intra-pulse modulation schemes such as binary phase shift keying (BPSK) and minimum shift keying (MSK) are introduced to transmit data symbols for frequency modulated radar waveforms.
2. **Symbol Construction through Transmit Beamforming:** The deployment of phased array radars in the operation field is under the limelight due to their advantages like digital beamforming at the receiver and flexible antenna beam pattern. Through proper manipulation of parameters of each antenna in the phased array radar system, information can be transmitted toward the desired direction. Various modulation schemes such as amplitude modulation (AM), phase shift keying (PSK), quadrature amplitude modulation (QAM) can be

applied to phased array radars. However, these modulation schemes allow to transmit only one symbol per radar pulse. MIMO radars can be used to increase the amount of transmitted information.

13.6.5.4 CommRad

In communication-centric or CommRad approach, communication waveforms, such as OFDM, are exploited to perform sensing functionalities. Most of the efforts in CommRad systems are concentrated around achieving reliable radar/sensing performance.

The frequency diversity feature of the multicarrier OFDM waveform equip it to encounter critical wireless channel impairments such as multipath effect and interference. Moreover, frequency diversity can also be exploited to achieve reliable sensing capability, as in radar communication a specific target behaves different for different frequencies. Therefore, OFDM allows to observe the target response over the ambiguity function.

The Matlab implementation for a simple OFDM based radar system is given below.

```
1  %%%%%%%%%%%%%% OFDM based Radar %%%%%%%%%%%%%%%%%%%%%%%%%
2
3  %%%%%%%%%%%%%% OFDM SIGNAL Generation %%%%%%%%%%%%%%%%%%
4  %%%%%%%%%%%%%% Parameters %%%%%%%%%%%%%%%%%%%%%%
5
6  N = 256 ;
7  CP_size = 1/4; % Cyclic Prefix
8  used_subcarriers_ind = 35:1: N-35;
9  symb = 2* (randn (1,N) > 0)-1; % BPSK symbol Generation
10 ifft_sgnl = zeros (1,N);
11 ifft_sgnl(used_subcarriers_ind)= symb (used_subcarriers_ind);
12 sgnl = ifft(ifft_sgnl);
13 sgnl_cp = [sgnl(end-N*CP_size+1:end) sgnl];
14 clf; subplot (3,1,1)
15 plot (10*log10(abs(sgnl_cp).^2))
16 title ('Transmitted Signal')
17 xlabel('sample time index in [nano-secs]')
18 ylabel('Power (dB)')
19
20 %%%%%%%%%%%%%% Delayed, Attenuated and Noisy Signal %%%%%%%%%%%%%%%%%%%%%%%%%
21 delay_tx_rx = 10;
22 sgnl_delay = [zeros(1, delay_tx_rx) sgnl_cp];  % Delayed Signal
23 sgnl_attenuated = 1 .* sgnl_delay;
24 noise_amplitude = 0.01;
25 noise = noise_amplitude * randn(1,length(sgnl_attenuated));
26 sgnl_rx = sgnl_attenuated + noise;
27 subplot (3,1,2)
28 plot (10*log10(abs(sgnl_rx).^2))
29 title ('Received Signal')
30 xlabel('sample time index in [nano-secs]')
31 ylabel('Power (dB)')
32
33 %%%%%%%%%%%%%%%%%%%%%%%%%%%% Cross Correlation %%%%%%%%%%%%%%%%%%%%%%%%
34
35 [Rxy,lags] = xcorr(sgnl_rx,sgnl_cp);
36 Rxy = Rxy/max(Rxy);
37 subplot (3,1,3)
38 plot (lags, Rxy)
39 title ('Cross-Correlation Transmit-Receive')
40 xlabel('sample time index in [nano-secs]')
```

In the above mentioned code, a simple OFDM signal is transmitted over the air interface or channel. In practical scenarios, a wireless channel contains numerous obstacles such as buildings, mountains, trees, etc. For simplicity, in this code example a single obstacle is modeled.

These obstacle are a source of interference/signal degradation for the communication signals, but from the radar sensing perspective, obtaining information about the environment is critical.

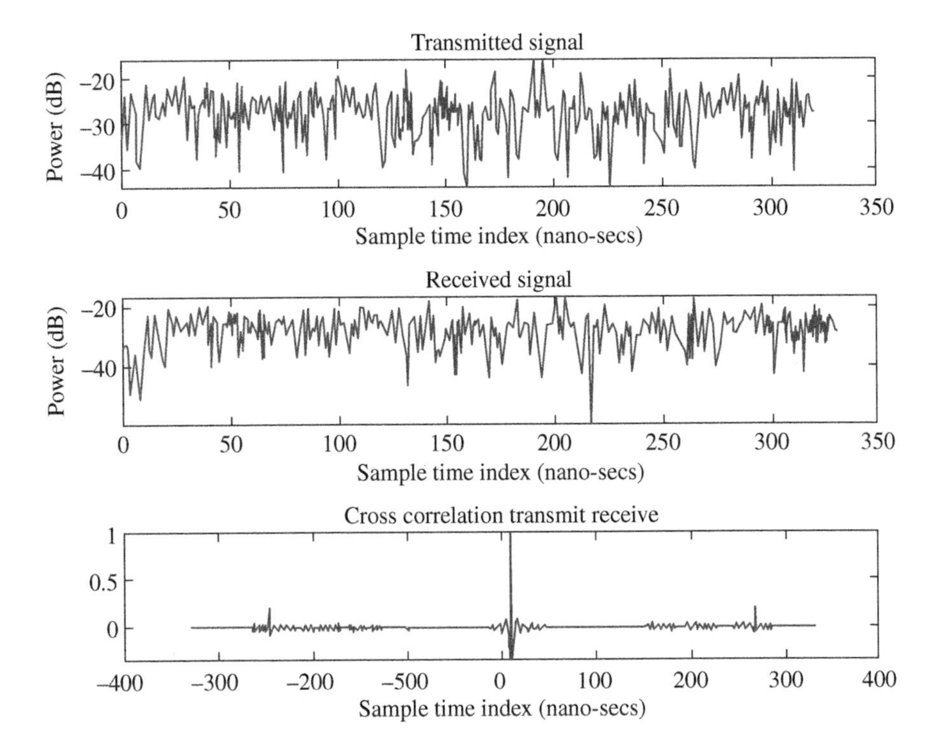

Figure 13.20 OFDM based radar system.

Therefore, to extract information about the range of the target, correlation processing is used at the receiver.

The peak of the cross-correlation plot in Figure 13.20 shows the presence of a target. The peak is present at 10 nano-seconds. Now, range of the target can be calculated from the formula

$$R = \frac{c\Delta t}{2} = \frac{3 * 10^8 * 10 * 10^{-9}}{2} = 1.5 \text{ m}. \tag{13.19}$$

The small peaks in the cross-correlation plot are due to the presence of the cyclic prefix in OFDM signal.

13.7 Mapping Methods

The data collected in the G-REM framework can be further processed to obtain more information. This is generally done by finding a relationship or function which relates changes in the observable data to high-level information or awareness. Therefore, the methods used in this process are referred to as mapping methods. The mapping methods mentioned in Figure 13.2 may not be specific to wireless communication, or signal processing, for that matter. In this section, these methods will only be explained briefly and two mapping case studies will be reviewed. Note that the methods here are not a complete list and may not apply to all sensing scenarios.

13.7.1 Signal Processing Algorithms

Signal processing algorithms do not map the extracted information to physical actions directly, but they are necessary to prepare the data for the mapping methods. There are many signal processing techniques, with entire books dedicated to this topic. Therefore, this chapter will explain three categorizations – noise reduction, signal transform, and signal extraction – rather than giving specific techniques and derivations.

The sensed information can contain noise due to hardware impairments, software limitations, and/or inconsistencies due to unexpected phenomena, hardware failures, or erroneous information. Hardware and software impairments could be sampling clock errors or nonlinearity due to power amplifiers (PAs). Such impairments cause phase offsets, which may be debilitating for a highly sensitive CSI-based sensing application. Inconsistencies in measurements result in outlying values that are not within the expected range. Examples could include spoofers feeding deliberately wrong information or unexpected interference. These erroneous measurements reduce the accuracy of the mapping method by either preventing the extraction and detection of the required pattern for an application or causing wrong patterns to be learnt. As a result, phase offset and outlier removing algorithms, such as averaging or thresholding, are used to reduce, if not remove, the noise.

For CSI based sensing, time-frequency domain analysis is required. In these scenarios, FFT, short time Fourier transform (STFT), discrete wavelet transform (DWT), etc., transforms can be used. Additionally, the received may require additional processing to extract the required signals from initial measurements. For example, radars can use filters to remove reflections from stationary objects or the environment. This additional processing can be in the form of filtering or thresholding, signal compression and signal composition. Signal compression can be used to reduce the size of the measurements by removing redundant measurements or reducing the resolution of the data to a tolerable extent. This increases the computational efficiency and can greatly reduce data size. Some radar and Wi-Fi sensing applications require the data or signals from multiple devices, like in the bistatic or multistatic sensing configurations explained in Section 13.6. In these cases, the measurements, for example CSI, from multiple sources need to be coherently combined. Interpolation techniques may also be required to fill missing gaps. This is done using signal composition techniques.

13.7.2 Interpolation Techniques

Because it is often unrealistic, if not impossible, to collect measurements from every location or time/scenario, interpolation algorithms are often used to estimate or predict the missing values from the measurements that can be collected. For applications related to the environment and coverage, spatial interpolation algorithms can be used. Here, two popularly used spatial interpolation algorithms are explained briefly.

13.7.2.1 Inverse Distance Weighted Interpolation

The inverse distance weighting (IDW) interpolation algorithm relies on the concept of spatial auto-correlation. In other words, it assumes that the values at unknown points are more similar to measured values at closer points than those further away. This interpolation technique can be mathematically formulated as

$$\hat{a} = \frac{\sum_{i=1}^{n}\left(\frac{a_i}{d_i^{\eta}}\right)}{\sum_{i=1}^{n}\left(\frac{1}{d_i^{\eta}}\right)}, \tag{13.20}$$

where $\hat{a}$ is the estimated value for a desired location, a_i are the measured values, d_i are their respective distances to the desired location and η represents the impact of d_i on the estimated value.

13.7.2.2 Kriging's Interpolation

The advantage of Kriging's interpolation method over others is that it is a geostatistics method – meaning that it extracts the spatial patterns of the environment and uses these patterns to find the weights for the interpolation. This results in spatially dependent estimated values, which is suitable for estimating the signal strength at different locations, as electromagnetic propagation phenomena are also affected by the physical environment.

In order to use the Ordinary Kriging (OK) interpolation technique, the measured data must be stationary, normally distributed, and not have spatial trends.[1] The OK estimator can be written as

$$\hat{a} = \boldsymbol{w}'\boldsymbol{A}, \tag{13.21}$$

where $\boldsymbol{w}'$ is the vector of weights for $\boldsymbol{A}$, the vector containing the measurements.

The basic steps for OK interpolation is as follows:

1. Compute the **experimental semivariogram** $\gamma_e(a, b)$:
   This is a measure of the spatial auto-correlation of two measured/known values, a and b, and can be calculated as

   $$\gamma_e(a, b) = \frac{(a - b)^2}{2}, \quad \forall a, b \in \boldsymbol{A}, a \neq b, \tag{13.22}$$

   where $\boldsymbol{A}$ is the set of measurements. The *semivariogram cloud plot* can be utilized to visualize this information by plotting the resulting values against their separation.
2. Fit a **semivariogram model** γ_m on the experimental semivariogram:
   A model can be fitted on the experimental semivariogram cloud plot to extract a spatial pattern. For this step, initially generic models, e.g. linear, exponential, spherical, periodic, etc., are used to find the one which best fits the data. Then, if necessary, certain parameters can be optimized to give a better fit. These parameters are as follows:
   - nugget: the difference between the 0 semivariance and the model's interception with the y-axis
   - partial sill: the difference between the nugget and the semivariance value at which the model curve levels off
   - range: the distance at which the model curve levels off

   For example, the exponential semivariogram has the form

   $$\gamma_m = \begin{cases} 0, & \textit{if } |\boldsymbol{d}| = 0 \\ \alpha + \left(\sigma^2 - \alpha\right)\left(1 - e^{\frac{-3|\boldsymbol{d}|}{r}}\right), & \textit{if } \boldsymbol{d} > 0 \end{cases}, \tag{13.23}$$

   where $\boldsymbol{d}$ is the lags or distances between the positions of the measured values, α is the nugget, r is the range and σ^2 is the sill.
3. Find the weights $\boldsymbol{w}$ using the Kriging semivariance equations:
   The weights for the measured values can be found using a system of linear equations, known as Kriging semivariance equations, given as

   $$\left[\begin{array}{c} \boldsymbol{w} \\ \hline \Lambda \end{array}\right] = \left[\begin{array}{c|c} -\boldsymbol{\Gamma} & \boldsymbol{1} \\ \hline \boldsymbol{1}' & \boldsymbol{0} \end{array}\right]^{-1} \left[\begin{array}{c} -\boldsymbol{\Gamma}_{\boldsymbol{0}} \\ \hline \boldsymbol{1} \end{array}\right], \tag{13.24}$$

   where λ is the Lagrange multiplier, $\boldsymbol{\Gamma}$ is the semivariance matrix of the measured positions, $\boldsymbol{\Gamma}_{\boldsymbol{0}}$ is the vector of semivariances between the measured positions and desired position The derivation of these equations can be found in [8].
4. Apply the **OK estimator** in Eq. (13.21) to get the estimated value at the desired position.

1 There are data processing techniques and variations of Kriging's interpolation technique which can be used if the data does not satisfy these conditions, but these will not be explained here.

13.7.3 Model-Based Techniques

Model-based mapping methods aim to mathematically define the changes in the measured values due to a desired action or phenomena. These methods can be theoretical and rely on physical theories, like Fresnel zone model, or statistical and be formed through the statistical analysis of numerous amounts of data. Widely used model-based techniques include thresholding, clustering and peak/trough detection algorithms. For example, the Fresnel zone model can be used to form thresholds for RSSI based motion recognition applications [9]. The drawback of model-based techniques is that they cannot always represent the environment or desired phenomenon accurately, either due to unknown contributions or the increase in complexity. This, in turn, limits the accuracy and feasibility of the sensing, allowing the extraction of only coarse-grained information. Additionally, model-based techniques require accurate measurements/estimations and signal processing. Furthermore, the models are usually generated for a specific environment or scenario, and cannot be used (effectively) in other environments/scenarios.

13.7.4 Learning-Based Techniques

These techniques utilize machine learning (ML) and deep learning (DL) algorithms in clustering or identification tasks. Their basic principle is to learn or fit functions to a collection of measurements and then utilize this function for sensing. Examples of popular learning algorithms include decision tree learning, *k*-nearest neighbor, convolutional neural networks, and support vector machines. The advantage of learning-based techniques over model-based is that they find more flexible functions which can be applied to a larger variety of environments and scenarios and they require little to no signal processing beforehand. However, there is are trade-offs between complexity-accuracy and generality-accuracy. More accurate results require a larger amount of data, which increases complexity. Additionally, more accurate results tend to mean that the learned model or function is over-fit, which may give good results for new measurements taken under the same conditions, but reduces the effectiveness of the function in different, but similar conditions.

13.7.5 Hybrid Techniques

These techniques consist of both model and learning-based algorithms and are used to remove or reduce the drawbacks of the techniques individually. Generally in these techniques, simpler model-based algorithms with less amount of signal processing are used to extract coarse information, followed by learning-based techniques to get finer grained information.

13.7.6 Case Study: Radio Frequency Map Construction

Radio frequency maps are sub-layers of the REM database. They are essentially radio signal coverage maps at different frequencies and are used for CR applications like spectrum sharing between primary and secondary network users. Here, the observed information is generally RSSI, but some mapping techniques also incorporate SNR to improve the performance in specific scenarios.

In early works, the system architecture consists of dedicated MCDs distributed randomly or equidistant from each other in the network. In a feedback-based approach, instantaneous, or periodic sensing modes are used to transmit sensing signals the base station or access point, which has REM capabilities. The MCDs measure the RSSI of the received sensing signals and forward this data to the REM capable device. Alternatively, they can also measure the power of the spectrum at a desired time instant and transmit this information. Because in these systems, one cannot have an infinite number of devices for measuring, the resulting map is sparse and is not useful for locations other than the measured ones. Therefore, interpolation methods must be used to estimate the signal strengths at other locations. A variety of interpolation and machine learning algorithms have been applied and modified in order to find the best way to do this [10]. The main criteria in choosing and developing these algorithms is the accuracy and computational complexity. In general, accuracy and computational complexity are inversely related; therefore, there is a trade-off between the accuracy of the estimated radio frequency map and the computational complexity and computation time due to the algorithm. The number of measurements also causes the same trade-off. It is important to note that the signal strength is affected by the physical environment and the presence of different obstacles. Therefore, it is overly naive to assume linear or uncorrelated interpolation techniques would be sufficient. Among all of the methods, geostatistical algorithms, like Kriging's interpolation method and its modifications, give the best results in terms of accuracy and relative complexity. This is because it incorporates the geospatial characteristics of the environment, therefore somewhat taking into account the propagation characteristics of electromagnetic waves. The resulting radio frequency map is a heatmap showing the estimated strength of the received signal at different locations.

Nowadays, there are so many measurement capable devices that may not be dedicated, but can still be utilized to acquire well distributed measurements. In this case, the defining interpolation criteria becomes complexity, as there is an abundance of measurements. Here, machine learning-based or hybrid algorithms are used to map the RSSI to locations [11].

13.7.6.1 Radio Frequency Map Construction Test-bed for CR

A simple test-bed for a RF-REM construction can be implemented using spectrum analyzers, signal generators and (dedicated) computers. The connection between the hardware components could be wired or wireless-most of the hardware these days have Internet capability. The test-bed set-up is given in Figure 13.21. The signal generator simulates the primary user (PU) and is made to transmit randomly for variable duration but fixed bandwidth and carrier frequency f_c to mimic

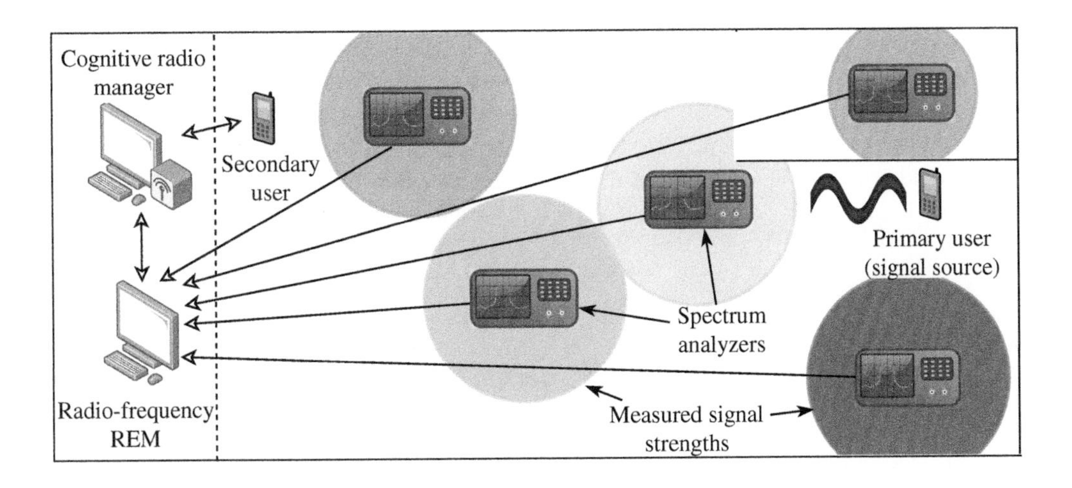

Figure 13.21 A simple test-bed set-up for RF-REM construction consisting of one secondary and primary users and a cognitive radio engine.

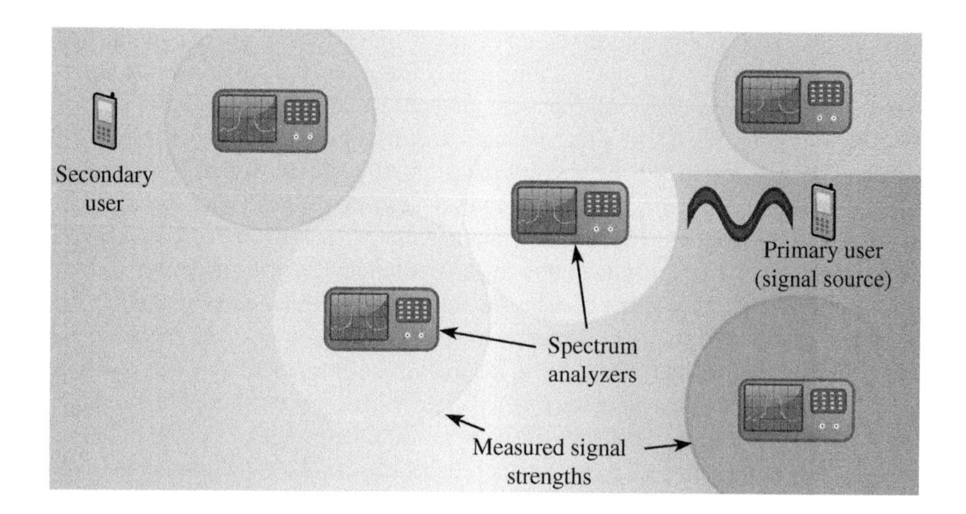

Figure 13.22 Expected RF-REM heatmap for example scenario.

the random nature of data transmissions. Spectrum analyzers are used to measure the power of the desired frequency at their locations. Another signal generator acts as the secondary user (SU). For more realistic scenarios, multiple bandwidths, carrier frequencies, PUs and SUs could be used, like in Wi-Fi networks, however, this would increase complexity. Three computers can be used in place of the CR manager, REM processor and REM manager and storage. Alternatively, if the capacity of the computer permits, a single computer can be used as well. References [12, 13] contain details for similar set-ups.

When the SU needs to transmit, it sends a request to the CR manager, which checks the availability of the spectrum resources from the RF-REM manager. The spectrum analyzers forward their measurements to the RF-REM manager either continuously or as requested. The RF-REM manager feeds this data to the RF-REM processor, which contains the mapping methods – or in this case, interpolation algorithms – and calculates spectrum power estimates for other locations. The expected heatmap produced as a result of this process for the scenario in Figure 13.21 is given in Figure 13.22. This information is given to the CR manager, which uses it to decide the transmitting frequency and power for the SU, in the simplest case.

The quality of the RF-REM depends on the number of measurement locations available. For the scenario depicted in Figure 13.21, let the collected measurements be as given in Table 13.1. The Kriging's interpolation can be applied using the Matlab function as given in the code below

Table 13.1 RSSI measurements for the scenario in Figure 13.21.

SA ID	*X*	*Y*	RSSI
1.	3.0	1.2	−75
2.	2.2	0.5	−65
3.	1.2	0.2	−70
4.	1.0	1.0	−80
5.	3.0	−0.2	−55

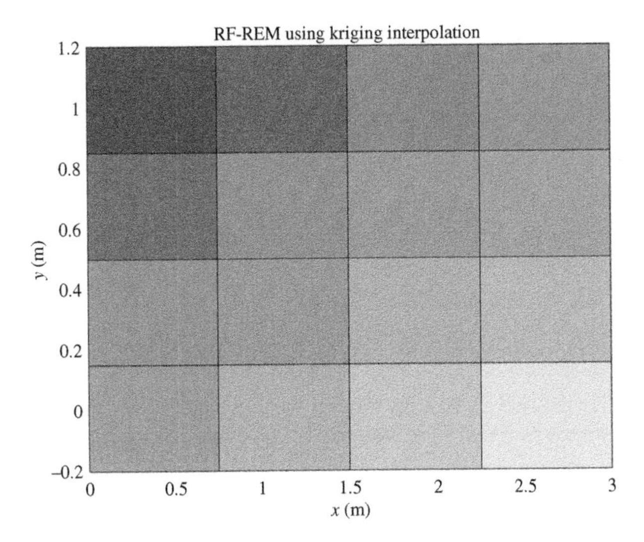

Figure 13.23 RF-REM constructed using Kriging's interpolation.

and the resulting RF-REM is given in Figure 13.23. The absence of more data has resulted in a low-resolution map which similar to, but not as accurate as the expected given in Figure 13.22.

```
1  A = [3.0, 1.2, -75;
2       2.2, 0.5, -65;
3       1.2, 0.2, -70;
4       1.0, 1.0, -80;
5       3.0, -0.2, -55];    % measurement locations and values [x,y,a]
6
7  a = kriging(A(:,1), A(:,2), A(:,3));
8
9  x_ = linspace(0,3,5);
10 y_ = linspace(-0.2, 1.2,5);
11 surf(x_, y_, a')
```

13.7.7 Case Study: Wireless Local Area Network/Wi-Fi Sensing

The principle of this method lies in the fact that changes in the physical environment causes observable changes in the radio environment. The RSSI and the CSI measurements are suitable measurements to monitor these changes. Depending on the measurement type, the mapping process varies slightly; therefore, a simplified outline will be given for each measurement. However, the common steps can be stated as follows:

1. Measurement Stage – Multiple measurements are taken at various base stages to form a training data set
2. Mapping Stage – Make a model or train a machine learning algorithm to map the measurements to the desired or labeled information

More primitive algorithms store every measurement in a look-up table instead of training or modeling. These algorithms have lower accuracy in complex environments where there are a lot of factors affecting the measurements. The resolution of these algorithms is also solely dependent on the number of measurements and difference between the various base stages. Other algorithms use interpolation, machine/deep learning, and signal processing methods to estimate missing values, extract and learn patterns corresponding to a change in the environment. Wireless local area network (WLAN) based sensing requires bi or multistatic configurations, as the channel between two communicating devices needs to be measured. The sensing can be with or without feedback. In bistatic configurations, the Rx can sense the environment using the transmitted signals or it can send the measurements to the Tx for it to construct the REM. Similarly, in multistatic configurations, a Tx will transmit communication or sensing signals, which will be used by the access points (APs) to make measurements. Once again, either the APs can use these measurements to construct their own REM or share them with the transmitting device.

The RSSI is a function of the distance between the communicating wireless devices. Therefore, mapping or learning the changes in the measured RSSI to the changes in the environment can be used to infer information about the UE with a given RSSI measurement. However, a direct mapping is usually not possible, due to the presence of external factors affecting the RSSI – obstacles, interference from devices in the network and interference from other devices. In order to work around this, artificial intelligence and machine learning algorithms are used to learn the relationship between the measurements and position. One application of this RSSI based sensing is localization and positioning through environment fingerprinting. Here, the RSSI is measured from a set of APs and labeled with the corresponding locations. Semisupervised or supervised learning algorithms are used to learn a function which maps the RSSI to a position and extracts features in the environment [14]. RSSI based sensing methods are limited in capability because RSSI provides limited information about the multipaths, which are the most effected in the radio propagation due to changes in the physical environment. For this reason, RSSI measurements are called *coarse-grained* measurements and are generally used to obtain coarse range measurements.

CSI, on the other hand, is a *fine-grained measurement*, since it is a reflection of how wireless signals propagate and the multipath produced. The schematic for CSI based WLAN sensing is given in Figure 13.24. The effective channel H for a MIMO system, contains the combined effects of the propagation environment (delay and Doppler), transmit/receive processing acting on the received signal at a given carrier frequency and RF impairments [15].

Using the relation in Eq. (13.6.4) and known data in the training symbols, the raw CSI is found. This CSI contains noise in the form of phase offsets and errors or outliers. These have to be removed, otherwise they would reduce the accuracy of mapping algorithms, as these impairments can be

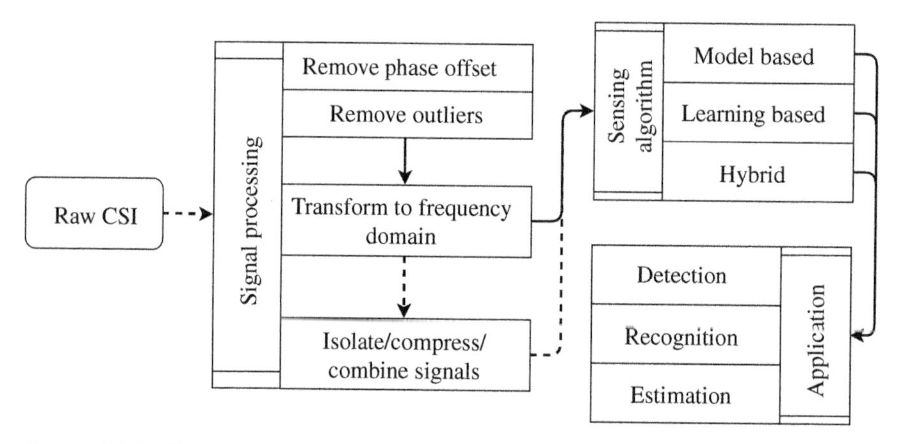

Figure 13.24 CSI based WLAN sensing framework.

considered random, where as channel changes are correlated. Next, signal transformation methods are applied. Depending on the time-frequency analysis required, FFT, STFT, DWT and other transforms can be used. In addition, the configuration of the system may result in signals from multiple devices or with multiple frequencies for sensing or contain interference. In these cases, the required signals have to be isolated, extracted or coherently combined. Similarly, irrelevant signals may need to be removed. Therefore, a combination of thresholding, compression, filtering or composition techniques may have to be used.

The preprocessed CSI is then fed to the mapping algorithm. The mapping algorithm essentially finds a function $f(.)$ that maps the CSI measurements to the desired result. Depending on the usage scenario and measurement features, such as rate of measuring and the frequencies measured, these algorithms can be model-based, learning-based or hybrid – a combination of the previous two. Model-based algorithms are based on theoretical modeling of physical concepts or statistical results. Learning-based algorithms are dependent on classical ML or DL algorithms, such as k-Nearest Neighbor or Deep Neural Networks. Hybrid models are used to eliminate the specific short comings of both the model-based and learning-based algorithms. For example, model-based algorithms are not able to accurately represent complex environments or become too complicated and take too long to converge if they are. Learning-based algorithms may require massive amounts of data, which results in longer training periods. There is also the risk of over-fitting. Hybrid models can compensate for these disadvantages by utilizing model-based algorithms to get coarse results, which are then refined using less data and training time with learning-based algorithms.

Based on the usage scenario, the resulting application can be classified as detection, recognition or estimation. Detection is a binary classification of "present" or "not present." Such applications could be human presence detection, object detection, etc. Recognition involves classifying the sensing results into nonbinary labels by comparing the results with previous or defined results. Example applications performing recognition would be person identification, object identification, gesture recognition, etc. Estimation, on the other hand, refers to inferring more specific information, such as localization and tracking or heart rate estimation.

13.7.7.1 WLAN Sensing Test-bed for Gesture Detection

A simple test-bed for WLAN gesture detection can be implemented using a Wi-Fi router and a computer with wireless cards. Both of them can have multiple antennas, which results in more CSI, the router, as it is in the WLAN standards, and the computer through the wireless cards. The scenario is depicted in Figure 13.25, where the waves indicate the sensing transmissions. In this basic scenario, a single responding node is present in a generic room filled with day-to-day objects. The computer is the G-REM capable node. A volunteer makes the desired hand gestures.

As per the G-REM process in Figure 13.3, the G-REM capable node initiates a sensing instance and the responding node begins transmitting OFDM based sensing signals. The computer uses these signals to calculate the CSI. At first, a base state with no hand gestures is measured. Then the volunteer is required to make the hand gestures – one at a time or in a specific sequence. These gestures cause a disturbance in the channel which is reflected in the CSI. Then, at the signal processing stage, noise and anomalies are removed using filtering techniques before defining features or attributes, such as phase difference and delays, are extracted from the raw CSI data. Finally, learning-based algorithms, like decision trees or support vector machines are trained to classify the hand gestures. More details on simulations and test-bed examples can be found in Refs. [9, 16, 17].

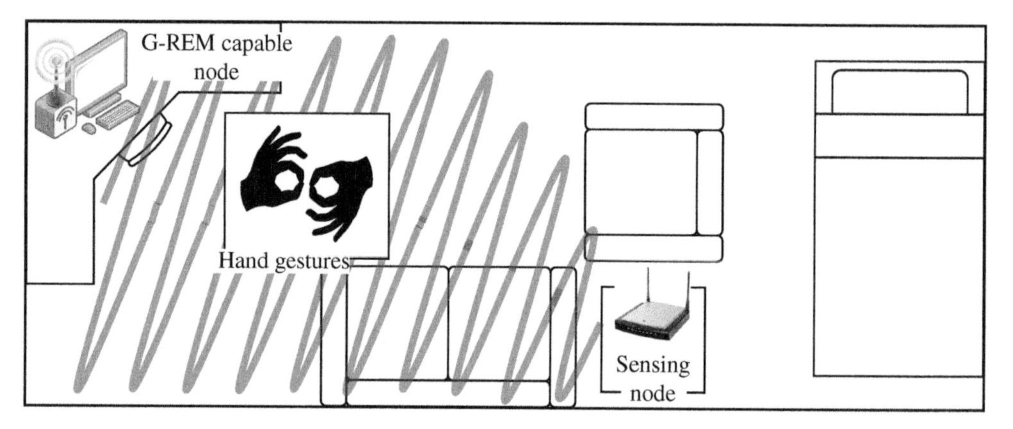

Figure 13.25 The WLAN sensing test-bed for gesture detection.

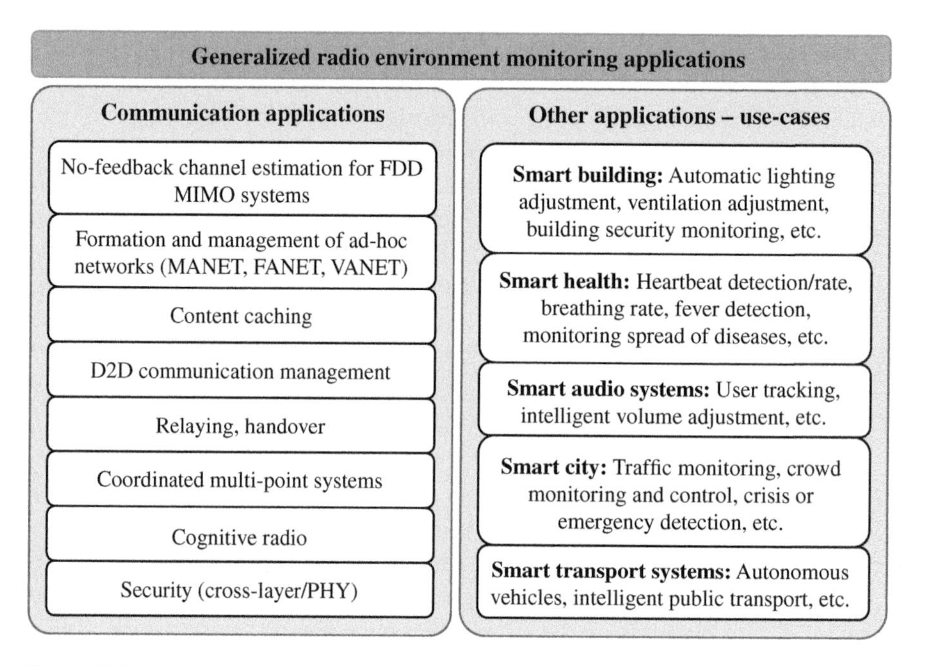

Figure 13.26 Applications and use-cases of G-REM.

13.8 Applications of G-REM

The radio environment is an information mine, containing multidimensional data from multiple domains. The G-REM framework enables information gathering and radio and physical environment awareness, behaving as an independent entity capable of feeding information to other entities. There are multiple ways G-REM can be utilized – whether for improving the quality of wireless communication or as a multisensor for other use-cases, as shown in Figure 13.26. This section will go over some of these applications.

13.8.1 Cognitive Radios

Perhaps the most obvious application of REM, CRs are intelligent devices capable of optimizing all aspects of communication through the adjustable communication parameters in order to satisfy UE quality of service (QoS) requirements and performance metrics [1]. CRs were prevalently studied for PU and SU coordination in opportunistic spectrum sharing between Wi-Fi devices and television white space (TVWS) devices. However, with efficiency across all domains – especially frequency and time – vital to the realization of 5G, 6G and beyond systems, CRs and CR based communication optimization techniques are once again a hot topic. G-REM can provide the information required by the cognitive engine in order to better manage the available resources. For example, spatio-temporal and spectrum usage relationships of UE can be used to predict future spectrum usage patterns, aiding in spectrum sharing, numerology selection, dynamic spectrum access techniques and much more.

13.8.2 Security

Technology is becoming more and more available, not only in terms of hardware, but also in terms of know-how. Additionally, the user-centric nature of emerging cellular requirements necessitate the sharing of more user information with other devices, and the increased demand for wireless communication services imply that there is a constant stream of user data. All of these reasons pose security threats and are signs that security will be critical in future wireless networks. Common attacks in wireless communications are briefly:

- Eavesdropping – a passive event where illegitimate nodes acquire data not intended for them.
- Jamming – an active event where an illegitimate node transmits signals with the aim of degrading communication performance for legitimate nodes.
- Spoofing – an active event where the illegitimate node transmits signals with the aim of misguiding the legitimate nodes.

REM can aid in preventing, disrupting or minimizing the effects of these threats.

13.8.2.1 PHY Layer Security

PHY layer security is a complementary technique to cryptography-based security techniques [18]. The main principle is exploiting channel information and characteristics to ensure complete secrecy and prevent the detection or decoding of the transmitted signals. REM is aware of UE state, objects in the environment and/or other nodes. This information can be used to add known affects to the signal or apply MIMO or beamforming techniques, preventing illegitimate nodes from understanding or receiving or even receiving the signal.

13.8.2.2 Cross-Layer Security

Cross-layer frameworks are recently being used to optimize communication performance. One aspect for optimization is security. Using information from multiple layers of the OSI network model, such as the network, traffic and MAC layers, more dynamic security techniques can be applied. REM contains information on the devices in the network and their spectrum usage patterns. Using these, traffic patterns can be predicted, which can be used for securing routing. Knowledge of UE in the environment, through the network and through RF fingerprinting-based UE detection, can be used for social reputation-based security techniques, where UEs are given weights representing their trustworthiness in order to determine a threat level for the environment or aid in multiuser security techniques [19].

13.8.3 Multi-Antenna Communication Systems

Multi-antenna communication systems, such as reconfigurable intelligent surfaces (RISs), MIMO, and massive MIMO (mMIMO), often perform spatial multiplexing or beamforming to provide diversity gain, increase reliability, increase data rate, for PHY security and/or to improve power efficiency.

13.8.3.1 UE and Obstacle Tracking for Beam Management

One common disadvantage of these systems is their susceptibility to obstructions. The associated wavelength is generally very small (mm-Waves), reducing their propagation distance, and the transmissions are narrow beams, leading to sparse channels with no or minimal multipath. This makes beam alignment crucial. Any error could prevent the signal from reaching the intended receiver. Therefore, environment awareness – specifically awareness of the UE location and state, as well as the location and mobility of objects in the environment – is crucial to exploit the full potential of these systems. The G-REM framework contains methods to aid in this. For example, each G-REM capable UE can share their specific mobility pattern with the required node. Common mobility models present in online databases can be used with the UE mobility pattern to predict the trajectory of the UE. Additionally, the G-REM framework can be used to collect CSI and map it to mobility in the environment, like in Figure 13.7.7.

13.8.3.2 No-Feedback Channel Estimation for FDD MIMO and mMIMO Systems

Channel estimation is a necessary load on communication systems. In time division duplexing (TDD) systems, the uplink and downlink channels are considered to have the same propagation path and therefore reciprocal. In frequency division duplexing (FDD) systems, on the other hand, the channel has to be measured for both uplink and downlink, which causes additional overhead. While this can be (and has to be) overlooked for single-input single-output (SISO) systems, it is crippling in MIMO and mMIMO systems, where the channel is an $M \times N$ dimension matrix. Here, M and N are the number of transmitting and receiving antennas, respectively. This is the case for narrow-band time-invariant channels. As such, calculating all the elements of this matrix compounds the induced overhead. While this is once again tolerated in TDD systems and reduced using geometric correlation property of closely placed antennas, the same cannot be said for FDD systems.

There are works in the literature, however, which exploit the relationship between radio wave propagation and the physical environment to try to map the channel at one frequency to the channel at another frequency [20]. They reconstruct the propagation paths from the measured channel information and map them to the channel at another frequency using neural networks. While this area has room for improvement, it is a start to reduce the computational overhead for these systems.

13.8.4 Formation and Management of Ad Hoc Networks and Device-to-Device Communication

Nowadays, specialized robots are being developed to carry out various tasks in cooperation. Controlling every action of the robots in a centralized fashion, whether autonomously or by trained professionals, is not always possible. For example, in military reconnaissance applications transmissions from the ground station may jeopardize the mission if detected or in devices underground or underwater, where the signals cannot propagate great distances. Now, with studies actualizing swarm intelligence – the decentralized and self-organized control of a multitude of unit robots with specific or basic functions - ad hoc networks should be of even more importance. Developments in autonomous vehicles have also led to the use of ad hoc networks for vehicular communications – vehicular ad-hoc networks (VANETs). Because of the decentralized nature of ad hoc networks, implementing them is not an easy task – with challenges such as scalability, security, cooperation, etc. [21].

Ad hoc devices are ideal examples for G-REM implementation. They have to make intelligent decisions about their communication parameters in order to maintain a connection with the other devices. Depending on the device capabilities, each device may have a dedicated G-REM architecture which can be used to infer information about the radio and physical environment. The gathered information can then be used to mitigate the challenges mentioned above. Additionally, the information on mobility in the environment and distance to other objects could be vital for autonomous vehicles.

13.8.5 Content Caching

Another task where REM could be useful is content-caching, where frequently desired data is stored at network edge nodes or the cloud to reduce data transfer overhead [22]. The main concern in this area is in deciding what data has sufficient popularity or will have popularity such that it should be cached. However, another issue is the location of the node where the data would be cached. Here, mobility patterns of UEs can be used to predict the area with heaviest traffic and the time at which that data is most requested.

13.8.6 Enabling Flexible Radios for 6G and Beyond Networks

The current 5G standards are the result of an evolution of the wireless communication requirements from vocal and data centric to user-centric communication. This evolution introduced communication services with different QoS requirements – enhance mobile broadband (eMBB), massive machine type communication (mMTC) and ultra-reliable, low-latency communication (URLLC) – tailored to meet the communication requirements of the users. eMBB is an extension of 4G LTE networks and is for communications requiring large data rates, and therefore more bandwidth, such as video streaming or virtual reality (VR). The mMTC service is for low power and data communications, typical to less complex devices, like sensors or machinery, while URLLC provides low latency and reliable communication for mission critical applications, like remote surgery [23]. To enable these services, the multiple numerology concept was introduced with the aim of choosing the most suitable numerology for the required service, therefore optimizing the available resources. However, as the number and variety of wireless applications, various communication applications along with sensing applications, increase, cellular networks are envisioned to evolve further – becoming even more user/use-case centric. Limitations on available spectrum resources will force a tight, yet not fully cooperative or coordinated, sharing of spectrum in other bands – licensed and unlicensed. As such, 6G networks will revolve around flexibility and adaptability [24]. With their ability to sense and infer information on the physical and radio environment G-REM and REM will be the enabling components for 6G radio devices and networks.

13.8.7 Non-Communication Applications

The G-REM framework can be used to obtain information for other applications. The received signal reflects the physical environment. Using the appropriate sensing and mapping techniques, these signals can be used as a futuristic "everything" sensor. WLAN sensing is one REM technique which has made a lot of progress in mapping the received signal features to actions or events in the environment. Such actions include gesture recognition, person authentication, person counting, temperature or smoke detection, and many more. Using this information, otherwise dumb devices can be "aware" of the environment. These devices can be used for home monitoring, crowd monitoring, smart health, smart city, etc., applications, increasing the quality of life with off-the-shelf equipment.

13.9 Challenges and Future Directions

The seamless integration of G-REM and sensing into current wireless communication networks is not without its challenges. In addition to complexity/accuracy trade-offs during sensing and REM construction, security is also an issue. Furthermore, new technologies are constantly being developed. The G-REM framework must evolve to include and take advantage of these technologies, or at the very least, be aware of their effects on the sensing process. New architectures, protocols and polices must be developed and standardized to provide common ground for a variety of devices to take advantage of this concept. This section will go over some challenges and future areas where research and/or standardization can take place.

13.9.1 Security

As mentioned throughout this chapter, the radio environment is a huge source of information which can be applied to enhance communication as well as quality of life. However, the sensing or communication signals could be intercepted by nodes other than the intended ones, meaning that illegitimate nodes can sense the environment passively, without transmitting any signals themselves. Therefore, these nodes could make their own REM and infringe on the UEs privacy or degrade the performance of the communication. For these reasons, studying threat scenarios and developing PHY and cross-layer security techniques to prevent the acquisition of valuable information on the environment from those who would misuse it is necessary [25].

REM gains awareness of the environment primarily from analyzing or processing the received signals. Because of this, it is vulnerable to spoofing attacks which cause misinformation on the state of the radio and physical environment. For example, a spoofer can transmit signals misleading the REM capable device on which or how many devices are in the environment, their locations and what resources they are using. Similarly, the illegitimate node may simulate motion or blockages, either by transmitting appropriate signals or causing these effects physically. This can trigger repeated sensing procedures, wasting the resources of the REM capable device and causing an additional load on the network. While it may not be possible for a radio device to physically prevent these attacks, methods should be developed to detect and minimize their affects on the REM and CR performance. Sensing signals can also be detected and a jammer can transmit noise, for example, preventing sensing completely. Therefore, developing security techniques to protect the REM integrity is also vital.

Additionally, the sheer number of MCDs for G-REM implementations – including the classic RSSI based RF REM construction – these days brings about new concerns as to how to verify that all the sensed information is legitimate. The G-REM framework is data dependent; therefore, any erroneous measurements could significantly impair the cognitive abilities of G-REM and thus G-REM dependent applications. For example, incorrect measurements fed to a machine learning algorithm could lead to the development of wrong models, which would be useless at best.

13.9.2 Scheduling

An example operation of the G-REM framework was given in Figure 13.3. In essence, communication transmissions may be used for sensing. However, the quality of the measured information may not be as desired by the requesting application, even if hybrid waveforms are utilized. As a result, different applications may require different radio environment monitoring technologies, e.g. radar, or different sensing parameters. In these cases, additional sensing transmissions may be necessary, which are an extra overhead to the network and may lead to interference to the communication transmissions and other sensing transmissions. As a result, proper scheduling techniques for sensing transmissions in the presence of communication traffic must be investigated. The trade-offs between communication and sensing performance must be investigated.

13.9.3 Integration of (New) Technologies

Some of the components of the REM framework may have been or are being studied, however, these studies are stand-alone and have not been related to REM as a concept. For example, radar is a well-studied method for target detection and tracking, especially in military research, but has not been studied jointly with communication until recently. Joining these and other separate concepts under one framework has created new opportunities and areas for research, which must be explored in order to fully realize the potential of REM. Additionally, it is an exciting period for the wireless communication field with new technologies being developed and explored rapidly. These technologies can either be placed under the REM framework or can utilize the framework for their performance. The seamless integration of the new technologies with the REM framework must also be studied.

13.9.3.1 Re-configurable Intelligent Surfaces

One such new technology is the RIS. These surfaces are made up of small elements which can collect the electromagnetic signals and reflect or scatter them in the desired manner. In this way, RIS enables communication around obstacles and allows the signal to reach otherwise unreachable areas. This capability of RIS can also be used to sense areas behind blockages [26]. However, there are some issues which remain unsolved, like given the signal attenuation and losses in the RIS, is this practical under real life scenarios? Given that it is, is the RIS aided channel reciprocal in time? In other words, can it be used for radar with the known radar relations or is there a need to derive new relations? Additionally, integrating the RIS into the G-REM framework is also an issue depending on whether the RIS is controlled by a central entity in the network, a G-REM capable device, or another entity (from another network or operator). The envisioned mass deployment of RIS all but ensures that not all RIS in the environment can be or will be controlled by one central entity. In this situation, differentiating between sensed changes due to reconfiguration of the RIS and changes in the environment becomes an issue. RIS can also be active and function as an independent, nontransmitting REM capable device utilizing REM to adjust its parameters in order to enhance the communication or sensing quality of other devices.

13.9.3.2 Quantum Radar

Quantum communication is trending, providing faster communication. Although many 6G and beyond papers envision quantum communications in the standards for the near future [27], hardware limitations may stall these predictions. However, progress in this area should be followed closely. Quantum radars are an interesting concept in this field [28]. Here, entangled particles are used to sense the environment. One particle is transmitted and the other is kept for observation. Because of the entanglement, everything that happens to the transmitted particle will be mirrored in the kept particle and can be observed instantly. Currently, quantum radars have not been commercially applied and are still in the research and development stage. Integrating quantum technologies into the G-REM framework can take place both to improve their performance and to enable radio environment monitoring.

13.10 Conclusion

While communication has been the main concern for previous generations of wireless networks; the future wireless applications also requires information about the environment. Vehicle-to-vehicle communication, human gesture recognition and drones that have ability to sense the environment and avoid obstacles while communicating with other devices at the same time are some of the examples of applications that support both communication and sensing features. The G-REM framework is a 6G and beyond enabler in this aspect because it incorporates sensing, communication and cognition in one structure. The sensing methods are briefly explained, along with the observable and extractable information, and two methods are simulated. The combination of these methods and concepts under one framework have led to new research areas, some of which are mentioned above. Exciting developments in wireless technologies, such as RIS or super-heterogeneous networks, also affect the REM framework. The evaluation of the REM framework and the integration studies of these new technologies to this framework are interesting, to say the least.

References

1 J. Mitola and G. Q. Maguire, "Cognitive radio: making software radios more personal," *IEEE Personal Communications*, vol. 6, no. 4, pp. 13–18, 1999.

2 Y. Zhao, B. Le, and J. H. Reed, *Network Support: The Radio Environment Map*. Burlington: Elsevier, 2006, pp. 337–363.

3 3GPP Task Group, "NR; Radio Resource Control (RRC) protocol specification (Rel-16)," 3rd Generation Partnership Project, Technical Report 38.331, 2020.

4 B. R. Mahafza, *Radar Systems Analysis and Design Using MATLAB*, 3rd ed. USA: Chapman and Hall/CRC, 2016.

5 M. A. Richards, *Fundamentals of Radar Signal Processing*, 2nd ed. New York: McGraw Hills Education, 2014.

6 L. Zheng, M. Lops, Y. C. Eldar, et al., "Radar and communication coexistence: an overview: a review of recent methods," *IEEE Signal Processing Magazine*, vol. 36, no. 5, pp. 85–99, 2019.

7 K. V. Mishra, M. B. Shankar, V. Koivunen, et al., "Toward millimeter-wave joint radar communications: a signal processing perspective," *IEEE Signal Processing Magazine*, vol. 36, no. 5, pp. 100–114, 2019.

8 T. Bailey and A. Gatrell, *Interactive Spatial Data Analysis*. USA: Longman/Wiley, 1995.

9 S. Ren, H. Wang, L. Gong, et al., "Intelligent contactless gesture recognition using WLAN physical layer information,"*IEEE Access*, vol. 7, pp. 92758–92767, 2019.

10 M. Pesko, T. Javornik, A. Kosir, et al., "Radio environment maps: The survey of construction methods," *KSII Transactions on Internet and Information Systems*, vol. 8, no. 12, pp. 3789–3809, 2014.

11 M. Akimoto, X. Wang, M. Umehira, et al., "Crowdsourced radio environment mapping by exploiting machine learning," in *2019 22nd International Symposium on Wireless Personal Multimedia Communications (WPMC)*, Lisbon, 24–27 Nov. 2019, pp. 1–6.

12 F. Casadevall and A. Umbert, "REM-based real time testbed: a proof of concept on the benefits of using REM for improving radio resource management capabilities," in *ICT 2013*, Casablanca, 6–8 Nov. 2013, pp. 1–5.

13 D. Denkovski, V. Rakovic, M. Pavloski, et al., "Integration of heterogeneous spectrum sensing devices toward accurate REM construction," in *2012 IEEE Wireless Communications and Networking Conference (WCNC)*, Paris, 1–4 Apr. 2012, pp. 798–802.

14 J. Yoo and J. Park, "Indoor localization based on Wi-Fi received signal strength indicators: feature extraction, mobile fingerprinting, and trajectory learning," *Applied Sciences*, vol. 9, no. 09, pp. 3930–3951, 2019.

15 Y. Ma, G. Zhou, and S. Wang, "Wi-Fi sensing with channel state information: a survey," *ACM Computing Surveys*, vol. 52, no. 3, 2019.

16 H. Abdelnasser, M. Youssef, and K. A. Harras, "WiGest: a ubiquitous Wi-Fi-based gesture recognition system," in *2015 IEEE Conference on Computer Communications (INFOCOM)*, Kowloon, 26 Apr.–1 May 2015, pp. 1472–1480.

17 W. He, K. Wu, Y. Zou, et al., "WiG: Wi-Fi-based gesture recognition system," in *2015 24th International Conference on Computer Communication and Networks (ICCCN)*, Las Vegas, 3–6 Aug. 2015, pp. 1–7.

18 Y. Shiu, S. Y. Chang, H. Wu, et al., "Physical layer security in wireless networks: a tutorial," *IEEE Wireless Communications*, vol. 18, no. 2, pp. 66–74, 2011.

19 I. Adam and J. Ping, "Framework for security event management in 5G," in *Proceedings of the 13th International Conference on Availability, Reliability and Security*, ser. ARES 2018. Hamburg: Association for Computing Machinery, 27–30 Aug. 2018.

20 D. Vasisht, S. Kumar, H. Rahul, et al.,"Eliminating channel feedback in next-generation cellular networks," in *Proceedings of the 2016 ACM SIGCOMM Conference*. Florianopolis: Association for Computing Machinery, 22–26 Aug. 2016, pp. 398–411.

21 S. Sharmila and T. Shanthi, "A survey on wireless ad hoc network: issues and implementation," in *2016 International Conference on Emerging Trends in Engineering, Technology and Science (ICETETS)*, Pudukkottai, 24–26 Feb. 2016, pp. 1–6.

22 J. Kwak, Y. Kim, L. B. Le, et al., "Hybrid content caching in 5G wireless networks: cloud versus edge caching," *IEEE Transactions on Wireless Communications*, vol. 17, no. 5, pp. 3030–3045, 2018.

23 S. Parkvall, E. Dahlman, and S. Johan, *5G NR: The Next Generation Wireless Access Technology*. Academic Press, 2018.

24 A. Yazar, S. Dogan-Tusha, H. Arslan, "6G Vision: An Ultra-Flexible Perspective," ITU Journal on Future and Evolving Technologies, 1(9), 1–20, Dec. 2020.

25 H. M. Furqan, M. S. J. Solaija, H. Türkmen, and H. Arslan, "Wireless Communication, Sensing, and REM: A Security Perspective," in IEEE Open Journal of the Communications Society, vol. 2, pp. 287–321, 2021, doi: 10.1109/OJCOMS.2021.3054066.

26 S. Gong, X. Lu, D. T. Hoang, et al., "Towards smart wireless communications via intelligent reflecting surfaces: a contemporary survey," *IEEE Communications Surveys and Tutorials*, pp. 1–1, 2020.

27 I. F. Akyildiz, A. Kak, and S. Nie, "6G and beyond: the future of wireless communications systems," *IEEE Access*, vol. 8, pp. 133995–134030, 2020.

28 S. Barzanjeh, S. Pirandola, D. Vitali, et al., "Microwave quantum illumination using a digital receiver,"*Science Advances*, vol. 6, no. 19, 2020.

Index

c

d

e

f

g

l

m

n

Printed and bound by CPI Group (UK) Ltd, Croydon, CR0 4YY